ADVANCES IN MATERIALS SCIENCE, ENERGY TECHNOLOGY AND ENVIRONMENTAL ENGINEERING

PROCEEDINGS OF THE INTERNATIONAL CONFERENCE ON MATERIALS SCIENCE, ENERGY TECHNOLOGY AND ENVIRONMENTAL ENGINEERING (MSETEE 2016), ZHUHAI, CHINA, 28–29 MAY 2016

Advances in Materials Science, Energy Technology and Environmental Engineering

Editors

Aragona Patty
Institute of Applied Industrial Technology Division, Portland Community College, Portland, OR, USA

Peijiang Zhou
School of Resource and Environmental Sciences, Wuhan University, Hubei, China

CRC Press is an imprint of the
Taylor & Francis Group, an **informa** business

A BALKEMA BOOK

Published by:
CRC Press/Balkema
P.O. Box 447, 2300 AK Leiden, The Netherlands
e-mail: Pub.NL@taylorandfrancis.com
www.crcpress.com – www.taylorandfrancis.com

First issued in paperback 2020

© 2017 by Taylor & Francis Group, LLC
CRC Press/Balkema is an imprint of the Taylor & Francis Group, an informa business

No claim to original U.S. Government works

Typeset by V Publishing Solutions Pvt Ltd., Chennai, India

ISBN 13: 978-0-367-73663-7 (pbk)
ISBN 13: 978-1-138-19668-1 (hbk)

Visit the Taylor & Francis Web site at
http://www.taylorandfrancis.com

and the CRC Press Web site at
http://www.crcpress.com

Advances in Materials Science, Energy Technology
and Environmental Engineering – Patty & Zhou (Eds)
© *2017 Taylor & Francis Group, London, ISBN 978-1-138-19668-1*

Table of contents

Energy science and environmental engineering

Materials science and materials processing

Preface

The 2016 International Conference on Materials Science, Energy Technology and Environmental Engineering (MSETEE 2016) took place on May 28–29, 2016 in Zhuhai City, China. MSETEE 2016 brought together academics and industrial experts in the field of materials science, energy technology and environmental engineering. The primary goal of the conference is to promote research and developmental activities in materials science, energy technology and environmental engineering and another goal is to promote scientific information interchange between researchers, developers, engineers, students, and practitioners working all around the world. The conference will be organized every year making it an ideal platform for people to share views and experiences in materials science, energy technology and environmental engineering and related areas.

In order to organize MSETEE 2016, we have sent our invitation to scholars and researchers from all around the world. Eventually, over 350 conference articles were submitted for publication. These articles have gone through a strict review process performed by our international reviewers. All the submissions were reviewed double-blind, both the reviewers and the authors remaining anonymous. First, all the submissions were divided into several chapters according to the topics, and the information of the authors, including name, affiliation, email and so on, removed. Then the editors assigned the submissions to reviewers according to their research interests. Each submission was reviewed by two reviewers. The review results should be sent to chairs on time. If two reviewers had conflicting opinions, the paper would be transmitted to the third reviewer assigned by the chairs. Only papers which were approved by all reviewers were accepted for publication.

With the hard work from the reviewers, only 91 papers were finally accepted for publication. These papers were divided into three chapters:

– Electrical and Electronic Engineering
– Energy Science and Environmental Engineering
– Materials Science and Materials Processing

To prepare MSETEE 2016, we have received a lot help from many people.

We thank all the contributors for their interest and support to MSETEE 2016. We also feel honored to have the support from our international reviewers and committee members. Moreover, the support from CRC Press/Balkema (Taylor & Francis Group) is also deeply appreciated; without their effort, this book will not be able to come into being.

The Organizing Committee of MSETEE 2016

Advances in Materials Science, Energy Technology
and Environmental Engineering – Patty & Zhou (Eds)
© *2017 Taylor & Francis Group, London, ISBN 978-1-138-19668-1*

Organizing committees

CONFERENCE CHAIRS

Prof. Aragona Patty, *Portland Community College, USA*
Prof. Peijiang Zhou, *Wuhan University, China*
Prof. Ke Chen, *Hefei University of Technology, China*

CO-EDITORS

Prof. Aragona Patty, *Portland Community College, USA*
Prof. Peijiang Zhou, *Wuhan University, China*

TECHNICAL PROGRAM COMMITTEES

Prof. Jeng-Haur Horng, *National Formosa University, Taiwan*
Prof. Mingqing Chen, *Jiangnan University, China*
Dr. Calogero Orlando, *Kore University of Enna, Italy*
Prof. Yingkui Yang, *School of Materials Science and Engineering, Hubei University, China*
Prof. Noureddine HASSINI, *University of Oran 1, Algeria*
A. Prof. Walailak Atthirawong, *King Mongkut's Institute of Technology, Thailand*
A. Prof. Chi-Wai Kan, *The Hong Kong Polytechnic University, Hong Kong*
Prof. Antonio Messineo, *University of Enna Kore, Italy*
Dr. Gajendra Sharma, *Kathmandu University, Nepal*
Dr. Ran Huang, *National Taiwan Ocean University, Taiwan*
Prof. Jianwei Liu, *Guilin University of Electronic Technology, China*
A. Prof. Xiurong Fang, *Xi`an University of Science and Technology, China*
Prof. Chunfeng Shi, *Research Institute of Petroleum Processing, SINOPEC, China*
Dr. Zeya Oo, *Yangon Technological University, Myanmar*
Prof. Yusri Yusof, *Universiti Tun Hussein Onn Malaysia (UTHM), Malaysia*
Prof. Yun Hae Kim, *Korea Maritime and Ocean University, South Korea*
Prof. Blednova Zhesfina e Mikhailovna, *Kuban State University of Technology, Russian Federation*
Prof. Yun-Hae KIM, *Korea Maritime and Ocean University, Korea*
Dr. Shaojun Cai, *Jianghan University, China*
Prof. Alokesh Pramanik, *Curtin University, Australia*
Dr. Ana Evangelista, *Federal University of Rio de Janeiro, Brazil*
Assistant Prof. Abdullahi Ali Mohamed, *University of Nottingham—Malaysia Campus (UNMC), Malaysia*
A. Prof. Famiza Binti Abdul Latif, *Universiti Teknologi MARA, Malaysia*

Electrical and electronic engineering

*Advances in Materials Science, Energy Technology
and Environmental Engineering – Patty & Zhou (Eds)
© 2017 Taylor & Francis Group, London, ISBN 978-1-138-19668-1*

Research on the friction vibration feature extraction based on the norm-coupling method in a running-in process

Ting Liu & Guobin Li
Marine Engineering College, Dalian Maritime University, Dalian City, Liaoning Province, P.R. China

Haijun Wei
Merchant Marine College, Shanghai Maritime University, Shanghai, P.R. China

ABSTRACT: A norm-coupling method of Tangential Friction Vibration (TFV) and Normal Friction Vibration (NFV) was proposed with the norm theory to solve the difficulty of completely reflecting the wear state feature of tribological pairs by friction vibration in a single direction. The feature of Norm-Coupling Friction Vibration (NCFV) was expressed with matrix 2-norms, and the variation in this feature was analyzed using a running-in process. Result indicates that the matrix 2-norms of friction vibration (M2NFV) can reflect the change in the wear state of tribological pairs in the running-in process. The M2NFV is large at the beginning of the running-in process, followed by a decline as the process continues, and finally, the process produces a stable and smooth fluctuation while reaching the running-in stage. The variation in M2NFV is in close agreement with the change in friction coefficient. Hence, the complete feature extraction of TFV and NFV by the norm-coupling method can reflect the variation in the wear state of the tribological pairs in the running-in process.

1 INTRODUCTION

It is well known that the friction vibration originates from the wear process of the tribological pairs, and thus, it can provide information about the wear states of the tribological pairs (Sun 2015). Therefore, the friction vibration can be used to monitor the changes in the wear states of the tribological pairs by complete feature extraction of the measured signals. Many tribology scientists performed thorough research for years from the aspect of the feature extractions of friction vibration (Li et al. 2003; Li & Peng 2009; Wan et al. 2009; Kang & Luan 2006; Chen & Zhou 2003; Chen & Zhou 2006; Zhu et al. 2007; Ji 2012). Chen & Zhou (2003, 2006) extracted the time-frequency characteristics of both TFV and NFV by methods of short-time Fourier transform, Zhao-Atlas-Marks Distribution, and db4 Wavelet Transform. Zhu et al. (2007) performed researches on the fractal feature of NFV, studied the variations in the correlation dimension in different wear processes, and found that the Correlation Dimension of normal friction vibration was in decline in the "divergence" wear process, while there was an increase in the running-in wear process. In order to discuss the wear state of a piston-cylinder, Ji (2012) extracted the chaos features and the fractal characteristics of NFV, discovered that all of the Correlation Dimension, the Largest Lyapunov Exponent (LLE), and the Kolmogorov Entropy could reflect the friction and wear information of tribological pairs. It was also observed that both the Correlation Dimension and the Kolmogorov Entropy showed a "Positive Bathtub Curve" from the running-in wear process to a stable wear stage and then to a severe wear state, while the LLE showed a "Inverted Bathtub Curve". As shown earlier, the friction vibration feature extraction was studied using the applications of wavelet transform, fractal, and chaos, and the friction vibration feature extraction was successfully achieved in a single direction. However, both the tangential friction vibration and the normal friction vibration can be excited simultaneously in the wear process, and they both contain the information of the variation in the wear state. Hence, it is incomplete when extracting the friction vibration feature in a single direction, and difficult to accurately reflect the wear state of the tribological pairs. Unfortunately, few have been done to research the coupling of TFV and NFV, and then extract the feature information, which can reflect the wear state of the tribological pairs.

In this report, the norm-coupling method of TFV and NFV was proposed with the norm theory, the feature of NCFV was extracted with matrix 2-norms, and the variation in the friction vibration feature was analyzed by the running-in wear test.

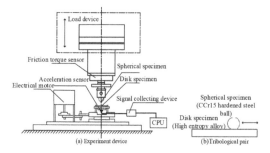

Figure 1.   Schematic diagram of CFT-I wear tester.

2   EXPERIMENTS

2.1   *Apparatus*

Running-in wear experiments were conducted on CFT-I friction and wear tester, which can measure the friction coefficient in real time, as shown in Figure 1. TFV and NFV were measured using a triaxial acceleration sensor (model 356 A16 ICP, PCB PIEZOTRONICS Company) with a sensitivity of 100 mv/g and a range of ± 50 g, which was fixed under the disk specimen. A data acquisition system (PXIe-1071, NI Company) with the sampling frequency of 25.6 KHz and the sampling interval of 6.25 Hz was used to collect vibration signals every 1 min.

2.2   *Parameters*

Spherical-disk specimen was used as tribological pairs in the test. The spherical specimen is made of CCr15 hardened steel ball with a hardness of 65 HRC and a diameter of 3 mm, and the chemical composition includes C, Si, Mn, Cr, Mo, P, S, Ni, Cu, and Fe. The disk specimen is composed of a high-entropy alloy with Ø30 mm × 10 mm and a hardness of 57.8 HRC, and the chemical composition includes Al, Co, Cr, Fe, and Ni. The disk specimen driven by an electric motor at a speed of 600 rpm covers a moving distance of 5 mm. The spherical specimen and the disk specimen slide at a calculated relative sliding velocity of 0.1 m/s. A continuous drip lubrication was conducted using CD40 (an ordinary marine lube oil) with a density of 0.8957 g/cm³, and viscosities of 139.6 cSt at 40°C and 12.5 cSt at 100°C. A spring loading system was used to apply a test load of 120 N to the disk specimen, resulting in a contact pressure of 20.65 MPa between the spherical specimen and the disk specimen.

3   NORM-COUPLING METHOD OF FRICTION VIBRATION

3.1   *Norm-coupling*

A distance cij is defined by a matrix norm $\| \cdot \|$ in a normed linear space $(C, \| \cdot \|)$ with the definition of normed space,

$$c_{ij} = \left\| x_{ij} - z_{ij} \right\|, x, z \in C;$$
$$i = 1, 2, 3, \cdots, 4096; \quad j = 1, 2, 3, \cdots, 60. \quad (1)$$

where c_{ij} is the amplitude of NCFV of point i in j min, x_{ij} is the amplitude of TFV of point i in j min, and z_{ij} is the amplitude of NFV of point i in j min.

3.2   *Matrix 2-norms*

M2NFV is introduced with NCFV by Eq. (1) in the normed space R_n and defines

$$\left\| d_j \right\|_2 = \left(\sum_{i=1}^{4096} \left| c_{ij} \right|^2 \Big/ 4096 \right)^{\frac{1}{2}},$$
$$i = 1, 2, 3, \cdots, 4096; \quad j = 1, 2, 3, \cdots, 60. \quad (2)$$

where $d_j \in R_n$, c_{ij} is the amplitude of NCFV of point i in j min, and d_j is the normalized M2NFV in j min, or M2NFV for short.

4   RESULTS AND DISCUSSION

4.1   *The analysis of NCFV*

With reference to the methods described in Sun et al. (2015), the Harmonic Wavelet Packet Transform (HWPT) was used to eliminate the influence of the noise of the acquired vibration signals. The acquired vibration signals were decomposed into seven layers and then were reconstructed with HWPT. Through the comparative analysis of reconstructed signals in different frequency bands, it was found that the reconstructed signals in the range of 3000–4000 Hz (31–40 frequency bands) showed a significant TFV and those in the range of 2000–3000 Hz (21–30 frequency bands) showed a significant NFV.

According to Eq. (1), the NCFV signals were acquired with TFV and NFV, and the waveforms of friction vibration signals at the beginning of the running-in process, and after 2 min, 4 min, and 6 min are shown in Fig. 2. TFV, NFV, and NCFV signals

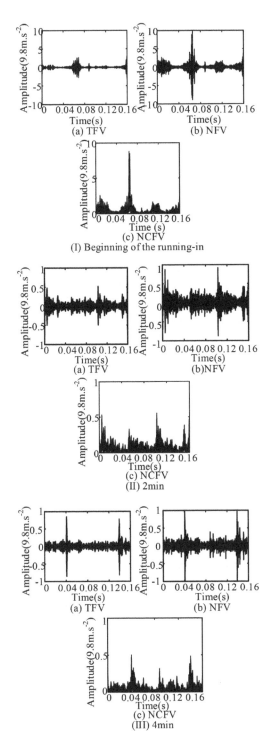

(a) TFV　　(b) NFV

(c) NCFV
(I) Beginning of the running-in

(a) TFV　　(b)NFV

(c) NCFV
(II) 2min

(a) TFV　　(b) NFV

(c) NCFV
(III) 4min

Figure 2. (*Continued*).

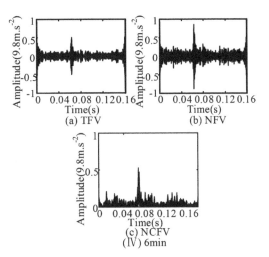

(a) TFV　　(b) NFV

(c) NCFV
(IV) 6min

Figure 2. Waveform diagram of friction vibration signals.

are shown in Figure 2(a), Figure 2(b) and Figure 2 (c), respectively. In Figure 2, NCFV signals were declining in the running-in wear process, which was in agreement with the variations in both TFV signals and NFV signals. Therefore, the NCFV signals also contain the information about the variation in the wear state of tribological pairs, which can preserve the information of both TFV signals and NFV signals.

4.2　The analysis of M2NFV

4.2.1　The variation of M2NFV

In order to analyze the variation in NCFV signals in the running-in process, the feature was extracted with the matrix 2-norms. According to Eq. (2), M2NFV was calculated, and the variation in M2NFV in the running-in process is presented in Figure 3(a). For comparative analysis, the Root Mean Square (RMS) values of TFV and NFV were also calculated, and the changes in RMS values of TFV and NFV are shown in Figure 3(b) and Figure 3(c), respectively. In Figure 3, all of M2NFV, and RMS values of TFV and NFV began to show a decreasing trend and then tended to be stable, while there were significant differences between them in different running-in times. All the features of the three signals began to decline for 2 min from the beginning of the running-in process. When the running-in continued to 5 min, M2NFV decreased slowly, while the RMS values of both TFV and NFV fluctuated, with that of TFV fluctuating to a greater extent. After another 5 min, M2NFV tended to be smooth and steady, while the RMS values of both TFV and NFV fluctuated to a greater extent.

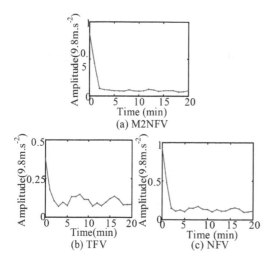

(a) M2NFV

(b) TFV

(c) NFV

Figure 3. Temporal variation of friction vibration features.

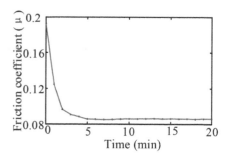

Figure 4. Temporal variation of the friction coefficient μ.

As discussed earlier, the general trend of the variations in M2NFV, and RMS values of TFV and NFV are similar, while there are significant differences between them in different times of the running-in wear process, which suggests that the reflections of the running-in wear state are different between them.

4.2.2 The relationship between M2NFV and the running-in wear state

The running-in wear is a self-adaptive process in which the tribological pairs tend to adapt and match gradually. As the friction coefficient can reflect different wear states of tribological pairs in the running-in process, the relationship between M2NFV and the running-in wear states of tribological pairs was also discussed utilizing friction coefficient shown in Figure 4. There were three stages of the friction coefficient in the running-in wear process. In the first stage, the friction coefficient was in decline from the beginning of the running-in for 2 min, which indicated that the running-in wear states changed greatly and tribological pairs worn severely. When the running-in continued for 5 min, the friction coefficient decreased slowly, which indicated that the running-in wear states tended to be steady at this stage and the tribological pairs worn less. In the last stage, the friction coefficient decreased to a small value and was stable after 5 min, which indicated that the wear state was smooth and steady and the tribological pairs formed the interoperable surface.

In Figure 3, there were also three stages in the running-in wear process. In the first stage, the wear states of tribological pairs changed greatly for 2 min from the beginning of the running-in process, and M2NFV was in decline. When the running-in continued for 5 min, the wear states of tribological pairs tended to be steady in this stage, and M2NFV decreased slowly. In the last stage, the wear state of tribological pairs was smooth and steady, and M2NFV decreased to a small value and changed gently. Therefore, M2NFV can accurately reflect the variation in the wear state of tribological pairs in the running-in process. However, RMS values of both TFV and NFV declined in the first stage, while fluctuated in the second and the last stages. Hence, the RMS values of both TFV and NFV can reflect the change in the running-in wear state of tribological pairs in the first stage, but not in the second and last stages. Hence, the RMS values of TFV and NFV cannot accurately reflect the variation in the wear state of tribological pairs in the running-in process.

The running-in wear process with submerged lubrication is extremely complex, which is influenced by different wear mechanisms, temperatures, lubricating media, etc. A single direction friction vibration is easy to be affected by many factors in its direction. Therefore, the feature extraction in a single direction friction vibration is difficult to reflect the wear state feature of tribological pairs completely, while M2NFV can effectively reduce the impact of the multi-variability and uncertainty of the friction vibration in a single direction. Hence, M2NFV can accurately reflect the variation in the wear state of tribological pairs in the running-in process.

5 CONCLUSION

A norm-coupling method of TFV and NFV is proposed with the norm theory, the feature of NCFV is extracted with matrix 2-norms, and the variation in friction vibration feature is analyzed in the running-in wear test. From the presented results, it can be concluded as follows:

1. The coupling of TFV and NFV is realized with the norm theory, and NCFV preserves the information of both TFV signals and NFV signals.
2. The feature of NCFV is extracted with matrix 2-norms, and M2NFV can accurately reflect the variation of the wear state of tribological pairs in the running-in process.

REFERENCES

Chen, G.X. & Zhou, Z.R. 2003. Investigation of the Modal Coupling Mechanism of Squeal Generation Based on Wavelet Transformation. *Tribology*. 23(6): 524–528.

Chen, G.X. & Zhou, Z.R. 2006. Time-frequency characteristics of friction-induced vibration. Chin. J. Mech. Eng. 42(2): 1–5.

Ji, C.C. 2012. *Study on the dynamic properties during the wear process of cylinder liner-piston ring based on chaos and fractal theories*. Xuzhou: China Univ. MIN. & Technol.

Kang, H.Y. et al. 2006. Diagnosis of bearing fault based on order envelope spectrum analysis. *J. Vib. & SHOCK* 25(5): 166–167, 174, 200.

Li, J.B. & Peng, T. 2009. Improved EMD and its application in rolling bearing fault diagnosis. *J Hunan Univ. Technol*. 23(6): 28–32.

Li, X.H. et al. 2003. Spectrum analysis and its application to gearbox fault diagnosis. J. Vib., Meas. & Diagn. 23(3): 168–170.

Sun, D. 2015. *Study on friction vibrations characteristics in the reciprocating motion running in process of the friction pairs*. Dalian: Dalian Maritime Univ.

Sun, D. et al. 2015. Investigation on friction vibration behavior of tribological pairs under different wear states. ASME J. Tribol. Trans. 137(2): 021606.

Wan, S.T. et al. 2009. Fault diagnosis method of rolling bearing based on undecimated wavelet transformation of lifting scheme. *J. Vib. & SHOCK* 28(1): 170–173.

Zhu, H. et al. 2007. The changes of fractal dimensions of frictional signals in the running-in wear process. *Wear*. 263(7–12): 1502–1507.

*Advances in Materials Science, Energy Technology
and Environmental Engineering – Patty & Zhou (Eds)*
© *2017 Taylor & Francis Group, London, ISBN 978-1-138-19668-1*

Fault diagnosis of a converter transformer based on the FOA-LSSVM

Yong Sun
Maintenance and Test Center, CSG EHV Power Transmission Company, Guangzhou, China

Qi Long
CSG EHV Power Transmission Company, Guangzhou, China

Gang Lv
Gui Yang Bureau, CSG EHV Power Transmission Company, Guiyang, China

Daikuan Huang
Da Li Bureau, CSG EHV Power Transmission Company, Dali, China

Yi Li & Youping Fan
School of Electrical Engineering, Wuhan University, Wuhan, China

ABSTRACT: In order to improve the accuracy of fault diagnosis of a converter transformer, a model has been proposed based on the Least Squares Support Vector Machine (LSSVM). Optimization of traditional LSSVM parameters is aimless and inefficient. In order to solve this problem, this paper uses fruit Fly Optimization Algorithm (FOA) to optimize the penalty parameter C and the kernel function parameter γ in LSSVM and build up the converter transformer fault diagnosis model based on the FOA-LSSVM. It is verified through a case analysis that FOA has higher precision on classification and global search ability compared with Genetic Algorithm (GA) and Particle Swarm Optimization (PSO). The simulation results indicate that the fault diagnosis model of the converter transformer based on the FOA-LSSVM has a faster convergence speed and a higher diagnostic accuracy compared with the traditional LSSVM, GA-LSSVM, and PSO-LSSVM.

1 INTRODUCTION

The fault diagnosis and life expectancy of a converter transformer have always been the key issues in the maintenance of electric power grid. This study focuses on the research on the new method for transformer fault diagnosis and life expectancy-oriented diagnosis. A large number of operating experience and actual research show that the dissolved gas analysis in oil (dissolved gas analysis, the DGA) can find the insulation of the converter transformer internal latent fault earlier, which is one of the most effective and most widely used methods of the current exchange converter transformer fault analysis.

With the development of the intelligent diagnosis technology, more and more artificial intelligence methods such as the Bayesian Classifier, the Artificial Neural Network, and the Support Vector Machine have been introduced into fault diagnosis. Owing to its advantageous nature, SVM has been applied to a wide range of classification tasks. In particular, SVM has been shown to perform very well on many fault diagnosis tasks. However, there is still a need for improving the performance of the SVM classifier.

This paper uses fruit Fly Optimization Algorithm (FOA) to optimize the penalty parameter C and the kernel function parameter γ in LSSVM and build up the converter transformer fault diagnosis model based on the FOA-LSSVM. It is verified through a case analysis that FOA has higher precision on classification and global search ability compared with GA algorithm and PSO algorithm. The simulation results indicate that the fault diagnosis model of converter transformer based on the FOA-LSSVM has a faster convergence speed and a higher diagnostic accuracy compared with the traditional LSSVM, GA-LSSVM, and PSO-LSSVM.

2 BACKGROUND

2.1 *Least Squares Support Vector Machines (LSSVM)*

The Least Squares Support Vector Machine (LSSVM) is an improvement of the standard SVM and performed well on the structural risk minimization principle of statistical learning theory. It has good generalization ability in little samples and can avoid falling into a local minimum. Therefore, it has become a powerful tool of intelligent fault diagnosis and prediction gradually. In the feature space, the LSSVM model is described as:

$$\min J(\omega, \xi) = -\frac{1}{2}\omega^T\omega + \frac{1}{2}C\sum_i^l \xi_i^2$$
$$y(x) = \omega^T \varphi(x) + b \qquad (1)$$

where ω is the weight vector, $\varphi(x)$ is the nonlinear mapping function from input space to high-dimension space, b is the offset, C is the penalty parameter, and ξ_i is the slack variable.

By introducing Lagrangian multipliers α_i,

$$L(\omega,b,\xi,\alpha) = \frac{1}{2}\omega^T\omega + \frac{1}{2}C\sum_i^l \xi_i^2$$
$$- \sum_{i=1}^l \alpha_i[\omega^T\varphi(x_i) + b + \xi_i - y_i] \qquad (2)$$

According to the KKT optimization condition:

$$\begin{cases} \partial L / \partial \omega = 0 \\ \partial L / \partial b = 0 \\ \partial L / \partial \xi_i = 0 \\ \partial L / \partial \alpha_i = 0 \end{cases} \qquad (3)$$

After solving equation (3), the following system of linear equations can be obtained:

$$\begin{bmatrix} 0 & y^T \\ y & \boldsymbol{\Phi} + C^{-1}\boldsymbol{I} \end{bmatrix}\begin{bmatrix} b \\ \alpha \end{bmatrix} = \begin{bmatrix} 0 \\ \boldsymbol{I}_l \end{bmatrix}$$
$$\boldsymbol{\Phi} = y_i y_j \varphi^T(x_i)\varphi(x_j) = y_i y_j K(x_i, x_j) \qquad (4)$$

where $K(x_i, x_j)$ is the kernel function; $y = [y_1,...,y_l]^T$; $\boldsymbol{I}_l = [1,...1,]^T$; $\boldsymbol{I}$ is the unit matrix; and $\alpha_i = [\alpha_i,...,\alpha_i]^T$. After obtaining vector α and b, the decision function of LSSVM is obtained as equation (5):

$$y(x) = \sum_{i=1}^l \alpha_i K(x, x_i) + b \qquad (5)$$

This paper chooses RBF function as the kernel function of LSSVM. It is expressed in equation (6):

$$K(x, x_i) = \exp\left(-\|x_i - x\|^2 / 2\gamma^2\right) \qquad (6)$$

where γ is the kernel function parameter. The penalty parameter C and the kernel function parameter γ are the key issues for establishing the LSSVM fault diagnosis model. This paper uses the fly optimization algorithm to optimize the two parameters, selecting an optimal solution.

2.2 Fruit Fly Optimization Algorithm

The fruit Fly Optimization Algorithm (FOA) was developed based on the foraging behavior of a fruit fly. The fruit fly is superior to other species in olfactory capabilities and visual sense. Therefore, they are capable of utilizing their instinct to locate food. According to the food searching characteristics of a fruit fly swarm, the FOA can be divided into several steps as follows:

Step 1: Initialization. Determine the initial population scale SIZEPOP, maximum iterations MAXGEN, population numbers N, and the initial position of the fruit fly population (X_0, Y_0).

Step 2: Random search. Set the initial iteration times $g = 0$. Then, set the search direction to the random number Rand ( ). Finally, set the search distance to a random value, and therefore, there are:

$$\begin{cases} X_i = X_0 + \text{Random Value} \\ Y_i = Y_0 + \text{Random Value} \end{cases} \qquad (7)$$

Step 3: Preliminary calculation. Calculate the distance of the food location from the origin position D_i. Then, compute the smell concentration judgment value S_i, which is the reciprocal of the distance of the food location from the origin. The specific formula is as follows:

$$\begin{cases} D_i = \sqrt{(X_i^2 + Y_i^2)} \\ S_i = 1 / D_i \end{cases} \qquad (8)$$

Step 4: Calculation of the odor concentration. According to the smell concentration judgment value S_i, the substitution taste concentration judging function (fitness function), calculate the taste of the density of individual fruit fly $S_{m,i}$, $S_{m,i} = f(S_i)$.

Step 5: Finding the best individual. According to the odor density, find out the best taste concentration of the fruit fly individual $S_{m,best}$, $S_{m,best} = \min/\max(S_{m,i})$.

Step 6: Visual positioning. Record and retain the best taste density $S_{m,best}$ and the coordinates of optimal individual fly at this time (X_{best}, Y_{best}).

Step 7. Iterative optimization. First, determine whether $g = $ MAXGEN. If $g < $ MAXGEN, repeat steps 2–5. Finally, decide whether the taste density at this time is the best when the best taste concentration is better than the previous iteration. If yes perform step 6; if otherwise continue to repeat steps 2–5 until the end or when $g = $ MAXGEN.

It can be observed from the above steps that the only parameters FOA algorithm needs to set are the population size, the initial position of the population, the search distance, and the maximum iterations. Compared with other optimization algorithms, it has the advantage of less number of calculations.

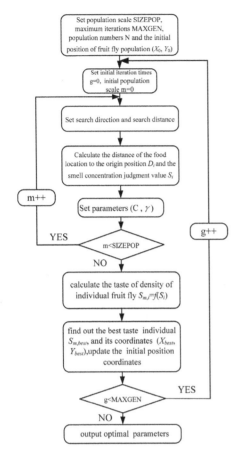

Figure 1. Flowchart of FOA-LSSVM.

2.3 Optimization of LSSVM parameters based on FOA

Parameters C and γ have a larger effect on LSSVM diagnosis accuracy, and this paper uses FOA to optimize the penalty parameter C and the kernel function parameter γ in LSSVM and build up the converter transformer fault diagnosis model based on the FOA-LSSVM. Specific optimization process is shown in Fig. 1

3 FAULT DIAGNOSIS MODEL BASED ON THE FOA-LSSVM

3.1 Fault sample pretreatment and fault classification

This paper takes gases dissolved in transformer oil, such as H_2, CH_4, C_2H_2, C_2H_4, and C_2H_6, as sample for fault diagnosis. In order to improve the stability and accuracy of fault diagnosis algorithm, scale change pretreatment should be carried out to the

original sample data, transforming original sample data into [0, 1]. The specific formula is as follows:

$$x_i' = \frac{x_i - x_{min}}{x_{max} - x_{min}} \qquad (9)$$

where x_i is the gas concentration of the original sample, x_{min} is the minimum gas concentration in the original sample, x_{max} is the largest gas concentration in the original sample, and x_i' is the sample data after pretreatment.

Converter transformer fault can be divided into six types: Partial Discharge (PD), Low-Energy Discharge (LE-D), High-Energy Discharge (HE-D), Low-Temperature overheating (LT), Medium-Temperature overheating (MT), and High-Temperature overheating (HT).

3.2 Fault diagnosis process

Fault diagnoses of the converter transformer based on the FOA-LSSVM process are as follows:

1. Collecting effective converter transformer fault DGA data samples;
2. Getting the training sample and test sample sets by preprocessing DGA data samples;
3. Optimizing the penalty parameter C and the kernel function parameter γ by using FOA algorithm to get the global optimal solution and establishing the FOA-LSSVM fault diagnosis model;
4. Feeding the processed sample data as input to obtain diagnostic results.

3.3 Case study

In this study, a large number of converter transformer DGA data with a clear fault cause were collected. All samples were divided into 150 groups: 120 groups were chosen as the training set and the remaining 30 groups as the test set. Converter transformer fault sample statistics is shown in Table 1.

Search range of parameters C and γ in FOA are set to [0.1, 100] and [0.1, 10], respectively; Maximum generation is 100, population number is 2, and the population size is 20. Fitness curves of FOA are shown in Fig. 2. The result for the optimal parameters is: C = 54.87 and γ = 0.43. Diagnostic results of FOA-LSSVM are shown in Fig. 3. Out of the 30 groups chosen as the test set, 28 groups gained right fault-type judgment with an accuracy of 93.33%. The outcome verified the accuracy and effectiveness of the method evidently.

3.4 Analysis and comparison

In order to verify the superiority of FOA-LSSVM, this paper compares the diagnosis results of

FOA-LSSVM with traditional LSSVM, GA-LSSVM, and PSO-LSSVM. In order to prove the advantage of FOA, the population size, population number, and maximum generation of the GA and

Table 1. Fault diagnosis sample statistics.

Fault type	Training set	Test set
PD	20	5
LE-D	20	5
HE-D	20	5
LT	20	5
MT	20	5
HT	20	5
Total	120	30

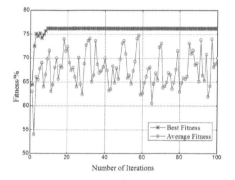

Figure 2.   Fitness curves of FOA.

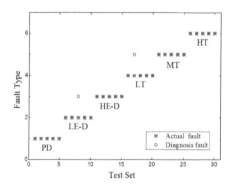

Figure 3.   Diagnostic results of FOA-LSSVM.

Table 2. Comparison of results of different diagnosis methods.

Diagnosis method	C	γ	Convergence generations	Accuracy /%
Traditional LSSVM	82.16	1.07	–	83.33
GA-LSSVM	46.72	0.76	24	86.67
POS-LSSVM	73.43	0.39	18	90
FOA-LSSVM	54.87	0.43	9	93.33

PSO were set as same as FOA. Comparison of results of different diagnosis methods is shown in Table 2.

As shown in the diagnosis result, compared with GA and PSO, FOA uses minimum convergence algebra and gains the obvious advantages in terms of convergence speed. In terms of diagnostic accuracy, the accuracy of traditional LSSVM was 83.33% which is the lowest, and the accuracies of GA-LSSVM and PSO-LSSVM are 86.67% and 90%, respectively, which gains a certain increase. The accuracy of FOA-LSSVM proposed in this paper is 93.33%, which is the highest. As the optimization of traditional LSSVM parameters is aimless and inefficient, compared with PSO and GA, FOA is used in this study. FAO converges rapidly, its global optimization ability is stronger, and its precision is higher under the same number of iterations, and hence, its fault diagnosis accuracy is higher.

4   CONCLUSION

FOA algorithm is very simple, convenient, and effective by using biological group cooperation system and information sharing principle to search for the global optimal solution. It has been proven that FOA has higher precision on classification and global search ability compared with GA algorithm and PSO algorithm by the simulation test.

This paper presents the FOA algorithm for optimization of the penalty parameter C and the kernel function parameter σ and put forward the fault diagnosis model of converter transformer based on the FOA-LSSVM. Examples prove that this method has a faster convergence speed and a higher diagnostic accuracy compared with the traditional LSSVM, GA-LSSVM, and PSO-LSSVM.

REFERENCES

Fei S W & Sun Y (2008). Forecasting dissolved gases content in power transformer oil based on support vector machine with genetic algorithm. *Electrical Power Systems Research, 3*, 507–514.
Meng K, Dong Z Y & Wang D H (2010). A self-adaptive RBF neural network classifier for transformer fault analysis. *IEEE Transactions on Power Systems. 3*, 1350–1360.
Pan W T (2012). A new fruit fly optimization algorithm: Taking the financial distress model as an example. *Knowl. Based Syst. 26*, 69–74.
Pan Q K (2014). An improved fruit fly optimization algorithm for continuous function optimization problems. *Knowl. Based Syst. 62*, 69–83.
Xue Haoran, Zhang Keheng & Li Bin (2015). Fault diagnosis of transformer based on the cuckoo search and support vector machine. *Power System Protection and Control. 8*, 8–13.
Zheng Ruirui (2011). New transformer fault diagnosis method based on multi-class relevance vector machine. *Proceedings of the CSEE . 7*, 56–63.

*Advances in Materials Science, Energy Technology
and Environmental Engineering – Patty & Zhou (Eds)*
© *2017 Taylor & Francis Group, London, ISBN 978-1-138-19668-1*

Study of the milling experiment and cutting parameter optimization of HT250

Yulong Wang, Shan Li, Yu Zhang & Shenzhen Li
Faculty of Mechanical and Electrical Engineering, Kunming University of Science and Technology, Kunming, China

ABSTRACT: By changing the cutting parameters while milling gray iron HT250, and designing an orthogonal experiment based on it, we can get different surface roughness values and cutting force values with different cutting parameters. Using the range analysis method, we analyzed the experimental data and got the optimal levels of each factor. Finally, the optimal parameters combination is obtained when taking the surface roughness or the cutting force as the target.

1 INTRODUCTION

Gray cast iron has good strength, abrasion resistance, heat resistance and shock absorption performance, as well as the casting performance. It can satisfy the requirement of design performance of a high-power diesel engine cylinder (Li, 2011; Zhang, 2010). At present, the global mainstream high-power engine cylinder is made of gray cast iron. In foreign countries, a lot of research and results on machining have been made (Wu, 1983; Iwata, 1984; Carroll, 1988; Jose, 2003). Moreover, the study of machining HT250 has been carried out domestically (Fang, 2001; Tang, 2007), but remains to be explored deeply. Therefore, studying the factors affecting the cutting performance of the gray cast iron is of important theoretical and practical value.

Taking a diesel engine cylinder from a company as the research object, this paper studies the optimization parameters during the finish machining process. Measuring the cutting force and surface roughness after completing milling experiments, this paper gets down to research from the following aspects. It studies, respectively, the influences of the cutting parameters on the cutting force or surface roughness by using the range analysis method, and it concludes the optimal group of cutting parameters named M when taking the cutting force as a target and the optimal group of cutting parameters named N when taking the surface roughness as a target.

2 EXPERIMENTAL DESIGN, IMPLEMENTATION AND DATA COLLECTION

This paper adopts the method of orthogonal test design for milling experiments. First, we need to confirm the factors that affect the experiment results, and design experiment plan according to these factors. Then, conduct the orthogonal experiment and evaluate the experiment results. Finally, we need a further experiment when necessary in order to make the experiment perfect and the results reliable.

Table 1. Orthogonal experiment list.

Group No.	A/(m/min)	B/(mm/z)	C/mm	Ra/μm
1	1 (140)	1 (0.04)	1 (0.2)	0.9390
2	1	2 (0.07)	2 (0.5)	1.3080
3	1	3 (0.09)	3 (0.8)	0.9927
4	1	4 (0.12)	4 (1.0)	2.0500
5	2 (196)	1	2	1.1767
6	2	2	1	1.1840
7	2	3	4	0.7260
8	2	4	3	0.6320
9	3 (252)	1	3	0.8497
10	3	2	4	0.7043
11	3	3	1	1.0210
12	3	4	2	0.9517
13	4 (308)	1	4	0.8223
14	4	2	3	0.5200
15	4	3	2	0.6890
16	4	4	1	0.7580
K_{1j}	5.2897	3.7877	3.9020	
K_{2j}	3.7187	3.7163	4.1254	
K_{3j}	3.5267	3.4287	2.9944	
K_{4j}	2.7893	4.3917	4.3026	
Optimal level	A4	B3	C3	
R	2.5004	0.963	1.3082	
Important order		$A > C > B$		

As a result of the limitation of experimental conditions, we choose a type of YG8 vertical milling cutter with a diameter of 20 mm and choose the engine cylinder block with HT250 materials as the workpiece, and type of KHC63 horizontal machining center as the machine tool. According to the actual cutting situation, the radial cutting depth is 20 mm, namely the whole cutter diameter. In this paper, the milling speed, feed per tooth, and axial depth of cut are chosen as the experiment factors, and are denoted by using A, B, and C, respectively. Depending on the cutting parameters during the finish machining process and the performance of the machining tool, an orthogonal experiment table $L_{16}(4^3)$ with three factors and four levels has been built, as shown in Table 1 (the roughness values in the table are filled after experiment implementation and data collection).

3 PROCESSING AND ANALYSIS OF EXPERIMENTAL DATA

The paper chooses roughness and cutting force as the experiment targets. The cutting force is measured by a force measurement system named Kistler made in Swiss and the roughness is measured by a portable roughness measuring instrument named Mahr made in Germany.

After collecting the data of roughness values and cutting force values, this paper analyzes, respectively, the cutting parameters that influence roughness or cutting force, and concludes the optimal parameter combination named M for improving the quality of the surface and the optimal parameter combination named N for reducing the cutting force. Then, by using the robustness design method, it analyzes the data and concludes the optimal parameter combination named P, under which the roughness and cutting force can get relative ideal results at the same time.

3.1 *Range analysis of experiment data*

For the range analysis, readers can refer to relevant books for detailed introduction. By using range analysis, we can get the following results: □confirming the influences of different levels under the same factor; □ confirming the important order of each factor that influences the experiment results; and □ getting the optimal parameters combination.

3.1.1 *Range analysis of roughness values*
1. Confirming the optimal level of each factor and the optimal combination

First, it is necessary to analyze the effects of factor A to roughness. Define level 1 of A as A1, then level 2 as A2, level 3 as A3, and level 4 as A4. It

can be concluded from Table 1 that the effect of A1 is reflected only in the no. 1, 2, 3, and 4 tests, the effect of A2 is reflected only in the no. 5, 6, 7, and 8 tests, the effect of A3 is reflected only in the no. 9, 10, 11, and 12 tests, and the effect of A4 is reflected only in the no. 13, 14, 15, and 16 tests. Here, we rule that X1-X16 represent the results of tests 1–16 in turn. Therefore, we can get the results as follows.

The total of experiment index of factor A under level 1 is as follows:

$$KA1 = X1 + X2 + X3 + X4 = 0.9390 + 1.3080 + 0.9927 + 2.0500 = 5.2897$$

In the same way, the total of factor A under level 2, level 3, and level 4 can be calculated as follows:

$$KA2 = 3.7187$$
$$KA3 = 3.5267$$
$$KA4 = 2.7893$$

Obviously, the four tests A1, A2, A3, and A4 are used to evaluate the effect of factor A on experiment indexes, namely roughness. In the four tests, each level of factor B and factor C appears only one time and there is no interaction between B and C. Therefore, it can be concluded from Table 1 that different combinations of the levels of B and C factors have no effect on the experiment indexes. Then, for the tests A1, A2, A3, and A4, the experimental conditions are exactly the same. If factor A has no effect on experiment indexes, then KA1, KA2, KA3, and KA4 should be equal. However, from the calculation results above, we can know that KA1, KA2, KA3, and KA4 are not equal. Clearly, this is caused by the change in levels of factor A. Therefore, it can be concluded that KA1, KA2, KA3, and KA4 reflect the influence of A1, A2, A3, and A4 on the experiment indexes, namely roughness. As it is known, smaller values of roughness are better. As KA1 > KA2 > KA3 > KA4, we can judge that level 4 of factor A is the superior level.

In the same way, we can get KB1 = 3.7877, KB2 = 3.7163, KB3 = 3.4287, KB4 = 4.3917.

As KB4 > KB1 > KB2 > KB3, it can be concluded that level 3 of factor B is the superior level, that is to say, the roughness values are minimum when the feed per tooth is 0.09 mm/z (602 mm/min).

In the same way, we can get KC1 = 3.9020, KC2 = 4.1254, KC3 = 2.9944, KC4 = 4.3026.

As KC4 > KC2 > KC1 > KC3, it can be concluded that level 3 of factor C is the superior level, that is to say, the roughness values are minimum when the axial depth of cut is 0.8 mm.

From the above analysis, it can be concluded that the superior levels of milling speed, feed per tooth, and axial depth of cut are, respectively,

A4, B3, and C3. Therefore, we can get the optimal parameter combination as M: A4B3C3, i.e. the milling speed, is 308 m/min, the feed per tooth is 0.09 mm/z, and the axial depth of cut is 0.8 mm. The calculated results given above can be filled in the corresponding positions in Table 1.

2. Confirming the important order of each factor

According to the range analysis, it is necessary to calculate the range in order to confirm the important order of each factor. The larger the range, the greater the influence of the factors on the target. According to the definition of range, the range of each factor can be calculated as follows:

$$R_A = KA1\text{-}KA4 = 5.2897 - 2.7893 = 2.5004$$

$$R_B = KB4\text{-}KB3 = 4.3917 - 3.4287 = 0.963$$

$$R_C = KC4\text{-}KC3 = 4.3026 - 2.9944 = 1.3082$$

For $R_A > R_C > R_B$, the important order of each factor is A, C, B.

3.1.2 Range analysis of cutting force

The results of the measured cutting force are given in Table 2.

1. Confirming the optimal level of each factor and the optimal combination

Table 2. The milling force measurements and range analysis for them.

No.	A/(m/min)	B/(mm/r)	C/mm	F/N
1	1 (140)	1 (0.04)	1 (0.2)	199
2	1	2 (0.07)	2 (0.5)	238
3	1	3 (0.09)	3 (0.8)	449
4	1	4 (0.12)	4 (1.0)	632
5	2 (196)	1	2	172
6	2	2	1	156
7	2	3	4	619
8	2	4	3	476
9	3 (252)	1	3	228
10	3	2	4	612
11	3	3	1	154
12	3	4	2	167
13	4 (308)	1	4	381
14	4	2	3	512
15	4	3	2	243
16	4	4	1	185
K_{1j}	1518	980	694	
K_{2j}	1423	1518	820	
K_{3j}	1161	1465	1665	
K_{4j}	1321	1460	2244	
Optimal level	A3	B1	C1	
R	357	538	1550	
Important order		C>B>A		

With the same method used previously, the total of each experiment target of every factor can be calculated and the optimal level of each factor can be confirmed. The results are as follows.

Level 3, namely 252 m/min, is the optimal level of the milling speed.

Level 1, namely 0.04 mm/z, is the optimal level of the feed per tooth.

Level 1, namely 0.2 mm, is the optimal level of the axial depth of cut.

The calculated results given above can be filled in the the corresponding positions in Table 2.

From the calculation and analysis, it can be concluded that the optimal parameter combination to reduce the cutting force is A3B1C1, which is denoted as N.

2. Confirming the important order of each factor

In the same way mentioned above, it can be concluded that $R_C > R_B > R_A$. Hence, we can judge the important order of factors, which affect the cutting force, given as follows: axial depth of cut, feed per tooth, and milling speed.

The results can be filled in the corresponding positions in Table 2.

4 CONCLUSION

We can get the following conclusions through the analysis of this paper.

1. Taking roughness as the only target, the best roughness can be obtained by using the optimal combination M-A4B3C3, namely 308 m/min of milling speed, 0.09 mm/z of feed per tooth, and 0.8 mm of axial depth of cut.

2. Taking cutting force as the only target, the smallest cutting force can be obtained by using the optimal combination N-A3B1C1, namely 252 m/min of milling speed, 0.04 mm/z of feed per tooth, and 0.2 mm of axial depth of cut.

ACKNOWLEDGMENTS

This work was supported by the National Key Science and Technology Projects (No. 2012ZX04012-031).

Corresponding author: Li Shan, an associate professor researching on the contemporary integrated manufacturing system. Email: 624814911@qq.com.

REFERENCES

Carroll J.T., Strenkowski J.S. Finite element models of orthogonal cutting with application to single point diamond turning[J]. Int. J. Mech. Sci., 1988, 30: 899–920.

Fang Gang, Zeng Pang. Advance in numerical simulation technology for cutting process[J]. Advances in mechanics, 2001, 31(3): 394–404.

Iwata, K., Osakada K., et al. Process modeling of orthogonal cutting by rigid-plastic finite element[J]. Journal of Engineering Materials and Technology, 1984, 106(2): 132.

Jose, L. Thermo mechanical analysis of a chip machining process[J]. ABAQUS Users' Conference, 2003.

Li Shaohua. The operation situation of car industry in China during 2010 and twelfth five-year automobile industry development goals. Tire industry, 31.6 (2011): 330–335.

Tang Zhitao, Liu Zhanqiang, Ai Xing. A study of Thermo-elastic-plastic Large-deformation Finite Element Theory and Key Techniques of Metal Cutting Simulation[J]. China Mechanical Engineering, 2007, 18(6): 746–750.

Wu De, Liu Jingan. Manual of alloy cast iron[M]. Chongqing: Science and technology literature press, 1983.

Zhang Cheng. The policy research of car industry of China[D]. Beijing: Minzu University of China, 2010.

*Advances in Materials Science, Energy Technology
and Environmental Engineering – Patty & Zhou (Eds)*
© 2017 Taylor & Francis Group, London, ISBN 978-1-138-19668-1

Light exoskeleton for civil use: A pragmatic carrying assistive apparatus

Peizhang Zhu
College of Mechanical and Electrical Engineering, Central South University, Changsha, China

Fenglin Han
*College of Mechanical and Electrical Engineering, State Key Laboratory of High Performance
Complex Manufacturing, Central South University, Changsha, China*

ABSTRACT: This paper presents the development of a civil-use carrying assistive apparatus, which is pragmatic for its high cost performance and flexibility. It functions in situations where carrying is common, filling up the gap of exoskeletons for daily tasks. Using motor-driven steel cables, this apparatus assists two frequently used parts of the operator: upper limbs and psoas. The assistance of the upper limbs concerns vertical actions, and is achieved by a windlass-like process. The assistance of the psoas concerns helping the operator bend and straightening, and is joined by a lower-limb structure that supports generating reactive force. To improve flexibility, a new clutch called electromagnetic jaw clutch was developed and used to help switch to a motion-free mode. The control system employs SCM STC89C52 as a core controller and contains force sensors, IMU and servo drivers. The test proved that the apparatus improves the carrying capability and has broad prospects.

1 INTRODUCTION

A traditional exoskeleton is an electromechanical structure worn by a human user and matching the shape and functions of human body. It is able to augment and/or treat the ability of human limbs that are weak or injured (Pons 2008, Perry et al. 2007, Low 2011). Moreover, it merges the machine power and the human intelligence in order to enhance the intelligence of the machine and to power the operator (Pons 2008). The US military has been developing several exoskeletons to augment and amplify the ability of the soldier for military purposes from early times (Cloud 1965).

Though invested by governments and well developed by military laboratories (Bogue 2009), the exoskeletons are still rare in civil markets largely because of the significantly high cost and difficulty to build a control system that is able to make the apparatus both easy to use and affordable. Many types of control systems are being developed for exoskeletons (Kazerooni et al. 2006), and the dynamic-model-based control system is one of them. In general, there are three ways to obtain the model (Anam & Al-Jumaily 2012): the mathematical analysis, the system identification, and the artificial intelligent method. The first one, mathematical modeling, is obtained theoretically by analyzing the system's physical properties; the BLEEX, a 6-DOF lower limb exoskeleton is a good example of this type (Kazerooni et al. 2005; Kazerooni et al. 2006; Zoss et al. 2006; Ghan et al. 2006).

Besides being viewed from the control system aspect, today's existing exoskeletons can also be viewed from the mechanical aspect. Owing to its futuristic appearance, its mechanical characteristics became very popular in recent years and have been reviewed and discussed by many authors: Bogue et al. (2009), who discussed the recent development of the exoskeleton; Gopura et al. (2009, 2011, 2016), who classified the mechanical designs of the existing active upper-limb exoskeleton robot and reviewed the mechanical design of upper extremity exoskeletons; Lo et al. and Xie et al. (2012), who presented the recent progress of upper limb exoskeleton robots for rehabilitation; Slavka et al. (2013), etc.

Many scholars focus on the applications of exoskeletons that can meet multiple requirements for human power augmentations, such as the full body exoskeleton HAL (Hybrid Assistive Limb) developed by the University of Tsukuba to augment the nurse's power to take care of the patient (Yamamoto et al. 2002) or/ and exoskeletons that help rehabilitation like the lower limb exoskeleton EXPOS (EXoskeleton for Patients and Old by Sogang) developed by Sogang University in Korea (Kong & Jeon 2006). However, the development and mechanical design of an affordable exoskeleton for ordinary daily use seem to be overlooked; except for necessary medical use like the upper limb exoskeleton type designed by John Hopkins University to help elbow flexion of paralyzed people (Schmeisser & Seamone 1973; Bogue 2009), ordinary people are not necessarily in need of such a specially designed and expensive exoskeleton. Among all the assistive functions of a power-augmentation exoskeleton, carrying assistance for upper limbs is one of the most used and concerned functions besides walking assistance: Perry et al. (2007) talked about human power amplifications of the upper limbs, Gopura et al. (2009, 2011, 2016) developed the hardware systems of an upper-limb exoskeleton, Wu et al. (2011) designed a device to strengthen upper-limb muscle, Kiguchi et al. (2008) developed exoskeleton robots for assisting upper-limb motion since, as he mentioned, "upper-limb motion is involved in a lot of activities of everyday life", etc.

In order to fill the gap of civil-use exoskeleton that is affordable and pragmatic, an assisting apparatus that focuses on carrying function is required. When carrying function of an exoskeleton is isolated, there is no need to use conventional structures seen in an exoskeleton to enable assistance. In a conventional structure, motors are installed at every joint to drive the upper limbs (Yan et al. 2014). These motors need to cooperate to mimic the actions of human limbs (Hassani et al. 2014), and since human actions are so complicated, fine sensors are necessary to collect accurate data to help motors work correctly. These sensors along with the structures and power system drive up the price too high. Additionally, the data collected from numerous sensors are usually oceanic for a simple control system to handle, thus researchers have to put a lot of efforts into developing the control system.

While powered exoskeletons require massive investment, a different structure design that focuses on practicality can cut down the expense and considerably raise the civil value.

2 FEATURES OF THE APPARATUS

2.1 Conception

When working in situations where moving or holding loads happens a lot, people often fail to endure due to an analogy of the cask effect: the upper limbs are more likely to become powerless faster than the lower limbs and shoulders or any other stronger parts involved in this action. Therefore, focusing on aiding upper limbs would add to the operator's productivity.

During the process of carrying, gravity is the main force that applies on the load and is applied in perpendicular direction; there are only minor loading forces existing in other directions—back and forth, left and right. Therefore, it seems that the problem is made complicated when an exoskeleton with conventional structure is used to carry things around: joint motors are no longer needed and the end-effector assistance (Meng et al. 2015) makes the whole carrying process easy and fast.

One method to carry out the end-effector assistance is to connect one end of a steel cable to the operator's wrist or a hook-like part that can help hold loads and the other end to the powered parts during carrying so that the main resistance—load on the upper limbs—that the operator confronts is ameliorated. Moreover, the use of steel cables adds to the flexibility of the upper limbs, which is vital for operators' efficiency of actions. Moreover, details about the assistance of the upper limbs will be covered later.

Although the force on the upper limbs is the main concern, the burden on psoas is also noticeable for that every gram the operator grabs generates resistant torque that goes to psoas. Therefore, it is creative but natural to take psoas assistance into consideration.

During the process of carrying weights around, analysis of human body's actions like bending down, straightening, and walking are often simplified into some two-dimensional movements with the side elevation as the plane. However, the fact is that the human body is so flexible if it is discussed in a two-dimensional way and the psoas assistance is achieved with mechanical transmission like gears or joint motors, the result would probably be either the operator being so limited in actions or the control system being too complicated to meet the requirement of degree of freedom. Therefore, as steel cables mentioned can improve the flexibility of the upper limbs and simplify the process, it is reasonable to also get steel cables involved in psoas assistance. The specific mechanism and structures for psoas assistance will be discussed later.

Moreover, in order to get a longer battery life not by merely enlarging its capacity, a way to save power needs to be worked out. Therefore, unlike the traditional way with which an exoskeleton works, this apparatus depends on an intermittent work to avoid consuming unnecessary energy and the mechanism behind this will also be discussed later.

2.2 Design and functions

The exterior of this light exoskeleton is shown in Figure 1. The apparatus which consists of an upper body part and a lower limb part is capable of performing the following four main functions: assisting the operator's upper limbs to lift up loads; aiding the operator to bend down and straightening up by assisting the operator's psoas; helping the operator hold the load while standing or walking around and transmit the loading force to the ground; and finally, switching to a motion-free mode to improve the operator's flexibility when assistance is not needed.

When the operator needs to carry a load, s/he could first make the action of bending down and grabbing the load in a motion-free mode, without being impeded by the resistance from the gears. Then, according to the relations between actions of the operator and reactions of the apparatus listed in Table 2, the DC servo motor A shown in Figure 1 would be responsible for assisting the upper limbs to lift up and hold the loads while motor B takes care of assisting the operator's psoas to help straighten

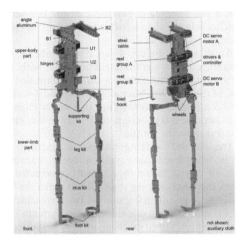

Figure 1.   Design and appearance.

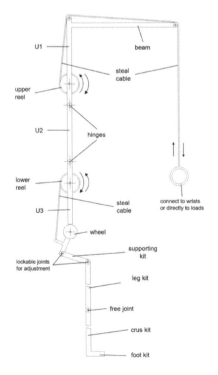

Figure 2.   Structure diagram.

up. Moreover, when the operator needs to lower down the load to a target place, s/he can lean over the upper body to certain degrees in order to inform the controller, and the latter would respond by reversing the motors at a certain speed so the operator could lower down the load and bend down the upper body simultaneously with assistance to both the upper limbs and the psoas. In addition, as mentioned in the last paragraph, the whole structure is designed to be able to transmit the loading force to the ground when the operator is standing or walking around. All the functions and the mechanism will be discussed in the following sections.

2.3   Structure

As is demonstrated in Figures 1 and 2, the upper body part can be divided into four components: the cantilever, U1, U2, and U3. The cantilever, consisting of two beams (B1, B2) and one angle of aluminum, is used to support the steel cables and produce a distance between the operator's body and the loads; U1, U2, and U3 are connected sequentially by the hinges; U1 and U3 both contain a DC servo motor and a group of two reels, which turn to winch the steel cables so that assistance is provided, while U2 contains mainly the servo drivers and the controller inside.

The lower limb part can also be divided into four components: the supporting kit, leg kit, crus kit, and foot kit. The supporting kit consists of two wheels used to support the upper body part and several lockable joints used to adjust the position of the supporting kit, so that it can fit the upper body part better according to different users. Both leg kit and crus kit respond to the supporting task and are adjustable in length; foot kit can be fixed to the operator's feet and help stabilize the whole structure when the apparatus is functioning, and it is also adjustable to fit various foot lengths.

Additionally, a changeable module is connected to the cable tail. It could be a hook that directly controls the loads or a bracelet that is wore by the operator as well as other functional modules. In addition, these changeable modules vary according to the situations and contribute to the practicality of the apparatus.

3   KEY MECHANISMS

3.1   Intermittent functioning

One core feature of this apparatus is the intermittent working mechanism, which is very important for reducing power consumption of the apparatus and is achieved by the structure called electromagnetic jaw clutch. It is a smart design that combines two common types of traditional clutches: the jaw clutch and the electromagnetic clutch.

For a traditional powered exoskeleton that can assist its operator to carry heavy loads, once the user puts it on, the whole system must keep functioning irrespective of the situation because if its dynamic system stops outputting power, resistance from those offline motors would make it difficult for the user to make any actions or even completely unmovable. Hence, when the user is between the actions of carrying loads, the continuous functioning wastes much energy, leading to the decrease in power endurance. On the contrary, using the EJC, this apparatus is able to switch between two states: enabling assistance and motion free. When jaw-meshing engagement builds, the motor outputs torque to assist carrying loads, and when disconnected, the whole power system stops working to save energy. This mechanism not only is confined to prolonging power endurance, but can also give the operator much freedom of movement when not carrying loads. Additionally, without complicated joint motor systems, it drastically reduces the size, weight, and cost of the whole apparatus to make it more practical for civilian use.

As EJC is a combination of two types of traditional clutches, two traditional jaw clutch parts (J1, J2) can be found in an EJC as shown in Figure 3. A traditional jaw clutch consists of two cylindrical semi-clutches, and each has circularly arranged teeth on one side or around the outside surface. One apparent advantage of jaw clutch is that the two semi-clutches engage each other by meshing, and therefore, they are capable of bearing significant torque without any relative rotation if properly sized. However, its connection and disconnection are usually controlled manually via a series of machine structures and this adds up to the use of excessive machine parts.

A typical electromagnetic clutch also has two cylindrical semi-clutches, which engage each other by friction instead of jaw mesh, and its way of control—using electromagnet—is an advantage of this type of clutch since anything that can be controlled by current is good for an automatic system.

Through frictional contact, the load is gradually accelerated to match the speed of the rotor, and this is a vital feature to avoid any rigid impact when actuating a high-speed rotor. However, in a power-assistance apparatus like this, high-speed and ever-rotating rotors do not exist, and thus, the rigid impact is out of concern. Additionally, friction contact consumes most part of the energy and wastes them through thermal dissipation. Therefore, a jaw-meshing engagement of the two semi-clutches is better than a friction contact in carrying assistance apparatus.

As shown in Figure 4, an EJC combines the advantage of an electromagnetic clutch with the advantage of a jaw clutch: it can be controlled by a SCM (Single Chip Microcomputer) and does not need continuous current inflow to keep the engagement because it is jaw coupling but friction contact that transmits the torque.

The detailed working process is described as below: when a reduction gearbox that is connected to a DC servo motor (not shown) is actuated, it drives the bevel gear on shaft (70) to make the latter rotate. A group (first group) of two semi-clutches (41,42) with teeth on one side and Neodymium ring magnets (61,62) fixed on them are mounted directly on shaft 70, and the shaft passes on torque to semi-clutches through a flat key on each side. Additionally, the semi-clutch is axially movable on the shaft, using the flat key as a rail. When a signal given by the operator reaches SCM, the connection between a DC power and the coils (51,52) that are wrapped around their round bases will switch on. Thus, the energized coil builds up a magnetic field that attracts the Neodymium ring magnets. Since the round bases are, respectively, fixed to another group (second group) of two semi-clutches (31,32), the magnetic attraction drags the two semi-clutches (41,42) of the first group to move, respectively, toward the two semi-clutches (31,32) of the second group. Thus, a jaw-meshing engagement built between semi-clutches and torque is able to pass on to the reels (21,22) connected, respectively, to the two semi-clutches (31,32)

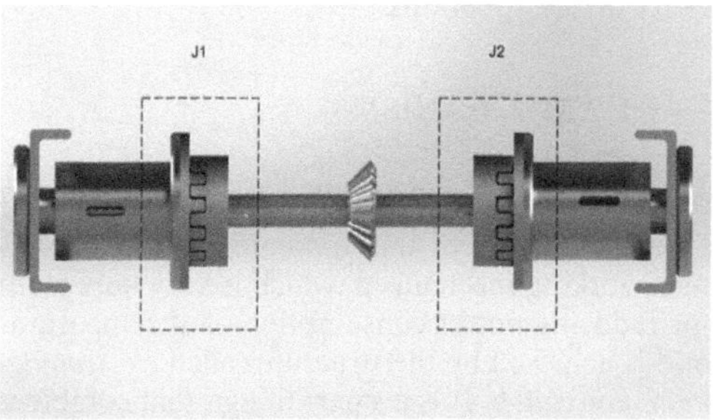

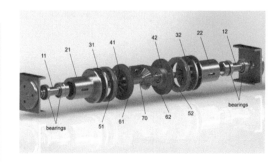

Figure 3.   The jaw clutch part in an EJC.

Figure 4.   Exploded view of the electromagnetic jaw clutch.

to output pulling force and assist carrying. More specifically, the reels are, respectively, mounted on shafts (11,12) using bearings shown in Figure 4. Shafts (11,12) are fixed to the end caps, and thus when jaw engagement is functioning, the reels along with shaft (70) will be revolving round the axis of the fixed shafts (11,12). Moreover, this means that the two ends of shaft (70) are supported, respectively, by one end of each of the fixed shafts. When another signal sent by the operator reaches SCM, the latter reverses the direction of the current flowing in the coils, and thus, the magnetic field is reversed to create a repulsive force on the Neodymium ring magnets. As a result, the repulsion drives semi-clutches (41,42) to move away from semi-clutches (31,32) to disconnect the jaw-meshing engagement.

The ability to switch swiftly and easily between the two states enables the apparatus to work longer and have less limitation on the users.

3.2   Upper limb assistance

The assistance of the upper limbs, as mentioned before, is a process of helping operators to resist the greatest force component produced by the loads. As gravity takes most part of the whole force exerted on the operator's upper limbs, even if the force of inertia is accounted when moving around, a structure to help lift loads vertically is in need. We can easily find such structures above a well or an elevator, and it is called a winch, and the reels driven by EJC function are two winches.

As Figure 5 shows, the detailed assisting process of the upper limbs is described as follows: each end of the two steel cables (C1, C2) is wrapped around two reels (R1, R2) of the upper EJC, respectively, with the tail of each cable stuck into a hole on the reel, so that the cable does not slip. The other end of the steel cable is wrapped around the inner side of shafts (Z1, Z2) with a few more rounds, and its tail also sticks into a hole on the shaft. On the outer side of each of the shaft (Z1, Z2), there is another group of two steel cables (P1, P2) wrapping, and the two cables are connected to the operator's wrists or the load hooks.

When actuated, EJC allows the torque to pass from motor to the reels, which function as the winches to pull back the two steel cables (C1, C2). This tension drives shafts (Z1, Z2) to start turning, and since cables (C1, C2) and cables (P1, P2) are wrapped and fixed to the same shafts, the torque makes cables (P1, P2) pull back and lift up the operator's wrist or the load hook, helping to carry the loads.

3.3   Psoas/waist assistance

Besides assistance to the operator's upper limbs, which are the most burdened parts involved, the load on psoas is also noticeable for that psoas generates the torque needed to straighten up the bent body when grabbing a weight from the ground or a lower place. Therefore, it would be helpful to involve some psoas assistance. Inspired by the flexible power transmission using winches and steel cables and the intermittent working mechanism enabled by EJC, we decided to also use an EJC and steel cables as the core of psoas assistance. The process needs the aid of a pair of lower limb structures similar to those of traditional exoskeletons, as shown in Figure 1, and these structures are tied to the operator's legs and ankles and are adjustable in length. On the top of the structures, as shown in Figure 6, there are the seats (S1, S2) for the tails of steel cables that come from the reels of the lower EJC to connect to. The wheels shown in Figure 6 are used to support the upper body part of the apparatus, which will be discussed later.

The detailed process of psoas assistance is described below: as shown in Figure 6, the actuation and working mechanism of lower EJC are the same with upper EJC, and the activation is controlled by a SCM that receives signal from the operator and some sensors comprehensively. Two cables, like the ones used in the upper EJC to assist upper limbs, are respectively wrapped around reels (R3, R4) with one tail of each fixed to the latter to avoid slip, but the difference is that the other tail of each cable comes to be connected to the seats

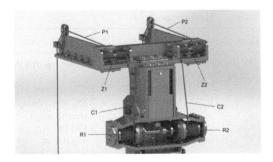

Figure 5.   The assistance of the upper limbs.

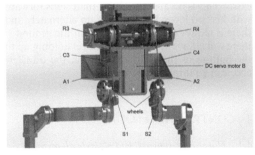

Figure 6.   The assistance of the psoas and top of the lower limb structures.

(S1, S2) mentioned before. Thus, when the operator bends down for the loads with the EJC disconnected, the steel cables are drawn out. After the load is ready, the operator actuates EJC to drive the reels (R3, R4) to turn, thus they pull the steel cables (C3, C4) back in. As the lower-limb structures are stabilized by the operator's lower limbs, the seats are stationary bases for the cables to exert force on, and according to the counterforce principle, the reels are now actually being pulled by the seats. Hence, this counterforce helps shorten the distance between the seats of the lower-limb structures and the reels of the lower EJC by pulling the upper body of the operator closer to the seats, and as a result, the operator is assisted to straighten up from the bend-down posture.

3.4 *Load transmission*

In this apparatus, as shown in Figure 1 (front view), the whole upper body part is supported by the lower limb structures, so that the force generated by the gravity of both the loads and the weight of the apparatus itself can be transferred to the ground, and thus, the operator does not have to suffer heavy loads just to alleviate upper limbs and psoas.

The specific transference path is described below: the cables that connect to the operator's wrists pass through some guide pulleys installed on the T-beams (B1, B2), thus the beams are the first group of apparatus units to bear loading force. Then, the force transfers to the main part of the upper body, which contains three sections (U1, U2, and U3) that can relatively rotate freely in a certain range, and this three-section design offers the operator a more comfortable and soft experience. Subsequently, wheels on the top of the lower-limb structures withstand two abutments fixed to section U3 as shown in Figures 6 and 1. Thereafter, the loading force transfers to the lower limb part and, apparently, it passes through the leg kit, crus kit, and foot kit in sequence, and finally, goes to the ground. Additionally, when the operator is walking with loads, this path of force transference still exists. This is because a walking man steps forward with his two lower limbs raising up alternately and there is always one limb standing on the ground—there will always be at least one wheel holding up to the bottom of the abutment, so that the transference can go through this contact.

4 CONTROL AND LOGIC

4.1 *Controller, sensors, and electric elements*

With the help of sensors listed in Table 1, the apparatus is able to assist the operator in performing the actions listed below: first, to lift up or lower down the loads; second, to bend down or straighten up the upper body; and third, to hold the loads and carry them around.

By using gloves or fingerstalls, force sensor A is mounted to closely fit the pad of the left index finger, so that it can detect whether or not the operator is carrying enough weight of loads (user set) for activating assistance and controlling the state of EJCs. Force sensor B is mounted on the left palm so that the operator can adjust the speed with which the load is lifted or lowered by pressing his/her palm against the load with different forces. IMU collects data on posture and movements of the operator and help decide which part of the apparatus functions and which does not: when the operator stands straight, the psoas assistance will not function and he or she can walk freely; when the operator leans by more than 10 degrees, the psoas assistance will function; and when the operator makes a swift bending-forward action, the motors will reverse and help the operator lower the load or bend down easily. The detailed relations between sensors and actions will be discussed later.

The DC servo motor A outputs power to assist the upper limbs. The DC servo motor B assists psoas. Electric relays control the current flowing through the electromagnetic coils and help revert the magnetic field of the coils. DC-DC voltage regulators provide proper voltage, which is tunable, to SCM and the electromagnetic coils. The electromagnetic coils generate magnetic field to attract or repulse the Neodymium ring magnets.

4.2 *Main circuit*

A schematic circuit diagram, as shown in Figure 7, contains all components that take power from the main battery. The voltage regulators help to avoid using more than one battery and the relays ([K1a], [K2a],...) are responsible for controlling

Table 1. Main components.

Controller	Sensors	Electric elements
STC89C52 SCM	Force sensor A	DC servo motor A
	Force sensor B IMU	DC servo motor B 2 DC servo drivers 8 electric relays 2 DC-DC voltage regulators 4 electromagnetic coils

the electromagnetic coils ([L1], [L2],...) of the EJC. The relays determine whether or not the coils are electrified and the direction of current: [K1a] and [K1b] share one signal output, and hence, they trigger on or off simultaneously, and so do the other pairs of relays ([K2a] and [K2b], [K3a] and [K3b], [K4a] and [K4b]). Importantly, relay pairs in the same sub-circuit do not switch on at the same time; for example, while [K1a] and [K1b] are on, [K2a] and [K2b] must stay switched off. The details are listed in Table 2.

4.3 Input signals and the actions

The control of the apparatus is based on three units: a controller that uses SCM STC89C52 mentioned before as the core, a sensing unit that contains force sensors and an IMU, and a driving unit that consists of servos and motors.

Based on the four main functions mentioned earlier, the logic relations are developed to link up the three units to make the apparatus work naturally. The controller responds to input signals from the sensing unit in those ways demonstrated in Table 2.

4.4 Function verification

In order to check whether the apparatus can be applied practically, as shown in Figure 8, a girl was

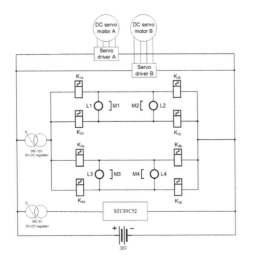

Figure 7.   A schematic circuit diagram.

Table 2.   Logic relations between actions, input signals, and their results.

Related actions	Involved sensors	Signal type	Signal content	Controlled objects	Results
(Upper limbs) Hold, lift, lower down	Force sensor A only	Analog voltage	$U(A)>U0$ (user set)	Relays [K1a], [K1b]	[K1a] and [K1b] trigger on, so the upper coils [L1] and [L2] generate attraction on ring magnets [M1] and [M2]; thus, the upper EJC is engaged
(Psoas) Hold, straighten, bend	Force sensor A and IMU	Analog voltage and serial	$U(A)>U0$ and $I(\theta)>10°$	Relays [K3a], [K3b]	[K3a] and [K3b] trigger on, so the lower coils [L3] and [L4] generate attraction on ring magnets [M3] and [M4]; thus, the lower EJC is engaged
(Upper limbs) Motion free	Force sensor A only	Analog voltage	$U(A)\leq U0$	Relays [K2a], [K2b]	[K1a] and [K1b] switch off, then [K2a] and [K2b] trigger on, so the upper coils [L1] and [L2] generate repulsion on ring magnets [M1] and [M2]; thus, the upper EJC is disengaged
(Psoas) Motion free	Force sensor A and IMU	Analog voltage and serial	$U(A)\leq U0$ and $I(\theta)>10°$	Relays [K4a], [K4b]	[K3a] and [K3b] switch off, then [K4a] and [K4b] trigger on, so the lower coils [L3] and [L4] generate repulsion on ring magnets [M3] and [M4]; thus, the lower EJC is disengaged
(Upper limbs) Lift up	Force sensor B only	Analog voltage	$U(B)>U1$ (user set)	DC servo driver A	DC servo motor A is actuated and the value of $U(B)$ determines its speed; thus, the upper limbs of the operator are being assisted
(Psoas) Straighten up	Force sensor B and IMU	Analog voltage and serial	$U(B)>U1$ and $I(\theta)>10°$	DC servo driver B	DC servo motor B is actuated and the value of $U(B)$ determines its speed; thus, the psoas of the operator is being assisted
(Upper limbs) Lower down	Force sensors A and B, and IMU	Analog voltage and serial	$U(A)>U0$ and $U(B)\leq U1$ and $I(\alpha)>A0$ (trigger, user set)	DC servo driver A	DC servo motor A is reversed and rotates at a uniform speed n0, which is customizable; thus, the load on the upper limbs is lowered with assistance
(Psoas) Bend down	Force sensors A and B, and IMU	Analog voltage and serial	$U(A)>U0$ and $U(B)\leq U1$ and $I(\alpha)>A0$ and $I(\theta)>10°$	DC servo driver B	DC servo motor B is reversed and rotates at a uniform speed n1, which is customizable; thus, the operator can bend down with assistance

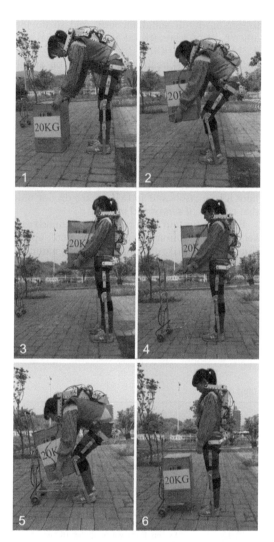

Figure 8.   The process of an operator carrying a load.

tor bent down and lowered the load to the cart with assistance; during this process, both upper and lower EJCs were engaged and motors A and B were both actuated. Finally, in picture 6, the operator had finished the whole process and the apparatus reverted to the motion-free mode.

5   CONCLUSIONS

As for both military and rehabilitative purposes, the powered exoskeleton is an ideal choice for human-power augmentation. Usually, the exoskeletons are designed to match all actions of human limbs, and such a design brings about the use of multiple sensors and a complex control system. These exoskeletons are among the most advanced technologies of big countries like USA, Japan, and China with significant military research investment because of the great potential of exoskeleton for military use. However, exoskeletons for civil use like walking assistance and carrying assistance are now often overlooked. Based on this notion, we designed and developed a compact exoskeleton that is mainly concerned with the assistance of carrying. It has an upper body part and a lower limb part and is driven by two DC servo motors.

The functioning of the apparatus is controlled by three sensors, as shown in Table 1, which follow the logic demonstrated in Table 2. With a view of prolonging the endurance and giving the operator more flexibility, a combination of two types of traditional clutches was developed and introduced into the apparatus: the electromagnetic clutch, as shown in Figure 4.

The assistance of the upper limbs is a basic function while the assistance of the psoas turns out to be practical when we tested the apparatus, because carrying consists of actions that rely on psoas much more like bending down and straightening up.

The apparatus can be implemented in most situations where carrying or/and holding weights are the main work, and is also a good choice for companies and enterprises that wish to improve the efficiency of their physical labor at a relatively low cost.

The future improvement of this apparatus will be revolving around reducing more weight and size, increasing portability, and the level of wearing comfort.

ACKNOWLEDGMENTS

This study was supported by the National Natural Science Foundation of China (51405518) and the Project of State Key Laboratory of High Performance Complex Manufacturing (zzyjkt2014-04).

invited to wear this apparatus to carry a box with books inside weighing 20 kilograms and load it onto a cart.

In picture 1, the operator was trying to bend down and the force sensor A was not engaged, so that the apparatus was in motion-free mode—EJCs were disengaged. Then, in picture 2, she grabbed the load with load hooks attached to her hands; during this process, she engaged the force sensors A and B, and an IMU, causing both upper limb assistance and psoas assistance to function. Subsequently, as picture 3 shows, when she had finished the action of straightening up, the lower EJC was disengaged while the upper EJC was not so, she could hold the load and walk. Then, in pictures 4 and 5, the opera-

REFERENCES

Anam, K. & Al-Jumaily, A.A. (2012). Active exoskeleton control systems: State of the art. *Procedia Engineering* 41: 988–994.

Bogue, R. (2009). Exoskeletons and robotic prosthetics: A review of recent developments. *Industrial Robot: An International Journal* 36(5): 421–427.

Cloud, W. (1965). Man amplifiers: Machines that let you carry a ton. *Popular Science* 187(5): 70–73.

Ghan, J., Steger, R., & Kazerooni, H. (2006). Control and system identification for the Berkeley lower extremity exoskeleton (BLEEX). *Advanced Robotics* 20(9): 989–1014.

Gopura, R.A.R.C., & Kiguchi, K. (2009). Mechanical designs of active upper-limb exoskeleton robots: State-of-the-art and design difficulties. *IEEE International Conference on Rehabilitation Robotics: Reaching Users & the Community (ICORR)* 2009: 178–187.

Gopura, R.A.R.C., & Bandara, D.S.V. (2011). A brief review on upper extremity robotic exoskeleton systems. *IEEE 6th International Conference on Industrial and Information Systems (ICIIS)* 2011: 346–351.

Gopura, R.A.R.C., & Bandara, D.S.V. (2016). Developments in hardware systems of active upper-limb exoskeleton robots: A review. *Robotics and Autonomous Systems* 75: 203–220.

Hassani, W., Mohammed, S., Rifaï, H., & Amirat, Y. (2014). Powered orthosis for lower limb movements assistance and rehabilitation. *Control Engineering Practice* 26: 245–253.

Kazerooni, H., Racine, Jean-Louis, Huang, Lihua, & Steger, R. (2005). On the control of the Berkeley lower extremity exoskeleton (BLEEX). *Proceedings of the 2005 IEEE International Conference on Robotics and Automation (ICRA)* 2005: 4353–4360.

Kazerooni, H., Steger, R., & Huang, Lihua (2006). Hybrid control of the Berkeley lower extremity exoskeleton (BLEEX). *International Journal of Robotics Research* 25(5–6): 561–573.

Kiguchi, K., Rahman, M.H., Sasaki, M., & Teramoto, K. (2008). Development of a 3DOF mobile exoskeleton robot for human upper-limb motion assist. *Robotics and Autonomous Systems* 56(8): 678–691.

Kong, Kyoungchul, & Jeon, Doyoung (2006). Design and control of an exoskeleton for the elderly and patients. *IEEE/ASME Transactions on Mechatronics* 11(4): 428–432.

Lo, Ho-Shing, & Xie, Sheng-Quan (2012). Exoskeleton robots for upper-limb rehabilitation: State of the art and future prospects. *Medical Engineering and Physics* 34(3): 261–268.

Low, K.H. (2011). Robot-assisted gait rehabilitation: From exoskeletons to gait systems. *Defense Science Research Conference and Expo (DSR)* 2011: 1–10.

Meng, Wei, Liu, Quan, Ai, Qingsong, Sheng, Bo, & Xie, Shengquan-Shane (2015). Recent development of mechanisms and control strategies for robot-assisted lower limb rehabilitation. *Mechatronics* 31: 132–145.

Perry, J.C. (2007). Upper-Limb Powered Exoskeleton Design. *IEEE/ASME Transactions on Mechatronics* 12(4): 408–417.

Pons, J.L. (2008). *Wearable Robots: Biomechatronic Exoskeletons*. Madrid: John Wiley & Sons.

Schmeisser, G., & Seamone, W. (1973). An upper limb prosthesis-orthosis power and control system with multi-level potential. *The Journal of Bone and Joint Surgery* 55(7): 1493–1501.

Slavka V., Patrik K., & Marcel J. (2013). Wearable lower limb robotics: A review. *Biocybernetics and Biomedical Engineering* 33(2): 96–105.

Wu, Tzong-Ming, Wang, Shu-Yi, & Chen, Dar-Zen (2011). Design of an exoskeleton for strengthening the upper limb muscle for overextension injury prevention. *Mechanism and Machine Theory* 46(12): 1825–1839.

Yamamoto, K., Hyodo, K., Ishii, M., & Matsuo, T. (2002). Development of power assisting suit for assisting nurse labor. *JSME International Journal Series C* 45(3): 703–711.

Yan, Tingfang, Cempini, M., Oddo, C.M., & Vitiello, N. (2014). Review of assistive strategies in powered lower-limb orthoses and exoskeletons. *Robotics and Autonomous Systems* 64: 120–136.

Zoss, A.B., Kazerooni, H., & Chu, A. (2006). Biomechanical design of the Berkeley lower extremity exoskeleton (BLEEX). *IEEE/ASME Transactions on Mechatronics* 11(2): 128–138.

Advances in Materials Science, Energy Technology
and Environmental Engineering – Patty & Zhou (Eds)
© 2017 Taylor & Francis Group, London, ISBN 978-1-138-19668-1

Screw extruder control system design and simulation using MATLAB

Yuanlou Gao & Lei Wang
Beihang University, Beijing, China

Jianli Zheng
Beifang Xing'an, Taiyuan, Shanxi Province, China

ABSTRACT: A pressure control system designed for a single-screw extruder is mainly studied in this paper. In order to achieve constant pressure control of the head in the extrusion system, we used fuzzy PID self-tuning techniques and Smith Predictor. Then, we used the Matlab Simulink toolbox to simulate the control result. Simulation results indicate that the method achieves good control effect and meets the design requirements.

1 INTRODUCTION

It was very difficult to establish a precise mathematical model as the single-screw extruder system in a big lag system, and the controlled object parameters change greatly. Fuzzy control does not need an explicit system model and has good dynamic performance. As the fuzzy control does not have high steady-state accuracy, we combined the fuzzy control and the PID control to form the FZ-PID controller. The structure of the Smith predictor control was devised to remove the delay effect from the closed-loop design and can be successfully applied together with PID controllers. Considering all these, we decided to use a predictive Smith FZ-PID controller, which is based on the PID controller combined fuzzy logic and the Smith predictive controller.

2 DIGITAL DEVICES MODELING

Configuration diagram for a single-screw extruder closed-loop pressure control system can be represented by Fig. 1.

In Fig. 1, PID controller, transmission, gearbox, and pressure sensor have accurate transfer function.

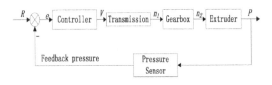

Figure 1.   Configuration diagram for closed-loop pressure control system.

2.1 Extruder

Our transfer function is based on the study on the relationship of speed and pressure. We always set the speed between 3 rmp and 5 rmp, the pressure varies linearly with speed, and the slop of the line is 5.23. Based on engineering experiments, we set the inertia time to be constantly 6 and lag time to be 2 s. Through the reduction, we get the transfer function as

$$W_1(s) = \frac{5.23}{6s+1} e^{-2s} \tag{1}$$

2.2 Pump-controlled motor

We use the Linde classic hydrostatic drive system HPV-02 + HMF55-02. Parameters: motor displacement is $D_m = 55$ ml/rev, volume of the chamber is $V_0 = 5.8 \times 10^{-4}$, effective bulk modulus is $\beta_e = 6.987 \times 10^8$ N/m³, leakage coefficient of pump is $C_{tp} = 2.3 \times 10^{-12}$ m⁵/N·s, pump speed is $\omega_p = 220$ rad/s, viscous damping coefficient is $B_m = 0.325$ N·m·s, and time constant of valve-controlled cylinder is T = 0.5 s. Then, we get the transfer function of the simplified model of pump-controlled motor as

$$W_2(s) = \frac{60}{2\pi} \cdot \frac{1.997}{5.45 \times 10^{-4} s^2 + 6.78 \times 10^{-3} s + 1}$$
$$\cdot \frac{1}{0.5s + 1} \tag{2}$$

2.3 Gearbox

The gear ratio is 80.357, and the transfer function is as follows:

$$W_3(s) = \frac{1}{80.357} \quad (3)$$

2.4 System

In conclusion, the transfer function of the system is

$$G(s) = W_1(s) * W_2(s) * W_3(s)$$
$$= \frac{2.484}{1.635 \times 10^{-3} s^4 + 2.388 \times 10^{-2} s^3 + 3.045 s^2 + 6.507 s + 1} \quad (4)$$

3 CONTROLLER DESIGN

Control strategy: We use fuzzy-PID, which works as a fuzzy control in big deviation scope and a PID control in small deviation scope. The Smith predictor control is a feedback control scheme that has a minor loop. The minor loop works to eliminate the actual delayed output as well as to feed the predicted output to the primary controller. This makes it possible to design the primary controller, assuming no time-delay in the control loop. The control system is shown in Fig. 2.

3.1 Smith predictive control

The Smith prediction algorithm is an effective control method to overcome the pure lag. It is through the prediction of the dynamic characteristics of the controlled object, using the estimated model to compensate for the time lag, that the controlled object and the common compensator constitute no time lag of the generalized controlled object. After compensation, the controller is equivalent to a system with no time lag to control.

The transfer function of the original system is $G(s) = G_0(s) \cdot e^{-\tau s}$, $G_0(s)$ is the part of $G(s)$ which does not contain pure hysteresis characteristics of the part, $G_m(s)(1 - e^{-\tau_1 s})$ is the Smith predictor. In the ideal case, $G_0(s) = G_m(s)$, $\tau = \tau_1$, $G_c(s)$, is the controller. The closed-loop transfer function of the system is

$$G(s) = \frac{G_c(s)G_0(s)e^{-\tau s}}{1 + G_m(s)G_c(s) + G_c(s)[G_0(s)e^{-\tau s} - G_m(s)e^{-\tau_1 s}]}$$
$$= \frac{G_c(s)G_0(s)e^{-\tau s}}{1 + G_m(s)G_c(s)} \quad (5)$$

After compensation of the closed-loop system, $e^{-\tau s}$ is equivalent to have been moved to the closed-loop characteristics of the hysteresis loop. Therefore, when the estimated model is in full agreement with the controlled object, the Smith estimation algorithm will contain the object with pure hysteresis, which does not contain pure hysteresis. The transition process of the system with the addition of the Smith predictor is only a delay in time, while the performance index and the object characteristics are all the same.

3.2 Structure of fuzzy controller

FZ-PID control is a fuzzy two-term control. The control input variables in the fuzzy two-term control are chosen as error (e), which is the voltage bias between the converter and the given derivatives of error (ė). The output variables are chosen as velocity output PID. E, EC, KP, KI, and KD are e, ė, P, I, and D linguistic variables after fuzziness (shown in Table 1 [1]).

They have the same set of fuzzy regions: {Negative Large (NL), Negative Medium (NM), Negative Small (NS), Zero (Z), Positive Small (PS), Positive Medium (PM), Positive Large (PL)}. They also have the same range:{-3, -2, -1, 0, 1, 2, 3}. Their membership function for each control is the triangle.

3.3 Fuzzy rules

The rules for FZ-PID control can be obtained as follows: KP = PL, KI = 0, KD = PS, when |E| = PL; KP = PM, KI = PS, KD = PM, when |E|,|EC| = PM; KP = PL, KI = PL, KD = PM, when |E| = PS. Based on these rules, we can obtain the rules for ΔK_p, ΔK_i, ΔK_d (in Tables 1 and 2).

3.4 Fuzzy inference and defuzzification

We use Mamdani (min-max) fuzzy inference and centroid defuzzification.

Table 1. The basic ranges.

Item	The basic range
e	[-10V,10V]
ė	[-10s,10s]
P	[-9,9]
I	[-0.6,0.6]
D	[-0.06,0.06]

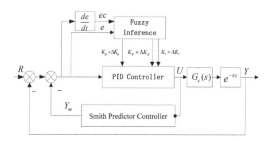

Figure 2. Smith Fuzzy-PID control system.

Table 2. Fuzzy-PID: The rule base.

	NL	NM	NS	ZR	PS	PM	PL
NL	PL/NL/PS	PL/NL/NS	PM/NM/NL	PM/NM/NL	PS/NS/NL	ZR/ZR/NM	ZR/ZR/PS
NM	PL/NL/PS	PL/NL/NS	PM/NM/NL	PS/NS/NM	PS/NS/NM	ZR/ZR/NS	NS/ZR/ZR
NS	PM/NL/ZR	PM/NM/NS	PM/NS/NM	PS/NS/NM	ZR/ZR/NS	NS/PS/NS	NS/PS/ZR
ZR	PM/NM/ZR	PM/NM/NS	PS/NS/NS	ZR/ZR/NS	NS/PS/NS	NM/PM/NS	NM/PM/ZR
PS	PS/NM/ZR	PS/NS/ZR	ZR/ZR/ZR	NS/PS/ZR	NS/PS/ZR	NM/PM/ZR	NL/PL/ZR
PM	PS/ZR/PL	ZR/ZR/PS	NS/PS/PS	NM/PS/PS	NM/PM/PS	NM/PL/PS	NL/PL/PL
PL	ZR/ZR/PL	ZR/ZR/PM	NM/PS/PM	NM/PM/PM	NM/PM/PS	NL/PL/PS	NL/PL/PL

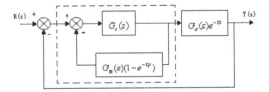

Figure 3. The Smith prediction control principle diagram.

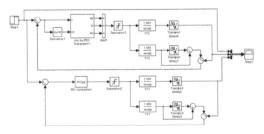

Figure 4. Predictive smith PID controller and predictive Smith Fuzzy-PID controller.

4 SIMULATION AND PERFORMANCE ANALYSIS

4.1 Simulation

In order to achieve a precise control of the prospective pressure conditions, there is a need to determine the PID parameters by analyzing its pressure closed-loop simulation and performance, which is based on the SIMULINK Tool of MATLAB.

We can establish the pressure closed-loop simulation model in SIMULINK, as shown in Fig. 4.

According to the actual production system control practices, the initial parameters have been set as follows. It has been assumed that Kp = 7.5, Ki = 0.43, Kd = 2, and G(s) is the closed-loop pressure model. The step response curve is shown in Fig. 5.

The blue line is the curve of step signal, the red line is the time domain response curve of the predictive Smith PID controller, and the green line is the time domain response curve of the predictive Smith FZ-PID controller. The rise time is the time required for the response curve increased from 10% of steady state value to 90%. The range of steady state is set at +1%.

4.2 Performance analysis

Overshoot is required to be less than or equal to 2%, and its settling time should be less than 5 s, in line with the producing condition. As can be seen by comparing the two curves, the predictive Smith

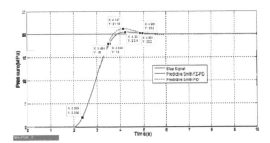

Figure 5. Time domain response curves of two controllers.

Table 3. Simulation results.

Item	Predictive Smith PID controller	Predictive Smith FZ-PID controller
Rise time[s]	1.125	1.149
Overshoot	5.90%	2.00%
Settling time[s]	4.961	4.851

FZ-PID controller has smaller overshoot and shorter settling time, and its rise time is close to the predictive FZ-PID. By comparison, the predictive Smith FZ-PID controller has better control effect, which fully meets our design requirements.

29

5 CONCLUSION

The analysis of simulation results above shows that the closed-loop is stable. After the predictive Smith FZ-PID correction, the rapidity and the accuracy of the system can meet the design requirements. Therefore, usage of the predictive Smith FZ-PID algorithm control can make the screw extruder system more effective and feasible. Hence, this device is suitable for mass production and can be widely used in actual production.

REFERENCES

Dingdu W, Ling H, 2009. Smith temperature control system based on fuzzy PID control algorithm, in Instrument technique and sensor, 2009-04-15.

Fang H, Weiming T, Qiang W, 2009. Joint control of Smith and Fuzzy in networked control systems, in Proceedings of the 2009, virtual instruments assembly (1).

Jing Z, Xianyun Z, 2002. Modeling and Simulation of fuzzy control of DC motor speed control system based on MATLAB/SIMULINK, in Modern electronic technology, 2002-04-25.

Shuai L, 2010. Characteristics of pump controlled motor closed loop speed regulating control system, in Zhejing University, 2010.

Tingting L, 2001. Research on constant speed control system of electro hydraulic proportional variable displacement pump motor. in Chang'an University, 2001.

Weizhe Z, Baorui Y, Yichong G, 2007. Study on transfer function of pressure speed closed loop control system of single screw extruder, in China Plastics, 2007(1).

Advances in Materials Science, Energy Technology and Environmental Engineering – Patty & Zhou (Eds)
© 2017 Taylor & Francis Group, London, ISBN 978-1-138-19668-1

Study on the optimizing design of a screw extruder system

Yuanlou Gao & Lei Wang
Beihang University, Beijing, China

Jianli Zheng
Beifang Xing'an, Taiyuan, Shanxi Province, China

ABSTRACT: This paper describes the optimizing design for a single-screw extruder. Extrusion molding plays an important role in the plastic product processing industry. In the plastic processing industry in our country, almost 1/3~1/2 plastic products are completed through the extrusion molding machine. In order to achieve better rapidity and constant pressure control of the head, we used the PID and the Smith Predictor to control the extrusion system and then we used Matlab Simulink toolbox to simulate control result. Simulation results indicate that the control system with a pump-controlled motor as the drive mechanism achieves better control effect and meets the design requirements than the control system with three-phase asynchronous motors.

1 INTRODUCTION

A traditional screw extruder system driven by three-phase asynchronous motors uses a frequency transformer and a gearbox to control the pressure of the head. Although the controller is included, the rising time of the system is too long and the rapidity is poor. To solve this problem, the pump-controlled motor system is used to replace the traditional three-phase asynchronous motor. Previously, we have completed the modeling and simulation of the motor-driven extruder system. In this paper, we completed the modeling and simulation of the closed-loop pump-driven pressure control system, and made comparison between the closed-loop pressure control system and the motor-driven system. The single-screw extruder system is a big lag system. The structure of the Smith predictor control was devised to remove the delay effect of the closed-loop design and can be successfully applied together with PID controllers. Therefore, we decide to use the predictive Smith-PID controller.

2 DIGITAL DEVICE MODELING

The configuration diagram of the single-screw extruder closed-loop pressure control system is presented in Fig. 1.

In Fig. 1, PID controller, transmission, gearbox, and pressure sensor have accurate transfer function.

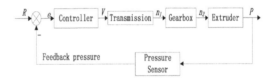

Figure 1. Configuration diagram for closed-loop pressure control system.

2.1 Extruder

Our transfer function is based on the study on the relationship between speed and pressure. We always set the speed between 3rmp and 5rmp, the pressure varies linearly with the speed, and the slop of the line is 5.23. Based on engineering experiments, we set the inertia time constant at 6 and the lag time at 2 s. Through the reduction, we get the transfer function as follows:

$$W_1(s) = \frac{5.23}{6s+1} e^{-2s} \tag{1}$$

2.2 Pump-controlled motor

We use the Linde classic hydrostatic drive system HPV-02 + HMF55-02. Parameters: motor displacement is D_m = 55 ml/rev, volume of the chamber is V_0 = 5.8 × 10⁻⁴, effective bulk modulus is β_e = 6.987 × 108 N/m³, leakage coefficient of pump is C_{tp} = 2.3 × 10⁻¹² m⁵/N · s, pump speed is ω_p = 220 rad/s, viscous damping coefficient is B_m = 0.325 N · m · s, and time constant of valve-controlled cylinder is T = 0.5 s. Then, we get the transfer function of simplified model of pump-controlled motor as follows:

$$W_2(s) = \frac{60}{2\pi} \cdot \frac{1.997}{5.45 \times 10^{-4} s^2 + 6.78 \times 10^{-3} s + 1}$$
$$\times \frac{1}{0.5s + 1} \tag{2}$$

2.3 Gearbox 1

The mathematical model of the speed reducer is a typical proportion. The maximum rotation speed of the plant is 5 rpm, but the maximum speed is set to 10 rpm in the production test and adaptability. In this paper, we choose the speed reducer to be Tianjin speed reducer with a transmission ratio of 143. The gear ratio is 80.357, and the transfer function is

$$W_3(s) = \frac{1}{80.357} \tag{3}$$

2.4 Frequency transformer

We use ABB ACS550 frequency transformer and set the acceleration time at 5 s, then get the transfer function of the simplified model of PWM control as follows:

$$W_4(s) = \frac{5}{3s + 1} \tag{4}$$

2.5 Motor

We choose the YB2-132M4A AC motor, which is directly connected with a reducer. Parameters: two pole pairs, rated speed is $n_e = 1440$ r/min, frequency is f = 50 Hz, rated power is $P_e = 7.5$ kW, and moment of inertia is $GD^2 = 0.167$ kg · m². Finally, we get the approximate transfer function as

$$W_5(s) = \frac{30}{0.01224288s + 1} \tag{5}$$

2.6 Gearbox 2

The maximum rotation speed of the plant is 5 rpm, but the maximum speed is set to 10 rpm in the production test and adaptability. In this paper, we choose the speed reducer to be Tianjin speed reducer with a gear ratio of 143, and the transfer function is

$$W_6(s) = \frac{1}{143} \tag{6}$$

2.7 Modeling of a pump-driven system

In conclusion, the transfer function of a pump-driven system is:

$$G_1(s) = W_1 * W_2 * W_3$$
$$= \frac{2.484}{1.635 \times 10^{-3} s^4 + 2.388 \times 10^{-2} s^3 + 3.045 s^2 + 6.507 s + 1} \tag{7}$$

2.8 Frequency transformer

In conclusion, the transfer function of the system is:

$$G_2(s) = W_1 * W_4 * W_5 * W_6$$
$$= \frac{5.486}{(3s + 1)(0.012s + 1)(6s + 1)} e^{-2s} \tag{8}$$

3 CONTROLLER DESIGN

Control strategy: We use the PID and the Smith predictor control. The Smith predictor control is a feedback control scheme that has a minor loop . The minor loop works to eliminate the actual delayed output as well as to feed the predicted output to the primary controller. This makes it possible to design the primary controller, assuming no time-delay in the control loop. The control system is shown in Fig. 2.

The Smith prediction algorithm is an effective control method to overcome the pure lag. It is through the prediction of the dynamic characteristics of the controlled object, using the estimated model to compensate for the time lag, that the controlled object and the common compensator constitute no time lag of the generalized controlled object. After compensation, the controller is equivalent to a system with no time lag to control.

The transfer function of the original system is $G(s) = G_0(s) \cdot e^{-\tau s}$, $G_0(s)$ is the part of $G(s)$ which does not contain pure hysteresis characteristics of the part, $G_m(s)(1 - e^{-\tau_1 s})$ is the Smith predictor. In the ideal case, $G_0(s) = G_m(s)$, $\tau = \tau_1$, $G_c(s)$ is the controller. The closed-loop transfer function of the system is

$$G(s) = \frac{G_c(s)G_0(s)e^{-\tau s}}{1 + G_m(s)G_c(s) + G_c(s)[G_0(s)e^{-\tau s} - G_m(s)e^{-\tau_1 s}]}$$
$$= \frac{G_c(s)G_0(s)e^{-\tau s}}{1 + G_m(s)G_c(s)} \tag{9}$$

After compensation of the closed-loop system, $e^{-\tau s}$ is equivalent to have been moved to the closed-loop characteristics of the hysteresis loop.

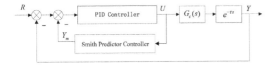

Figure 2.   Smith-PID control system.

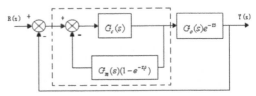

Figure 3.   Smith predictive control principle diagram.

Therefore, when the estimated model is in full agreement with the controlled object, the Smith estimation algorithm will contain the object with pure hysteresis, which does not contain pure hysteresis usually. The transition process of the system with the addition of the Smith predictor is only a delay in time, while the performance index and the object characteristics are all the same (which does not contain pure lag object part).

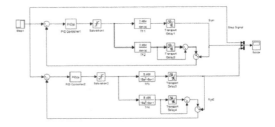

Figure 4.   Predictive Smith-PID controller.

4   SIMULATION AND PERFORMANCE ANALYSIS

4.1   Simulation

In order to achieve a precise control of prospective pressure conditions, there is a need to determine the PID parameters by analyzing its closed-loop pressure simulation and performance, which is based on the SIMULINK Tool of MATLAB.

We can establish the closed-loop pressure simulation model in SIMULINK as shown in Fig. 3.

According to the actual production system control practices, the initial parameters were set as follows. It can be assumed that Kp1 = 7.5, Ki1 = 0.45, Kd1 = 1.1 for PID controller 1 and Kp2 = 9, Ki2 = 0.06, Kd2 = 3.2 for PID controller 2. The step response curve is shown in Fig. 4.

In Fig. 5, the blue line is the curve of step signal, the red line is the time domain response curve of the pump-driven pressure control system, and the green line is the time domain response curve of the motor-driven extruder system. The rise time is the time required for response curve increased from 10% of the steady state value to 90%. The range of steady state value is set at +1%.

4.2   Performance analysis

As can be seen by comparing the two curves, the pump-driven pressure control system has shorter rise time and settling time, and the overshoot is the same as the motor-driven pressure control system. By comparison, the pump-driven pressure control system has better control effect, which fully meets our design requirements.

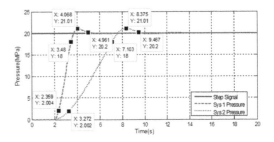

Figure 5.   Time domain response curves of two systems.

Table 1.   Simulation results.

Item	Pump-driven pressure control system	Motor-driven pressure control system
Rise time [s]	1.121	3.831
Overshoot	5.05%	5.05%
Settling time [s]	4.961	9.467

5   CONCLUSION

The analysis of simulation results given above shows that the closed-loop model is stable. After using the pump-controlled motor, we replaced the traditional three-phase asynchronous motor; the rising time of the system is much shorter and the rapidity is good. The rapidity and the accuracy of the system can meet the design requirements. Therefore, being driven by the pump can make the screw extruder system more effective and feasible.

REFERENCES

Bin H, 2001. Design and Realization of the computer control system of twin screw extruder.

Lina D, 2008. Research on variable frequency speed regulating synchronous transmission algorithm of AC asynchronous motor, JiLin University, 2008.

Long P, Yuxia Z, Zhiming J, 2009. Research Progress on extrusion theory of single screw extruder, in China Plastics. 2009(5).

Wenqi Z, 2006. Application of frequency converter in precision extrusion machinery. 2006(4): 117–118.

Weizhe Z, Baorui Y, Yichong G, 2007. Study on transfer function of pressure speed closed loop control system of single screw extruder, in China Plastics 2007(1).

Xiaodong Z, Jun W, Hong W, 2004. Control of pure delay system based on Smith predictor, in Journal of Zhengzhou University (Engineering Science Edition) 2004.

Yang L, Tinghua L, 2004. Research and development of plastic extrusion machinery.

Yongli Z, Lihua L, 2008. Principle and development of variable frequency speed regulation of three phase asynchronous motor, in Heilongjiang Science and Technology Information. 2008(5): 033.

Zhang J. and Morris A, 2004. PID controller, in IEEE transaction on Fuzzy Systems, 2004, 6(4): 449–463.

Zhizhong W, 2006. Design of closed loop control system for induction motor. TC, 2006, 47: 3.

Advances in Materials Science, Energy Technology
and Environmental Engineering – Patty & Zhou (Eds)
© 2017 Taylor & Francis Group, London, ISBN 978-1-138-19668-1

Constant speed control system of a test-bed for a hydraulic energy recovery turbine

G.Q. Yang

College of Information Science and Engineering, Northeastern University, Shenyang, China

ABSTRACT: The hydraulic energy recovery turbine is a kind of equipment which is used to recover the energy from the high-pressure fluid produced in a chemical process. In view of the problems that any fluctuation in the flow rate of the tested turbine leads to an instability of the rotational speed, a new control strategy was proposed to achieve a constant output of the rotational speed of the turbine, using throttling speed regulation technology of hydraulic servo valve. A transfer function of the control system was established based on the analysis of the dynamic process. In addition, an emulation analysis on the turbine constant speed control system example was performed using the MATLAB. The research results and the example show that the rotational speed of the turbine can be controlled in an acceptable range of errors, and the dynamic response is also good by using an ordinary PID controller and adjusting the controller parameters properly.

1 INTRODUCTION

At present, energy recycling is increasingly considered by people. The hydraulic turbine, as an energy recovery device for high pressure in the chemical process, has attracted much attention. There are three ways of recycling energy (Yang, 2011), namely the hydraulic turbine driving a pump or fan, helping the motor to act, and driving the generator. However, in order to avoid the fluctuation in the flow affecting the output speed, the hydraulic turbine requires a constant output speed, no matter which method can be used.

Hydraulic transmission control technology can realize flexible control of the power, realize stepless speed regulation easily, and have large adjustment speed range. Moreover, it can be operated by a remote control and an automatic control (Guan, 2011; Wang, 2011). The problems of unstable output speed caused by high-pressure fluid flow fluctuation in the hydraulic turbine can be solved by using hydraulic transmission control technology for constant speed control of the hydraulic turbine. In addition, it can realize the constant speed work of the equipment driven by the hydraulic turbine.

2 CONSTITUTION TEST RIG OF CONSTANT SPEED HYDRAULIC CONTROL OF THE HYDRAULIC TURBINE

The test rig schematic diagram of constant speed hydraulic control of the hydraulic turbine, including the hydraulic turbine, speed sensor, hydraulic pump, safety valve, PID controller, electro-hydraulic servo valve, cooler, coupling, etc., is shown in Figure 1. The speed sensor is used for measuring the output rotational speed of the hydraulic turbine, whose input terminal is connected to the hydraulic turbine shaft, and output terminal is connected to the input shaft of the hydraulic pump. The input circuit of the hydraulic pump connects with the oil tank, and the output oil circuit connects with the hydraulic valve block, and the valve block is equipped with a safe valve, an electro-hydraulic servo valve, etc. It can be realized the adjustment of the pressure difference at both ends of the valve port of the electro-hydraulic servo valve is achieved by adjusting the opening size of the electro-hydraulic servo valve. The electro-hydraulic servo valve is equivalent to the load of the hydraulic pump. Adjusting the pressure difference at both ends of the electro-hydraulic servo valve means adjusting the actual output pressure of the hydraulic pump from the load depending on the hydraulic system pressure. Therefore, through adjusting the output oil pressure of the hydraulic pump, it can achieve the matching of output torque and speed between the hydraulic pump and the hydraulic turbine, and realize the constant speed control of the hydraulic turbine at a given flow and head. As the output energy of the hydraulic pump is consumed in the electro-hydraulic servo valve and pipeline, this process will produce a lot of heat. As a result of this, installing the oil circuit cooling device compelled cooling the oil so as to guarantee the normal system operation. At the same time, the safe valve

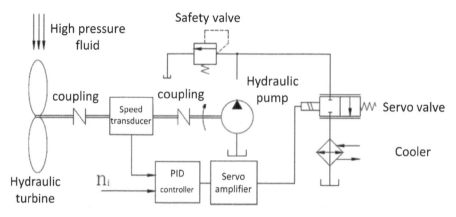

Figure 1. A constant speed hydraulic control of hydraulic turbine test rig.

is installed at the pump outlet in order to prevent the fast speed of the hydraulic pump from causing damage to the hydraulic system.

3 CONSTANT SPEED CONTROL

Hydraulic turbine reclaims that the high-pressure fluid energy in the chemical process makes the energy of certain pressure and flow of the fluid into the rotating mechanical energy, which can drive other machinery actions. Generally, it requires a constant speed. Taking the hydraulic turbine drives the generator for an example. It is well known that the induction electromotive frequency force depends on the speed n and the number of the magnetic poles P of the synchronous AC generator, which is $f = \frac{n}{60}p$ (Wei, 2013). The hydraulic turbine drives the AC generator, which can lead to the instability of the frequency of the AC generator due to the speed instability caused by high-pressure fluid fluctuation of hydraulic turbine. Therefore, it needs to make the inverter rectification when in use. Therefore, the constant speed control of the hydraulic turbine is the key to determine whether or not the generator can be connected to the grid.

3.1 Hydraulic energy recovery turbine constant speed hydraulic control principle

The test rig of the constant speed hydraulic control of the hydraulic turbine is shown in Fig 1. A high-pressure fluid produced in the chemical process overcomes the friction and inertia of the hydraulic turbine and, the high-pressure fluid energy is converted into rotary mechanical energy by the hydraulic turbine. At the same time, the hydraulic pump output shaft drives the hydraulic pump through the shaft coupling. The hydraulic pump

oil outlet connects with the servo valve directly. The oil generates pressure drop through the servo valve mouth, which determines the hydraulic pump outlet pressure, i.e. through the adjustment of the differential pressure of the servo valve. It can realize the adjustment of the load torque of the hydraulic pump works as the load hydraulic turbine. The specific work process is as follows: the hydraulic turbine actual rotational speed signal forms a closed loop in the real-time feedback using a speed torque sensor, and the speed signal is compared with a given speed signal to produce a deviation signal. It controls the servo valve action directly through the disposal by the PID controller to adjust the pressure difference between the two ends of the servo valve, i.e. the load hydraulic pump. Then, the constant speed control of the hydraulic turbine is achieved. It is clear that the high-pressure fluid flow must be in a certain range. When the flow is too high, the servo valve will be fully closed and the hydraulic pump output pressure will increase to a safe valve setting. Simultaneously, the oil from the safety valve will flow back to the tank, speed adjustment function will fail, the flow will become too slow, it will not drive the work of the hydraulic turbine, and the speed adjustment function will fail.

3.2 Transfer function model of speed control loop

With the flow of high-pressure fluid increase, the speed of hydraulic turbine will accelerate, and torque and rotational speed sensor will feedback speed signal and will comprise a given speed signal by PID regulation to control the servo valve. As the hydraulic pump is loaded, the servo valve controls the output pressure of the hydraulic pump by adjusting the valve port size; the larger the output pressure of the hydraulic pump, the more the energy consumed by the hydraulic pump as the hydraulic turbine. This leads to the energy reduction in the hydraulic turbine distribution, and the hydraulic turbine speed will be reduced correspondingly; until the speed achieved the given speed signals, the speed will achieve balance. When the high-pressure fluid flow decreases, the speed regulation process is like this.

From the above, changes in the high-pressure fluid flow will cause changes in the rotation speed of the hydraulic turbine. Through the torque and speed sensors, the actual working speed of the hydraulic turbine will be sent to the PID controller in the system. According to the deviation of the turbine speed signal and a given speed signal, the controller will adjust the servo valve opening size. The servo valve opening size will adjust the output pressure of the hydraulic pump, thereby regulating the speed of hydraulic turbine.

In this process, the PID controller, servo valve, hydraulic pump, hydraulic turbine, speed sensor, etc., together constitute a closed-loop control system, and the control system block diagram is shown in Figure 2.

PID controller transfer function (Xue, 2002; Hu, 2011; Yang, 2011):

$$\frac{u(s)}{e_n(s)} = K_p \left(1 + \frac{1}{T_i s} + T_d s\right) \tag{1}$$

where u is the controller output voltage; e_n is the deviation signal; K_p, T_i, and T_d are the controller ratio factor, the integral time constant, and the differential time constant, respectively.

Servo amplifier transfer function:

$$\frac{i(s)}{u(s)} = K_A \tag{2}$$

where i is the servo amplifier output current and K_A is the servo amplifier amplification factor.

The servo valve frequency response is much higher than the natural frequency of the system, after linearization processing can be regarded as the proportional component at the steady-state

operating point $(Q_0, n_{o0}, \Delta P_{v0})$, and its transfer function is

$$\frac{\Delta P_V(s)}{i(s)} = K_{pv} \tag{3}$$

where ΔP_V is the hydraulic pump output pressure and K_{PV} is the pressure gain at a steady operating point.

The servo valve is regarded as the load to adjust the output pressure of the hydraulic pump, and this pressure reacts with the hydraulic pump load torque. The hydraulic pump outputs the oil at a certain flow rate, assuming that the modulus of elasticity of oil and the oil temperature is a constant, without considering the friction and leakage of oil, hydraulic pump transfer function:

$$\frac{T_P(s)}{\Delta P_V(s)} = K_B \tag{4}$$

where T_P is the load torque of the hydraulic pump acting on the hydraulic turbine, K_B is the hydraulic pump ratio coefficient, $K_B = 2\pi \cdot V_p$, and V_P is the displacement of the hydraulic pump.

Speed sensor transfer function:

$$\frac{n_f(s)}{n_o(s)} = K_f \tag{5}$$

where K_f is the speed sensor ratio coefficient.

The hydraulic pump is used as a hydraulic turbine load, without considering the friction moment, elastic moment, etc., by the hydraulic turbine and the load torque balance equation steady-state operating point $(Q_0, n_{o0}, \Delta P_{v0})$ can be

$$Jsw(s) = T_T(s) - T_P(s) \tag{6}$$

$$w(s) = 2\pi n_o(s)/60 \tag{7}$$

$$T_T(s) = K_Q \cdot Q(s) - K_n \cdot n_o(s) \tag{8}$$

where J is the equivalent moment of inertia of the shaft; w is the angular velocity of hydraulic

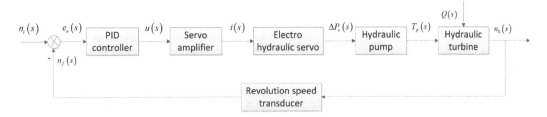

Figure 2.   Block diagram of hydraulic turbine speed control system.

turbine; T_T is the load torque of high-pressure fluid acting on the hydraulic turbine; K_Q is the hydraulic turbine flow coefficient, $K_Q = \frac{30\rho g H \eta}{\pi} \cdot \frac{1}{n_{o0}}$; K_n is the rotation speed coefficient of hydraulic turbine, $K_n = \frac{30\rho g H \eta}{\pi} \cdot \frac{Q_0}{n_{o0}}$; H is the head of hydraulic turbine; ρ is the fluid density; and η is the hydraulic turbine efficiency.

Combined with the above formulas (1–8), the speed output response characteristics of the hydraulic turbine under the combined action of the input command n_i and Q can be obtained as follows:

$$n_o(s) =$$

$$\frac{K_P \cdot K_A \cdot K_{PV} \cdot K_B \cdot \left(1 + \frac{1}{T_i s} + T_d s\right)\left(\frac{-1}{\frac{2\pi J}{60}s + K_n}\right)}{1 + f(s)}$$

$$\cdot n_i(s) + \frac{\frac{K_Q}{\frac{2\pi J}{60}s + K_n}}{1 + f(s)} \cdot Q(s),$$

where $f(s) = K_P \cdot K_A \cdot K_{PV} \cdot K_B \cdot K_f \cdot (1 + \frac{1}{T_i s} + T_d s)\left(\frac{-1}{\frac{2\pi J}{60}s + K_n}\right)$.

In the experiment, the hydraulic turbine head (H), flow rate (Q), efficiency (η), rotation speed (n_{o0}), and density of working medium (ρ) are 70 m, 90 m³/h, 73%, 3000 r/min, and 1000 kgf/m³, respectively, and the experiment medium is water. The Rexroth A4V series hydraulic pump was selected, and its displacement is 56 ml. The electric hydraulic servo valve is QDY3-400-10-21, and the steady-state operating point is $(Q_0, n_{o0}, \Delta P_{v0}) = (90, 3000, 3.47)$, $K_P = 2$, $K_A = 1$ mA/v, $K_{PV} = 0.7$ MPa/mA, $K_B = 8.9$ Nm/MPa, $K_f = 0.00167$ V/r/min, $J = 2.178$ kgm², $K_Q = 1.55 \times 10^4$, $K_n = 1.356 \times 10^{-2}$.

The system model was built in MATLAB, as shown in Figure 3.

In the experiment, the actual requirement for rotation speed of the hydraulic turbine is 3000 r/min. The corresponding voltage of the rotation speed n_i

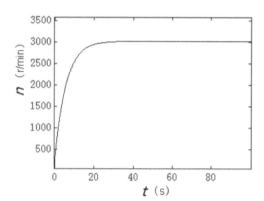

Figure 4. The constant speed control system of hydraulic turbine's input flow rate order step response.

is 5V. The flow step signal is regarded as the system input. In addition, the PID controller parameters are $K_P = 0.1$, $T_i = 0.0001$, and $T_d = 0.1$, and the ideal simulation results were obtained as shown in Figure 4. As can be seen from Figure 4, the output rotation speed of the hydraulic turbine is gradually accelerated, and it achieves a stable output rotation speed of 3000 r/min at a time of 25 s. As the hydraulic turbine can drive the hydraulic pump to rotate as long as it is rotated, according to the rotation speed of the hydraulic pump, the servo valve will immediately adjust the hydraulic pump load pressure to produce the load torque. When this load torque and the torque generated by the fluid acting on the hydraulic turbine are equal, the rotation speed will no longer rise and achieve the required values (as shown in Figure 4). When the input flow rate changes, the servo valve makes the corresponding adjustments immediately, to control the size of the load torque, to maintain the output rotation speed of the hydraulic turbine within the specified range.

4 CONCLUSION

Constant rotation speed control technology is the key of the hydraulic turbine as a kind of energy recovery device. With respect to the instability of the output rotation speed of the hydraulic turbine caused by the flow rate fluctuation, hydraulic transmission control technology is proposed in this paper. This technique is based on the method of changing the torque of the hydraulic turbine to realize the constant rotation speed control of the hydraulic turbine. Through the actual simulation, this method can effectively solve hydraulic turbine output speed instability. At the same time, the servo valve and the hydraulic pump are used

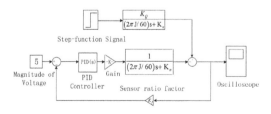

Figure 3. Constant speed control system of hydraulic turbine by MATLAB.

as the load to form the load torque to carry on the energy dissipation control, and the rotation speed of the hydraulic turbine is adjusted by reversing. This method can provide the theoretical basis for similar control systems, and can provide reference for the research of the hydraulic turbine energy recovery device.

ACKNOWLEDGMENTS

The authors specially thank Professor Wei Liejiang for his guidance.

REFERENCES

Guan Zhong-fan. Hydraulic transmission system [M]. Beijing: Mechanical Industry Press, 2011.
Hu Shou-song. Automatic control principle [M]. Beijing: Science Press, 2011.
Wang Chun-xing. Hydraulic control system [M]. Beijing: Mechanical Industry Press, 2011.
Wei Lie-jiang, Wang Dong-liang, Hu Xiao-min. Constant Speed Control of Hydraulic Transmission Wind Turbine [J]. Machine Tool & Hydraulics, 2013, 41(11): 77–79.
Xue Ding-yu. System simulation technology and application based on MATLAB/Simulink [M]. Beijing: Tsinghua University Press, 2002.
Yang Guo-lai, Zhang Cheng-cheng, Zuo Gang-yong, et al. Research of Electric-hydraulic Servo System in Continuous Casting Machine Based on Fuzzy-PID Control [J]. Journal of Gansu Sciences, 2011, 23(2): 120–123.
Yang Jun-hu, Zhang Xue-ning, Wang Xiao-hui, et al. Overview of Research on Energy Recovery Hydraulic Turbine [J]. Fluid Machinery, 2011, 39(6): 29–33.
Yang Jun-hu, Zhang Xue-ning, Wang Xiao-hui, et al. Progress in Energy Recovery Hydraulic Turbine Research [J]. Chemical Engineering & Machinery, 2011, 38(6): 655–658.

Advances in Materials Science, Energy Technology and Environmental Engineering – Patty & Zhou (Eds)
© 2017 Taylor & Francis Group, London, ISBN 978-1-138-19668-1

Isolated RS485 communication interface for the metal working fluids monitoring system

T. Škulavík
Faculty of Materials Science and Technology in Trnava, Institute of Applied Informatics, Automation and Mechatronics, Slovak Technical University in Bratislava, Trnava, Slovak Republic

K. Gerulová
Faculty of Materials Science and Technology in Trnava, Institute of Integrated Safety, Slovak Technical University in Bratislava, Trnava, Slovak Republic

ABSTRACT: This paper deals with the proposal of the communication interface of the Real Time Monitoring and Control System (RMACS) for metalworking fluids. The RMACS will be a complex decentralized system, of which some components will be mounted in different sections of the 5 axis machines. The user interface of the RMACS will be located at the greater distance. The industrial environment and the greater distances between the components have an important role during the design of the communication interface and the network topology.

1 INTRODUCTION

1.1 *Metalworking Fluids monitoring*

Metalworking Fluids (MWF) are used in many metalworking technologies. Despite the introduction of so called dry machining the MWFs could not be excluded from some machining processes. Especially for today's challenging machining processes, such as milling titanium for lighter aircraft or boring compacted graphite iron for next generation vehicles powered by biofuel, metalworking fluids are required (Skleros et al. 2008). The quality of MWFs has to be monitored continuously during the process of machining. The quality is obtained from parameters as concentration, pH, conductivity, residual oil, corrosion and the amount of microorganisms. To monitor and control some of the mentioned parameters a Real-Time Monitoring and Control System (RMACS) is being developed. The RMACS will be deployed in real conditions in the Centre of excellence of 5 axes machining in MTF Trnava.

2 RMACS NETWORK PHYSICAL LAYER AND TOPOLOGY

Industrial applications, as it is in our case, require transmission of data between multiple nodes often over very long distances. The RS-485 bus standard is one of the most widely used physical layer bus designs in industrial applications. The key features of RS-485 bus standard that make it ideal for use in the industrial communication applications are as follows: Long distance links—up to 1200 m; Bidirectional communications possible over a single pair of twisted cables; Differential transmission increases noise immunity and decreases noise emissions; Multiple drivers and receivers can be connected on the same bus; Wide common-mode range allows for differences in ground potential between the driver and receiver; TIA/EIA-485-A allow for data rates of up to 10 Mbps. Devices meeting the TIA/EIA-485-A specifications do not have to operate over the entire range and are not limited to 10 Mbps. Applications for RS-485 include process control networks; industrial automation; remote terminals; building automation, such as Heating, Ventilation, Air Conditioning (HVAC), security systems; motor control; and motion control (Marais 2008). Precisely because of above mentioned advantages, we decided to use the RS-485 as a physical layer for the communication network of the RMACS. Physical layer of the RS-485 interface is normally used with higher-level protocols, such as Profibus, Interbus, Modbus. This allows for robust data transmission over relatively long distances. In our case the communication protocol was also developed as a part of the system and will be described later. The proposed network topology of the real-time monitoring and control system is shown in Figure 1. As it was mentioned before, the benefit of the RS-485 is that multiple devices can be connected on the same bus. As an

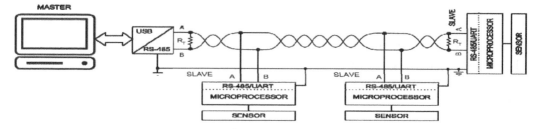

Figure 1. Network topology of RMACS.

example shown in Figure 1, three slave devices and one master device are connected to half-duplex RS-485 bus. In our case the master device is a Human Machine Interface (HMI), which can be a basic computer with USB port. Slave devices are microprocessor-controlled sensors (Pitel' et al. 2005) or microprocessor-controlled actuators (in our case mostly microprocessor-controlled valves). Usually none of the connected devices have a RS-485 interface. Because of this an RS-485 to USB converter must be used with the HMI. The before mentioned microprocessor-controlled devices are developed exactly for the purposes of the RMACS. The RS-485 interface is therefore a part of the development.

3 RS-485 INTERFACE

Many facts should be taken in to consideration during the development of the communication interface for the microprocessor-controlled devices and the HMI—physical layer, half duplex/full duplex communication, the quantity and distance between connected devices, data rates and communication packet length, type of environment and not least the price. These facts have impact on the design of the electrical circuit and on the selection of the electronic components.

RS-485 is specified as a multipoint standard, which means up to 32 transceivers can be connected on the same bus. The input impedance of the RS-485 receiver is specified as larger than or equal to 12 kΩ. This impedance is defined as having one Unit Load (UL). The RS-485 specification specifies the capability to sustain up to 32 ULs. Some RS-485 receivers are specified as having ¼ UL or ⅛ UL. A receiver specified to have ¼ UL means that the receiver only loads the bus by ¼ of the standard UL and, therefore, 4 times as many of these receivers can be connected to the bus (4 × 32 = 128 nodes). Similarly, if an RS-485 receiver is specified to have ⅛ UL, the receiver only loads the bus by ⅛ of the standard UL and, therefore, 8 times as many of these receivers can be connected to the bus (8 × 32 = 256

nodes) (Marais 2008). For the proposed system the 32 transceivers are quite sufficient. An RS-485 system must have a driver that can be disconnected from the transmission line when a particular node is not transmitting. The DE (RTS) pin on the RS-485 transceiver enables the driver when a logic high is set to DE (DE = 1). Setting the DE pin to low (DE = 0) puts the driver in a tristate condition. This effectively disconnects the driver from the bus and allows other nodes to transmit over the same twisted pair cable. RS-485 transceivers also have an RE pin that enables/disables the receiver. With respect to the communication method structure and the final price the half duplex communication was selected. Half-duplex communication allows for data transmission in both directions, but only in one direction at a time. Half-duplex RS-485 links have multiple drivers and receivers on the same signal path. This is the reason why the proposed RS-485 transceivers must have driver/receiver enable pins enabling only one driver to send data at a time. For reliable RS-485 communications, it is essential that the reflections in the transmission line be kept as small as possible. This can only be done by proper cable termination. Reflections happen very quickly during and just after signal transitions. On a long line, the reflections are more likely to continue long enough to cause the receiver to misread logic levels. On short lines, the reflections occur much sooner and have no effect on the received logic levels. RS-485 applications require termination at the master node and the slave node furthest from the master (Marais 2008). The termination RT is shown in Figure 1. The data rates are strongly dependent on the cable lengths. In general when high data rates are used, the application is limited to a shorter cable. It is possible to use longer cables when low data rates are used. In our application a 9600 kbps data rates are used and a maximum cable length is approximately 20 m to 30 m.

Long links in communication applications can cause the ground potential at different nodes on the bus to be slightly different. This causes ground currents to flow through the path of least resistance through either the common earth ground or the ground wire. If the same electrical system is used to connect the power supplies of all nodes to the same earth ground, the ground connection may have reduced noise. Note, however, that motors, switches, and other electrically noisy equipment can still induce ground noise into the system. When different nodes are situated in different buildings, different power systems are required. This is likely to increase the impedance of the earth ground and the ground currents from other sources are more likely to find their way into the link's ground wire. Isolating the link reduces or even eliminates these problems. Galvanic isolation is a perfect solution if there is no guarantee that the potential at the earth grounds at different nodes in the system are within the common-mode range of the transceiver. Galvanic isolation allows information flow, but prevents current flow (Marais 2008). Note that the signal lines, as well as the power supply, must be isolated.

As an RS-485 interface suitable for the developed system the ADM2486 was selected. The ADM2486 differential bus transceiver is an integrated, galvanically isolated component designed for bidirectional data communication on multipoint bus transmission lines. It is designed for balanced transmission lines and complies with ANSI EIA/TIA-485-A and ISO 8482: 1987(E). The device employs Analog Devices iCoupler technology to combine a 3-channel isolator, a three-state differential line driver, and a differential input receiver into a single package. The logic side of the device is powered with either a 5 V or a 3 V supply, and the bus side uses an isolated 5 V supply. The ADM2486 driver has an active-high enable feature. The driver differential outputs and the receiver differential inputs are connected internally to form a differential input/output port that imposes minimal loading on the bus when the driver is disabled or when VDD1 or VDD2 = 0 V. Also provided is an active-high receiver disable feature that causes the receive output to enter a high impedance state. The device has current-limiting and thermal shutdown features to protect against output short circuits and situations where bus contention may cause excessive power dissipation. The part is fully specified over the industrial temperature range and is available in a 16-lead, wide body SOIC package.

In the ADM2486, electrical isolation is implemented on the logic side of the interface. Therefore, the part has two main sections: a digital isolation section and a transceiver section (Figure 2).

Driver input and request-to-send signals, applied to the TxD and RTS pins, respectively, and referenced to logic ground (GND 1), are coupled across an isolation barrier to appear at the transceiver section referenced to isolated ground (GND 2). Similarly, the receiver output, referenced to isolated ground in the transceiver section, is coupled across the isolation barrier to appear at the RxD pin

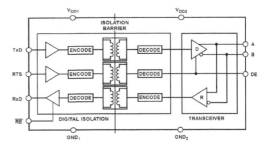

Figure 2. Digital isolation and transceiver section of ADM2486 (Anon 2013).

referenced to logic ground (Anon 2013). The receiver input includes a fail-safe feature that guarantees a logic high RxD output when the A and B inputs are floating or open-circuited. This means that, the differential input threshold Voltage (VTH) has been adjusted from ±200 mV to −200 mV to −30 mV during the bus idle condition, VIA−VIB = 0 and therefore is larger than −30 mV, resulting in the receiver output being high (RO = 1). This means that if all transceivers connected to the bus have true fail-safe features, the receiver output is always defined. The ADM2486 requires isolated power capable of 5 V at up to approximately 75 mA (this current is dependent on the data rate and termination resistors used) to be supplied between the VDD 2 and the GND 2 pins. Supply currents for iCoupler are impacted by the values of supply voltage, output load, and data rate of the isolation channels. IDD 1 and IDD 2 are determined by performing separate calculations for each channel and summing the results. The values for I_{DDO} and I_{DDI} for a given channel are calculated using Equation 1 and Equation 2 (Anon 2013):

$$I_{DDO} = \left(I_{DDO(D)} + \left(0.5 \times 10^{-3}\right) \times C_L \times V_{DDO}\right) \times \left(2f - fr\right) + I_{DDO(Q)} \tag{1}$$

$$I_{DDI} = \left(I_{DDO(D)}\right) \times \left(2f - fr\right) + \left(I_{DDI(Q)}\right) \tag{2}$$

where: IDDI(D), IDDO(D) are the dynamic input and output supply current per channel (mA/Mbps), CL is the output load capacitance (pF), f is the input logic frequency (MHz, half of input data rate, NRZ signaling), fr is the input stage refresh rate (Mbps), IDDI(Q) and IDDO(Q) are the input and output quiescent supply currents (mA), VDDO is the output supply value (V).

Figure 3 shows the designed schematic of UART/RS-485 converter using the ADM2486. The ADM2486 isolated RS-485 transceiver requires no external interface circuitry for the logic interfaces. The power supplying is ensured from two galvanically isolated 3.3 V and 5 V power supplies. The isolation is made by sub-miniature isolated

DC/DC converter TME0505S. The TME0505S drives the 5 V side of the RS-485 transceiver. The 3.3 V side is driven from CPU (UART communication) power supply. Power supply bypassing is done by capacitors C29 and C30. Resistor R74 is used as a termination on the Master node and on the farthest Slave node. On the other slave nodes the termination resistor is unplaced. Transient voltage suppressors TS5 and TS6 protect the component from high voltage transients. Signal pins RX (Receiver Output Data), RTS (Receiver Enable Input) and TX (Transmit Data Input) are connected to CPU and provide an UART communication between the two connected components (CPU and ADM). Signals 485_B1 and 485_A1 are the inverting and noninverting driver output/ receiver input. The circuit shown in Figure 3 is used as the communication interface in all Slave nodes—micro processor controlled devices (Figure 1). From Figure 1 could be seen that the Master node is equipped with an USB communication interface.

The communication between the Master node and Slave nodes is carried over the USB/RS-485 converter. The designed USB/RS-485 converter (Figure 4) is based on the circuit shown in Figure 3. The circuit shown in Figure 4 illustrates an isolated USB/ RS-485 converter. The conversion between USB and

RS-485 standards are done by the before mentioned ADM2486 and an FTDI chip FT232RL. The FT232R is a USB to serial UART interface device which simplifies USB to serial designs and reduces external component count by fully integrating an external EEPROM, USB termination resistors and an integrated clock circuit which requires no external crystal, into the device. It has been designed to operate efficiently with a USB host controller by using as little as possible of the total USB bandwidth available (Anon 2010). Figure 5 shows the FT232R in a typical USB bus powered design configuration. A ferrite bead L1 is connected in series with the USB power supply to reduce EMI noise from the FT232R and associated circuitry being radiated down the USB cable to the USB host. The coupling capacitors are connected between every VCC and GND pin to bypass the power supply. Any of the CBUS I/O pins of the FT232RL can be configured to drive a LED. The FT232R has 3 configuration options for driving LEDs from the CBUS. These are TXLED#, RXLED#, and TX&RXLED#. In our application of the CBUS pins one is used to indicate transmission of data (TXLED#) and another one is used to indicate receiving data (RXLED#). When data is being transmitted or received the respective pins will drive from tri-state to low in order to provide indication on the LEDs of data transfer. A digital one-shot is used so that even a small percentage of data transfer is visible to the end user. In this application, a TTL to RS485 level converter ADM2486 is used on the serial UART interface of the FT232R to convert the TTL levels of the FT232R to RS485 levels. With RS485, the transmitter is only enabled when a character is being transmitted from the UART. The TXDEN signal CBUS pin option on the FT232R (RTS) is provided for exactly this purpose and so the transmitter enable is wired to CBUS2 which has been configured as TXDEN. Similarly, CBUS3 has been configured as RXDEN. This signal is used to control the receiver enable RE.

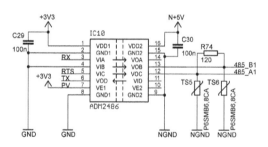

Figure 3.   UART/RS-485 converter schema.

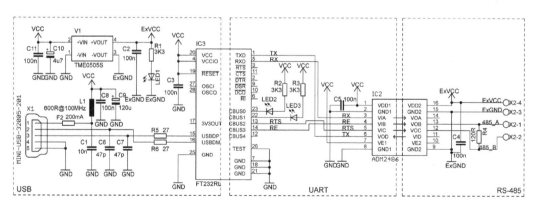

Figure 4.   USB/RS-485 converter schema.

44

Table 1. Synchronization, Command and Acknowledge packet format.

P_TYPE	START		ADR		SIZE	TYPE	D12	D11	D10	D9	D8	D7	D6	D5	D4	D3	D2	D1	D0	CRC XR	CRC OR	STOP
Synch	FF	0	0-FF	FF-0	0-FF	0	-	-	-	-	-	-	-	-	-	-	-	-	-	0-FF	0-FF	0|FF
Cmd	FF	0	0x18	0x31	0x0C	1	0-FF	0-FF	0-FF	0-FF	0-FF	0-FF	0-FF	0-FF	0-FF	0-FF	0-FF	0-FF	-	0-FF	0-FF	0|FF
Ack	FF	0	0x31	0x18	0x0D	1	0-FF	0-FF	0-FF	0-FF	0-FF	0-FF	0-FF	0-FF	0-FF	0-FF	0-FF	0-FF	0-FF	0-FF	0-FF	0|FF

4 COMMUNICATION PROTOCOL

All communication between the nodes of the RMACS network requires that the nodes agree on many physical aspects of the data to be exchanged before successful transmission can take place. The set of rules defining transmissions are called a communication protocol and was developed as a part of the RMACS. The communication protocol commonly defines the packet size, transmission speed, error correction types, handshaking and synchronization techniques, address mapping, acknowledgement processes, flow control, packet sequence controls, routing and address formatting (Franeková et al. 2007, Franeková et al. 2014). As it was mentioned above, the communication protocol defines the packet format which could be transmitted between the communication nodes (Lojka & Zolotová 2014). Table 1 shows the proposed format of three types (P_TYPE) of packets used in communication between nodes in RMACS. The communication speed is 38400 bit/s and the three packets are the Synchronization packet, Command packet and Acknowledge packet. All of them have the same 2 byte start Sequence (START). The address field consists from the address of the node which is sending the message and the address of the recipient. The Master in the communication should have an address between 0×0 and 0×19 (0x stands for HEX). The Slave nodes should have an address between 0×20 and 0xFF. After the address field, the size of the data space (SIZE) and the packet type (TYPE) are given. The data space is marked as D0 to D13. After the transmitted data the control bytes (CRC RX and CRC OR) and a 2 byte STOP are given.

5 RESULTS AND CONCLUSION

The proposed communication interface and the communication protocol were successfully tested on the first prototype of microprocessor controlled sensor and PC based HMI. Using a complicated, but sophisticated method for communication interface and protocol testing would lead to another scope of the research and extend the time of developing the RMACS. Therefore relatively simple but still effective test was executed. The test was executed in real conditions in three week uninterrupted operation. During the operation of the device the measured values and the timestamps sent from the microprocessor controlled sensor to HMI were monitored and compared with an etalon measuring device data. Also the number of measured samples were counted and compared with the result of measurement time and sampling time ratio. The above mentioned test proved the robustness and reliability of the developed RMACS communication interface.

The Real Time Monitoring and Control System will be a complex decentralized system, for metalworking fluids monitoring on 5-axis machines. The user interface of the RMACS will be located at the greater distance. The industrial environment and the greater distances between the components have an important role during the design of the communication interface and of the network topology. Taking into account the before mentioned facts an RS-485 Master-Slave communication network with appropriate communication protocol and interfaces was developed. The Master node is an HMI (standard PC) with USB connection and the Slave nodes are microprocessor controlled sensors and actuators. These nodes together form a system which will support a project focused on the study of advance oxidative processes for metalworking fluids lifetime extension and for their following acceleration of biological disposal at the end of the life cycle.

ACKNOWLEDGMENT

This work was supported by the Grant Agency VEGA of the Slovak Ministry of Education, Science, Research and Sport via project no. 1/0640/14: "Studying the use of advanced oxidative processes for metalworking fluids lifetime extension and for their following acceleration of biological disposal at the end of the life cycle".

REFERENCES

Anon (2010). Future Technology Devices International Ltd, FT232R USB UART IC Datasheet, Version 2.10, Document No.: FT_000053, United Kingdom.
Anon (2013). Analog Devices, ADM 2486 data sheet, REV. E, Analog Devices.
Franeková, M., F. Kállay, P. Peniak & P. Vestenický (2007). Communication security of industrial networks. Monography, EDIS ŽU, Žilina 2007, ISBN 978-80- 8070-715-6.

Franeková, M., R. Pirník & Ľ. Pekár (2014). Modelling of data transfer in MATLAB, Simulink and in the Communications System Toolbox. EDIS ŽU, Žilina, ISBN 978-80- 554-0896-5.

Lojka, T. & I. Zolotová (2014). Distributed sensor network—data stream mining and architecture. Advances in Information Science and Applications—Volume 1: Proceedings of 18th International Conference on Computers (part of CSCC `14), Santorini Island, Greece, ISBN 978-1-61804-236-1—ISSN 1790-5109, 98-103.

Marais, H. (2008). RS-485/RS-422 Circuit Implementation Guide. AN-960 REV. 0, Application note.

Piteľ, J., I. Polanecká & K. Žídek (2005). Sensors and sensor systems—A Guide for Exercises. Košice: TU, ISBN/ISSN: 80-8073-451-8, p. 106.

Skerlos, S.J., K.F. Hayes, A.F. Clarens & F. Zhao (2008). Current Advances in Sustainable Metalworking Fluids Research. *Int. J. of Sustainable Manufacturing 1*, No.1/2, 180–202.

*Advances in Materials Science, Energy Technology
and Environmental Engineering – Patty & Zhou (Eds)*
© 2017 Taylor & Francis Group, London, ISBN 978-1-138-19668-1

Theoretical modeling and simulation analysis of ultra-precision turning of ZnSe crystals

S.P. Li & Y. Zhang
Kunming University of Science and Technology, Kunming, Yunnan, China

Q.M. Xie & W.Q. Zhang
Yunnan KIRO-CH Photonics Co. Ltd., Kunming, Yunnan, China

ABSTRACT: In order to explore the machining mechanism of ultra-precision turning of ZnSe polycrystalline materials and explain its processing phenomena, in view of the characteristics of ultra-precision turning of brittle materials, a theoretical turning model with the characteristics of crack propagation is established. According to the results of theoretical modeling and simulation analysis, critical processing conditions of ZnSe crystals lens turning are explored, and the critical undeformed chip thickness, "brittle-ductile transition" depth and critical turning thickness in machining are researched.

1 INTRODUCTION

In recent years, with the ever-increasing application or requirements in space science and technology, the national defense and military fields, the preparation and properties of infrared materials and the optical components manufacturing are concerned in many countries. In addition, processing crystal optical elements by the Single Point Diamond Turning technology (SPDT) can ensure high precision, improve the laser damage threshold and avoid embedding in polishing abrasive, so as to effectively reduce subsurface damage on the crystal surface, so the SPDT technology is widely used in engineering. In fact, the infrared materials like Zinc Selenide (ZnSe) crystals currently have been able to conduct a more mature manufacture by SPDT in other countries. But in view of its important military scientific value, the process has heavily security features and is reported more rarely. Therefore, the key technical problems of these materials in ultra-precision machining have to be solved independently.

The ultra-precision oblique turning models of brittle material which are coupled with material properties, tool geometrical characteristics and machining parameters of oblique turning are established in this paper. Based on critical turning conditions of "brittle-ductile coupled material removal", the turning models and brittle-ductile transition depth of the critical feed rate, cutting depth and tool rake angle, and the tool tip's arc radius and the tool oblique turning angle are predicted, which provides a theoretical foundation

for revealing the brittle-ductile coupled material removal mechanism of soft brittle materials.

2 ULTRA-PRECISION OBLIQUE TURNING PROCESS MODELING OF ZNSE CRYSTALS

2.1 *Geometric model of oblique turning*

The oblique turning processing models will be established mainly from three aspects, including cutting process parameters, tool geometry characteristics and the inherent characteristics of the material. In the actual turning experiments, the tool cutting edge blunt round radius can be controlled in less than 30 nm, which is far less than the undeformed chip thickness of ZnSe criticals. Therefore, the tool edge can be approximated infinitely sharp. In addition, three-dimensional model for ultra-precision turning coupled with tool oblique angle is put forward, as shown in Figure 1.

According to turning conditions of "brittle-ductile coupled material removal", the processing of SPDT for brittle materials is studied. That is to say, when the tool undeformed chip thickness is smaller than a critical value or the critical depth of "brittle-ductile transition", and cracks in machining process has not yet spread to the machined surface, ZnSe crystals can be turned by ultra-precision turning to obtain fine machined surface, as shown in Figure 1 a). Workpiece coordinate system is supposed C_1: O-$X_1Y_1Z_1$. The direction of X_1-axis is opposite to the direction of the tool feed rate f_1. The direction of Y_1-axis is the same as the tool

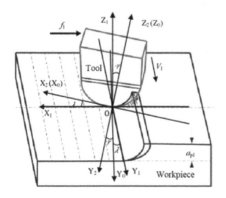

a) The transform relationship between orthogonal cutting and oblique cutting

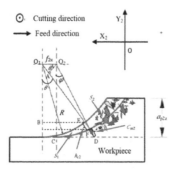

b) The solution of critical turning thickness in the coordinate system

Figure 1. Geometric models of ultra-precision oblique turning of ZnSe crystal.

cutting speed $\vec{V_1}$. The direction of the Z_1-axis is opposite to the direction of the tool cutting depth $\overline{a_{p1}}$.

In addition, based on the geometric features of the tool, the tool coordinate system C_2: O-$X_2Y_2Z_2$ on the front surface of the tool is established. The X_2-axis is located in the same plane with the tool cutting edge, which is tangent to the tip point O of the tool. The Y_2-axis is through the tip point of the tool and is perpendicular to the front surface of the tool. The Z_2-axis is located in the same plane with the tool cutting edge and is perpendicular intersection on the tip point of the tool with the X_2-axis. The directions of the coordinate axes of the tool coordinate system C_2 can be determined by the coordinate transformation of the workpiece coordinate system C_1. In order to establish the quantitative transform relationship between C_1 and C_2, the intermediate coordinate system C_0: O-$X_0Y_0Z_0$ is introduced. The specific transformation process is as follows: C_1 is counterclockwise rotated λ

around the Z_1-axis by the right hand rule to obtain C_0, and C_0 counterclockwise rotated a rake angle -γ around the X_0-axis by the right hand rule to obtain C_2. The transformation relation can be expressed by the formula (1):

$$C_2 = M_{20} \times M_{01} \times C_1 = M_{21} \times C_1 \qquad (1)$$

The matrix M_{21} can be obtained from the formula (2):

$$M_{21} = M_{20} \times M_{01}$$

$$= \begin{bmatrix} 1 & 0 & 0 & 0 \\ 0 & \cos\gamma & \sin\gamma & 0 \\ 0 & -\sin\gamma & \cos\gamma & 0 \\ 0 & 0 & 0 & 1 \end{bmatrix} \times \begin{bmatrix} \cos\lambda & -\sin\lambda & 0 & 0 \\ \sin\lambda & \cos\lambda & 0 & 0 \\ 0 & 0 & 1 & 0 \\ 0 & 0 & 0 & 1 \end{bmatrix}$$

$$= \begin{bmatrix} \cos\lambda & -\sin\lambda & 0 & 0 \\ \sin\lambda\cos\gamma & \cos\lambda\cos\gamma & \sin\gamma & 0 \\ -\sin\lambda\sin\gamma & -\cos\lambda\sin\gamma & \cos\gamma & 0 \\ 0 & 0 & 0 & 1 \end{bmatrix}$$

$$(2)$$

The vector processing of machining parameters in the Figure 1 a) is carried out according to C1: cutting depth is $\overline{a_{p1}} = [0 \quad 0 \quad -a_{p1} \quad 0]$, tool feed rate is $\overline{f_1} = [-f_1 \quad 0 \quad 0 \quad 1]$. In the same way, the processing parameters in C_2 are also processed: cutting depth is $\overline{a_{p2}} = \left[-a_{p2x} \quad -a_{p2y} \quad -a_{p2z} \quad 1 \right]$, tool feed rate is $\overline{f_2} = \left[-f_{2x} \quad -f_{2y} \quad -f_{2z} \quad 1 \right]$. Therefore, the processing parameters in C_2 after the transformation can be obtained by the formula (3):

$$\begin{cases} a_{p2z} = a_{p1} \cdot \cos\gamma \\ f_{2x} = f_1 \cdot \cos\lambda \end{cases} \qquad (3)$$

In the formula, a_{p2z} is the cutting depth component along the Z_2-axis. f_{2x} is the component of feed rate along the X_2-axis, as shown in Figure b).

2.2 Calculation of critical turning thickness

The model of the critical turning thickness of ZnSe crystal mirror surface machining in C_2 is established, as shown in Figure 1 b). The point O_1 and O_2 are the natural diamond tools tip's arc center before and after the f_{2x}, and the tool tip's arc radius is R. When the a_{p2z} is larger, ZnSe crystals can be in the complicated turning state of plastic deformation and brittle fracture. Before the stage that the turning thickness h_2 is increased from zero to the critical value h_{c2}, the material is in the stage of plastic deformation process, corresponding area is S_1, and with the further increase of turning thickness, materials will be in the stage of brittle fracture, corresponding area is S_2. At the same time, the crack is gradually extended to the machined surface along the radius of the cutting edge. In addition, part of the cracks disappear or are cut off with the removal of the chips. So the intermediate crack can be extended to the machined surface when the critical turning thickness is assumed as h_{c2}, which is reasonable. Therefore, the critical crack length C_{m2} of the in initial point F of Figure 1 b) can be obtained by the formula (4):

$$C_{m2} = \left[\eta \bullet \left(\frac{E}{H} \right)^{1/2} \bullet \frac{4 \bullet F_c}{K_{ID}} \right]^{2/3} \qquad (4)$$

In the formula, η is a coefficient, which is configured as 0.016 ± 0.004. The dynamic fracture toughness $K_{ID} = 30\% \bullet K_{IC}$. F_c is the main cutting force of the tool rake face, which can be obtained from the formula (5):

$$F_c = \sigma_s \bullet S_1 + \sigma_b \bullet S_2 \qquad (5)$$

In the formula, σ_s is the yield limit of the workpiece, σ_b is the fracture strength of the workpiece.

Formula (6) is the solution formula of the plastic deformation cutting area S_1 and the brittle fracture removal area S_2:

$$\begin{cases} S_1 = \int_{X_a}^{X_b} \left| \sqrt{R^2 - (x - f_{2x})^2} - \sqrt{R^2 - x^2} \right| dx \\ S_2 = \int_{X_b}^{X_c} \left| \sqrt{R^2 - (x - f_{2x})^2} - (R - a_{p2z}) \right| dx \end{cases} \qquad (6)$$

In the formula, $X_a = \frac{f_{2x}}{2}$, $X_b = \sqrt{R^2 - (R - a_{p2z})^2}$, $X_c = \sqrt{R^2 - (R - a_{p2z})^2} + f_{2x}$.

According to the geometric relations in Figure 1 b), formula (7) and (8) are obtained:

$$\cos\varphi = \frac{R}{R + C_{m2}} \qquad (7)$$

$$\cos\theta = \frac{R - C_{m2} \bullet \cos\varphi - h_{c2} \bullet \cos\theta}{R} \qquad (8)$$

According to formula (7) and (8), the formula (9) can be obtained:

$$h_{c2} = \frac{R^2}{(R + C_{m2}) \bullet \cos\theta} - R \qquad (9)$$

In addition, according to the geometric relations in Figure 1 b), formula (10) is obtained:

$$h_{c2} = L_{O_1A} + L_{AF} - L_{O_1E}$$
$$= f_{2x} \bullet \sin\theta + \sqrt{R^2 - (f_{2x} \bullet \cos\theta)^2} - R \qquad (10)$$

2.3 Critical conditions of mirrors cutting

Research shows that the critical depth of brittle materials is not intrinsic parameter of the material. The parameter is related to not only the intrinsic characteristics of the material but also material stress distribution and the characteristics of loading tool. However, when specific workpiece materials are processed by SPDT, the intrinsic characteristics of materials, cutting tool and material stress distribution, etc. are basically stable, so the critical plastic deformation depth d_c of the workpiece materials can be treated as an approximate invariant.

Therefore, this paper presents the ZnSe crystals mirror cutting conditions: the critical turning thickness is less than the critical depth of the "brittle-ductile transition", $h_{c2} \le d_c$. That is to say, when workpieces are processed, brittle cracks may appear, but machined surface is not yet damaged, which has significantly reduced the requirements for the surface machining of brittle materials, and provides a new theoretical basis for the exploration of efficient processing technology.

3 STUDY ON CRITICAL PROCESS PARAMETERS OF ZNSE CRYSTALS MIRROR TURNING

Table 1 shows the material properties of ZnSe crystals, these parameters are put into the model to obtain the critical turning thickness h_{c2}. In this paper, the process parameters of ZnSe crystals mirror turning are studied, such as the feed rate, cutting depth, tool rake angle, tool tip's arc radius and tool oblique turning angle. Then, the critical turning conditions are simulated and analyzed, and the critical turning range or the optimal range of each process parameter is determined.

3.1 Critical the feed rate

The initial cutting depth is $a_{p1} = 1.0$ μm. The tool tip's arc radius is $R = 750$ μm. The tool rake angle is $\gamma = -30°$. The range of tool oblique turning angle λ is from $0°$ to $-50°$, and the step length is $-10°$. The range of feed rate f_1 is from 1 $\mu m/r$ to 6 $\mu m/r$. The material parameters in Table 1 are substituted into the model of oblique turning, and the crack growth length C_{m2} and the critical turning thickness h_{c2} are solved, as shown in Figure 2 and Figure 3.

Figure 2 shows that the cutting crack length C_{m2} of ZnSe crystal is longer with increase of f_1, but C_{m2} is decreased with increase of λ. And Figure 3 shows that the change rule between critical cutting thickness h_{c2} and cutting crack length C_{m2} is same.

Table 1. The material properties of ZnSe crystals.

Material type	Vickers hardness HV_0 (kg/mm^2)	Elastic modulus E (GPa)	Fracture toughness K_{IC} (MPa·m$^{1/2}$)	Yield stress σ_s (MPa)	Fracture strength σ_b (MPa)	Critical depth d_c (nm)
ZnSe	107.4	70.3	0.5	52	55	200

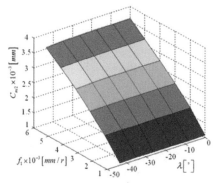

a) Relationship between C_{m2} and f_1, λ

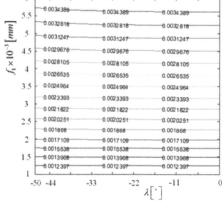
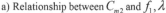

b) Contour relationship between C_{m2} and f_1, λ

Figure 2. Relationship between feed rate, tool oblique turning angle and crack growth length.

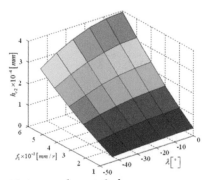

a) Relationship between h_{c2} and f_1, λ

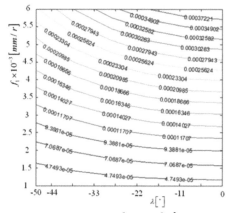

b) Contour relationship between h_{c2} and f_1, λ

Figure 3. Relationship between feed rate, tool oblique turning angle and critical turning thickness.

In addition, Figure 3 b) shows that when $\lambda = 0°$ and f_1 is from 3.3 $\mu m/r$ to 3.7 $\mu m/r$, the h_{c2} will change from 186.66 nm to 209.85 nm, while the critical plastic deformation depth $d_c = 200$ nm is contained in this interval. According to the critical cutting conditions, it can be inferred that the critical condition of the ZnSe crystal mirror cutting is from 3.3 $\mu m/r$ to 3.7 $\mu m/r$. However, when the critical feed rate is 3.5 $\mu m/r$, observing trends of contour lines of h_{c2} in 186.66 nm and 209.85 nm by the change rule of λ, it can be inferred that when λ is from $-10°$ to $-20°$, plastic removal of ZnSe crystal can be achieved, which shows that the processing of brittle materials is effectively improved by oblique turning.

3.2 Critical the cutting depth

The feed rate is $f_1 = 2.0$ $\mu m/r$. The tool tip's arc radius is $R = 750 \mu m$. The tool rake angle is $\gamma = -30°$. The range of tool oblique turning angle λ is from $0°$ to $-50°$, and the step length is $-10°$. The range of cutting depth a_{p1} is from 0.5 μm to 3.0 μm. The material parameters in Table 1 are substituted into the model of oblique turning, and the crack growth length C_{m2} and the critical turning thickness h_{c2} are solved, as shown in Figure 4 and Figure 5.

Figure 4 shows that the cutting crack length C_{m2} of ZnSe crystal is longer with increase of a_{p1}, but C_{m2} is decreased with increase of λ. And Figure 5

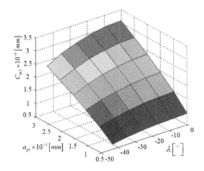

a) Relationship between C_{m2} and a_{p1}, λ

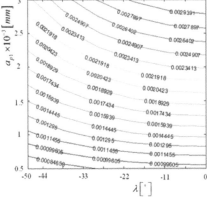

b) Contour relationship between C_{m2} and a_{p1}, λ

Figure 4. Relationship between cutting depth, tool oblique turning angle and crack growth length.

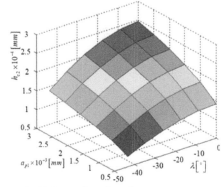

a) Relationship between h_{c2} and a_{p1}, λ

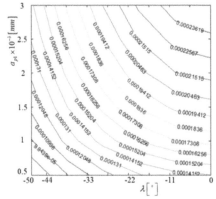

b) Contour relationship between h_{c2} and a_{p1}, λ

Figure 5. Relationship between cutting depth, tool oblique turning angle and critical turning thickness.

shows that the change rule between critical cutting thickness h_{c2} and cutting crack length C_{m2} is same. In addition, Figure 5 b) shows that when $\lambda = 0°$ and a_{p1} is from 1.35 μm to 1.65 μm, the h_{c2} of ZnSe crystal will change from 194.12 nm to 204.63 nm, while the critical plastic deformation depth $d_c = 200$ nm is contained in this interval. According to the critical cutting conditions, it can be inferred that this range of a_{p1} is the critical condition of the ZnSe crystals mirror cutting. However, observing trends of contour lines of h_{c2} in 194.12 nm to 204.63 nm by the change rule of λ, it can be inferred that when λ is from −10° to −15°, plastic removal of ZnSe crystal can be achieved, which will improve the cutting efficiency and further optimize the processing parameters.

3.3 Critical the tool rake angle

The cutting depth is $a_{p1} = 2.0$ μm. The tool tip's arc radius is $R = 750$ μm. The feed rate is $f_1 = 3.0$ $\mu m/r$. The range of tool oblique turning angle λ is from 0° to −50°, and the step length is −10°. The range of tool rake angle γ is from 0° to −50°. The mate-

rial parameters in Table 1 are substituted into the model of oblique turning, and the crack growth length and the critical turning thickness are solved, as shown in Figure 6 and Figure 7.

Figure 6 shows that the cutting crack length C_{m2} of ZnSe crystal is shorter with increase of γ, and C_{m2} is decreased with increase of λ. And Figure 7 shows that critical cutting thickness h_{c2} is gradually decreased with increase of γ and λ. In addition, Figure 7 b) shows that when $\lambda = 0°$ and γ is from −28° to −40°, the h_{c2} of ZnSe crystal will change from 205.45 nm to 198.74 nm, while the critical plastic deformation depth $d_c = 200$ nm is contained in this interval. According to the critical cutting conditions, it can be inferred that this range of γ is the critical condition of the ZnSe crystals mirror cutting. However, when the tool rake angle $\gamma = -30°$, observing trends of contour lines of h_{c2} in 205.45 nm to 198.74 nm by the change rule of λ, it can be inferred that when λ is from −5° to −15°, plastic removal of ZnSe crystal can be achieved, which will further optimize the processing parameters.

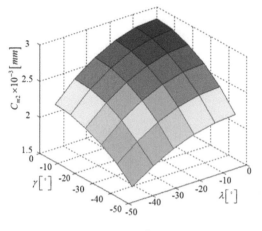

a) Relationship between C_{m2} and γ, λ

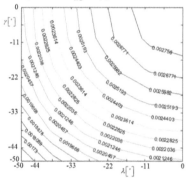

b) Contour relationship between C_{m2} and γ, λ

Figure 6. Relationship between tool rake angle, tool oblique turning angle and crack growth length.

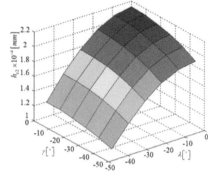

a) Relationship between h_{c2} and γ, λ

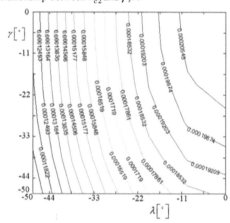

b) Contour relationship between h_{c2} and γ, λ

Figure 7. Relationship between tool rake angle, oblique turning angle and critical turning thickness.

3.4 The best tool tip's arc radius

The cutting depth is $a_{p1} = 1.0 \ \mu m$. The feed rate is $f_1 = 2 \ \mu m/r$. The tool rake angle is $\gamma = -30°$. The range of tool oblique turning angle λ is from $0°$ to $-50°$, and the step length is $-10°$. The range of tool tip's arc radius R is from $50 \ \mu m$ to $5000 \ \mu m$. The material parameters in Table 1 are substituted into the model of oblique turning, and the crack growth length and the critical turning thickness are solved, as shown in Figure 8 and Figure 9.

Figure 8 shows that the cutting crack length C_{m2} is decreased with increase of λ. However, when λ is more than $-15°$, the change of C_{m2} was not significant with increase of R. When the tool is in the state of orthogonal cutting, the cutting crack length C_{m2} is decreases rapidly with increase of R, but it tends to be stable soon. Figure 9 shows that critical cutting thickness h_{c2} is dramatically decreased with increase of R, and it tends to be

stable finally. At the same time, when R is relatively small, h_{c2} decreases rapidly with increase of λ. While R is large, the change of h_{c2} was not significant with increase of λ. Figure 9 b) shows that when tool tip's arc radius is $R = 700 \ \mu m$, the critical turning thickness is $h_{c2} < 200 \ nm$, that is $h_{c2} < d_c$. According to the cutting conditions and changing trend of h_{c2}, it can be inferred that when tool tip's arc radius is $R < 700 \ \mu m$, Mirror machining of ZnSe crystals can be realized. When tool tip's arc radius is R > 3000 μm, the critical turning thickness is almost stable. In addition, Figure 8 and Figure 9 show that the effect of the oblique angle on cutting crack is significant, but there is a small effect for the critical turning thickness, especially when the tool tip's arc radius is very small, the phenomenon is more prominent. In summary, when R is from 700 μm to 2500 μm, Mirror machining of ZnSe crystals can be realized, so this range of the parameters is reasonable.

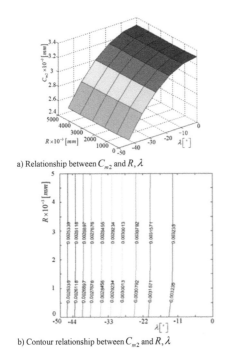

a) Relationship between C_{m2} and R, λ

b) Contour relationship between C_{m2} and R, λ

Figure 8. Relationship between tool tip's arc radius, oblique turning angle and crack growth length.

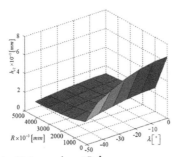

a) Relationship between h_{c2} and R, λ

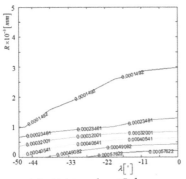

b) Contour relationship between h_{c2} and R, λ

Figure 9. Relationship between tool tip's arc radius, oblique turning angle and critical turning thickness.

4 CONCLUSIONS

The ultra-precision oblique turning models of brittle materials which are coupled with material properties, tool geometrical characteristics and machining parameters of oblique turning were established in this paper, and critical conditions of mirrors cutting of ZnSe crystals is put forward: $h_{c2} \le d_c$. Based on the critical conditions and oblique turning models, the critical feed rate, cutting depth, tool rake angle, tool tip's arc radius for ZnSe crystals are forecasted and analyzed. In addition, the positive effect of oblique turning on the ultra-precision turning process of ZnSe crystals is explored, which provides a theoretical guide for the analysis of the brittle-ductile coupling removal process in mirror surface machining.

REFERENCES

Arif M., Zhang X. & Rahman M. 2013. A predictive model of the critical undeformed chip thickness for ductile–brittle transition in nano-machining of brittle materials. *International Journal of Machine Tools & Manufacture* 64(4): 114–122.

Cheng X., Wei X.T. & Yang X.H. 2014. Unified Criterion for Brittle–Ductile Transition in Mechanical Microcutting of Brittle Materials. *Journal of Manufacturing Science & Engineering*: 136(136).

Klocek P. 1991. Handbook of infrared optical materials. *New York: Dekker CRC Press.*

Venkatachalam S., Li X. & Liang S.Y. 2009. Predictive modeling of transition undeformed chip thickness in ductile-regime micro-machining of single crystal brittle materials. *Journal of Materials Processing Technology* 209(7): 3306–3319.

Yu H.Z. 2007. Infrared Optical Material (Second Edition). *Beijing: National Defense Industry Press.*

Zong W.J., Cao Z.M. & He C.L. 2015. Critical undeformed chip thickness of brittle materials in single point diamond turning. *International Journal of Advanced Manufacturing Technology* 81(5–8): 1–10.

*Advances in Materials Science, Energy Technology
and Environmental Engineering – Patty & Zhou (Eds)*
© *2017 Taylor & Francis Group, London, ISBN 978-1-138-19668-1*

Site-preference effects in proton conductors of $BaCe_{1-x}Eu_xO_{3-\delta}$

J.X. Wang
School of Electronic and Information Engineering, Ningbo University of Technology, Ningbo, P.R. China

L.F. Lai
School of Materials Science and Engineering, Ningbo University of Technology, Ningbo, P.R. China

J.W. Jian & F.A. Li
College of Information and Science Engineering, Ningbo University, Ningbo, P.R. China

ABSTRACT: Perovskite oxides $BaCe_{1-x}Eu_xO_{3-\delta}$ (x = 0.05, 0.10, 0.15, 0.20) using starting ratios of Ba/(Ce + Eu) = 1.0 and 1.1 were prepared by solid state reactions. X-ray diffraction analysis shows that all XRD data are quite similar to those of the standard diffraction data for $BaCeO_3$ with the exception of the peak shifts. The samples synthesized from Ba/(Ce + Eu) = 1.1 preparations are shown to have different cell volume trends versus x and higher proton conductivities in water-containing atmospheres between 80 and 700°C, compared to samples synthesized from Ba/(Ce + Eu) = 1.0 preparations. These observations demonstrated that site-preference effects in proton conductors of $BaCe_{1-x}Eu_xO_{3-\delta}$.

1 INTRODUCTION

ABO_3-based perovskite oxides (Li et al. 2002; Fu & Weng 2014) have received great attention due to their great applications in fuel cells and other solid-state ionic devices that are required to have higher conductivity with negligible electronic component, lower thermal expansion, and higher structural stability. One of the effective routes to higher conductivity is to dope trivalent rare earth ions at B-sites via the introduction of the oxygen vacancies process (in Kröger-Vink notation (Kröger & Vink 1956)):

$$2B_B^x + O_o^x + R_2O_3 \rightarrow 2R_B' + V_o^{..} + 2BO_2 \tag{1}$$

When these doped-oxides are exposed to the humid atmosphere, the protons would incorporate into lattice via the reaction:

$$H_2O(g) + V_o^{..} + O_o^x \rightarrow 2(OH)_o^{.} \tag{2}$$

Such protons are free to migrate from one ion to the next, resulting in higher proton conduction.

$BaCeO_3$-based oxides are a kind of compounds that exhibit the promising proton conduction features such as relatively high conductivity and low activation energy (Kreuer 1999). However, these materials have to be prepared by high temperature solid-state reactions above 1500°C, which would lead to somewhat barium loss. To prevent the bar-

ium loss, one has to put more barium species before the formation reactions. This has led to some compositional uncertainties in the final products and difficulties in optimizing the proton conduction property (Peng et al. 2006; Chi et al. 2013). To design and search for the new proton conductors, it is very important to make clear the site-preference effects during the compositional variations. However, to the best of our knowledge, there is no experimental evidence on such site-preference effects. In this work, we report on the single-phase synthesis and structural/property characterization of proton conductors of $BaCe_{1-x}Eu_xO_{3-\delta}$.

2 EXPERIMENTAL

Chemical reagent pure $BaCO_3$, CeO_2, and Eu_2O_3 were used as the starting materials and were preheated in air at 900°C for 2 h so as to remove completely the carbonate species absorbed. For group A, the initial molar ratio was fixed as Ba: Ce: Eu = 1.0:1–x: x, while for group B, the initial molar ration was 1.1:1–x: x (x = 0.05, 0.10, 0.15, 0.20). After weighing according to these ratios, the mixtures were ball-milled for 4 h and calcined in air at 1250°C for 10h. The powders were then pressed into pellets under a pressure of 800 MPa and sintered in air at 1600°C for 10 h. After cooling to room temperature, the samples of groups A and B were obtained.

3 RESULTS AND DISCUSSION

3.1 *Structural characterization*

Fig. 1 illustrates X-ray diffraction patterns of the typical samples from group A Similar XRD patterns and group B. X-ray diffraction analysis shows that all XRD data are quite similar to those of the standard diffraction data for $BaCeO_3$ with the exception of the peak shifts. These results demonstrated the formation of the orthorhombic perovskite structure, while the peak shifts were associated with the variations of the lattice parameters under doping.

Assuming the initial atomic positions for $BaCeO_3$ (orthorhombic Pbnm with a = 6.216, b = 6.236, c = 8.777 Å) (Knight 2001), the lattice parameters for the samples of both groups were refined using the Rietica Rietveld refinement program. Fig. 1 illustrates the indexing results. The dopant content dependence of the cell volume is shown in Fig. 2. For group A, the cell volume increased with dopant content up to x = 0.10, and then decreased (Fig. 2a). The presence of the cell volume maximum can be understood in terms of the barium loss and site—preference of Eu^{3+} at Ba^{2+} sites. At lower dopant content than x = 0.10, the samples of group A probably have extremely low concentration of barium loss, Eu^{3+} substitution at Ce^{4+} would lead to lattice expansion since Eu^{3+} ion are larger than Ce^{4+} in 6-fold coordination. At higher dopant contents, barium loss might become significant; some Eu^{3+} ions would substitute at the Ba^{2+} sites by reducing the lattice energy. Even though part of larger Eu^{3+} at Ce^{4+} would yield somewhat lattice expansion, Eu^{3+} ions are much smaller than Ba^{2+}, Eu^{3+} substitution at Ba^{2+} sites is expected to produce a pronounced lattice shrinkage. This assumption is highly reasonable since the samples of group A did not show any EPR signals associated with the barium vacancies at A-site (Li et al. 2001) (see inset of Fig. 2). For the samples of group B, however, the excess of barium in the starting compositions would compensate the barium loss during the high temperature reactions. Therefore, the compositions probably match well the nominal ratios of $BaCe_{0.85}Eu_{0.15}O_{3-\delta}$ with no barium vacancies occurred. This is also strongly confirmed by our EPR measurements (inset of Fig. 2), where no EPR signals were observed for barium loss. Further, the site-preference of Eu^{3+} at Ba^{2+} sites is greatly suppressed since no extra Ba^{2+} sites are available. Substitution of Ce^{4+} by Eu^{3+} accounts for the nearly linear increase of the cell volume

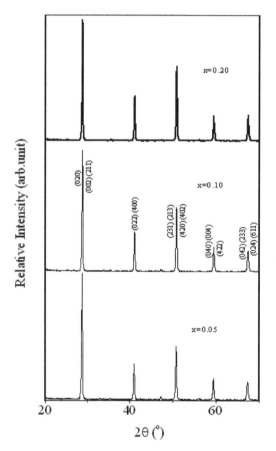

Figure 1. XRD patterns for the samples $BaCe_{1-x}Eu_x$ from group A.

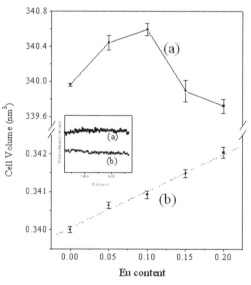

Figure 2. Relationship between cell volume of the samples $BaCe_{1-x}Eu_xO_{3-\delta}$ from (a) group A and (b) group B. The inset shows the corresponding EPR signals.

with the dopant content in Fig. 2b. Our previous Raman results also showed that all Eu^{3+} ions are doped onto the Ce-sites, the oxygen vacancy concentration would increase with increasing Eu content (Wang et al. 2006).

3.2 Electrical properties

The conductivity for all of the samples were measured by AC complex impedance technique over the temperature range from 200 to 750°C in a water containing atmosphere. Figure 3 shows a typical impedance spectrum of $BaCe_{0.85}Eu_{0.15}O_{3-\delta}$ from group B that was recorded at 200°C in water containing atmosphere. The impedance data consisted of two arcs and one spike with decreasing frequencies that are corresponding respectively to the proton conduction in lattice and along grain boundaries, as well as the electrode effects. Similar spectra have been found for all samples of groups A and B. With increasing temperature, the effect of electrode became obviously and the sizes of semicircles from bulk and grain boundary decreased dramatically, which indicated that the resistance decreased with the temperature. To get the conductivity of grain boundary, more information about the microstructure is needed. Here only the bulk conductivity was emphasized. The bulk conductivity data were obtained by the intersection of the bulk arc on the real part of the impedance axis. The dependence of conductivity on temperature exhibits Arrhenius behavior over the temperature regions studied; the activation energies (E_a) and pre-exponential terms (A) were determined by

fitting the expression, $\sigma = A/T\exp(-E_a/KT)$, to the data.

As shown in the inset of Fig. 3, the conductivity data for the sample $BaCe_{0.85}Eu_{0.15}O_{3-\delta}$ from group A were smaller than those from group B, while the activation energy was nearly the same. This result can be understood in terms of the site-preference-related oxygen vacancy concentration. As our XRD analysis, the samples of group A has some amount of barium loss, site-preference of Eu^{3+} at Ba^{2+} sites would reduce the oxygen vacancy concentration and lattice shrinkage. The similar activation energy data indicated that site-preference of Eu^{3+} at Ba^{2+} sites did not have a big influence on the proton hopping in lattice, which is however different from a lattice simulation where activation energies increased with a decrease in lattice parameter. For the samples from group B, all Eu^{3+} would substitute at Ce^{4+} sites, and a higher concentration of oxygen vacancies can be the main reason for the enhanced proton conduction in perovskite lattice.

The temperature dependences of bulk conductivity for all the samples from group B in humid atmosphere is shown in Fig. 4. The $Ba_{1.04}Ce_{0.85}Gd_{0.15}O_{3-\delta}$ data (dot line) was also shown for comparison. The fitting results in Table 1, along with 300°C bulk conductivity values, yield two interesting observations. First, the conductivity at 300°C increases with Eu content and reached the maximum for x = 0.15, then decreases for x = 0.20. Second, the conductivity of x = 0.15 composition is nearly the same as previously reported for $Ba_{1.04}Ce_{0.85}Gd_{0.15}O_{3-\delta}$ (Shima & Haile 1997).

We used the defect chemistry to explain these results. According to Eqs. (1) and (2), one oxygen

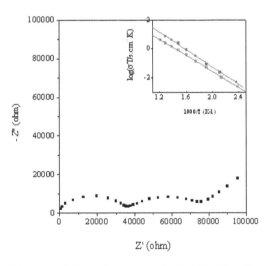

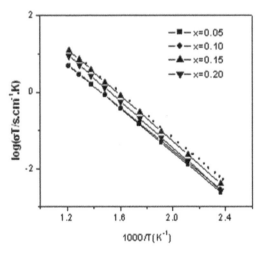

Figure 3. AC impedance spectra for $Ba_{1.1}Ce_{0.85}Eu_{0.15}O_{3-\delta}$ recorded at 200 in water containing atmosphere. The inset shows the conductivity data for the samples from group A (open square) and group B (solid square).

Figure 4. Arrhenius plots of bulk conductivity for samples $BaCe_{1-x}Eu_xO_{3-\delta}$ from group B in humid air and the data of $Ba_{1.04}Ce_{0.85}Gd_{0.15}O_{3-\delta}$ (black dot) for comparison.

Table 1. Data of conductivity, activation energy and pre-exponent term at 300°C.

$BaCe_{1-x}Eu_xO_{3-\delta}$	σ ($\times 10^{-4}$ S·cm^{-1})	E_a (eV)	Log (A) (S·cm^{-1}·K)
0.05	2.53	0.56	4.19
0.10	2.69	0.55	4.09
0.15	5.29	0.60	4.77
0.20	3.65	0.60	4.66

vacancy was created by two Eu^{3+} ions doping at Ce-site and two protons can incorporate into the lattice at the humid atmosphere. Thus, proton conductivity should be increased with the increasing of Eu content. However, when Eu content was larger than $x = 0.15$, the defect association $Eu_{Ce}'\cdot V_{\ddot{O}}$ may be formed. Further Eu doping may increase the lattice distortion because of a lager ionic size of Eu^{3+} ions relative to that of Ce^{4+}. These two factors prohibit further incorporation of proton, and hence after $x \geq 0.15$, the conductivity decreased. Furthermore, by using excessive Ba $BaCe_{1-x}Eu_xO_{3-\delta}$ ($x = 0.15$) and $Ba_{1.04}Ce_{0.85}Gd_{0.15}O_{3-\delta}$, both were prepared and no trivalent ions were doped at Ba-sites. Since Eu^{3+} and Gd^{3+} have very similar ionic size, the Eu and Gd doped $BaCeO_3$ samples should show almost same lattice distortion. It can be explained the very close conductivities for the sample $BaCe_{0.85}Eu_{0.15}O_{3-\delta}$ and $Ba_{1.04}Ce_{0.85}Gd_{0.15}O_{3-\delta}$ (Shima & Haile 1997).

4 CONCLUSIONS

In summary, this work reports on the site-preference in proton conductors of $BaCe_{1-x}Eu_xO_{3-\delta}$ by using a precise XRD analysis in combination with EPR measurements. It is clearly demonstrated that excessive Ba species compensated the Ba loss during high temperature sintering, in which Eu^{3+} ions would substitute at Ce^{4+} sites by producing higher concentration of oxygen vacancies and promoting the proton conduction. Instead, in the case of barium loss, Eu^{3+} would prefer substituting at Ba^{2+} sites by reducing the oxygen vacancy concentration. The result is that the conductivity of the sample $BaCe_{0.85}Eu_{0.15}O_{3-\delta}$ from group B reached the maximum at the same temperature.

ACKNOWLEDGEMENTS

This work was financially supported by The National Natural Science Foundation of China (51472126, 61471210), Natural Science Foundation of Ningbo City (2014A610153).

REFERENCES

Chi, X., J. Zhang, M. Wu, Y. Liu & Z. Wen (2013). Study on Stability and Electrical Performance of Yttrium and Bismuth Co-Doped $BaCeO_3$. Ceramics International. 39, 4899–4906.

Fu, Y.P. & C.S. Weng (2014). Effect of rare-earth ions doped in $BaCeO_3$ on chemical stability, mechanical properties, and conductivity properties. Ceramics International. 40, 10793–10802.

Knight, K.S. (2001). Structural phase transtions, oxygen vacancy ordering and protonation in doped $BaCeO_3$: results from time-of-flight neutron power diffraction investigation. Solid State Ionics 145, 275–294.

Kreuer, K.D. (1999). Aspects of the formation and mobility of protonic charge carriers and the stability of perovskite-type oxides. Solid State Ionics, 125, 285–302.

Kröger, F.A. & H.J. Vink (1956). Relations between the Concentrations of Imperfections in Crystalline Solids. Solid State Phys. 3, 307–435.

Li, L, P., G.S. Li, R.L. Smith & H. Inomata (2002). Activation of oxide-ion conduction in $KNbO_3$ by addition of Mg^{2+}. Appl. Phys. Lett. 81, 2899–2901.

Li, L., P. G.S. Li, J.P. Miao, W.H. Su & H. Inomata (2001). Valence variations in titanium-based perovskite oxides by high-pressure and high temperature method. J. Mater. Res. 16, 417–424.

Peng, R.Y., L. Wu & Z. Mao (2006). Electrochemical properties of intermediate-temperature SOFCs based on proton conducting Sm-doped $BaCeO_3$ electrolyte thin film. Solid State Ionics. 177, 389–393.

Shima, D. & S.M. Haile (1997). The influence of cation non-stoichiometry on the properties of undoped and gadolinia-doped barium cerate. Solid State Ionics. 97, 443–445.

Wang, J.X., W.H. Su, D.P. Xu & T.M. He (2006). Electrical properties of Solid Solutioms $Ba_{1.1}Ce_{1-x}Eu_xO_{3-\delta}$. J. Alloy Compd. 421, 45–48.

*Advances in Materials Science, Energy Technology
and Environmental Engineering – Patty & Zhou (Eds)*
© 2017 Taylor & Francis Group, London, ISBN 978-1-138-19668-1

Parameter analysis of the spiral generator of rotary stream for sand reclamation equipment based on FLUENT

Wenlong Lu, Shan Li & Qiang Chen
Mechanical and Electrical Engineering College, Kunming University of Science and Technology, Kunming, China

ABSTRACT: The arc plate blade applied in the guide blade rotation method of rotary generator is analyzed in this study. From the analysis, it has been found that the radian of the blade, the mounting angle, the number of blades, the length of the chord, and the height of the blade all can make strong rotary stream, and especially, the radian, the mounting angle, and the number of leaves can produce strong effect for rotation.

1 INTRODUCTION

The design of the spiral generator is the core content of the design of the rotary stream for sand reclamation equipment, though the spiral generator can produce strong rotary stream that is a key part of the design. With the increase in curl of rotary stream, the turbulence intensity will increase and the mixing capacity will be strengthened. In addition, the rotary stream can cause a collision of sand and the top cover with a certain deflection angle, decreasing the impact force without reducing the impact velocity, so that the sand is not easy to break. Owing to the existence of the deflection angle, sand and the top cover occur in linear contact, increasing the damage area of the inert membrane on the surface of sand particles (Fiore, S., & Zanetti, M. C. 2008).

2 RESEARCH OF BLADE PARAMETERS

Get strong rotary stream by the guide blade rotation method, so that it just carries on a design to the blade shape, position, etc. As the axial-compression machinery typically used an arc plate blade, the shape of the arc plate blade can be selected within the spiral generator (J. P. Van Doormaal, & G. D. Raithby. 1984).

3 ANALYSIS OF BLADE PARAMETERS BASED ON FLUENT

Depending on the actual needs to rule the diameter of the guide vane shaft and vane wheel, every spiral generator has the same diameter as that of the guide vane shaft and vane wheel. The longest chord length is selected to be 100 mm, the diameter of the guide vane shaft is 180 mm, the maximum height of the blade is 100 mm, the outer diameter of the blade wheel is 380 mm, the diameter of the outer tube is 400 mm, and the inlet wind velocity is 40 m/s, as research of the unified regulations. For different radian airfoils, the mounting angle, number of blades, length of the chord, and height of the blade were analyzed.

3.1 *Analysis of the radian of the blade*

For a constant chord length of the arc plate blade, the radian is not uniform. It is necessary to study using blades of different radians of the spiral generator to generate the curl effect of the rotary stream. Two blades of different radians are analyzed, as shown in Figure 1 and Figure 2.

The spiral generator no. 1: Arc plate blade no. 1 is installed, radian 1.4; the spiral generator no. 2: arc plate blade no. 2 is installed, radian 0.8 (mounting angle is 60°, chord length is 100 mm, height of blade is 100 mm, guide vane journal is 180 mm, vane wheel shaft journal is 380 mm, diameter of the outer tube 400 mm, and number of blades is 8).

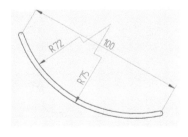

Figure 1.   Cross-section of arc plate blade no. 1.

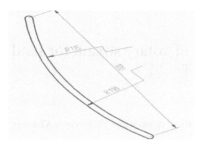

Figure 2. Cross-section of arc plate blade no. 2.

Figure 3. Arrangement of the arc plate blade of the spiral generator.

Figure 4. Diagram of arrangement of the spiral generator.

Figure 3 shows the arc plate blade of the spiral generator stereogram. Figure 4 is a schematic view of the pipe containing the spiral generator. The direct current stream enters from the air inlet at the left, though the spiral generator can produce rotary stream. At the entrance, the wind speed was set at 40 m/s, blowpipe length 1000 mm, and the distance of three intercepted sections from the nozzle are 150 mm, 200 mm, and 400 mm to compare wind speed distribution.

By comparing Figures 5 and 6, it has been found that arc plate blade no. 1 is equipped with the spiral generator no. 1, rotary stream from the spiral generator nozzle distance of 150 mm, the outermost layer of the annulus velocity major is distributed between 70 m/s and 80 m/s, the speed of the intermediate annulus ranged from 50 m/s to 70 m/s, and the speed of the inner surface ranged from 10 m/s to 50 m/s. However, the spiral generator is provided with arc plate blade no. 2, rotary stream outer ring speed distribution mainly between 47 m/s and 57 m/s, which is obviously to blow with the spiral generator no. 1 generating the degree of rotation. The same phenomenon can be found through the observation of the rotational speed of the airflow in the 200 mm and 400 mm.

3.2 Arc plate blade mounting angle

Application of Arc plate blade no. 1, respectively, is used for the design of the mounting angle of 20°, 40°, and 60° of the spiral generator simulation analysis (other parameters are: chord length is 100 mm, height of the blade is 100 mm, guide vane journal is 180 mm, vane wheel shaft journal is 380 mm, diameter of the outer tube is 400 mm, number of blades is 8, and inlet wind velocity is 40 m/s), and using the above method to compare analysis of the speed at the spiral generator nozzle 150 mm distribution.

From Figure 7 and Figure 8, it can be found that when the blade mounting angle is small, a low

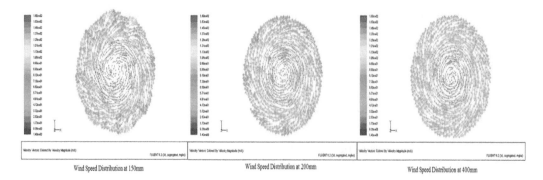

Figure 5. Wind speed distribution at 150 mm, 200 mm, and 400 m in the spiral generator no. 1.

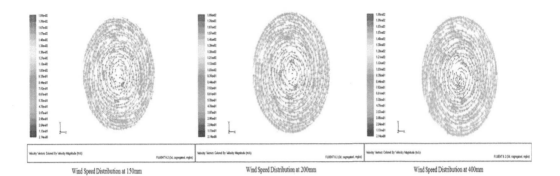

Figure 6. Wind speed distribution at 150 mm, 200 mm, and 400 mm in the spiral generator no. 2.

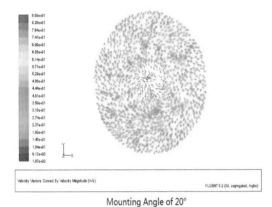

Mounting Angle of 20°

Figure 7. Mounting angle of 20° in wind speed distribution at 150 mm.

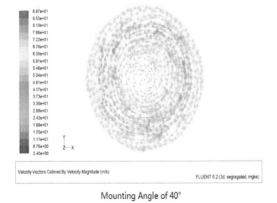

Mounting Angle of 40°

Figure 8. Mounting angle of 40° in wind speed distribution at 150 mm.

degree of rotary stream of rotation produced by the spiral generator and also the maximum speed of rotary stream have not increased significantly.

3.3 Chord length analysis

Through the above simulation, it can be determined that to select a larger radian and larger mounting angle of the arc plate blade is conducive to getting strong rotary stream. Therefore, through the same radian and mounting angle of the two kinds of chord length comparative analysis, it has been found that different chord lengths affect the rotation degree. Setting parameters: mounting angle is 20°, height of blade is 100 mm, guide vane journal is 180 mm, vane wheel shaft journal is 380 mm, diameter of the outer tube is 400 mm, number of blades is 8, and inlet wind velocity is 40 m/s, respectively, 100 mm and 40 mm chord length of the blade for analysis.

By comparing Figure 1 and Figure 9, it has been found that compared with installation 100 mm chord length arc plate blade of the spiral generator with installation, 40 mm chord length arc plate blade of the spiral generator and 100 mm chord length arc plate blade of spiral generator are significantly better.

3.4 Analysis of the height of the blade

Through the above simulation, it can be found that the other parameters are the same for spiral generator, and longer chord length of the blade has produced a greater degree of rotation. Here, it is equipped with: mounting angle is 20°, guide vane journal is 180 mm, vane wheel shaft journal is 380 mm, diameter of the outer tube is 400 mm, number of blades is 8, inlet wind velocity is 40 m/s, two spiral generators with heights of blade is 50 mm and 100 mm, respectively, are compared and analyzed, to determine the effects of rotary stream due to height of blade.

By comparing Figure 1 and Figure 9, it has been found that comparing 100 mm height of blade with 50 mm height of blade, the 100 mm height of blade of the spiral generator is significantly better.

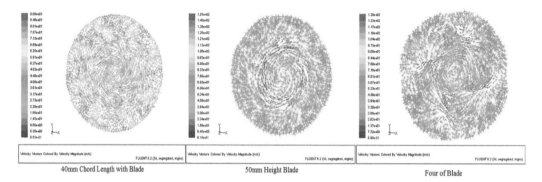

| 40mm Chord Length with Blade | 50mm Height Blade | Four of Blade |

Figure 9. Representative 40 mm chord length with blade, 50 mm height of blade, and four of blade in wind speed distribution at 150 mm.

3.5 Blade number analysis

Here, it is equipped with mounting angle 20°, guide vane journal is 180 mm, heights of blade are 50 mm and 100 mm, vane wheel shaft journal is 380 mm, diameter of the outer tube is 400 mm, and inlet wind velocity is 40 m/s, Two spiral generators with 4 blades and 8 blades, respectively, are compared and analyzed, to determine the effects of rotary stream due to blade number.

By comparing Figure 1 and Figure 9, it has been found that comparing 4 blades of spiral generator with 8 blades of spiral generator, 8 blades of spiral generator is significantly better.

4 CONCLUSIONS

Through the above analysis, it can be found that the number of blades, the blade mounting angle, and the radian of blade are important factors to produce strong rotary stream. However, under the guide vane journal certain conditions, the number of blades, chord length, and blade mounting angle are mutual restraint. It is necessary to optimize the three parameters in dissimilar environments.

REFERENCES

(1972). A Calculation Procedure for Heat, Mass and Momentum Transfer.Three-Dimensional Parabolic Flows," International Journal of Heat and MASS Transfer.

Fiore, S., & Zanetti, M.C. (2008). Industrial treatment processes for recycling of green foundry sands. International Journal of Cast Metals Research, 21(6), 435–438.

Van Doormaal, J.P. & G.D. Raithby. (1984). Enhancements of the simple method for predicting incompressible fluid flows.Numerical Heat Transfer Applications, 7(7), 147–163.

Advances in Materials Science, Energy Technology and Environmental Engineering – Patty & Zhou (Eds)
© 2017 Taylor & Francis Group, London, ISBN 978-1-138-19668-1

Designing of an adjustment device for a semi-feeding peanut picking roller

Xiaorong Lv
College of Machinery and Electronics, Sichuan Agricultural University, Ya'an, China

Xiaolian Lv, Xiaoqiong Zhang & Wei Wang
College of Machinery and Automotive Engineering, Chuzhou University, Anhui, China

ABSTRACT: The installation position of a semi-feeding peanut picking roller is fixed, which cannot meet the diverse needs of the relative position of picking parts. It comes from the differences in region and peanut variety and leads to lower efficiency of picking and poor quality. Designing of an adjustment device for a semi-feeding peanut picking roller is discussed in this paper. By changing the position of the bolts in the fixed slot, the adjustment device achieves the modifications of picking roller installation position and installation angle. During the process of peanut picking, the adjustment device effectively alters the relative position of picking components according to the actual situation. It will make sense in improving the work efficiency and the quality of a semi-feeding peanut picker.

1 INTRODUCTION

A peanut picking machine is a symbol of peanut picking mechanization. It can effectively replace manpower to harvest dry and wet peanut pods. The adjustment device has also improved the efficiency of peanut harvesting.

Depending on the feeding methods, the peanut picking machine can be divided into two types: full-feeding and semi-feeding (Ma, 2007). Full-feeding peanut pickers have many defects, such as high consumption of power, residue of peanut, unclear separation of peanut and vine, and high percentage of peanut damage. They are more suitable for dry vine peanut picking jobs. Semi-feeding peanut picking machines have strong job adaptability. They can work well on wet and dry peanut vine picking operations. The machine consumes less power and can work with a small tractor. Especially in the southern rainy region, the machine is portable enough for field work (Yu, 2011, Lv, 2015). Owing to the difference in peanut varieties and growth characteristics, the peanuts have greater differences in the peanut seedling length and the range of pod distribution. Therefore, at the time of picking peanut pods, it has a greater impact on picking that affects the relative position between the picking rollers and clamping conveyor chains. However, the relative position among the picking rollers, the transportation equipment, and the holding part is relatively fixed in the existing semi-feeding peanut picker. It cannot be adjusted according to the actual work situation.

The assembly location between the picking-transporting part and the peanut picking parts is relatively fixed (Lv, 2015; Guo, 2015). In order to overcome the problem, this paper provides an adjustment device for the semi-feeding peanut picking roller, which adds the support bracket, the mounting bracket, the adjust brackets, and the adjust screw. The device can achieve that the adjustment of the relative height and angle between picking roller and picking-transporting part, and the adjustment of the relative distance and angle between two picking rollers.

2 STRUCTURE AND WORKING PRINCIPLE OF ADJUSTMENT DEVICE

2.1 Overall structure

In this paper, the semi-feeding peanut picking roller adjustment device consists of the support bracket, the mounting bracket, the adjust brackets, the adjust screw, and other components. The overall structure of the semi-feeding peanut picking roller adjustment device is shown in Figure 1 (a) and (b).

2.2 Working principle

When the peanut harvester is picking in the field, the machine's picking rolls should adjust the irrelative position according to the actual situation to achieve high efficiency. Picking rollers are

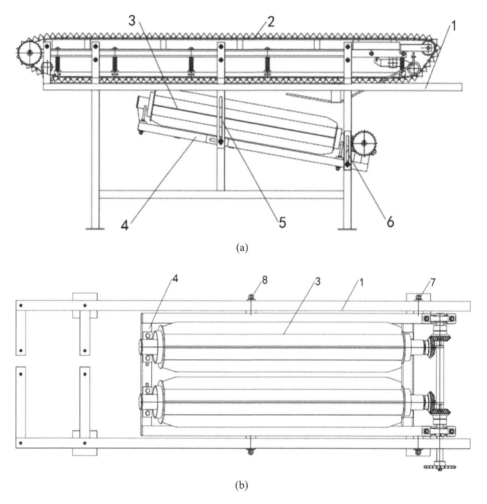

(a)

(b)

Figure 1. Overall structure of the adjustment device: (a) lateral view and (b) plan view. 1. Support bracket; 2. Clamping conveyor chains; 3. Peanut picking roller; 4.Mounting bracket; 5. Adjust bracket I; 6. Adjust bracket II; 7. Adjust screw I; 8. Adjust screw II.

mounted on the mounting bracket. The mounting bracket is connected to the adjust brackets I and II by the adjust screws I and II. In this way, the picking rollers are mounted on the support bracket. In addition, the mounting bracket is designed with adjust holes and adjust slots. Moreover, the adjust brackets are designed with oval long-slots in the installing locations. Through the cooperation of adjust-holes and adjust-slots, the mounting bracket can achieve the adjustment of installation height and angle, and then the adjustment device can equitably adjust the relative position between clamping conveyor chains and peanut picking rollers. Similarly, the adjustment device can regulate the relative position between two peanut picking rollers, through oval fixed holes I and II, which are

designed on the mounting bracket, to realize the distance adjustment and angle adjustment between two picking rollers.

3 DESIGN OF AN ADJUSTMENT DEVICE OF A SEMI-FEEDING PEANUT PICKING ROLLER

3.1 Design of the mounting bracket

The mounting bracket is composed of a support seat, adjust hole, adjust slot, and fixed holes I and II. The detailed structure is shown in Figure 2 (a) and (b). Both ends of the peanut picking roller are installed in the bearing seat. The front and the back bearing seats of each peanut picking

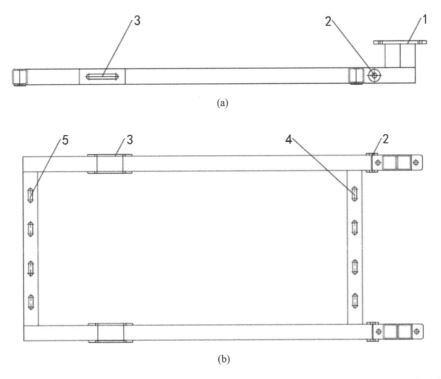

(a)

(b)

Figure 2. Structure of the mounting bracket: (a) lateral view and (b) plan view. 1. Support seat; 2. Adjust hole; 3. Adjust slot; 4. Fixed hole I; 5. Fixed hole II.

roller are arranged on the mounting bracket by the adjust screw and the fixed hole. The fixed hole is an oval long slot. Through the different fastening positions of the front and the back bearing seats in the fixed hole, the adjustment device can achieve the distance regulation and the angle regulation of two peanut picking rollers.

Adjust holes are arranged in the front of the mounting bracket, and the adjust slots are arranged in the back of the mounting bracket. The adjust slot is an oval long slot. With the cooperation of adjust holes and adjust slots, the adjustment device can easily change the installation angle of the mounting bracket. The adjustment device can easily adjust the relative installation position of the mounting bracket and clamping conveyor chains by the cooperation of the mounting bracket and the adjust brackets.

The support seats in the front of the mounting bracket are used to fix the bearing seats of the bevel gears.

3.2 *Design of the adjust bracket*

Adjust brackets mainly consist of support arms and oval long slots. The structure characteristics are shown in Figure 1 (a) of 5 and 6. Adjust bracket I and adjust bracket II are parts of the

support bracket, which can support the weight of the part of the machine. Moreover, the adjust brackets have an effect on supporting the clamping-conveyor chains.

Adjust bracket I and adjusting bracket II have oval long slots in the installation position. With the cooperation of them and the mounting bracket's holes and slots, the mounting bracket will possess a variety of installation positions and angle. When the relative position of the adjust bracket and the mounting bracket is fixed, they will be clamped by the adjust screw.

4 CONCLUSIONS

This paper describes the designing of an adjustment device for a semi-feeding peanut picking roller. When the peanut harvester is working in the field, the adjustment device can effectively adjust the relative position between two peanut picking rollers and the relative position between the peanut picking roller and the clamping conveyor chains according to the actual situation.

Peanut picking rollers are mounted on the mounting bracket. The mounting bracket and the adjust bracket are connected by the adjust screws.

65

In this way, the peanut picking rollers are mounted on the frame of the peanut picking machine. The adjustment device can adjust the installation height and the angle of the mounting bracket with the collaboration of the adjust hole and adjust slot, which are designed on the mounting bracket and the oval long slots, which are designed on the adjust bracket's installation position. Furthermore, it can adjust the relative position between the picking roller and the clamping conveyor chains. With the cooperation of the fixed hole I and fixed hole II, which are designed on the mounting bracket, the adjustment device can adjust the installation distance and angle of two peanut picking rollers.

Compared with the existing technology, the adjustment device, which is designed in this paper has a simple structure and design. It can ensure excellent adaptability of the peanut harvesting machine, and greatly improve the picking quality. The adjustment device of a semi-feeding peanut picking roller has been successfully applied for a national invention patent, which provides a new way for the design of the adjustment device of a semi-feeding peanut picking roller.

ACKNOWLEDGMENTS

This study was supported by the Natural Science Foundation of Anhui province of china (1408085ME103).

REFERENCES

Guo, N. & Guo F.H. 2015. A kind of fruit picking roller mechanism for peanut combine harvester. *China's intellectual property office:* CN201420812794.7.

Lv, X.L. & Hu, Z.C. 2015. Design and performance test of semi-feeding peanut picker. *Journal of Huazhong Agricultural University* 34(3): 124–129.

Lv, X.L. & Lv, X.R. 2015. A kind of longitudinal semi-feeding peanut picking roller. *China's intellectual property office:* CN201510680562.X.

Ma, T. & Bi, J.J. 2007. Development trend and countermeasures of peanut production in Shandong Province. *Anhui Agricultural Science Bulletin* 13(16): 118–119.

Yu, X.T. & Gu, F.W. 2011. The general situation and development of peanut fruit picking machine. *Journal of Chinese Agricultural Mechanization* (3): 10–12.

Advances in Materials Science, Energy Technology
and Environmental Engineering – Patty & Zhou (Eds)
© 2017 Taylor & Francis Group, London, ISBN 978-1-138-19668-1

Research on an intelligent algorithm diagnosis system

B.M. Chen
College of Information Science and Engineering, Central South University, Changsha, Hunan, China
The Second Xiangya Hospital, Central South University, Changsha, Hunan, China

X.P. Fan
College of Information Science and Engineering, Central South University, Changsha, Hunan, China

X.R. Li
The Second Xiangya Hospital, Central South University, Changsha, Hunan, China

H.W. Zhou
Changsha Tianlei Network Technology Co. Ltd., Changsha, Hunan, China

Z.M. Zhou
Changsha Environmental Protection College, Changsha, Hunan, China

ABSTRACT: Research on an intelligent algorithm diagnosis system is based on artificial neural network and an expert system. It combines the diagnosis standard of ICD 10, DSM IV with 40 years experiences, and knowledge. The learning samples come from the data nationwide. The correct rate of system diagnosis is 99%.

1 INTRODUCTION

Based on a large-scope epidemiological investigation, more than 60 million children with mental health disorders need mental health care in China. However, the child psychiatrists in China are not enough in number to serve so many children.

At the same time, the artificial neural network is becoming more and more important in many fields along with computer technology development. It comes from that computer which simulates the structure and function of the human brain neural system. It originates from modern biology research on human brain organization.

Therefore, we wish that applying the artificial neural network into child's mental health disorders intelligent diagnosis. The National Natural Science Foundation of China supports the research (39270262). By computer technology, we make an intelligent diagnosis system.

2 BP NEURAL NETWORKS

The human brain neural system consists of $10^{10} \sim 10^{11}$ neural cells. Artificial neural network Processing Elements (PEs) are a type of simulation to the human brain neural cell.

Artificial neural network processing element consists of many inputs, xi, i = 1, 2..., n, and an output y. As shown in (1) and (2):

$$y_j(t) = f\left(\sum_{i=1}^{n} w_{ji} x_i - \theta_j \right) \qquad (1)$$

$$f(x) = \frac{1}{1 + e^{-ax}}, \ 0 < f(x) < 1 \qquad (2)$$

where θj is the threshold of the neural network cell, wji is the connection weight, n is the input number, yj is the output of the neural network cell, and f (.) is the output transform function.

The neural network processing element has three types of transform functions. We use S shape function as an output transform function of the neural network processing element as shown in Figure 1.

Neural network has two learning algorithms: directed learning algorithm and undirected learning algorithm. We apply (back-propagation) the algorithm into the intelligent diagnosis system. BP is one type of directed learning algorithm. It can adjust neural network weights according to the difference between expectation target dj and practical output yj. In BP, the difference of the error is measured by the mean square error, as in (3) and (4):

Figure 1. Transform functions.

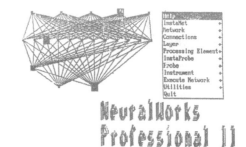

Figure 2. Neural expert diagnosis system structure.

$$E_k = \frac{1}{2}\sum_{j=1}^{q}(d_j - y_j)^2 \qquad (3)$$

$$E = \sum_{k=1}^{m} E_k = \frac{1}{2}\sum_{k=1}^{m}\sum_{j=1}^{q}(d_j - y_j)^2 \qquad (4)$$

3 NEURAL NETWORK COMBINED WITH AN ARTIFICIAL INTELLIGENT EXPERT SYSTEM

An artificial intelligent expert system is based on the logic reasoning thought. It is usually used to simulate the left brain of the human being. However, it faces many obstacles, such as knowledge acquisition bottleneck, matching conflict and combinatorial explosion, infinitude recursion in the reasoning process, etc.

The artificial neural network overcomes these obstacles. It is based on the visual thought and is used to simulate the right brain of the human being. It enhances the system intelligent level, management ability at real time, and strong ability. It has many available abilities, such as: learning ability by itself, organizing by itself, adapting by itself, allowing error data ability, and synthesis reasoning ability to not learned data. Though the neural network system has more advantages than the expert system, it cannot solve excluding diagnosis problems and so on.

As the single system cannot solve some problems, we combine the neural network with the artificial intelligent expert system to make a neural intelligent expert system. It is a beneficial practice to simulate the full brain of the human being. The system structure of child mental health disorder diagnosis is shown in Figure 2.

Besides using Pro*C language, C language, and oracle database, we use neural works professional and many others in our system as shown in Figure 3.

We select a full connect network mode, a hidden layer, and a directed learn algorithm for our diagnosis system. Similarly, we build 17 types of suitable neural networks to diagnose different child mental health disorders.

Comparison of the 17 types of networks training times number and network training times is shown in Figure 4 and Figure 5:

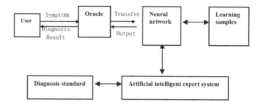

Figure 3. Neural works Professional II software.

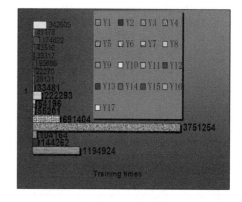

Figure 4. Network training times number y1–y17.

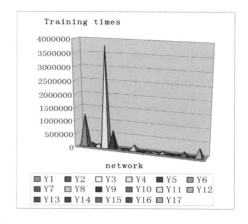

Figure 5. Comparison of network training times.

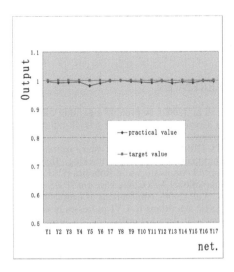

Figure 6. Network practical output comparison with the target value.

Table 1. Network (Y8) parameter setting.

Name	Parameter
Type	bkpcum
Input PE	7
Hidden PE	7
Output PE	2
Control Strategy	backprop
Learning Rule	Cum-Delta (but to input layer is none)
Summation Function	Sum (i.e. $I_i = \sum_k W_{ij}X_j$)
Transfer Function	Sigmoid Function (but to input layer is Linear)
Output Function	Direct
Error Function	Standard
Learn Source	File Rand
Recall Source	File Seq.
Probe and Instrument	Out-err

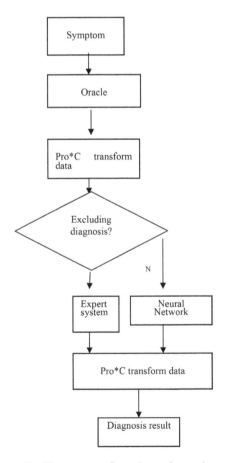

Figure 7. The program flow chart of neural expert diagnosis.

Comparison of the practical output with the target value of the 17 types of neural networks is shown in Figure 6:

The program flow chart of neural expert diagnosis system is shown in Figure 7:

For example, we give parameter setting of the eighth neural network. Other networks (Y1, Y2,..., Y17) are set similarly (as shown in Table 1).

4 RESULTS

Research on diagnosis and treatment system of child mental health disorders is based on the artificial neural network and the expert system. It is an applied basic research of interdisciplinary subject. This research can help children with mental health disorders all over the country. The intelligent diagnosis system combines the artificial intelligent expert system with the neural network. It combines the diagnosis standard of ICD 10, DSM IV with 40 years of clinical experiences, and knowledge of senior child psychiatrists. It also combines computer science with child psychiatry, child psychology, psychological estimate, and psychological therapy, etc. The learning samples come from the clinical data and the epidemiological data in more than a dozen nationwide hospitals. It is a very beneficial practice that uses the computer simulating the human brain too. The correct rate of system diagnosis is 99%. The system can diagnose 61 types of child mental health disorders and give a suggestion on the treatment methods.

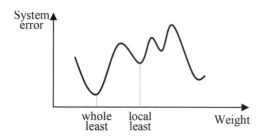

Figure 8. Local minimum value.

5 CONCLUSIONS

Through research, we find out that using only one method of artificial intelligence expert system or neural network cannot solve all problems in the diagnosis system. It does better to simulating function of the human brain that combined the artificial intelligence expert system with the artificial neural network.

Though BP arithmetic is better than others for our system, BP arithmetic has a big disadvantage that is easy to drop into the local minimum value rather than the whole minimum value as shown in Figure 8:

At the present time, many intelligent algorithms spring up like mushrooms in certain types of fields, such as support vector machine, ant colony algorithm, particle swarm optimization, immune computation, etc. For finding a better intelligent algorithm, we still have to perform many researches. For example, the Support Vector Machine (SVM) has a big advantage compared with the neural network, i.e. no local minimum value problem in the support vector machine. Therefore, SVM is becoming a new research focus after the neural network research. It will push machine learning theory and significant technology development.

ACKNOWLEDGMENTS

Foundation item: This project (39270262) was supported by the National Natural Science Foundation of China. The research wins the third prize of Hunan Province Medical Science and Technology Progress.

REFERENCES

Han et al. (2007). A pruning algorithm for RBF neural network based on rough sets [J]. Information and Control. 36: 604–609.

Jin, H. (2010). Application of fuzzy neural network in multi-maneuvering target tracking. Informatics in Control Automation and Robotics (CAR). 2010 2nd International Asia Conference on: 92–95.

Li, A. et al. (2004). Applications of neural networks and genetic algorithms to CVI processes in carbon composites [J]. Acta Materialia.52: 299–305.

Liu, J. et al. (2008). A novel recurrent neural network forecasting model for power intelligence center. Journal of Central South University of Technology. 15: 726–732.

Ma, T.H. & Zhao Y.W. (2007). Attribute reduction algorithm based on reduction pruning [J]. Computer Engineering.18: 56–58.

Maojun, C. et al. (2010). Double Chains Quantum Genetic Algorithm with Application in Training of Process Neural Networks. Education Technology and Computer Science (ETCS). 2010 Second International Workshop on: 19–22.

Quteishat, et al. (2010). A Modified Fuzzy Min–Max Neural Network With a Genetic-Algorithm-Based Rule Extractor for Pattern Classification. Systems. Man and Cybernetics. Part A: Systems and Humans. IEEE Transactions on. 40(3): 641–650.

Xu, X. H. et al. (2006). Decision tree dynamic pruning method based on minimum description length in speech recognition [J]. Acta Acustica. (4): 370–375.

Zhang, H. et al. (2008). Flame image recognition of alumina rotary kiln by artificial neural network and support vector machine methods [J]. Journal of Central South University of Technology. 15(1): 39–43.

Zhao, J. et al. (2010). Study on the Ripple Effects of Requirement Evolution Based on Feed forward Neural Network. Measuring Technology and Mechatronics Automation (ICMTMA). 2010 International Conference: 677–680.

Zhiwei, W. et al. (2004). Particle Swarm Optimization and Neural Network Application for QSAR. Proceedings of the 18th International Parallel and Distributed Processing Symposium (IPDPS'04). 0-7695-2132-0/04 (C) 2004 IEEE.

Advances in Materials Science, Energy Technology and Environmental Engineering – Patty & Zhou (Eds)
© 2017 Taylor & Francis Group, London, ISBN 978-1-138-19668-1

Image restoration based on structure and texture decomposition

Qiong Zhang
Shantou University Medical College, Shantou, Guangdong, China

Minfen Shen
Shantou Polytechnic, Shantou, Guangdong, China

Bin Li
Shantou Institute of Ultrasonic Instruments Co. Ltd., Shantou, Guangdong, China

ABSTRACT: A new method for structure and texture filling-in of complex images with missing information is proposed. The presented algorithm relies on edge-based region segmentation. The segmented regions are used both to reconstruct a structure component and to guide the restoration of a texture component. The contributions of this paper are two-fold. First, we propose an efficient method to prevent the edge-blur in filling-in complex image. Second, the texture can be quickly and nicely fixed in our method. Examples with real images show the advantages of the proposed scheme.

1 INTRODUCTION

The technique of automatic digital inpainting is to fill in the regions of missing information with available information from their surroundings of the image. The technique can be used in old image restoration, special effects, coding, and wireless image transmission. The algorithms proposed in the literature fill in missing data from different points of view.

As a common method, the Partial Differential Equations (PDEs) are widely used for image inpainting (Caselles, 2014), but it is only fitted for low-resolution images with light scratches or small holes and the computation is time consuming. Furthermore, the drawback of PDEs method for filling-in larger regions is lack of consideration for the extension of image textures. Recently, some new procedures for image inpainting are proposed (Zhou, 2004). For example, the algorithm converts the traditional 2D image inpainting problem to 3D implicit surface reconstruction using a Radial Basis Function (RBF). This approach can restore large smooth areas and can be computed quickly with a fast solution algorithm. However, it cannot get reasonable results for large variation regions because of the isotropic character of the RBF. On the other hand, in order to remedy the results of PDEs algorithm for larger regions, the concept of local texture analysis has been considered by several researchers (Bonet, 1997; Efros, 1999). They analyze the textures of adjacent zones and upgrade the inpainting algorithms to promote the performance of image inpainting. Nevertheless, this approach has failed to ensure the edge continuity and the synthesis algorithm is time consuming because it must be computed in the entire image. M. Bertalmio *et al.* (Betalmio, 2003) decompose an image into structure and texture images and restore them separately. This method cannot restore complex images because of the drawback of the used structure and texture inpainting algorithms.

To obtain an efficient method for preventing the edge-blur in filling-in complex image, we present an algorithm in which both the structure and the texture are restored separately. The rest of this paper is organized as follows: we present the main steps of the algorithm in section 2. Section 3 concentrates on image decomposition. Section 4 concentrates on region segmentation. Section 5 describes our structure reconstruction method. Section 6 describes our adaptive texture restoration method. The experiment results are given in section 7. Finally, section 8 draws conclusions and outlines the future work.

2 OUTLINE OF OUR ALGORITHM

The presented algorithm consists of five main steps including image decomposition, edge-based region segmentation, structure reconstruction based on Normalized Basis Function (NRBF) (Cowper, 2002), texture restoration based on adaptive texture matching, and adding back these two sub-images. The input is an image with missing areas.

In this paper, we assume that the missing areas have already been detected.

In the first step, an image is decomposed into structure and texture components using total variation minimization and oscillating patterns algorithm (Vese, 2003). This decomposition algorithm produces images that are very well suited for the image structure reconstruction and texture synthesis techniques described in the following.

In the second step, we try to segment the structure and texture images into different regions according to their edges. The purpose of region segmentation is to ensure that the segmented regions in the structure component u are smooth enough for NRBF surface reconstruction and, in the texture component v, different textures belong to different segmented regions, i.e. the searching region is small in the texture synthesis step. Fortunately, we observed that, in most situations, the segmented regions in both structure and texture components are the same. Therefore, we only need to segment the structure image based on edges. However, edge information is missing in the hole. In order to segment a structure image, we have to infer the broken edges as accurately as possible. This can be achieved with the help of tensor voting (Medioni, 2000).

In the third step, we simply use NRBF surface interpolation to restore the structure image in every segmented region. Accordingly, the two-dimensional image inpainting problem is transformed into three-dimensional surface reconstruction problem. The suitable choose is Radial Basis Function (RBF). However, in order to enhance the universality of our algorithm, NRBF is used.

In the fourth step, an adaptive texture matching is used to synthesize the missing texture information in each region because the same texture belongs to the same region. In our algorithm, the searching area for a matching block is not the whole image but a segmented region and the size of seed-block can be changed automatically according to the complexity of the segmented region, i.e. the missing texture information can be quickly and nicely fixed.

Finally, the image is reconstructed adding back these two sub-images.

3 DECOMPOSITION

In this section, we review the image decomposition approach proposed in Vese (2003), which is considered to be one of the most suitable decomposition approaches for image inpainting. The two main ingredients of the decomposition developed in Vese (2003) are the total variation minimization of Rudin (1992) for image denoizing and restoration,

and the space of oscillating functions introduced in Meyer (2002) to model texture or noise.

Let $f : R^2 \to R$ be a given observed image, $f \in L^2(R^2)$. In real applications, the observed image f is just a noisy version of a true image u, or it is a textured image, and u would be a simple sketchy approximation or a cartoon image of f. In the presence of additive noise, the relation between u and f can be expressed by the linear model, introducing another function v, such that

$$f(x,y) = u(x,y) + v(x,y) \qquad (1)$$

In Rudin (1992), the problem of reconstruction u from f is posed as a minimization problem in the space of function of bounded variation $BV(R^2)$, allowing for edges:

$$\inf_{u \in BV} \left\{ F(u) = \int |\nabla u| + \lambda |v|_{L^2}^2, \quad f = u + v \right\} \qquad (2)$$

where $\lambda > 0$ is a tuning parameter. The second term in the energy is a fidelity term, while the first term is a regularizing term, to remove noise or small details, while keeping important features and sharp edges.

Meyer (2002) proved that, for small λ, the model will remove the texture. To extract both the $u \in BV$ and the v component as an oscillating function (texture or noise) from f, Meyer proposed the use of a different space of functions, which is in some sense the dual of the BV space. He introduced and proved the following definition.

4 REGION SEGMENTATION

If the structure image were restored by RBFs directly, the restored region would be blurry in large variation areas, Figure (2-(B), (D)). In addition, the texture restoration algorithm would be time consuming in restoring large image, if the texture synthesis algorithm is directly used to v. In order to overcome these drawbacks, an image inpainting based on area segmentation algorithm has been proposed.

The purpose of area segmentation was to make sure that the variation of every area is small in the structure component u and all the texture of the same area are the same in the texture component v in order that the searching region would be small when implementing texture synthesis.

How to segment an image is one of the most important problems of our algorithm. Fortunately, the conclusion that the segmented regions in both structure image and texture image are the same can be drawn from the analysis of the image decomposition algorithm. Therefore, we only need to

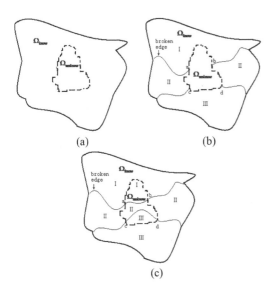

(a)　　　　　　(b)

(c)

Figure 1.   Region segmentation.

segment u into different smooth areas. There are two main processes in segmentation. One is to segment the surrounding region of missing information. The other is to segment the region of missing information. The first process can be carried out with k-means class proposed in Shinichi (1998). Moreover, the problem of the second process was to connect the broken edges naturally that pass though the missing region. The tensor voting (Medioni, 2000) was used to solve that problem.

Assume that there is an image with a damaged region Ω_{unknow} and the correlative region Ω_{know} as shown in Figure 1(a).

5 STRUCTURE RECONSTRUCTION

Given a set of zero-valued surface points and non-zero off-surface points, we have a scattered data interpolation problem: we want to approximate the signed-distance function $f(x)$ by an interpolant $s(x)$. The problem can be stated formally as follows:

Problem 1. Given a set of distinct nodes $X = \{x_i\}_{i=1}^{N} \subset R^3$ and a set of function values, $\{f_i\}_{i=1}^{N} \subset R$ finds an interpolant $s: R^3 \to R$ such that

$$s(x_i) = f_i, i = 1, \cdots, N \qquad (3)$$

Note that we use the notation $x = (x, y, z)$ for points $x \in R^3$.

The interpolant will be chosen from $BL^{(2)}(R^3)$, the Beppo Levi space of distributions on R^3 with square integrable second derivatives. This space is sufficiently large to have many solutions to Problem 1, and therefore, we can define the affine space of interpolants:

$$S = \{s \in BL^{(2)}(R^3) : s(x_i) = f_i, \quad i = 1, \cdots N\} \qquad (4)$$

The space $BL^{(2)}(R^3)$ is equipped with the rotation invariant semi-norm defined by

$$\|s\|^2 = \int_{R^3} s_{xx}^2(x) + s_{yy}^2(x) + s_{zz}^2(x) + 2s_{xy}^2(x) + 2s_{zx}^2(x) + 2s_{yz}^2(x) \qquad (5)$$

This semi-norm is a measure of the energy or "smoothness" of functions: functions with a small semi-norm are smoother than those with a large semi-norm. Duchon (1977) showed that the smoothest interpolant, i.e.

$$s^* = \arg\min_{s \in S} \|s\| \qquad (6)$$

where p is a linear polynomial, the coefficients λ_i are real numbers, and $|\cdot|$ is the Euclidean norm on R^3.

This function is a particular example of a *Radial Basis Function* (RBF). In general, an RBF is a function of the form

$$s(x) = p(x) + \sum_{i=1}^{N} \lambda_i \varphi(|x - x_i|) \qquad (7)$$

where p is a polynomial of low degree and the basic function φ is a real valued function on $[0, \infty)$, usually unbounded and of non-compact support (see, e.g., Cheney & Light (1999)). In this context, the points x_i are referred to as the centers of the RBF.

Popular choices for the basic function include the thin-plate spline $\varphi(r) = r^2 \log(r)$, the Gaussian $\varphi(r) = \exp(-cr^2)$, and the multiquadric $\varphi(r) = \sqrt{r^2 + c^2}$. For fitting functions of three variables, good choices include the biharmonic and triharmonic splines. In order to enhance the universality of our algorithm, normalized radial basis function $\varphi(|x - x_i|) = \frac{|x - x_i|}{\sum_{j=1}^{N} |x - x_j|}$ was chosen (Mario, 2005; Cherrie).

An arbitrary choice of coefficients λ_i in Equation (12) will yield a function s^* that is not a member of $BL^{(2)}(R^3)$. The requirement that $s^* \in BL^{(2)}(R^3)$ implies the orthogonality or side conditions

$$\sum_{i=1}^{N} \lambda_i = \sum_{i=1}^{N} \lambda_i x_i = \sum_{i=1}^{N} \lambda_i y_i = \sum_{i=1}^{N} \lambda_i z_i = 0$$

More generally, if the polynomial in Equation (7) is of degree m, then the side conditions imposed on the coefficients are

$$\sum_{i=1}^{N} \lambda_i q(\mathbf{x}_i) = 0, \text{ for all polynomials } q \text{ of degree}$$

at most m (8)

These side conditions along with the interpolation conditions of Equation (3) lead to a linear system to solve for the coefficients that specify the RBF.

Let $\{p_1, \cdots, p_l\}$ be a basis for polynomials of degree at most m and let $c = (c1, \cdots, c_i)$ be the coefficients that give p in terms of this basis. Then, Equations (3) and (8) may be written in the matrix form as

$$\begin{bmatrix} A & P \\ P^T & 0 \end{bmatrix} \begin{bmatrix} \lambda \\ c \end{bmatrix} = B \begin{bmatrix} \lambda \\ c \end{bmatrix} = \begin{bmatrix} f \\ 0 \end{bmatrix} \quad (9)$$

where

$$A_{ij} = \phi\big(|\mathbf{x}_i - \mathbf{x}_j|\big), \quad i, j = 1 \cdots N,$$
$$P_{i,j} = p_j(\mathbf{x}_i), \quad i = 1, \cdots, N, \quad j = 1, \cdots, N$$

where P is the matrix with ith row $(1, x_i, y_i, z_i)$, $\lambda = (\lambda_1, \cdots, \lambda_N)^T$ and $c = (c_1, c_2, c_3, c_4)^T$.

Solving the linear system (9) determines and c, and hence $s(x)$.

6   ADAPTIVE TEXTURE RESTORATION

For the general texture synthesis approaches, the size of seed-block is fixed and the searching region is the whole image. These introduce two drawbacks:

1. There will be mistake when an image contains many different textures, because different textures should choose different sizes of seed-block.
2. It will be time consuming to search the fittest filling-in block when the image is large.

In order to overcome these drawbacks, we developed the texture adaptive synthesis technique based on segmentation. It is only necessary to search the fittest filling-in block in the corresponding segmented region. Therefore, it is suitable for large image texture synthesis. In addition, our approach can choose the size of seed-block automatically according to the variation in the region in restoring different segmented regions.

The variation of region is measured as following:

$$\bar{x}^k = \frac{\sum_{\forall i} \sum_{\forall j} x_{ij}}{N^k}, \quad (i,j) \in \Omega_{know}^k . \quad (10)$$

$$\text{var}^k = \frac{\sum_{\forall i} \sum_{\forall j} (x_{ij} - \bar{x}^k)^2}{N^k}, \quad (i,j) \quad (11)$$

where Ω_{know}^k is the available domain in the kth region, N^k is the number of know points in the kth region, $\bar{x}^k$ is the mean of know points in the kth region, var^k is the variation of know point in the kth region.

The larger the value of var^k is, the more complication the texture is. Therefore, the size of the block should be smaller. In addition, for the purpose of reducing error diffusion, the block that contains more know points should be restored earlier.

7   RESULTS

The restoration steps of our method are shown in Figure 2. In the decomposition step, the damaged area is replaced by block color. Figure 2 (b)-(g)

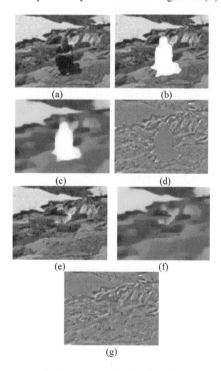

Figure 2. Object removal. Results from different algorithms and the progress of our algorithm with $\lambda = 0.02, \mu = 0.04$ are shown: (a) an original image and (b) the image with a selected region (white region) to be filled in. Then (b) is decomposed into (c) a structure image and (d) a texture image. These two images are reconstructed via structure reconstruction and adaptive texture synthesis proposed in our paper, respectively ((f) and (g)). (e) The final result of our algorithm.

74

shows the filling-in process of our algorithm. The main component of structure is presented in (c) and the main component of texture is presented in (d). The missing areas in both structure and texture images are naturally restored (Figure 2(f) and (g)). The final result (Figure 2(e)) of our method indicated that the image is restored without detection.

8   CONCLUSIONS AND FUTURE WORK

An algorithm for structure and texture image inpainting based on region segmentation was presented. The proposed algorithm can ensure the edge continuity, which is one of the most important characters in image inpainting, and make the restored regions less blurry. Additionally, region-segmentation-based structure reconstruction ensures that the surface can be nicely reconstructed by NRBF. Moreover, adaptive texture restoration based on region segmentation can achieve fast and accurate results. Therefore, our method is more appropriate for the restoration of large damaged regions.

We are currently looking at improved methods for image inpainting, including a more accurate segmentation method and a compactly supported radial basis function for surface reconstruction.

ACKNOWLEDGMENTS

This work is supported by the Science and Technology Planning Project of Guangdong Province (No. 2012B050300024 and No. 2015B020233018), the National Natural Science Foundation of China (No. 61302049), and Science and Technology Planning Project of Shantou City.

REFERENCES

Betalmio, M., L. Vese, G. Sapiro, and S. Osher, "Simultaneous structure and texture image inpainting," *IEEE Computer Society Conference on Computer Vision and Pattern Recognition*, 2003.

Caselles, V. "Variational models for image in painting," *European Congress of Mathematics Kraków*, 58, pp. 227–242, 2014.

Cheney E.W. and W. A. Light: A Course in Approximation Theory. Brooks Cole, Pacific Grove (1999).

Cherrie, J.B., R.K. Beatson and G.N. Newsam: "Fast Evaluation of Radial Basis Functions: Methods for Generalized Multiquadrics in Rn," SIAM J. Sci. Comput., Vol. 23, pp. 1549–1571.

Cowper, M.R., B. Mulgrew, and C.P. Unsworth, "Nonlinear prediction of chaotic signals using a normalized radial basis function network," *Signal Processing, ELSEVIER*, 82, pp. 775–789, 2002.

De Bonet, J.S. "Multiresolution sampling procedure for analysis and synthesis of texture images," *Proceedings of ACM*, SIGGRAPH, Jul. 1997.

Duchon, J. Splines Minimizing Rotation-invariant Seminorms in Sobolev spaces. In W. Schempp and K. Zeller, editors, Constructive Theory of Functions of Several Variables, number 571 in Lecture Notes in Mathematics, Berlin Springer-Verlag (1977) 85–100.

Efros, A. and T. K. Leung, "Texture synthesis by nonparametric sampling," *IEEE International Conference on Computer Vision*, Corfu, Greece, pp. 1033–1038, Sept. 1999.

Mario Botsch and Leif Kobbelt: "Real-time shape editing using Radial Basis Functions," EUROGRAPHICS Vol. 24 (2005), Num. 3.

Medioni, G., M.S. Lee, and C.K. Tang, *A computational framework for feature extraction and segmentation*, Elsevier, New York, 2000.

Meyer, Y. "Oscillating Patterns in Image Processing and Nonlinear Evolution Equations, " AMS University Lecture Series 22, 2002.

Rudin, L., S. Osher, and E. Fatemi: "Nonlinear total variation based noise removal algorithms," Physica D, no. 60, pp. 259–268, 1992.

Shinichi, Yoshiaki Shishikui, Yutaka Tanaka, and Ichiro Yuyama: "Image segmentation by region integration using initial dependence of the k-means algorithm," Systems and Computers in Japan, Vol. 29, No. 14, 1998.

Vese, L. and S. Osher, "Modeling textures with total variation minimization and oscillating patterns in image processing," *Journal of Scientific Computing*, Vol. 19, Nos. 1–3, Dec. 2003.

Zhou Tingfang, Tang Feng, Wang Jin, Wang Zhangye and Peng Qunsheng, "Digital image inpainting with radial basis functions," *Journal of Image and Graphics*, 9(10), pp. 1190–1196, 2004.

*Advances in Materials Science, Energy Technology
and Environmental Engineering – Patty & Zhou (Eds)*
© *2017 Taylor & Francis Group, London, ISBN 978-1-138-19668-1*

Preparing the FE model update of a rolling stock car body under servicing conditions based on profile plate equivalence

Xiaojun Deng
School of Mechanical, Electronic and Control Engineering, Beijing Jiaotong University, Beijing, China
CRRC Qingdao Sifang Co. Ltd., Qingdao, Shandong Province, China

Qiang Li
School of Mechanical, Electronic and Control Engineering, Beijing Jiaotong University, Beijing, China

Yanju Zhao
CRRC Qingdao Sifang Co. Ltd., Qingdao, Shandong Province, China

ABSTRACT: Structural dynamics analysis of the high-speed train is an important means to improve the vehicle vibration, noise, and comfort level, but there are challenges in vehicle modeling with a certain accuracy as a much elaborate model may lead to waste of time in structural dynamics analysis. This paper details a study carried out based on the substructure modal test and the equivalent finite element model, from the equivalent model of single block extrusion plate, to the equivalent finite element model of the whole car body, then considering the influence of the other equipment and interior. We updated the finite element model of the car body, referred to the modal test results through sensitivity analysis method, and got the correct dynamics model of it. The updated model has an average frequency error of 3.9%, the MAC of four orders are over 70% when compared with the modal test in servicing conditions, which provided a accurate model for dynamic analysis of the whole car body.

1 INTRODUCTION

At present, the total operating length of high-speed railways in China has exceeded 10,000 km and an additional line of the same mileage is under construction. More and more new EMUs resistant to environmental conditions such as freezing cold and moist heat have been developed and put into operation since the 5-year service of CRH380 A 380 km/h EMU developed by CRRC corporation. Rapid modeling of car body dynamics finite element model is of great importance to shorten the development time of EMU and improve the performance.

Many scholars at home and abroad have conducted outstanding researches. Jiang Yanqing (2006) studied the finite element modeling method for connecting structures between car bodies, Zhao Jinsen (2006) studied the equivalent dynamic model of an aluminum honeycomb sandwich panel using the dynamic equivalent method. Chen Feifei (2010) studied equivalent elastic constants of a car body corrugated profile plate. Ribeiro et al. (2013) calculated the elasticity modulus and the shear modulus of a single corrugated profile plate of EMU vehicles in two directions and developed the car body dynamics model using a static equivalent formula.

On the other hand, FE model updating technology has obtained considerable development. It examines the FE model with test data for the accurate dynamics model. This technology proves to be reliable and effective by the successful application of several engineering cases. The literature provides key parameter identification for a high-speed train car body under servicing conditions using model updating technology. The literature provides model updating for the connecting parameters of similar car body structures based on modal test results. Yu Jinpeng (2015) studied the car body dynamics model updated using the response surface method according to modal test results. However, the car body model used is the detailed FE model instead of a simplified equivalent model, and hence, large-scale calculation is required. The vibration model is not subject to quantization processing.

Based on several single block corrugated profile plates and modal test results of a complete car, this paper has obtained some ideal results by developing an equivalent model of a single block profile plate and then an equivalent FE model of a complete vehicle using model updating technology.

2 FE MODEL UPDATING THEORY

2.1 *FE model updating based on sensitivity analysis*

As shown in the following equation, optimization is the nature of FE model updating:

$$Min\|R(\mathbf{p})\|^2, R(\mathbf{p}) = \{f_E(\mathbf{p})\} - \{f_P(\mathbf{p})\} \quad (1)$$
$$s.t \quad VLB \leq \mathbf{p} \leq VUB$$

where p is the design parameter vector, $\{fE(p)\}$ is the test result, $\{fP(p)\}$ is the finite element simulation result, $R(p)$ is the residual error for simulation and test results, and VUB and VLB are the upper and lower limits for design parameters, respectively.

$$R(p) = G\Delta p \quad (2)$$

where Δp is the perturbation for design parameters and G is the sensitivity matrix of the characteristic against design parameters.

The realization of FE model updating of parameter type depending on sensitivity analysis is based on the process of constant iterating. The main steps are as follows:

1. Experimental modal identification of the research object;
2. Initial finite element modeling and analysis;
3. Sensitivity analysis results based on characteristic values (modal frequency, vibration model, correlation coefficient of vibration model, etc.; choose the parameters to be updated as p, initial value being p0);
4. Calculate characteristic residual R between the FE simulation value and the test value;
5. Calculate the sensitivity matrix G when the design parameter is p0;
6. Inversion of the sensitivity matrix G by equation (2) generates the changed parameter Δp;
7. Back to step 3, calculate the new design parameter pi; the cycle until the convergence condition is satisfied.

Exact match between the test modal order and the FE simulation modal order as well as correlation calculation are of great importance to the FE model updating process based on test modal identification results.

2.2 *Vibration model-related coefficient*

The Modal Assurance Criterion (MAC), also called the vibration model-related coefficient, can be used to determine the correlation of the vibration model with the second-order modal.

The value of MACij is a scalar in [0, 1]. It indicates a completely linear correlation between i and j, when MACij equals 1, and they are modals of the same order; it indicates that i is completely unrelated to j, when MACij equals 0. We generally calculate the MAC matrix for the test model and the simulation model in actual engineering practice. It can be concluded that the two models are in good correlation of same-order vibration and unrelated in non-order vibration when off-diagonal element of MAC matrix is no more than 0.1 and the diagonal element is no less than 0.7; thus, the simulation modeling is accurate.

3 FE MODELING FOR CORRUGATED PROFILE PLATE AND SELECTION OF PARAMETERS TO BE UPDATED

3.1 *Vibration model-related coefficient*

As shown above, the extruded aluminum profile plate has been widely used in current high-speed train car bodies, and its section features corrugated the hollow structure. Compared with the traditional frame skin structure, it is characterized by light weight, corrosion resistance, high sectional rigidity, low cost, etc. As the wall structure is thin (approx. 2 mm) and the dimension ratio between the corrugated structure and the entire car is too small (corrugated section dimension is approx. 30 mm), the entire car FE model will be huge when modeling according to traditional actual element or shell element. Take the profile plate (1500 mm × 1200 mm) in Figure 1, for example, when modeling with 3D solid elements, it will involve 540,000 nodes and 320,000 elements; when modeling with traditional shell elements, it will involve approximately 6,000 elements and 3,000 nodes. A huge model will be required when it is calculated to be 25 m long car body. The dynamics equivalent model in this paper significantly reduces the FE model size and improves the calculation speed.

The equivalent FE model for the profile plate with 121 nodes and 100 shell elements will be built with reference to the point measuring position of the sensor during the modal test, as shown below:

3.2 *Selection of parameters to be updated*

From Figure 1, we see that the sectional corrugated structure of the profile plate is direction based, and the length direction and width direction have different equivalent elasticity moduli when the plate is equivalent to one-layer shell element, thus the properties of 2D orthotropic materials are assigned to an equivalent shell element in this paper. Details of constitutive parameters for 2D orthotropic materials are shown in Table 1.

The profile plate in Figure 1 is considered as the research object, and the sensitivity matrix of

Figure 1.   Corrugated profile plate.

Figure 2.   The FE model of a profile plate.

Table 1.   Constitutive parameters schedule of 2D orthotropic material.

Name of variable	Meaning of variable
E1	Direction 11 elasticity modulus
E2	Direction 22 elasticity modulus
NU12	Poisson's ratio in 12 directions
G12	Shear modulus in 12 directions
G1z	Shear modulus in 23 directions
G2z	Shear modulus in 13 directions
Name of variable	Meaning of variable

its modal frequency in the first 14 orders against the 6 design parameters in Table 1 is shown in Figure 3. It can be observed that the sensitivity values for design parameters 1, 2, and 4 are large, and hence, we select E1, E2, and G12 of a 2D ortho-tropic material as the parameters to be updated in a consequent model updating process.

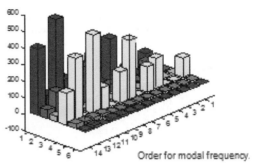

Figure 3.   The sensitivity matrix histogram of eigenval-ues to the design parameters.

4   FE MODELING AND SELECTION OF PARAMETERS TO BE UPDATED FOR A CAR BODY UNDER SERVICING CONDITIONS

For comparison of accuracy between the updated car body equivalent model and the traditional shell element FE model, two FE models of the entire car are developed in this paper. 1) The traditional shell element is used to simulate the corrugated profile plate and the entire car FE model is developed with-out model updating. 2) A single-layer 2D orthotropic shell element is used to simulate the corrugated pro-file plate and an entire car equivalent FE model is developed meanwhile by reference to modal test identification results, and carry out parameter-type FE model updating for car body based on sensitivity analysis. The car body equivalent FE model is shown in Figure 4 (no equivalent processing for cab area).

4.1   FE modeling for the traditional shell element

Figure 5 is the sectional view of the car body model with the traditional shell element which simulates the thin wall in a corrugated structure. The scale of the entire car (excluding cab area) involves approx-imately 270,000 nodes and 610,000 elements.

4.2   Equivalent FE modeling

Figure 6 is the sectional view of the entire car equiv-alent model with a single-layer shell element and each shell element is assigned with 2D orthotropic materials. As shown in the sectional view of the car body in Fig. 5, due to the sectional dimension error of the corrugated profile plate and uneven distribution of interior decoration and seat, the model will be divided into 6 areas as shown in Figure 6 and the constitutive parameters for 2D orthotropic materials in each area are mutually independent. The scale of the entire car (excluding cab area) involves approximately 3,500 nodes and 5,500 elements.

Figure 4.   The FE model of car body.

Figure 5.   Sectional view of the vehicle model with the shell element.

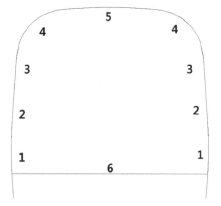

Figure 6.   Sectional view of the equivalent complete car model.

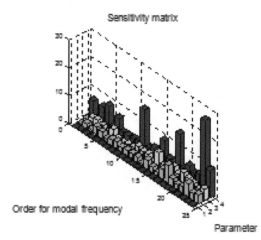

Figure 7. Sensitivity matrix of body eigenvalues to design variables.

4.3   Selection of the equivalent model parameter to be updated

The mass of parts such as interior decoration, air conditioning unit, and seat in the car under servicing conditions are simulated as concentrated mass points distributed across the profile plate; meanwhile, the influence of additional rigidity on car body dynamic performance should be equivalent to the profile plate, thus equivalence updating of additional rigidity should be considered after the development of a single-block profile plate equivalent model. The car body is divided into 6 areas as shown in Figure 6 and E1, E2, and G12 of 2D orthotropic materials in each area are selected as the parameters to be updated for a total of 18 parameters. FE model pre-updating will be performed by dividing the 18 parameters into 4 groups according to the change rule during the iterative process, and formal updating will be performed according to the fact that the parameters change along with the same percentage coefficient, namely 4 variables.

The sensitivity matrix of the car body equivalent model in the first 25 orders modal frequency against the 4 variables is shown in Figure 7.

5   FE MODEL UPDATING OF THE CAR BODY UNDER SERVICING CONDITIONS

5.1   Modal test results of the car body under servicing conditions

The modal test takes the air spring of the train as elastic support to simulate the free state. The random burst signals generated by four 50 kg electromagnetic vibrators are used as exciting sources.

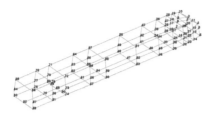

Figure 8. Distribution of modal test measuring points.

Figure 9. Scene of modal test.

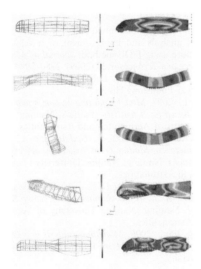

Figure 10. Car body modal vibration types.

The entire car is provided with 88 measuring points as shown in Figure 8 and each measuring point will collect transverse and vertical acceleration signals. The test scene is shown in Figure 9.

According to test modal identification results, select the typical 5 orders as the reference vibration mode for model updating. As shown in Figure 10, the left is modal test result and the right is FE model simulation result.

5.2 Car body FE model updating based on sensitivity analysis

As shown in Figure 11, convergence is responded when updating iterates to step 100, the error between updated fifth-order modal frequency and modal frequency to the corresponding order identified by the test is less than 10%.

5.3 Accuracy comparison of two FE models

The comparison of calculation results between the traditional FE model and the equivalent FE model is shown in Table 2. The maximum frequency error between traditional FE model calculation results and modal test is 46.3% and the average error is

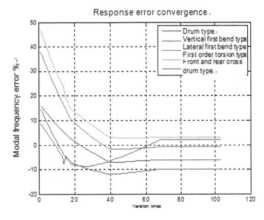

Figure 11. Convergence figure of modal frequency error.

Table 2. Accuracy comparisons of two FE models.

	Traditional model			Equivalent model			
Test/Hz	Frequency/Hz	Error/%	MAC	Frequency/Hz	Error/%	MAC	Description of vibration mode
12.8	17.3	42.2	0.27	12.1	−6.2	70.93	Inner drum
13.3	16.6	29.6	0.83	13.4	0.6	86.43	Car body vertical first bend
13.8	19.1	44.7	0.35	13.7	−0.8	49.83	Car body lateral first bend
16.2	25.2	46.3	0.44	16.0	−2.0	83.32	Car body torsion first order
20.3	22.4	9.1	0.69	18.4	−10.1	70.69	Front and rear cross drum
Average		34.4	0.52		3.9	72.24	

81

34.4%, the minimum error with test vibration model is 0.27, and the average error is 0.52; the maximum frequency error between the updated equivalent model and the modal test is 10.1% and the average error is 3.9%, except that with test vibration MAC for order one, which is lower than 70%, the error with the other is higher than 70%.

6 CONCLUSION

1. The high-speed train equivalent FE modeling method is studied and the entire car model with sound accuracy is obtained through updating combined with test modal identification results.
2. Corrugated profile plate from the car body is selected as the research object and 2D orthotropic materials have their constitutive parameters analyzed by sensitivity analysis for the selection of appropriate parameters to be updated.
3. The equivalent FE model developed with FE model updating technology based on sensitivity analysis can significantly improve model accuracy (modal frequency and MAC) and reduce model scale compared with the traditional FE model.

REFERENCES

Baoqiang, Zhang., Qintao, Guo & Guoping, Chen. 2011. Maglev Bearing Support Parameter Identification Based on Complex Model Updating. *Journal of Nanjing University of Aeronautics and Astronautics* 42(6): 748–752.

Feifei, Chen. 2010. *Equivalent Elastic Constants Analysis for Corrugated Sandwich Structure*. Wuhan University of Technology.

Guangming, Wu., Wenku, Shi & Wei, Liu. 2013. Carbody Optimization for Passenger Car Based on Modal Sensitivity Analysis. *Vibration and Impact* 32 (3): 41–45.

Jingzhe, Yang., Zhaoxiang, Deng & Shuna, Gao. 2011. White Carbody Structural Lightweight Design Based on Sensitivity Analysis. *Modern Manufacturing Engineering* 2: 103–106.

Jinsen, Zhao. 2006. *Equivalent Model Analysis for Mechanical Property of Aluminum Honeycomb Sandwich Panel*. Nanjing: Nanjing University of Aeronautics and Astronautics.

Jinpeng, Yu., Weihua, Zhang & Xuefei Huang. 2015. Carbody Structural Dynamics Model Updating Research for High Speed Trains Based on Test Modal. *Noise and Vibration Control*, 3 (03): 73–77.

Mottershead, J.E. & Friswell M.I. 1993. Model Updating in Structural Dynamics: A Survey. *Journal of Sound and Vibration* 167 (2): 347–375.

Qintao Guo., Dandan Yang & Haitao Li. 2015. Modal Testing and Parameter Identification of an Overall Rolling Stock Carbody of a High-speed Train for FE Model Validation. *ECCOMAS Thematic Conference on Multibody Dynamics*. Barcelona, Catalonia, Spain.

Qin, Shi., Chengming, Wang & Zhao, Liu. 2009. Carbody Structural Lightweight Design Based on Sensitivity Analysis. *Journal of Hefei University of Technology* 32 (7): 955–958.

Ribeiro D., Calçada R & Delgado R. 2013. Finite element model calibration of railway vehicle based on test modal parameters. *Vehicle System Dynamics* 51 (6): 821–856.

Shahram Azadi., Mohammad Azadi & Farshad Zahedi. NVH analysis and improvement of a vehicle body structure using DOE method. *Journal of Mechanical Science and Technology* 23 (2009): 2980–2989.

Shuguang, Zhang. 2007. *CRH2 EMU*. Beijing: China Railway Publishing House.

Xiaofa, Li. 2007. *Model Updating Research and Realization Based on Sensitivity Analysis*. Nanjing: Nanjing University of Aeronautics and Astronautics.

Yanhe, Tao. 2015. *FE Model Updating For Connecting Parameters of Similar Train Structures under Servicing conditions*. Nanjing: Nanjing University of Aeronautics and Astronautics.

Yunqin, Hu. 2008. *Equivalent Model Research and Numeric Analysis for Aluminum Honeycomb Sandwich Panel*. Nanjing: Nanjing University of Aeronautics and Astronautics.

Yanqing, Jiang. 2006. *Modeling Analysis and Optimization for High Speed Metro Vehicle*. Nanjing: Southeast University.

Advances in Materials Science, Energy Technology and Environmental Engineering – Patty & Zhou (Eds)
© 2017 Taylor & Francis Group, London, ISBN 978-1-138-19668-1

Description of the spatial effect of ground motion on multi-support excitation

Baochu Yu & Jiasong Wang
Dalian Ocean University, Dalian, China

ABSTRACT: This paper first gives a comprehensive review of the spatial effect of ground motion on the basis of fundamental types and propagation characteristics of seismic waves. The spatial effect of ground motion includes conduction effect, partial correlation effect, and local site effect. The common mathematical expressions of the ground acceleration power spectral density model, the traveling wave effect model, and the coherence effect model are presented here. Finally, this paper establishes matrix formulas of power spectrum of seismic ground motion, analyzes its physical significance, and points out the current problems in the seismic motion model.

1 INTRODUCTION

At present, long-span structures are widely applied to the field of engineering. As these structures have a large plan view size, the spatial variation of ground motion has an important influence on them. Therefore, the spatial variation effect of ground motion should be considered in seismic response analysis of this kind of problem. Ground motion is a complicated process of temporal and spatial variations. Previous researches place emphasis on the characteristics of temporal variation and mostly study from the aspects of intensity and spectrum structures, while neglecting the characteristics of its spatial variation. These researches mostly assumed that the excitation of each fulcrum is consistent in the structural analysis. When analyzing the seismic response of a long-span bridge, the distance of the fulcrum is larger; hence, the seismic waves arrive each fulcrum at different instants of time with time-delay. Moreover, the influence of seismic waves' reflection, refraction, dissipation in the medium, and other factors leads to the spatial effect of ground motion. For lack of actual observation records of ground motion of multi-point earthquake, some theoretical assumptions were presented only according to the degree of structural damage after the earthquake. In the past two decades, with the establishment of a high-density seismic observation network (SMART-1), seismic acceleration records of some intensive measured points have been obtained. Based on these data, seismic multi-point input ground motion models were proposed successively. These models, which are only for a particular observation station or a particular earthquake, do not have universality, but its random characteristics of ground motion are incomparable in all deterministic models for ground motions.

2 TYPES OF SEISMIC WAVES AND TRANSFORMATION OF ACCELERATION

When the earthquake occurs, the source rock stratum breaks and moves, and the deformation that rock stratum accumulates can be released suddenly. Rock stratum spreads all around in the form of seismic waves, which cause ground motion and then vibrate the engineering structure. According to the location of the seismic waves in the earth's crust, the seismic waves can be divided into body waves and surface waves. The body waves contain longitudinal waves and transverse waves; surface waves include Rayleigh waves, Love waves, etc. Strong-motion earthquake observation shows that no matter how complicated the recording waveforms are, the longitudinal waves with a higher velocity reach first, then transverse waves follow, and surface waves reach the last. From the direction of propagation and vibration, seismic waves can fall into two categories:

a. Pressure wave (P wave): the movement direction of the particle is consistent with the horizontal propagation direction of the wave.
b. Shear wave (S wave): the movement direction of the particle is perpendicular to the propagation direction of the wave. If all particles do horizontal motions, this kind of shear wave is the Shear-Horizontal wave (SH wave); if all particles do vertical motions, this kind of shear wave is the Shear-Vertical wave (SV wave);

The influence of these three types of seismic waves (P wave, SH wave, and SV wave), propagating along the horizontal direction to self-anchored suspension bridge will be discussed in this paper.

The below will present the specific form of $[E_{mN}]$ for the three types of seismic waves propagating along the horizontal direction mentioned above, given that the included angle between propagation direction of the wave and X-axis is β.

The relations between the acceleration components of any support along the wave propagation directions and each axis are as follows:

P wave

$$\ddot{x}_i = \ddot{u}_i \cos\beta, \quad \ddot{y}_i = \ddot{u}_i \sin\beta, \quad \ddot{z}_i = 0 \tag{1}$$

SH wave

$$\ddot{x}_i = -\ddot{u}_j \sin\beta, \quad \ddot{y}_i = \ddot{u}_j \cos\beta, \quad \ddot{z}_i = 0 \tag{2}$$

SV wave

$$\ddot{x}_i = 0, \quad \ddot{y}_i = 0, \quad \ddot{z}_i = \ddot{u}_k \tag{3}$$

Hence, for P wave, $[E_{mN}]$ can be expressed as

$$[E_{mN}] = \begin{bmatrix} \cos\beta & 0 & & 0 \\ \sin\beta & 0 & \vdots & 0 \\ 0 & 0 & & 0 \\ 0 & \cos\beta & & 0 \\ 0 & \sin\beta & \vdots & 0 \\ 0 & 0 & & 0 \\ \vdots & \vdots & \vdots & \vdots \\ 0 & 0 & & \cos\beta \\ 0 & 0 & \vdots & \sin\beta \\ 0 & 0 & & 0 \end{bmatrix} \tag{4}$$

while, for SH wave and SV wave, meta matrix in the above formula $\begin{bmatrix} \cos\beta \\ \sin\beta \\ 0 \end{bmatrix}$ should turn to $\begin{bmatrix} -\cos\beta \\ \sin\beta \\ 0 \end{bmatrix}$ and $\begin{bmatrix} 0 \\ 0 \\ 0 \end{bmatrix}$, respectively.

3 SPACE STATE OF SEISMIC GROUND MOTION

Since the 1960s, Taiwan Luodong, Japan Arakawa, and some other places establish plenty of successively compact arrays, especially SMART-1 array in Taiwan Lotung with the largest scale and the best effect. The records of seismic motion ground from these arrays provide valuable information for the study of spatial variation of ground motion. The seismic observations of this year (such as the Prieta Loma earthquake in the United States in 1989) clearly show that the process, which motion grounds propagate along the long-span structures, has changed significantly. The support leading to different structures do different movements. Generally, the following three factors are considered as the main reasons leading to vibration with the space changes, which should be considered in the analysis of seismic response of long-span bridges:

a. Conduction effect (also called travelling wave effect). As a result of the seismic velocity which is finite, when the distance between the fulcrums is large, the arrival time to different supporting points and different phase positions must be considered.

b. Partial correlation effect. Seismic wave is formed by superposition of the wave groups with different properties; each wave group consists of composition parts of different frequencies. In the conduction process, due to the influence of reflection, refraction in inhomogeneous soil medium, and interference between waves, the fulcrum may have different spectrum structures, even the different amplitude; the excitation of fulcrum is not completely coherent.

c. Local site effect. The uneven structure of the soil, the diverse period, and amplification effect of the rock stratum and the attenuation in the propagation process of wave give rise to diverse displacement of structural bearings.

The spatial correlation of ground motion can be described by the power spectral density function matrix of ground acceleration of each point.

When considering partial correlation effect and local site effect of motion ground, cross power spectral density function matrix for the different sites of ground can be expressed as:

$$[S(i\omega)] = \begin{bmatrix} S_{\ddot{x}_1\ddot{x}_1}(i\omega) & S_{\ddot{x}_1\ddot{x}_2}(i\omega) & \cdots & S_{\ddot{x}_1\ddot{x}_N}(i\omega) \\ S_{\ddot{x}_2\ddot{x}_1}(i\omega) & S_{\ddot{x}_2\ddot{x}_2}(i\omega) & \cdots & S_{\ddot{x}_2\ddot{x}_N}(i\omega) \\ \vdots & \vdots & \vdots & \vdots \\ S_{\ddot{x}_N\ddot{x}_1}(i\omega) & S_{\ddot{x}_N\ddot{x}_2}(i\omega) & \cdots & S_{\ddot{x}_N\ddot{x}_N}(i\omega) \end{bmatrix} \tag{5}$$

In the formula:

$$S_{\ddot{x}_k\ddot{x}_l}(i\omega) = \rho_{kl}(i\omega)\sqrt{S_{\ddot{x}_k}(i\omega)S_{\ddot{x}_l}(i\omega)} \tag{6}$$

thereinto,

$$\rho_{kl}(i\omega) = |\rho_{kl}(i\omega)|\exp[i\theta_{kl}(\omega)] \tag{7}$$

known as coherence function, and the modulo must be satisfied:

$$|\rho_{kl}(i\omega)| \le 1 \tag{8}$$

The phase angle is:

$$\theta_{kl}(i\omega) = \tan^{-1}\left[\operatorname{Im}\rho_{kl}(i\omega)/\operatorname{Re}\rho_{kl}(i\omega)\right] \quad (9)$$

Therefore, the spectral density function of each point and the coherence function between two points are given, what means that the random field of ground motion is given. Applying formula (7) into formula (6) results in:

$$S_{\ddot{x}_k\ddot{x}_l}(i\omega) = |\rho_{kl}(i\omega)|\exp\left[i\theta_{kl}(\omega)\right]\sqrt{S_{\ddot{x}_k}(i\omega)S_{\ddot{x}_l}(i\omega)} \quad (10)$$

From the above formula, we can see that the acceleration cross spectrum $S_{\ddot{x}_k\ddot{x}_l}(i\omega)$ of any two points on the ground, k and l, is related to the self-power spectrum $S_{\ddot{x}_k}(i\omega)$ and $S_{\ddot{x}_l}(i\omega)$, the modulo and phase angle of coherence function $\rho_{kl}(i\omega)$. In the formula, $\exp\left[i\theta_{kl}(\omega)\right]$ reflects the traveling wave effect; $\theta_{kl}(\omega) = 0$ shows what is not included in the traveling wave effect; $|\rho_{kl}(i\omega)|$ reflects the partial correlation effect, when $|\rho_{kl}(i\omega)| = 1$, it represents a complete coherence between the various incentives; and $S_{\ddot{x}_k}(i\omega)$ and $S_{\ddot{x}_l}(i\omega)$ reflect the local effect of points k and l on site.

4 THE MODEL OF SPATIAL EFFECT OF GROUND MOTION

4.1 Ground acceleration power spectral density model

The acceleration self-power spectrum density function $S_{\ddot{x}_k}(i\omega)$ of any point on the ground reflects the spectral characteristics of the local vibration of the point. At present, many kinds of models have been proposed. As long as the statistical parameters of the model can be determined reasonably, these models can imitate the main features of ground motion in varying degrees. Several power spectral density models are listed below.

4.1.1 Housner model

In 1947, Housner first proposed to simplify the seismic ground motion to stationary pulse series, and to model acceleration power spectrum density function as white noise:

$$S(\omega) = S_0 \quad (11)$$

4.1.2 Kanai-Tajimi model

Kanai and coworkers who assumed bedrock ground motion as white noise proposed a filtered white noise model of the stochastic process, which has obvious physical significance according to the characteristics of the site,

$$S(\omega) = \frac{1 + 4\xi_g^2(\omega/\omega_g)^2}{\left[1 - (\omega/\omega_g)^2\right]^2 + 4\xi_g^2(\omega/\omega_g)^2}S_0 \quad (12)$$

where ω_g and ξ_g are, respectively, the circle frequency damping ratio of the site soil and S_0 is the white spectral intensity of the bedrock ground motion.

4.1.3 Ou Jinping-Niu Ditao model

Since the Kanai-Tajimi Model cannot reflect the spectral characteristics of the bedrock ground motion, Ou Jinping et al. assumed that the bedrock acceleration spectrum is a Markov chain and proposed a filtered chromatographic model of ground acceleration as:

$$S(\omega) = \frac{1 + 4\xi_g^2(\omega/\omega_g)^2}{\left[1 - (\omega/\omega_g)^2\right]^2 + 4\xi_g^2(\omega/\omega_g)^2}\frac{1}{1 + (\omega/\omega_h)^2}S_0 \quad (13)$$

In the formula, ω_h reflects the parameters of bedrock spectral characteristics, advise that $\omega_h = 8\pi\, rad/s$, and the expressions of ω_g, ξ_g and S_0 are the same as above.

This model exaggerated the energy of the low-frequency ground motion, which may cause unreasonable results for the seismic response analysis of flexible structures. Moreover, it does not meet the condition that the speed and displacement of the ground motion are limited when $\omega = 0$; it cannot be used for analyzing the stochastic response of the structure under a multi-point seismic input. Therefore, Hu Jinxian, Ruiz, et al. proposed an improved model.

4.1.4 Hu Jinxian-Zhou Yuanxi model

As early as 1962, Zhou Yuanxi and Hu Jinxian introduced the parameters to reduce the ultra-low-frequency component of Kanai-Tajimi spectrum and proposed a modified model:

$$S(\omega) = \frac{1 + 4\xi_g^2(\omega/\omega_g)^2}{\left[1 - (\omega/\omega_g)^2\right]^2 + 4\xi_g^2(\omega/\omega_g)^2}\frac{\omega^6}{\omega^6 + \omega_c^6}S_0 \quad (14)$$

In the formula, ω_c is the parameter controlling low-frequency contents; the larger the value ω_c is, the smaller the low-frequency content of the motion ground is.

4.2 Calculation of equivalent power spectrum by code response spectrum

Compared with the code response spectrum of the anti-seismic bridge, ground acceleration power spectrum model mentioned above is not enough to describe the earthquake. Some scholars studied the conversion relation between ground motion response spectrum and power spectrum, the literature via the conversion relation between the two builds a power spectrum curve, which is equivalent to the response spectrum curve and used to calculate the response of acceleration, velocity, and displacement of single-degree-of-freedom system by the methods of response spectrum and power spectrum, and the results are very close which achieved much effect. Based on this, it is very convenient to adopt the method of random vibration, analyzing the seismic response whose fortification lever is the same as the method of response spectrum.

4.3 Travelling wave model

The expression of $\exp[i\theta_{kl}(\omega)]$ is generally presented in the following forms:

$$\exp[i\theta_{kl}(\omega)] = \exp[-i\omega d_{kl}^L / v_{app}] \qquad (15)$$

or further written as

$$\exp[i\theta_{kl}(\omega)] = \exp\left[-i\omega\frac{\bar{v}_{app} \cdot \vec{d}_{kl}}{v_{app}^2}\right] \qquad (16)$$

In the formula, d_{kl} is the horizontal distance between two points; d_{kl}^L is the projection of d_{kl} along the propagation direction; and v_{app} is the apparent wave velocity of seismic ground.

The apparent wave velocity is an important quantity, which is difficult to determine. The literature (Oliveira, 1991) shows that the discreteness of the apparent wave velocity changing with the frequency is very large. In practical application, it is a practical method to take the apparent wave velocity as a constant according to different conditions. Suppose the movement time differences between the ground sites and the origin of coordinate are $T_1, T_2, \ldots, T_N$, then

$$d_{kl}^L / v_{app} = T_l - T_k \qquad (17)$$

Therefore, formula (13) and formula (14) can be written as

$$\exp[i\theta_{kl}(\omega)] = \exp[-i\omega(T_l - T_k)] \qquad (18)$$

5 CONCLUSION

This paper first gives a comprehensive review on the spatial effect of ground motion on the basis of fundamental types and propagation characteristics of seismic waves. The spatial effect of ground motion includes the conduction effect, the partial correlation effect, and the local site effect. The common

mathematical expressions of the ground acceleration power spectral density model, the traveling wave effect model, and the coherence effect model are presented here. At last, this paper establishes matrix formulas of power spectrum of seismic ground motion. As a result of inadequate measured data and incomplete development of ground motion model, there are still some factors which are not take into account, for example, the frequently used coherence function model at present is obtained based on one or more ground motion records of one site. As is known to all, the seismic source characteristics, propagation path, geological condition of site, etc., all have daedal and strong randomness. Therefore, how to establish the coherence function model with statistical significance is an urgent problem to be solved.

REFERENCES

Dong K.K., Wieland M. Application of response spectrum method to a bridge subjected to multiple support excitation. IN: Proc. *9th World Conference. Earthquake Engineering*, Tokyo, Japan. 1988, 6: 531~536.

Harichandran R.S., Vanmarcke E.H. Stochastic variation of earthquake ground motion in space and time. J. *Journal of Engineering Mechanics*, ASCE, 1986, 105(2): 217~231.

Housner G.W. Characteristic of strong motion earthquakes. Bull. Seism. Soc. Am., 1947, 37: 17~31.

Hu Yuxian. Earthquake engineering M. *Beijing Seismological Press*, 1991.

Kanai K. Semi-empirical formula for seismic characteristic of the ground. J. *Earthquake Research Institute of University of Tokyo*, 1957, 35(2).

Kiureghian A.D., Neuenhofer A. Response spectrum method for multi-support seismic excitations. J. *Earthquake Engineering and Structural Dynamics*, 1992, 21: 713~740.

Lin Jinghao. Fast algorithm for random seismic response. J. *Earthquake Engineering and Structural Dynamics*, 1985,5(1): 89~94.

Loh C.H., Lin S.G. Directionality and simulation in spatial variation of seismic waves. J. ASCE, *Engineering Structures*, 1990, 12: 1~27.

Loh C.H., Yeh Y.T. Spatial variation and stochastic modeling of seismic differential ground movement. J. EESD, 1988, 16: 583~596.

Oliveira C.S., Hao H., Penjien J. *Ground motion modeling for multiple-input structural analysis, Structural Safety*, 1991, 10: 79~93.

Ruiz P., Penzien. J. *Probabilistic study of the behavior of structures during earthquakes. Earthquake engineering*. C.,UCB, CA, Report No. EERC69-03,1969.

Yamamura N., Tanaka H. Response analysis of flexible MDF system for multiple-support seismic excitations. J. *Earthquake Engineering and Structure Dynamics*. 1990; 19: 345~357.

Zhang Yahui. Buckling and dynamic analysis of complex structures under multiple loading conditions. D. *Doctoral Dissertation of Dalian University of Technology*, 1999.6.

Zhao Canhui. Seismic response of long span CFST Arch Bridges D. *Doctoral Dissertation of Southwest Jiao Tong University*, 2001.10.

*Advances in Materials Science, Energy Technology
and Environmental Engineering – Patty & Zhou (Eds)*
© 2017 Taylor & Francis Group, London, ISBN 978-1-138-19668-1

Study on the development process of a vehicle chassis

Lixin Jing, Liguang Wu & Fei Li
China Automotive Technology and Research Center, Tianjin, P.R. China

Minghui Liu
School of Mechanical Engineering, Tianjin Polytechnic University, Tianjin, P.R. China

ABSTRACT: During the development process of a light bus, due to the limitations of double wishbone front suspension with an air brake system, low floor, suspension system, and steering system were unable to refer the benchmarking bus, and hence, chassis hard points and components had to be redesigned. By analyzing the advantages and disadvantages of the benchmarking bus, in the development of the new product, we optimized the front suspension kingpin parameters, K & C characteristics, layout of steering transmission system, and rear suspension K characteristics, which can improve the vehicle handling stability performance. In late chassis turning, we achieved the goal of the vehicle performance. Through these, we have studied the development process of the vehicle chassis, which is very helpful for future development of the chassis.

1 INTRODUCTION

Chassis development plays a very important role in the development of the whole vehicle, and it is mainly divided into chassis layout, performance optimization, structural design of components, validation of sample vehicle, chassis tuning, etc. In this paper, we have studied the development process of the vehicle chassis through the development of a light bus.

A light bus chassis development process must meet the following design requirements:

1. Front double wishbone suspension with an air brake system;
2. Rear leaf spring non-independent suspension;
3. Two steps floor (630 mm height of the front floor);
4. Handling stability and ride comfort performance aiming at a listed vehicle.

As no light bus has all configurations above on the market, it requires a new chassis design and development, ensuring chassis performance while also ensuring the durability and reliability requirements of the chassis components.

2 CHASSIS LAYOUT

Owing to the adoption of an air brake system, a larger air chamber is needed at the steering knuckle position, which needs to rotate with steering knuckle during steering and keep a distance from the upper and lower control arms. Therefore, it presents a huge challenge for chassis assignment, checking of movement clearance and structural strength design of components.

Figure 1. Double wishbone front suspension of a light bus.

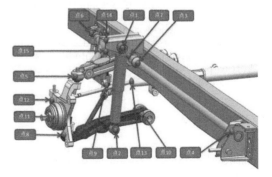

Figure 2. Double wishbone front suspension of the benchmarking bus.

Table 1. The Kingpin parameters.

	The benchmarking bus	A light bus	Target
The kingpin inclination angle, deg	7.4	13	8–14
Caster angle, deg	0	5	3–7
Mechanical trail, mm	0	33	20–35

Considering the requirements of two-step floor and frame manufacturing, the torsion bar spring can be installed only on the lower arm of double wishbone suspension, and therefore, a new structure design is required to meet the strength aims.

For the double wishbone suspension, the kingpin is the line connecting two points of lateral ball joints on the upper and lower arms. The kingpin inclination angle is the relative angle between kingpin's projection on the Y-Z plane with Z-axis. The caster angle is the relative angle between kingpin's projection on the X-Z plane with Z-axis. The distance between the intersection of kingpin's extension and ground with tire contact patch has projections on both the Y-Z plane and the X-Z plane, which are kingpin offset and mechanical trail. The distance between wheel center and the kingpin's projection on the Y-Z plane is the longitudinal force arm.

The kingpin inclination angle has significant effects on vehicle's tire aligning torque with gravity. The caster angle and mechanical trail affect tire aligning torque with lateral force and vehicle high-speed stability. The kingpin offset of the vehicle has a significant impact on the braking stabilization, which usually ranges from –10 mm to 5 mm, and longitudinal force arm influences a lot on the vehicle steering torque (acceleration and deceleration condition), which is usually less than 75 mm.

The results of subjective test and objective test show that the target vehicle is not good at steering return and high-speed stability. To promote it, we increased the caster angle and mechanical trail, made the kingpin offset and longitudinal arm as short as possible to satisfy the space requirement of the air chamber.

3 PERFORMANCE OPTIMIZATION

The objective test also shows that the suspension K&C property is not totally satisfactory, for example, the front roll camber rate is greater than 1 (the camber angle rise with wheel bumping), the rear roll steer is too large, and the lateral wheel

stiffness of front suspension is too low. Therefore, we improved the K&C property by multi-body kinematic analysis and adjust the suspension hard points and bushing stiffness to solve these problems.

The suspension K&C is optimized following the principles below: minimizing the rear axle's oversteer tendency and increasing the equivalent cornering stiffness to improve the vehicle lateral response characteristics and stability, improving the lateral wheel stiffness to improve the vehicle lateral response characteristics, reducing longitudinal wheel stiffness of the suspension to improve the comfort of the vehicle, optimizing front and rear suspension vertical stiffness to match the front and rear suspension natural frequency, optimizing front and rear suspension roll center height and roll stiffness to make the vehicle's roll gradient and the Tire Lateral Load Transfer Distribution (TLLTD) reasonable, etc.

For double wishbone suspension, the bump camber depends on swing arms' relative length and slope angle. As shown in the figure, if the upper control arm's lateral ball joint moves in against the lower arm's lateral ball joint, the camber angle decreases with wheel bumping. If upper control arm's lateral ball joint moves out against the lower arm's lateral ball joint, the camber angle increases with wheel bumping. In order to satisfy the former case, the swing arm length and slope angle were adjusted in the design.

The serious axle roll steering of the standard vehicle's rear suspension reduces the vehicle's stability and transient response. To improve these performances, the front installation position of leaf

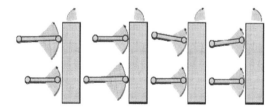

Figure 3. Bump camber of double wishbone suspension.

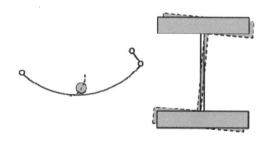

Figure 4. Rear axle roll steering of the benchmarking bus (turn left).

Figure 5. The ADAMS model of wishbone suspension.

Figure 6. The ADAMS model of leaf spring suspension.

Table 2. Optimized suspension characteristics.

	The benchmarking bus	A light bus	Target
Front Roll camber deg/deg	1.159	0.83	<0.85
Rear Roll steer deg/deg	0.1847	0.06	<0.1
Lateral Compliance mm/1000 N	2.35	0.6	<1

spring became lower in the design process. However, in order not to affect the road clearance, the position point cannot be too low, which makes the oversteer still remained.

By building Adams models of both the front and rear suspensions and adjusting hard points positions and stiffness of bushings, the K&C characteristics of the suspensions can be optimized to a reasonable level.

The optimized suspension characteristics are given in Table 2.

4   THE STRUCTURAL DESIGN OF COMPONENTS

As a result of chassis assignment, the components required a new design. Except for reaching the motion and clearance of components and satisfying chassis performance, the strength is also a requirement. It is checked by empirical static intensity analysis in design, and the specific conditions are shown below. In the intensity analysis, the entire suspension system is built in the finite

Table 3. Chassis strength conditions.

	Conditions	The inertia load		
		x/g	y/g	z/g
1	Vertical bump	0	0	4
2	Brake	1.1	0	1
3	Cornering impact	0.316	0.58	3
4	Through a deep pit	1.5	0	4
5	Rebound	0	0	−2
6	Cornering	0	1	1

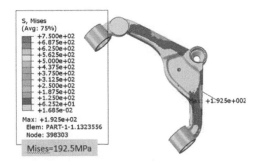

Figure 7. Analysis result of the upper control arm of front suspension.

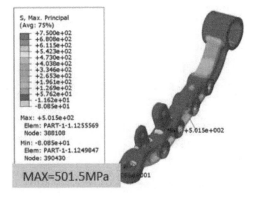

Figure 8. Analysis result of the lower control arm of front suspension.

89

element modeling software, where the constraint relationships between the various components are settled, and static forces are loaded on the wheel to analyze the stress level of each condition. Only strength requirements are reached, the structural components can be initially identified, and the fatigue durability needed a further verification at a later stage.

5 THE VALIDATION OF SAMPLE VEHICLE

The suspension hard points are strictly limited in trial-manufacture phase, and next is the verification work after the completion of the sample vehicle:

1. four-wheel alignment;
2. suspension K&C characteristic test

The four-wheel alignment ensures correct initial alignment parameters of the vehicle, which is the basis of the follow-up work. The suspension K&C test can check whether the sample vehicle satisfies the design objects, and it can detect the symmetry of suspension during the vehicle manufacturing process.

The suspension K&C test result validated the related parameters in the design phase, included the roll camber of front suspension that reduced the camber roll angle (the sample vehicle is shown in red curve, target vehicle in blue curve, the same below) and increased tire equivalent cornering stiffness, the roll steer of rear suspension that reduced the oversteering trend of rear axle and benefited the stability and responsiveness of vehicle, the lateral wheel stiffness of front suspension that increased the stiffness and benefited the vehicle's lateral responsiveness, the kingpin caster angle and mechanical trail adjusted to the target value by hard points, the kingpin offset and longitudinal force arm which are in proportion to the design value, and benefited the braking stability and characteristics of steering torque.

Figure 9. The suspension K&C test.

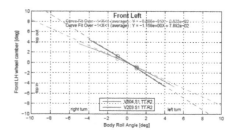

Figure 10. Roll camber of front suspension.

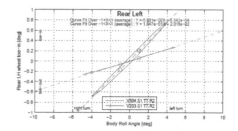

Figure 11. Roll steer of rear suspension.

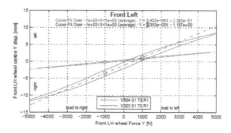

Figure 12. Front suspension lateral stiffness.

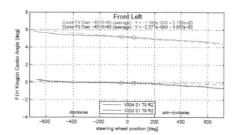

Figure 13. Caster angle of front suspension.

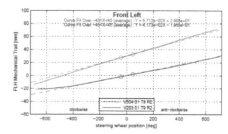

Figure 14. Mechanical trail of front suspension.

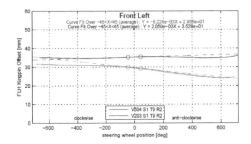

Figure 15. Kingpin offset of front suspension.

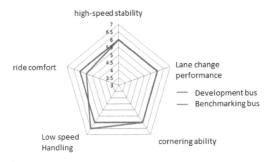

Figure 16. Subjective evaluation results.

6 CHASSIS TUNING

The verification work proved that the sample vehicle met the design objects well, and the next work is chassis tuning. Chassis tuning work is mainly based on the subjective evaluation results of experts to conduct several rounds of adjustments to the suspension springs, stabilizer bar, rubber bushings, and damper characteristics, and ultimately make the handling stability and ride comfort increased and balanced.

This chassis was tuned by international experts in National Automobile Quality Supervision and Test Center (Xiangyang), while the dampers and the steering system supplier cooperate the performance adjustment of the sample vehicle.

Subjective evaluation results after the Chassis tuning are as follows.

The sample vehicle is superior to the target vehicle in terms of the handling stability, especially in the high-speed stability and a steering aligning performance. The comfort is a little lower than the target vehicle, mainly due to a 40% increase in the unsprung mass relative to the target vehicle, which decreases the ride quality.

The reliability test is conducted with the sample vehicle after the chassis tuning, and there is no obvious fatigue failure during the test, proving that the components' structural design meets the reliability requirements.

7 SUMMARY

In the development process of this light bus, the primary condition is to meet the space of air brake systems, causing a new design of chassis hard points and components. With reference to the characteristics of target vehicle, there are still some parameters that can be further optimized, which can improve some performance. By using the method of multibody dynamics, the suspension K&C characteristics get optimized by changing hard points position and rubber bushing stiffness. By using the method of finite element analysis, chassis components get preliminarily improved to meet the chassis strength conditions, which have been proved effectively by past reliability tests. By adjusting the elastic component through the chassis tuning, the handling stability and ride comfort are increased and balanced. The vehicle performance of sample car achieved development goal, which proved that the effectiveness of the development process and provided a new thought for the chassis development in the future.

REFERENCES

Fulu Sun, Jiansheng Zhu. Discussion of Specification Setting for Chassis Tuning Sample on a Specific Car. Agricultural Equipment & Vehicle Engineering, 2013, 51(10): 12–14, 22.

Haibo Yu, Youde Li, Yuzhuo Men. Modeling and Kinematic Characteristics Analysis of Double-Wishbone Independent Suspension on ADAMS. Automobile Technology, 2007, 3: 5–8.

Jerson Rhyme Pal. The Automotive Chassis. Popular Science Press, 1992.

Konghui Guo. Automobile chassis platform integration and calibration technology. Changchun: Automobile chassis integrated matching design senior workshop handout, 2007.

Lingge Jin, Studies on the Design Optimization Method of Suspension K&C Characteristics of C-Class Vehicle. Jilin: Jilin University, 2010.

Lixin Jing, Konghui Guo, Dang Lu. Study on Parameter Identification of 3-link Leaf Spring Model. Automobile Technology, 2010, (12): 10–13, 54.

Mingjun Luo, Welin Wang. Finite element analysis and structural optimization on double-wishbone independent suspension with torsion bar. Machinery Design & Manufacture, 2009(7): 19–21.

Weixin Liu. Automobile design. Beijing: Tsinghua University Press, 2001, 462–479.

Yuanyi Huang, Guanneng Xu. Subjective Evaluation and Chassis tuning for Handling and ride of M150. Popular Science & Technology, 2008(12): 117–119.

*Advances in Materials Science, Energy Technology
and Environmental Engineering – Patty & Zhou (Eds)*
© 2017 Taylor & Francis Group, London, ISBN 978-1-138-19668-1

Study on the thermal expansion coefficient of athermalization infrared lens' thermal compensation element

F. Zhu & Y. Zhang
Kunming University of Science and Technology, Kunming, Yunnan, China

Z.W. Wang
North University of China, Taiyuan, Shanxi, China

ABSTRACT: A material suitable for the thermal compensation element is selected, which is important for the athermalization design of infrared lens. Using the infrared lens temperature range, the thermal expansion coefficients of three materials were measured by experiments, and the simulation of the thermal expansion process was carried out using the finite element software ANSYS. The simulation result was compared with the data of the experiment, and the suitable thermal expansion material was selected. The analysis results indicate that the thermal expansion amount of polyformaldehyde (POM) is relatively stable, and the simulation result is similar to the data of the experiment. POM is selected as the suitable material for the thermal expansion element, and thus, the design error of the thermal expansion structure can be reduced, and the infrared lens can have a better effect on thermal compensation.

1 INTRODUCTION

The infrared optical system is usually required to be able to work in a wide range of temperature (Yang et al. 2006). Any changes in the temperature will cause the optical system to produce the phenomenon of defocusing (Wu et al. 2002). Generally, the athermal design can be divided into: (1) mechanical passive athermalization; (2) mechanical and electrical active athermalization; and (3) optical passive athermalization. Mechanical passive athermalization is the special performance of the infrared lens, which can compensate the defocusing amount and improve the optical performance of the optical system in the thermal environment. Choosing the appropriate material is the key to thermal compensation ability of the structure (Li et al. 2006). The coefficient of thermal expansion is the most important index to choose the thermal compensation of a material (Rayces & Lebich, 1990). The larger thermal expansion coefficient can adapt to a wide range of temperature change (Zhang et al. 2008). Owing to the defocusing of the optical system and temperature that showed a linear relationship, the thermal expansion coefficient of the material needs to have certain stability. Infrared lens is a precision machinery, so the thermal expansion coefficient of the material has a direct impact on the results of the compensation (Tang et al. 2013). At present, when it comes to lens athermal design, we refer to the thermal expansion coefficient of the thermal compensation structure design, which is given by the manual, but in the

real environment, with changes in temperature, the thermal expansion coefficient of the material is also changed. In order to reflect the thermal expansion characteristics of the real material, accurate measurement of the thermal expansion coefficient of the selected material is necessary.

According to the operating temperature range and the mechanical structure of infrared lens, the thermal expansion coefficient of the material at $100 \times 10^{-6}/°C$ is selected. Initially, according to the manual of relevant materials, three kinds of heat compensation materials were selected. They respectively are polyformaldehyde (POM), polytetrafluoroethylene (PTFE), and PA1010. In this paper, the thermal expansion coefficients of the selected materials were measured, and the finite element software ANSYS was used to analyze the thermal expansion of materials. The simulation results of the three kinds of materials were obtained. Through analysis and comparison, we finally choose the appropriate material for the thermal compensation element.

2 TESTING OF THE THERMAL EXPANSION COEFFICIENT OF MATERIALS

2.1 *Experimental principle*

The equipment used in this experiment is a PCY-type super high-temperature horizontal expansion device. The structure of the test system is shown in Figure 1. The experimental principle is that the deformation of the measuring device is subtracted from the total deformation. A schematic diagram of thermal expansion analysis of the internal structure of the test system is shown in Figure 2.

ΔL, readings of electronic dial indicator; ΔL_1, sample expansion; ΔL_2, corundum sample tube expansion.

When the temperature is raised, the reading on the thermal expansion indicator is:

$$\Delta L = \Delta L_1 - \Delta L_2 \tag{1}$$

The net elongation of the sample is:

$$\Delta L_1 = \Delta L + \Delta L_2 \tag{2}$$

According to the definition of the coefficient of linear expansion, the linear expansion coefficient of the sample is:

$$\alpha = \frac{\Delta L_1}{L \cdot \Delta T} = \frac{(\Delta L + \Delta L_2)}{L \cdot \Delta T} = \frac{\Delta L}{L \cdot \Delta T} + \frac{\Delta L_2}{L \cdot \Delta T} \tag{3}$$

Therefore, the coefficient of thermal expansion for the test sample is:

$$\alpha = \alpha_{\overline{\mathcal{F}}} + \frac{\Delta L}{L \cdot (T_2 - T_1)} \tag{4}$$

$\alpha_{\overline{\mathcal{F}}}$ is the average coefficient of thermal expansion of corundum, L is the original length of the specimen, T_2 is the final temperature, and T_1 is the initial temperature.

As the temperature range measured in this experiment is 30°C to 80°C, the coefficient of thermal expansion of corundum is $5.7 \times 10^{-7}/°C$ within this range, and hence, it can ignore the thermal expansion of corundum. The coefficient of thermal expansion of the sample is $\alpha = \frac{L}{L \cdot (T_2 - T_1)}$.

2.2 Experimental samples

Owing to the size of the experimental samples, a structure with a diameter of 6–10 mm and a length of greater than or equal to 40 mm is needed, and therefore, three sets of samples to be tested were processed into a φ10 × 50 (mm) cylinder. The processed experimental samples are shown in Figure 3.

2.3 Experimental data and analysis

In order to reduce any experimental error, each material was made up of four specimens, whose thermal expansion coefficients were measured, and then averaged. Three materials per degree of thermal expansion are shown in Figure 4.

(a)PA1010　　　　(b)POM　　　　(c)PTFE

Figure 3.　Experimental samples.

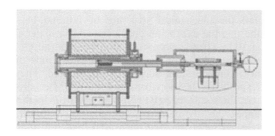

Figure 1.　The diagram of test system structure.

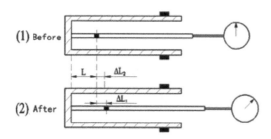

Figure 2.　Diagram of thermal expansion analysis.

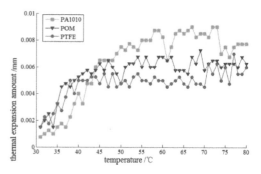

Figure 4.　Comparison chart of thermal expansion amount of three materials at each degree Celsius.

94

As we can see, Figure 4 shows that as the temperature increased, the thermal expansion of the three materials increased gradually, even when the temperature changes by 1°C. In the range of test temperature, the average values per degree amount of thermal expansion changes in different ranges are, for example, 0~0.010 mm for PA1010, 0~0.008 mm for POM, and 0~0.006 mm for PTFE. The maximum variation range of thermal expansion is that of PA1010, and hence, it has better thermal compensation capability. The average per degree amount of thermal expansion of PA1010 changes greater, and hence, its stability is poor. Therefore, the above experimental data show that when selecting the thermal compensation element material, we can prioritize POM and PTFE.

3 FINITE ELEMENT ANALYSIS OF THERMAL COMPENSATION MATERIALS

3.1 Finite element model

In this paper, the finite element software ANSYS is used to simulate the thermal expansion process of the three materials.

The structure of the sample in the cylinder is simple, and therefore, they are analyzed using the 2D and 3D model. According to the experimentally measured thermal expansion, this paper calculated the three kinds of materials in the experimental temperature range for every 10°C of the average thermal expansion coefficient. As the simulation analysis of the input values and the material thermal expansion coefficient are given in Table 1, the other performance parameters are shown in Table 2.

The finite element software is used to establish a material model with the same size as the experiment. The model is simplified to be a half-rectangular plane along the longitudinal section of the

Table 2. The parameters of material.

Material	Thermal conductivity /(W/(m·K))	density /(Kg/m³)	Elastic modulus /Gpa	Poisson ratio
PA1010	0.24	1040	1.3	0.35
POM	0.26	1410	2.6	0.35
PTFE	0.25	2180	0.45	0.4

(a)Two-dimensional model   (b)Three-dimensional model

Figure 5. The finite element model of the sample material.

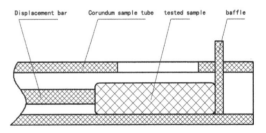

Figure 6. Chart of the installation location of the sample.

cylinder axis, using the plane coupling element PLANE223. When 3D analysis is performed, the model does not make any simplification, using the solid coupling element SOLID226. The 2D and 3D finite element model of the specimen is shown in Figure 5.

3.2 Applying constraint and load

In the process of material thermal expansion analysis, the accuracy of the results is largely affected by the constraint and load. In the analysis, the way of constraint should be closer to the actual installation situation of the sample. The installation position of the sample in the test system is shown in Figure 6, and the displacement on the bar exerts a very small pre-tightening force, which can be ignored during analysis. The right side of the block slice can limit its degrees of freedom on the horizontal direction, and the bottom is supported by an alumina tube and finally exerts temperature load.

Table 1. Thermal expansion coefficient of the experimental material.

Temperature/°C	Coefficient of thermal expansion / × 10-6°C-1		
	PA1010	POM	PTFE
30	14.9	15.0	30.0
40	36.4	69.6	53.5
50	118.3	114.6	100.0
60	158.7	127.1	100.5
70	165.2	124.1	96
80	156.2	126.1	115.5

In order to get the thermal expansion amount of the sample under different temperatures, we set the reference temperature to 30°C, applying the model from 30°C to 80°C load. The whole process is divided into fifty steps, and the temperature rises for each load step by 1°C. The picture of sample displacement distribution cloud is shown in Figure 7 (PA1010 sample, for example, the temperature is 80°C).

From Figure 7 (a) and Figure 7 (b), it can be seen that the trends of results of the 2D and 3D models are the same, and the sample on one end of the thermal expansion amount of pre-tightening force is the largest. By comparing with the analysis results, two methods of analysis in the setting temperature range of the material are the same about the axial thermal expansion displacement, and the analysis results of the other two materials were the same with PA1010. In order to simplify the calculation, the thermal expansion process of three kinds of materials can be analogously simulated by the 2D model.

3.3 Result analysis and discussion

By analyzing the simulation of the thermal expansion process of the three kinds of sample materials, we can get the sample amount of thermal expansion at a given temperature range, the two-dimensional sample analysis data, and compare with the experimental thermal expansion (as shown in Fig. 8).

With the increase in temperature, thermal expansion of the three kinds of material quantity simulation value increased gradually. At the range of 30 °C to 45 °C, the simulation results coincide with the experimental data. As the temperature rise further, the error of the experiment and simulation is also increased. Seeing from the simulation data curve, the stability of PA1010's thermal expansion process is poorer. Although the thermal expansion of PTFE is very stable in the early stage, when the temperature is greater than 70 °C, the thermal expansion amount rise sharply. However, the proc-

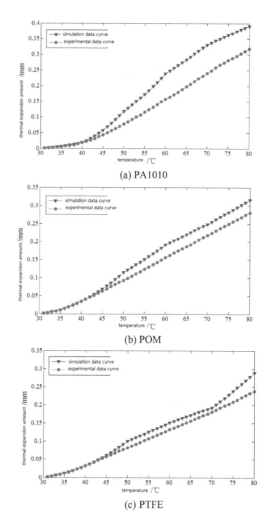

(a) PA1010

(b) POM

(c) PTFE

Figure 8. Comparison chart of simulation and experiment of the thermal expansion amount of three materials.

ess of POM thermal expansion in the whole process analysis performance is very stable. There is error between the simulation data and experimental data, because the software simulation of the thermal expansion process is ideal, and the actual materials of thermal expansion are influenced by many factors, such as the inside structure of the material, which the software cannot take into account. In order to compare the approximation degree of three kinds of the material simulation curve with the experimental data curve, we calculate the residual sum of squares of Q, the residual standard deviation between S and the related index R2, depending on the experimental thermal expansion, which acts as raw data (in Table 3,).

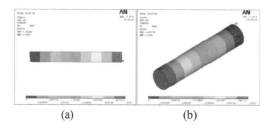

(a)   (b)

Figure 7.   Displacement distribution of PA1010.

Figure 9.　Mechanical athermalization infrared lens.

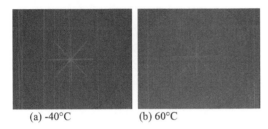

(a) -40°C　　　　　　　(b) 60°C

Figure 10.　Target imaging at –40°C and 60°C.

According to linear regression of related knowledge of the statistics, the residual sum of squares of Q and the residual standard deviation S are more close, the related index R2 is close to 1, and two curve similarity is best. From Table 3, it can be observed that among the three kinds of materials, the simulation results of POM and PTFE are close to experimental data. As the POM thermal expansion process is relatively stable and easy to control the quantity of heat expansion while designing the thermal compensation device, it is reasonable to choose POM as the thermal compensation element material from overall consideration. The machining of the mechanical athermalization infrared lens is shown in Figure 9.

In the high-low temperature experiment, the target imaging at –40°C and 60°C of the lens are shown in Figure 10.

It can be seen from the figure that the infrared lens has good imaging quality at high and low temperatures, which meets the requirement of the application.

4 CONCLUSION

In this paper, in order to realize the infrared lens design with athermalization, we selected three kinds of materials for the thermal compensation device to work on measuring the material thermal expansion coefficient. We input the experimental result as the value of software simulation and use the finite element software ANSYS to simulate the process of thermal expansion material to analyze the simulation data curve, which does not coincide with the experimental data. Analysis is based on the similarity of the experimental data and simulation data. Among the three kinds of materials, the POM and PTFE simulation result are closer to the experimental measured data. What is more, POM has a larger thermal expansion coefficient and its thermal stability is better than the other two materials. Therefore, POM has been selected as the suitable material for the thermal compensation device, and it can help the infrared camera get better athermalization effect.

REFERENCES

Li Juan, Wang Yingrui & Zhang Hong (2006). New passive compensating mechanism for athermalisation. Infrared and Laser Engineering, 35(4): 476–479.
Rayces L, Lebich L (1990). Thermal Compensation of Infrared Achromatic Objectives With Three Optical Materials. Proc. SPIE, 1990, 1354: 752–759.
Tang Liying, Gao Dongliang & Xiang Guang (2013). Effect of BaO and Li_2O Content on the Thermal Expansion Coefficient and Chemical Stability of Na_2O-CaO-SiO_2 Glass[J]. Materials Review, 27(z1): 195–197.
Wu Xiaojing, Sun Chiquan & Meng Junhe (2002). Relationship between athermalizing infrared optical system and zoom lens. Infrared and Laser Engineering, 31(3): 249–252.
Yang Yi. (2006). Study on the thermal optics property of primary mirror applied on a space[J]. Optical Technique, 32(1): 144–147.
Zhang Wenxue, Yang Juan & Cheng Xiaonong (2008). Research Advances in Thermal Expansion Properties of Epoxy Molding Compounds. Materials Review, 22(z1): 291–294.

Advances in Materials Science, Energy Technology
and Environmental Engineering – Patty & Zhou (Eds)
© 2017 Taylor & Francis Group, London, ISBN 978-1-138-19668-1

Study on the dynamics of the electronic equipment in a self-propelled artillery system

Jing Gao
The 63985 Unit of PLA, Beijing, China

Ben Li, Shengxu Wang, Qiang Wang & Shuang Li
The 63981 Unit of PLA, Wuhan, China

ABSTRACT: In order to study the dynamic characteristics of the electronic equipment in a self-propelled artillery system under the vibration and shock environment, the study takes the fire control computer as an example, for which an installation system dynamics model is established in the ADAMS. By applying the corresponding excitation, the dynamic response of the whole system is obtained, and then the vibration isolation effect of the whole system is determined. Then, the modal analysis of the fire control computer chassis is carried out, and the natural frequency of the chassis is obtained, which provides reference to the vibration isolation design.

1 INTRODUCTION

With the rapid development of the modern military technology, information technology played a significant role in the war. In order to improve the operational performance of the weapon system, more and more electronic equipment were installed to the different weaponries, and they occupy an irreplaceable position. For example, the fire control computer installed inside a self-propelled artillery turret is vital for the whole fire control system (Pflegl, 2000).

The self-propelled artillery electronic equipment must suffer vibration and shock in motor or shooting process. In this harsh environment, the reliability of the electronic equipment will be greatly affected, which would make weapon systems unable to maximize the operational effectiveness or even failure in use process, in order to reduce combat. Therefore, the dynamic characteristics of the electronic equipment of the gun are generally paid more attention (Bundy, 2001; Li, 1991).

In this paper, the dynamic simulation analysis of the artillery electronic equipment (fire control computer) installation system is carried out, and the vibration isolation effect of the whole system is determined by the response of the chassis under the vibration and shock environment. In addition, the natural frequency of the chassis is obtained by the modal analysis, which can provide reference for its vibration isolation design.

2 DYNAMIC ANALYSIS OF FIRE CONTROL COMPUTER CHASSIS

The fire control computer is the core part of the fire control system, and its main function is to complete the communication data and command with gunner fire control operation display platform, aiming in military display, navigation system, initial velocity radar, and main control device; conduct the data acquisition and conversion of the attitude angle, elevation angle, and azimuth angle sensor; conduct the operating and aiming solution and control the artillery by output adjustment artillery control signals; and exchange information with other systems. The reliability of the fire control computer is of great significance for artillery to play a war efficiency.

2.1 *Simulation and analysis of the installation system*

In the ADAMS, the dynamic simulation model of the fire control computer installation system is established. The model consists of three parts: the base, the chassis, and the vibration damper connecting the base and the chassis (defined by BUSHING, the setting of the parameters is referred to as the stiffness and damping of the vibration isolator). The excitation of the whole system is applied at the base position, including three acceleration along the coordinate axis direction and three angular acceleration around the coordinate axis. The excitation derives from the vibration response of the fire control computer installation position of the gun when firing under a certain working condition. The virtual system model is shown in Figure 1.

The excitation of six degrees of freedom is added by "General Point Motion (Prescribe all coordination)" in the ADAMS. After importing the vibration response data file of the fire control computer, the AKISPL function is used to realize the adding of the excitation of each group. The six groups of excitation functions are shown in Table 1.

The excitation function curves are shown in Figure 2 and Figure 3, and the acceleration of

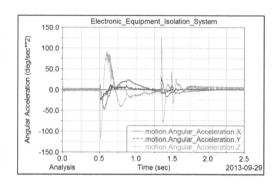

Figure 3. Angular acceleration curves around the X-, Y-, and Z-axis.

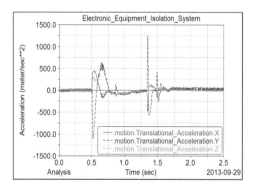

Figure 1. The model of the system to install the fire control computer.

Table 1. Activation function.

DoF/ Freedom	Type	f(time)/Function
TraX	acce(time) =	AKISPL(1000*time,0,Spline_1,0)
TraY	acce(time) =	AKISPL(1000*time,0,Spline_2,0)
TraZ	acce(time) =	AKISPL(1000*time,0,Spline_3,0)
RotX	acce(time) =	AKISPL(1000*time,0,Spline_4,0)
RotY	acce(time) =	AKISPL(1000*time,0,Spline_5,0)
RotZ	acce(time) =	AKISPL(1000*time,0,Spline_6,0)

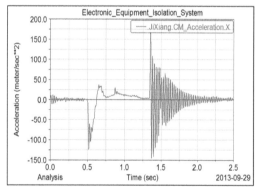

Figure 4. The box along the X-direction acceleration curve.

three directions along the X, Y, and Z directions is shown in Figure 2, and the angular acceleration excitation around the X-, Y-, and Z-axis is shown in Figure 3.

By setting, the simulation time was 2.5 s and the computational step was 0.001 s. The calculation results were obtained as follows: the acceleration response curve of the chassis is shown in Figure 4~Figure 6, and the angular acceleration response curve is shown in Figure 7~Figure 9.

Extracting the corresponding data list as follows:

It can be observed from the simulation results, that the peak acceleration along the X, Y, and Z directions of the chassis response (output) declined compared with the input of the base. Namely, the vibration isolator reduced the peak acceleration and reached the effect of vibration reduction. Also from input and output angle acceleration peak, contrast can be seen that they have no too big change basically, indicating that the vibration isolator

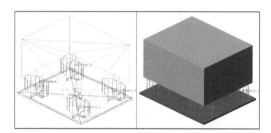

Figure 2. Acceleration curves along the X, Y, and Z directions.

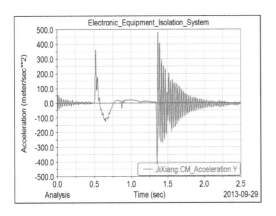

Figure 5. The box along the Y-direction acceleration curve.

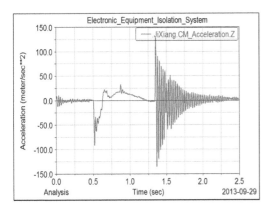

Figure 6. The box along the Z-direction acceleration.

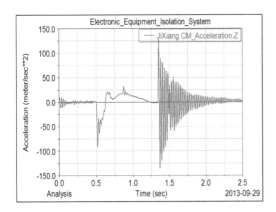

Figure 7. The box around the X-axis angular acceleration curve.

has a high stiffness of the torsional rigidity, and the angular acceleration of the base is basically synchronous attenuation transferred to the chassis.

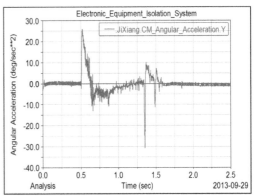

Figure 8. The box around the Y-axis angular acceleration curve.

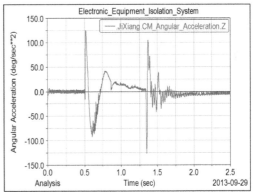

Figure 9. The box around the Z-axis angular acceleration curve.

3 MODE ANALYSIS OF THE CHASSIS

When analyzing the vibration response of the fire control computer installed system, the base and the chassis were considered as a rigid body. As the whole system is taken into account, the stiffness of the vibration isolator was much lower than the base and the chassis. Otherwise, it not only simplifies the calculation, but also ensures the correctness of the simulation results. However, in the actual vibration environment, the chassis tends to produce elastic deformation, sometimes even coupling resonance appears with the internal printed circuit board, which has a negative impact on the electronic equipment. Therefore, it is necessary to conduct mode analysis of the chassis to find out the natural frequency, thus avoiding local resonance and coupling phenomenon (Pitarresi, 1992).

Table 2. Simulation results of the installed system.

Simulation results		Acceleration peak (m/s²)			Angular acceleration peak (deg/s²)		
		X direction	Y direction	Z direction	Around X-axis	Around Y-axis	Around Z-axis
Input	Positive	451.68	1248.12	300.96	23.39	30.72	126.19
	Negative	−501.51	−1107.98	−378.11	−47.71	−25.95	−119.03
Output	Positive	167.35	481.30	128.35	47.61	25.93	124.88
	Negative	−145.88	−420.34	−135.25	−39.68	−30.66	−126.76

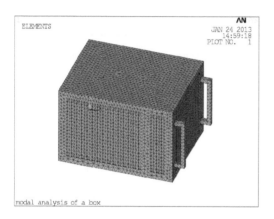

Figure 10. Finite element model of the chassis.

Using the established model function of Pro/E, the model of the entity of fire control computer chassis is established. Then, the model is imported into ANSYS and mode analysis of the chassis is conducted. The characteristics of the chassis material is 2A12, the density is 2700 kg/m³, the elastic modulus is 7.0E10 Pa, the Poisson's ratio is 0.3, and the grid is divided by SOLID 186 unit. The finite element model for the chassis is shown in Figure 10.

The first six orders of natural frequencies of the chassis are shown in Table 3.

Order	1	2	3	4	5	6
f(Hz)	946.99	999.06	1191.1	1225.6	1379.8	1387.6

In order to avoid local resonance and coupling phenomena in the fire control computer system, according to the frequency double rule, for the chassis, its first-order natural frequency ratio between vibration isolator and the printed circuit board were greater than 2. At this point, the mutuality between the resonance coupling effects is smaller.

From the above analysis, it was found that the first-order natural frequency of the chassis is 946.99 Hz, and it varies greatly with the general vibration isolator natural frequency (for the rubber isolator, the first-order natural frequency is generally in the range from a few hertz to several tens of hertz), which is enough to meet the requirements. For printed circuit board, according to the frequency double rule (or inverse frequency double rule), the internal printed circuit board of the first-order natural frequency could be less than 473.5 Hz or greater than 1893.98 Hz.

4 CONCLUSION

Taking the fire control computer as an example, this paper carries out the dynamic simulation analysis of the artillery electronic equipment. Through the analysis of the whole installation system simulation, the results indicate that the connection box and the base isolator can effectively attenuate the peak acceleration, but has a smaller effect on attenuating the diagonal peak acceleration. Through the modal analysis of the chassis, the results indicate that its first-order natural frequency is up to 946.99 Hz, with which the vibration isolator natural frequency is enough to meet the requirements of the frequency double rule. Based on this, it could provide reference to its internal printed circuit board anti-vibration design.

REFERENCES

Bundy Mark L, Erline Thomas F, Marrs Terry. (2001). Marrs T [C]. Proceedings of The 10th Gun Dynamics Symposium, Austin: AD-A393007.
Li Zhe (1991). The Amplitude-frequency Response and Balancing of the Shaking Force and Shaking Moment of Elastic Linkages [J]. Journal of Mechanical Engineering, 4(1): 38–43.
Pflegl G. A. (2000). Adaptive Control of Nonlinear System with Applications to Flight Control Systems and Suspension Dynamics [C]. AD-A375229: Proceedings of the 9th U.S. Army Symposium on Gun Dynamics.
Pitarresi J. M. and Primavera A. A. (1992). Comparison of Modeling Techniques for the Vibration Analysis of Printed Circuit Cards [J]. Electronic Packing, 1992(114): 378–383.

Advances in Materials Science, Energy Technology and Environmental Engineering – Patty & Zhou (Eds)
© 2017 Taylor & Francis Group, London, ISBN 978-1-138-19668-1

Bending testing apparatus with a welding plate

Cunjun Li, Hua Zhao & Hairong Wang
Zhoushan Institute of Calibration and Testing for Quality and Technology Supervision, Zhoushan, China

Shijie Su & Jitao Liu
College of Mechanical Engineering, Jiangsu University of Science and Technology, Zhenjiang, China

ABSTRACT: Plate bending testing machine is the main device used for metal plates bending test. There are many problems in bending testing samples existing on the current market, mainly such as the deviation and single function in the process of experiment. This paper describes the designing of a device to achieve the bending through clamping various kinds of plate specifications. Finally, this device can effectively solve the above problems.

1   INTRODUCTION

Bending machines are a widely used bending test apparatus with a range of applications, whose function is to be bent according to different needs of different products. Chr. Haeusler, a Swedish company, produces the world's biggest four-roll bending equipment. The three roller and four roller machines of **PROMAU DAVI, MG**, an Italian company, are very common as well as the German **SCHAFER** and other Italian companies. In recent years, Bender development level in China has been greatly improved, as adjustable three-roll plate bending machine and four-roll plate bending machine have been increasingly used instead of the traditional plate bending device, symmetrical three-roll plate bending machine, in the structure. However, more powerful two-roll bending machine is not common at home and abroad. Besides, less bending machines are used in the field of scientific research.

This paper provides a kind of plate bending test device with a weld, which can detect the mechanical properties of sheets of different thicknesses with weld flexibly.

2   PROJECT DESIGN

In current techniques, the quality of welding seam should be tested by bending the same plates with different specifications through welding joint. Generally, all are bent by the finished plate benders. These bending machines often have the following problems: when the welding steel plate is bent, the parts welded may be raised. Therefore, it is difficult to control any angle of the

given bend, which will cause inaccurate data of the experiment. Thus, welding qualities cannot be detected effectively and test requirements will not be completed by the universal Bender. In an existing principle of bending test equipment, apart from the hot-bending bend, Jack bend with the use of fire for bending, and finished pipe bending machines were mostly applied to bend tubes. The test of welding and requirements cannot be met easily.

In order to meet the requirements of the bend test of metal materials under GB/T 232–2010, the neck bending test is used in this paper, namely, dynamic rolled sheet does circular motion around the fixed roller. This project can effectively avoid the problems of the raised welding parts and can also accurately control any angle of the bending. Bending schematic is shown in Figure 1.

The bending machine according to functional design consists of several parts, including racks, benches, hydraulic systems, power transmission

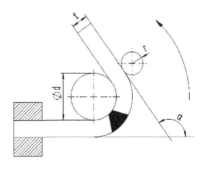

Figure 1.   Schematic diagram of bending.

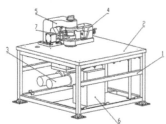

1, frame; 2, table; 3, power transmission and detection devices; 4, bending device; 5, fixture device; 6, hydraulic system; 7, sheet.

Figure 2.   Diagram of bending structure.

and detection devices, bending device, and fixture device. Structural diagram is shown in Figure 2.

3 STRUCTURAL DESIGN AND STRENGTH CHECK OF MAIN PARTS OF EACH DEVICE

The bending machine designed includes the following: rack, bench, hydraulic systems, power transmission and detection devices, bending device, and fixture device.

3.1 Design of the frame

The frame mainly is welded by square steel, only holding the entire work table-board and the above-gravity.

3.2 Design of power transmission and detection devices

Power transmission and detection devices include a swing cylinder, a steering angle sensor, fuel tank fixation, torque sensors, and a drive shaft. The source of transmission power is the swinging cylinder. Depending on the thickness of the sheet, the swinging cylinder can output different torques. The torque is transmitted by the torque sensor and the torque size can also be read directly by the torque sensor. Finally, the torque sensor transmits the torque to the drive shaft and the upper end of drive shaft is transmitted by a spline torque.

3.3 Design of bending device

The bending device includes a rocker seat, a rocker panel, fixed rollers, a movable roller frame, movable rollers, movable roller pins, a trapezoidal lead screw, and handwheel. The force of major parts of mechanical bending device are mainly caused by the force transmission in the process of the

bending. Analysis of bending strength of main parts and finite element is as follows.

3.3.1 Stress analysis and strength check of the movable roller

According to torque and force change, we know that the maximum force to the movable roller is 124 kN. The working diameter of the movable roller is 100 mm; the length is 165 mm and ID is 70 m. Force analysis is shown in Figure 3.

$$\begin{cases} F_a + F_b = 124\ KN \\ 0.165\ m \cdot F_a - 124KN \times 0.055m = 0 \end{cases} \quad (1)$$
$$F_a = 41KN \quad F_b = 83KN$$

The maximum bending moment that the movable roller withstands is at the center of the contact with the specimen

$$\begin{aligned} M_{max} &= 41\ KN \times 0.11\ \text{m} \\ &= 4.51\ KN \cdot m \end{aligned} \quad (2)$$

If the primary material of the movable roller is 45 steel, then the tensile strength is

$$\sigma_b \geq 630 MPa$$
Yield strength $\sigma_s \geq 370 MPa$ $\quad (3)$
as $\dfrac{\sigma_s}{\sigma_b} = \dfrac{370 MPa}{630 MPa} \approx 0.59,$

A safety factor $[S] = 1.8$
Allowable bending stress,

$$[\sigma] = \frac{300 MPa}{1.8} \approx 166.7 MPa \quad (4)$$

Movable roller diameter $D = 100\ mm$,

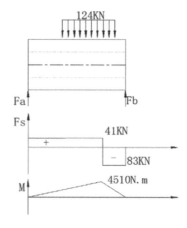

Figure 3.   Stress analysis of the movable roller.

Inside diameter $d = 70\ mm$

The section modulus in bending of the movable roller

$$W_1 = \frac{\pi D^3}{32}\left(1 - \alpha^4\right) = 7.5 \times 10^{-5}\ m^3, \tag{5}$$

For which $\alpha = \frac{d}{D}$.

The maximum bending stress in the work area:

$$\sigma_{1max} = \frac{M_{max}}{W_1} = \frac{4.51 KN \cdot m}{7.5 \times 10^{-5}\ m^3} \approx 60 MPa < [\sigma] \tag{6}$$

Therefore, the face stress of the movable roller is 124 KN, the diameter is 100 mm, and the inner diameter is 70 mm to meet the requirements

3.3.2 Stress analysis and strength check of the fixed roller

Stress analysis when the sheet starts to be bent is shown in Figure 4.

$$\begin{cases} F_1 + 124\ KN = F_d \\ 0.16 m \cdot F_d - 124\ KN \times 0.2375\ m = 0 \\ \qquad F_d = 184\ KN\quad F_1 = 60\ KN \end{cases} \tag{7}$$

(Force in specimen of fixture clamping)

The diameter of the fixed roller is 120 mm, and the length is 165 mm. Force analysis is shown in Figure 5.

$$\begin{cases} F_a + F_b = 184\ KN \\ 0.165\ m \cdot F_a - 184\ KN \times 0.055\ m = 0 \\ \qquad F_a = 61.3\ KN\quad F_b = 122.7\ KN \end{cases} \tag{8}$$

The maximum bending moment that the fixed roller withstands is at the center of the contact with the specimen,

$$\begin{aligned} M_{max} &= 61.3\ KN \times 0.11\ m \\ &= 6.743\ \text{KN.m} \end{aligned} \tag{9}$$

If the primary material of the fixed roller is 45 steel, then the tensile strength is

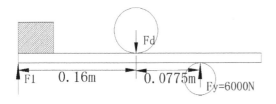

Figure 4. Stress diagram of bending.

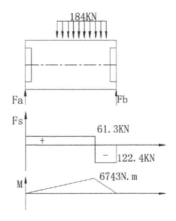

Figure 5. Stress analysis of Fixed roller.

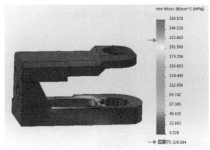

(a)Stress diagram of the rocker and the rocker plate

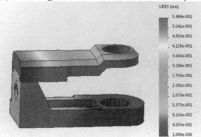

(b)Deformation maps of the rocker and the rocker plate

Figure 6. Rocker and finite element analysis of the rocker plate.

$$[\sigma] = \frac{300 MPa}{1.8} \approx 166.7 MPa \tag{10}$$

The diameter of the fixed roller, $D_1 = 160\ mm$

Bending the fixed roller face of the section modulus,

$$W_1 = \frac{\pi D_1^3}{32} \approx 4.02 \times 10^{-4}\ m^3 \tag{11}$$

The maximum bending stress in the work area,

$$\sigma_{1max} = \frac{M_{max}}{W_1} = \frac{6.743KN \cdot m}{4.02 \times 10^{-4} m^3} \approx 16.77 MPa < [\sigma]$$

(12)

to meet the requirements.

3.3.3 *Rocker and finite element analysis of the rocker plate*

Results of analysis of finite element of stress and strain of the rocker and the rocker plate are shown in Figure 6.

Fro Figure 6 (a) and Figure 6 (b), it can be observed that the maximum stress of the rocker and the rocker plate at concentrated stress is 268.6 MPa; other stresses are relatively small, hence stress meets requirements; 0.55 mm for maximum deformation is acceptable. As the requirements of the rocker and the rocker plate on deformation are not high, 0.55 mm of deformation can meet the requirements.

3.4 *Design of the fixture device*

The fixture device includes a fixture, a linear guide and a thin hydraulic cylinder, a front plywood, a

(a) Stress diagram of fixture

(b) Deformation maps of fixture

Figure 7. Fixture finite element analysis.

plywood, a pad, and a fixed roller block. The main parts of the fixture device involve force transmission in the process of bending and fixture clamping force. Finite element analysis of the main components of the fixture device is as follows:

3.4.1 *Finite element analysis on fixture*

Results of finite element analysis of stress and strain of fixture are shown in Figure 7.

From Figure 7 (a) and Figure 7 (b), it can be observed that the maximum stress is 127 MPa and the maximum deformation is 0.15 mm, which are within the permissible stress and strain, and hence, the intensity of the fixture is qualified.

4 CONCLUSIONS

First, this paper introduces the research status of bending machines. Then, in order to avoid the deficiencies of the current bending machines, this paper presents a powerful, adaptable sheet metal bending test machine. Finally, practical designs of the whole bending test machines finished by general program and structure design specific working parts as well as strength check of the main parts in bending machines. Therefore, it provides the basis for the production of bending machines with the above function.

REFERENCES

Li Xiang-yi. Study on Control System of Multi-Press Head Sheet Forming Equipment. Wuhan University of Technology. 2011
Wang Shi-ke. Sheet metal bending machine. China Academic Journal Electronic Publishing House, 1994–2016
Yan Hai-wen. Research on auto-control system of three-roll bender. Wuhan University of Technology, 2013: 1–2.
Yao Jiu-jun. Adjustable plate bending test machine. Industrial Technology [J], 2012: 64–65.
Zhang Jian-hui. Development and Application of Automatic Plate Bending Testing Machine. 2012
Zhou Hua. Research on 3D Reconstruction of Large Size Plate in Line Heating System. Guangdong University of Technology. 2014

*Advances in Materials Science, Energy Technology
and Environmental Engineering – Patty & Zhou (Eds)*
© 2017 Taylor & Francis Group, London, ISBN 978-1-138-19668-1

An inexact smoothing Newton algorithm for NCP based on a new smoothing approximation function

Chunlei Shang, Zhiyuan Tian & Ling Liu
School of Mathematics and Statistics, Qingdao University, Qingdao, China

ABSTRACT: In this paper, an inexact smoothing algorithm based on new smoothing approximation function has been proposed for solving the nonlinear complementarity problem with a non-monotone line search. Under appropriate conditions, a globally and locally superlinearly convergence of the proposed algorithm is established. Numerical experiments show that the algorithm is effective.

1 INTRODUCTION

In this paper, we consider the NCP that finds a vector $x \in R^n$ such that

$$x \geq 0, F(x) \geq 0, x'F(x) = 0 \qquad (1)$$

where $F: R^n \to R^n$. Throughout this paper, we assume that F is a continuously differentiable P_0 function and solution (1) is nonempty and bounded. The nonlinear complementarity problem has some important applications in engineering and economic. Scholars have proposed many effective algorithms according to the problem (Maetal, 2008; Yu, 2011). Recently, there have been many algorithms proposed based on smoothing Newton algorithm for solving NCP (Qi, 2000; Tseng, 2000; Ma, 2008; Zhang, 2009). The smoothing Newton method uses a smoothing function and a smoothing parameter reformulates the problem considered as a family of parameterized smoothing equations to be solved approximately by using the Newton method per iteration. By reducing the smoothing parameter to zero, it is believed that a solution to the original problem can be found.

We propose a new inexact smoothing algorithm with a non-monotone line search based on a new approximation function; we view the smoothing parameter as an independent variable. Numerical experiments will be employed to show the efficiency of the algorithm.

The rest of this paper is organized as follows: In section 2, we will present a new approximation function and inexact smoothing algorithm; then, we prove the feasibility of the algorithm. Globally and locally superlinearly convergence of the proposed algorithm is established in section 3. In section 4, the numerical behavior of the algorithm will be reported. Some conclusions are drawn in the last section.

2 NEW APPROXIMATION FUNCTION AND ALGORITHM

In this section, we propose a new smoothing function about NCP. First, we need to understand some definitions in this paper, which are as follows:

Definition 2: (Mangasarian, 1976): function $\phi: R^2 \to R$ is said NCP function, if

$$\phi(0, a, b) = 0 \Leftrightarrow a \geq 0, b \geq 0, ab = 0 \qquad (2)$$

Definition 3: Function $F: R^n \to R^n$ is said to be a P_0 function, if $\forall x, y \in R^n$ and $x \neq y$, and there is an index $i_0 \in \{1, 2, \cdots, n\}$

$$x_{i_0} \neq y_{i_0}, (x_{i_0} - y_{i_0})\left[F_{i_0}(x) - F_{i_0}(y)\right] \geq 0, \qquad (3)$$

Definition 4: Function $H_\mu(\bullet): R^n \to R^n (\mu > 0)$ is said to be a smoothing approximation function, for a function $H: R^n \to R^n$, if $\exists \kappa > 0$ such that $\forall x \in R^n$

$$\|H_\mu(x) - H(x)\| \leq \kappa \mu, \forall \mu > 0 \qquad (4)$$

Function $H_\mu(\bullet): R^n \to R^n (\mu > 0)$ is said to be a smoothing coincident approximation function, if κ has nothing to do with x.

In many algorithms for solving NCP, the most intuitive method is that based on min function converting NCP to the following equations:

$$\Phi(x) := \begin{pmatrix} \phi_{\min}(x_1, F_1(x)) \\ \phi_{\min}(x_2, F_2(x)) \\ \vdots \\ \phi_{\min}(x_n, F_n(x)) \end{pmatrix} = 0 \qquad (5)$$

where $\phi_{\min}(a,b) = \min\{a,b\}$, $\phi_{\min}(a,b)$ is a NCP function. If $F(x)$ is a group smooth function, then it is direct that $\Phi : R^2 \to R^n$ is a semismooth function.

We can get a semismooth equation based on the NCP function, the common method is smoothing NCP function, and finally, we can approximate the solution of NCP by solving the smoothing equation.

The following is the new smoothing approximation function according to min function:

$$\phi_\mu(a,b) = \begin{cases} a, & a \le b \\ b + \mu\left(1 - e^{-\frac{a-b}{\mu}}\right), & a > b \end{cases} \qquad (6)$$

where $\mu > 0$, $(a,b) \in R \times R$.

Proposition 2.1

For any $\mu > 0$ and $(a,b) \in R \times R$, $\phi_\mu(a,b)$ is continuously differentiable and a smoothing coincident approximation function of min function.
Proof: for any $\mu > 0$, $(a,b) \in R^2$

$$\partial_a \phi_\mu(a,b) = \begin{cases} 1, & a \le b \\ e^{-\frac{a-b}{\mu}}, & a > b \end{cases} \qquad (7)$$

$$\partial_b \phi_\mu(a,b) = \begin{cases} 0, & a \le b \\ 1 - e^{-\frac{a-b}{\mu}}, & a > b \end{cases} \qquad (8)$$

For $\forall \mu > 0$, when $a \le b$ $|\phi_\mu(a,b) - \phi_{\min}(a,b)| = 0$,

$a > b$ $|\phi_\mu(a,b) - \phi_{\min}(a,b)| = \mu\left(1 - e^{-\frac{a-b}{\mu}}\right) \le \mu$,

Then $\phi_\mu(a,b)$ is the smoothing coincident approximation function of min function. It is obvious that $\phi_\mu(a,b)$ is continuously differentiable.

Let $z := (\mu, x) \in R_+ \times R^n$ and $H(z) := \begin{pmatrix} \mu \\ \Phi(\mu, x) \end{pmatrix}$
$$\qquad (9)$$

there $\Phi(\mu, x) = \begin{pmatrix} \phi_\mu(x_1, F_1(x)) \\ \phi_\mu(x_2, F_2(x)) \\ \vdots \\ \phi_\mu(x_n, F_n(x)) \end{pmatrix} \qquad (10)$

since $H(z) = 0 \Leftrightarrow \mu = 0$ and x is the solution of NCP, $H(z) = 0$ is equivalent to $\|H(z)\|^2 = 0$, then defined the following value function:

$$\Psi(z) = \|H(z)\|^2 \qquad (11)$$

We can prove that $H : R^{n+1} \to R^{n+1}$ has the following property:

Proposition 2.2

1. Assume that $\mu \ne 0$, then H is continuously differentiable at $z := (\mu, x)$, and the *Jacobian* matrix has the following form:

$$H'(\mu, x) = \begin{pmatrix} 1 & 0 \\ C & D + EF'(x) \end{pmatrix} \qquad (12)$$

where $C \in R^n$, D and E are the diagonal matrix.
2. H is semismooth at R^{n+1}. If $F'(x)$ is *Lipschitz* continuously at R^n, then H is semismooth strongly at R^{n+1}.
3. $\nabla \Psi(z)$ is continuously differentiable at R^{n+1} and $\nabla \Psi(z) = 2H'(z)^T H(z)$.

Lemma 2.1 (Qi, 1993)

Suppose that $\varphi : R^n \to R^n$ is a locally Lipschitzian function and semismooth at x, then

a. $\forall V \in \partial \varphi(x+h)$, $h \to 0$,
   $Vh - \varphi'(x; h) = o(\|h\|)$.
b. $\forall h \to 0$, $\varphi(x+h) - \varphi(x) - \varphi'(x; h) = o(\|h\|)$.

Algorithm 2.1

Step 0: choose constants $\delta \in (0,1)$ and $\sigma \in (0,1)$, let $z^0 = (\mu^0, x^0) \in R_{++} \times R^n$ be an arbitrary point. Take $\gamma \in (0,1)$ such that $\gamma \mu^0 < \frac{1}{2}$, choose a sequence $\{\beta^k\}$ such that $\beta^k \in [0, \beta]$, where $\beta \in [0, 1 - \gamma \mu^0]$ is a constant, take $\rho_0 = \rho(z^0) := \gamma \min\{1, \psi(z^0)\}$, $C_0 := \Psi(z^0)$, $Q_0 = 1$, $\eta_{\min}$ and $\eta_{\max}$ is a constant satisfying $0 \le \eta_{\min} < \eta_{\max} < 1$, set $\eta_0 \in [\eta_{\min}, \eta_{\max}]$, take $k := 0$.
Step 1: if $\Psi(z^k) = 0$, stop.
Step 2: compute $\Delta z^k := (\Delta \mu^k, \Delta x^k)$ by

$$\begin{pmatrix} \mu^k \\ \Phi(\mu^k, x^k) \end{pmatrix} + \begin{pmatrix} 1 & 0 \\ C & D + EF'(x) \end{pmatrix}\begin{pmatrix} \Delta \mu^k \\ \Delta x^k \end{pmatrix} = \begin{pmatrix} \rho_k \mu^0 \\ \gamma^k \end{pmatrix} \qquad (13)$$

where $\gamma^k = \beta^k \Phi(x)$.
Step 3: let θ^k be the maximum of the values $1, \delta, \delta^2 \dots$, such that

$$\psi(z^k + \theta^k \Delta z^k) \le [1 - 2\sigma(1 - \gamma\mu^0 - \beta_k)\theta^k]C_k \qquad (14)$$

108

Step 4: set $z^{k+1} = z^k + \theta^k \Delta z^k$, $k = k+1$ and

$$\rho(z^k) := \min\{\gamma, \gamma\psi(z^k), \rho(z^{k-1})\} \qquad (15)$$

Step 5: choose $\eta_k \in [\eta_{\min}, \eta_{\max}]$, take

$$Q_k := \eta_{k-1}Q_{k-1} + 1, \ C_k = \frac{\eta_{k-1}Q_{k-1}C_{k-1} + \Psi(z^k)}{Q_k} \qquad (16)$$

Go back to step 1.

A similar algorithmic framework has been extensively studied by many researchers (see, for example, Huang, 2004; Qi, 2000). In Algorithm 2.1, we used a non-monotone line search technique introduced by Zhang and Hager (Zhang, 2004). If $\eta_k = 0$ for each k, then the line search is the usual monotone Armijo line search. Some basic results involving Algorithm 2.1 are included in the following remark.

Remark 2.1: let the sequence $\{C_k\}$ and $\{z_k\}$ be generated by Algorithm 2.1.

1. $C_{k+1} \leq C_k \ \forall k$.
   it follows from (7) and (5).
2. $\Psi(z^k) \leq C_k \ \forall k$
3. $\rho(z^{k+1}) \leq \rho(z^k) \ \forall k$.

which holds from the definition of $\rho(z^k)$.

4. $\mu^0 \rho(z^k) \leq \mu^k \ \forall k$.
5. $\mu_k > 0$, $\mu_{k+1} \leq \mu_k \ \forall k$

Pro. since

$$\mu_{k+1} = \mu_k + \alpha_k \Delta \mu_k = (1-\alpha_k)\mu_k + \alpha_k\mu^0\rho(z^k) > 0$$
$$\mu_k > 0$$

By above result (5)

$$\mu_{k+1} = \mu_k + \alpha_k\Delta\mu_k$$
$$= (1-\alpha_k)\mu_k + \alpha_k\mu^0\rho(z^k)$$
$$\leq (1-\alpha_k)\mu_k + \alpha_k\mu_k$$
$$= \mu_k$$

Therefore, the conclusion holds.
Theorem 2.1. Suppose that $F(x)$ is a continuously differentiable P_0 function. Then, Algorithm 2.1 is well defined.

3 CONVERGENCE ANALYSIS

In this section, we discuss the global and local superlinear convergence of Algorithm 2.1.

Theorem 3.1

Assume that F is a continuously differentiable P_0 function and the solution set of NCP (1) is nonempty and bounded, and let $\{z^k\}$ be gener-

ated by Algorithm 2.1, then $\lim_{k\to\infty} \rho(z^k) = 0$ and $\lim_{k\to\infty} \mu_k = 0$; every accumulation point of $\{z^k\}$ is a solution to the NCP (1).

Pro. by remark 2.1, we know consequence $\{C_k\}, \{\rho(z^k)\}$ and $\{\mu^k\}$ is convergent, so we can assume $\lim_{k\to\infty} C_k = \bar{C}$, $\lim_{k\to\infty} \rho_k = \bar{\rho}$ and $\lim_{k\to\infty} \mu_k = \bar{\mu}$. It is visible that $\bar{C} \geq 0$, $\bar{\rho} \geq 0$, and $\bar{\mu} \geq 0$. From remark 2.1, we can get

$$0 \leq \Psi(z^{k+1}) \leq C_{k+1} \leq C_k \leq C_0 \qquad (17)$$

Hence, consequence $\{\Psi(z^k)\}$ is bounded.

Assume that $\bar{\rho} \neq 0$, since consequence $\{\Psi(z^k)\}$ is bounded, it has a convergent subsequence, denoted by $\{\Psi(z^{k_n})\}$, defining $\bar{\Psi} := \lim_{k_n\to\infty} \to \Psi(z^{k_n})$, by the definition of $\rho(z^k)$ and $\bar{\rho} \neq 0$, we can get $\bar{\Psi} > 0$; moreover, based on (17), we know that $\bar{C} > 0$, $\bar{\mu} > 0$. By lemma 3.1, we obtain that the sequence $\{z^{k_n}\}$ is bounded. For subsequence, we can assume that $\lim_{k_n\to\infty} z^{k_n} = \bar{z}$; hence,

$$\lim_{k_n\to\infty} C_k = \bar{C}, \qquad \lim_{k_n\to\infty} \rho(z^{k_n}) = \bar{\rho}, \qquad \lim_{k_n\to\infty} \mu_{k_n} = \bar{\mu},$$
$$\lim_{k_n\to\infty} \Psi(z^{k_n}) = \bar{\Psi}.$$

For $\lim_{k\to\infty} \rho(z^k) = 0$ and $\lim_{k\to\infty} \mu_k = 0$, we divide the proof into two parts:

1. Assume that $\alpha_{k_n} \geq c > 0$, for $\forall k_n$, where c is a constant. In this case, we have

$$C_{k_n+1} \leq C_{k_n} - \frac{\sigma(1-\mu^0-\beta_{k_n})c}{Q_{k_n+1}} C_{k_n} \ \forall k_n, \lim_{k_n\to\infty} C_k = \bar{C},$$

we can get

$$\sum_{k_n=k_1}^{\infty} \frac{\sigma(1-\mu^0-\beta_{k_n})c}{Q_{k_n+1}} C_{k_n} < \infty.$$

Furthermore, by $\eta_{\max} \in [0,1]$ and the definition of Q_k, we have

$$Q_{k_n+1} = 1 + \sum_{i=0}^{k_n}\prod_{j=0}^{i}\eta_{k_n-j} \leq 1 + \sum_{i=0}^{k_n}\eta_{\max}^{i+1} \leq \sum_{i=0}^{\infty}\eta_{\max}$$
$$= \frac{1}{1-\eta_{\max}} \ \forall k_n$$

Therefore, we can get $\lim_{k_n\to\infty} C_{k_n} = 0$, which is in contradiction with $\bar{C} > 0$.

2. Assuming that $\lim_{k_n\to\infty} \alpha_{k_n} = 0$, the step-size $\bar{\alpha}_{k_n} = \alpha_{k_n}/\theta$ does not satisfy (14) for sufficiently large k_n, i.e. $\psi(z^{k_n} + \bar{\alpha}_{k_n}\Delta z^{k_n}) > [1-\sigma(1-\mu^0-\bar{\beta})\bar{\alpha}_{k_n}]C_{k_n}$ hold for sufficiently large k_n, since $\Psi(z^{k_n}) \leq C_{k_n}$, and the above inequality becomes

$$\left[\psi\left(z^{k_n}+\bar{\alpha}_{k_n}\Delta z^{k_n}\right)-\Psi\left(z^{k_n}\right)\right]/\bar{\alpha}_{k_n}$$
$$>-\sigma\left(1-\gamma\mu^0-\bar{\beta}\right)C_{k_n},\bar{u}>0,$$

and $\Psi\left(z^{k_n}\right)$ is continuously differentiable at $\bar{z}$. We have

$$-\sigma\left(1-\gamma\mu^0-\bar{\beta}\right)\bar{\Psi}=$$
$$-\sigma\left(1-\gamma\mu^0-\bar{\beta}\right)\bar{C}\leq\Psi'\left(\bar{z}\right)\Delta\bar{z}$$
$$=H\left(\bar{z}\right)^{\mathrm{T}}\left(-H\left(\bar{z}\right)+\binom{\rho\mu^0}{r}\right) \quad (18)$$
$$=-\Psi\left(\bar{z}\right)+\mu\rho\mu^0+\Phi\left(\mu,x\right)^{\mathrm{T}}r$$
$$\leq-\Psi\left(\bar{z}\right)+\gamma\mu^0\Psi\left(\bar{z}\right)+\bar{\beta}\Psi\left(\bar{z}\right)$$
$$=\left(-1+\gamma\mu^0+\bar{\beta}\right)\Psi\left(\bar{z}\right)$$

By the definition of ρ and β, we have $1-\gamma\mu^0-\bar{\beta}>0$, the equation is in contradiction with the known.

In above two cases, we obtain

$$\lim_{k\to\infty}\rho\left(z^k\right)=0,\lim_{k\to\infty}\mu_k=0.$$

Based on the definition of $\rho(z^k)$ and $\lim_{k\to\infty}\rho\left(z^k\right)=0$, we can get $\lim_{k_n\to\infty}\Psi\left(z^{k_n}\right)=0$; therefore, every accumulation point of $\left\{z^k\right\}$ is a solution to NCP (1).

Theorem 3.2

Assume that, for any $z\in\left(\mu,x\right)$, $H'(z)$ is nonsingular and $z^*=\left(\mu^*,x^*\right)$ is an accumulation point of the iteration sequence $\left\{z^k\right\}$ generated by Algorithm 2.1. If $H(z)$ is semismooth at z^*, then $\left\{z^k\right\}$ converges to z^* superlinearly.

Pro. since $H'(z)$ is nonsingular, then

$$\left\|H'\left(z^k\right)^{-1}\right\|\leq C.$$

$C>0$ is a constant, z^k converges to z^*, we get

$$\|z^k+\Delta z^k-z^*\|$$
$$=\left\|z^k+H'\left(z^k\right)^{-1}\left[-H\left(z^k\right)+\left(\rho\left(z^k\right)\mu^0,r^k\right)\right]-z^*\right\|$$
$$\leq\left\|H'\left(z^k\right)^{-1}\right\|\left\|H'\left(z^k\right)\left(z^k-z^*\right)-H\left(z^k\right)+\left(\rho\left(z^k\right)\mu^0,r^k\right)\right\|$$
$$\leq C\left(\left\|H'\left(z^k\right)\left(z^k-z^*\right)-H\left(z^k\right)\right\|+\left\|\left(\rho\left(z^k\right)\mu^0,r^k\right)\right\|\right)$$
$$\leq C\left(o\left(\|z^k-z^*\|\right)+\rho\left(z^k\right)\mu^0+\|r^k\|\right)$$
$$\leq C\left(o\left(\|z^k-z^*\|\right)+\mu^0\gamma\Psi\left(z^k\right)+\beta_k\left\|H\left(z^k\right)\right\|\right)$$
$$(19)$$

Since $H(z)$ is semismooth at z^* and locally *Lipschitz* continuous near z^* for all z^k sufficiently close to $z*$

$$\Psi\left(z^k\right)=\left\|H\left(z^k\right)\right\|^2=\left\|H\left(z^k\right)-H\left(z^*\right)\right\|^2$$
$$=O\left(\|z^k-z^*\|^2\right) \quad (20)$$

Furthermore, because $\beta_k\to0$ and from (20)

$$\beta_k\left\|H\left(z^k\right)\right\|=o\left(\|z^k-z^*\|\right) \quad (21)$$

Hence, from (19)–(21), we can get

$$\|z^k+\Delta z^k-z^*\|=o\left(\|z^k-z^*\|\right) \quad (22)$$

By the following proof of Theorem 3.1 of [12],

$$\|z^k-z^*\|=O\left(\left\|H\left(z^k\right)-H\left(z^*\right)\right\|\right)$$

hold for all z^k close z^*

$$\Psi\left(z^k+\Delta z^k\right)=\left\|H\left(z^k+\Delta z^k\right)\right\|^2$$
$$=O\left(\|z^k+\Delta z^k-z^*\|^2\right)$$
$$=o\left(\|z^k-z^*\|^k\right) \quad (23)$$
$$=o\left(\left\|H\left(z^k\right)-H\left(z^*\right)\right\|^2\right)$$
$$=o\left(\Psi\left(z^k\right)\right)$$

Hence, (23) implies that $\theta^k=1$ holds for all z^k close to z^*, i.e. $z^{k+1}=z^k+\Delta z^k$, it together with (22) proving $\left\{z^k\right\}$ converges to $\left\{z^*\right\}$ superlinearly.

The same can be obtained

$$\mu^{k+1}=\mu^k+\Delta\mu^k=\rho\left(z^k\right)\mu^0=\gamma\left\|H\left(z^k\right)\right\|^2\mu^0$$
$$\lim_{k\to\infty}\frac{\mu^{k+1}}{\mu^k}=\lim_{k\to\infty}\frac{\left\|H\left(z^k\right)\right\|^2}{\left\|H\left(z^{k-1}\right)\right\|^2}=0$$

this proves $\mu^{k+1}=o\left(\mu^k\right)$, thus we complete our proof.

4 NUMERICAL RESULT

In order to verify that the smoothing approximation function and the algorithm are effective, we will show the following numerical experiments.

Throughout the experiments, the parameters used in Algorithm 2.1 are as follows.

$$\delta=0.3,\ \sigma=0.1,\ \mu^0=0.1,\ \gamma=0.001,\ \beta_k=0.5^k.$$

We choose $k\leq2000$ and $\Psi(z)\leq10^{-12}$ as the termination conditions. We will regard the η_k as a fixed value, $\eta_k=0.5$.

Table 1.

x0	It	Val	x0	It	Val
[1,2,3,4]	11	1.3356e-16	[10,10,10,10]	8	2.0255e-17
[1,1,1,1]	9	2.6618e-16	[4,4,4,4]	9	5.9795e-15
[1,3,5,7]	9	1.6252e-14	rand(4,1)	7	2.5595e-17
[1,0,2,0]	8	6.1168e-17	50*rand(4,1)	20	2.3802e-24
[0.5,0.5,0.5,0.5]	7	1.8793e-13	[1,1,0,0]	7	3.0066e-14

Table 2.

	M1			M2		
n	It	val	cpu	It	val	cpu
100	7	1.7845e-15	0.452466	7	1.9149e-15	0.064856
200	7	2.2682e-15	0.321881	7	3.8492e-15	0.192716
400	7	3.2354e-15	1.235723	7	7.7177e-15	0.812828
800	7	5.1699e-15	8.202265	7	1.5454e-14	6.227407
1600	7	9.0388e-15	79.284173	7	3.0929e-14	52.825999

Example 1. (*Kojima–Shindo problem*) considering the following nonlinear complementarity problem, the test function is

$$F(x) = \begin{pmatrix} 3x_1^2 + 2x_1x_2 + 2x_2^2 + x_3 + 3x_4 - 6 \\ 2x_1^2 + x_1 + x_2^2 + 10x_3 + 2x_4 - 2 \\ 3x_1^2 + x_1x_2 + 2x_2^2 + 2x_3 + 9x_4 - 9 \\ x_1^2 + 3x_2^2 + 2x_3 + 3x_4 - 3 \end{pmatrix}$$

The above problem has a degenerate solution $x^* = (\sqrt{6}/2, 0, 0, 1/2)^T$ and a nondegenerate solution $x^{**} = (1, 0, 3, 0)^T$. For different initial points, we can get different iteration numbers. Numerical results are shown in Table 1.

Example 2. Considering the following two linear complementarity problems, $F(\cdot)$ are linear mapping, and $F(x) = Mx + q$, where $q = (-1, -1, ..., -1)^T$ M are two cases, respectively, $M1$ and M_2 are two different matrices:

$$M1 = \begin{pmatrix} 4 & -2 & 0 & \cdots & 0 & 0 \\ 1 & 4 & -2 & \cdots & 0 & 0 \\ \vdots & \vdots & \vdots & \vdots & \vdots & \vdots \\ 0 & 0 & 0 & \cdots & 4 & -2 \\ 0 & 0 & 0 & \cdots & 1 & 4 \end{pmatrix}$$

$$M2 = \begin{pmatrix} 1 & 2 & 2 & \cdots & 2 & 2 \\ 0 & 1 & 2 & \cdots & 2 & 2 \\ \vdots & \vdots & \vdots & \vdots & \vdots & \vdots \\ 0 & 0 & 0 & \cdots & 1 & 2 \\ 0 & 0 & 0 & \cdots & 0 & 1 \end{pmatrix}$$

We use $x0 = (1, 1, ..., 1)^T$ as the initial point and the dimensions were 100, 200, 400, 800, and 1600. Finally, the test results are given in Table 2.

5 CONCLUSIONS

In this paper, we present an inexact smoothing method with a new smooth approximation function. Solving NCP in our approach is feasible, and has a good value effect, and we establish a global convergence and superlinear convergence of the algorithm, and the algorithm is capable of solving large NCP and can have a very good value effect.

REFERENCES

Huang, Z.H. et al. 2004. Sub-quadratic convergence of a smoothing Newton algorithm for the P0-and monotone LCP, Math. Program. 99:423–441.
Ma, C.F. et al, 2008. A globally and superlinearly convergent smoothing Broyden-likemethod for solving nonlinear complementarity problem, Applied Mathematics and Computation 198:592–604.
Ma, C.F. & X.H. Chen, 2008. The convergence ofa one-step smoothing Newton method for P0-NCP base on a new smoothing NCP-function, Journal of Computation and Applied Mathematics 216(1):1–13.
Mangasarian, O.L, 1976. Equivalence of the complementarity problem to a system of nonlinear equations. SIAM J, Applied Mathematics, 31:89–92.
Qi, L. & G. Zhou, 2000. A smoothing Newton method for minimizing a sum of Euclidean norms, SIAM J. Optimize. 11:389–410.
Qi, L. & J. Sun, 1993. A nonsmooth version of Newton's method, Mathematical Programming 58(3):353–367.
Qi, L. 1993.Convergence analysis of some algorithms for solving nonsmooth equations, Mathematics of Operations Research 18(1):227–244.
Qi, L. & D. Sun, 2000. Improving the convergence of non-interior point algorithms for nonlinear complementarity problems, Mathematics of Computation 69 (229):283–304.
Tseng, P. 2000. Error bounds and superlinear convergence analysis of some Newton-type methods in optimization, in: G. Di Pillo, F. Giannessi (Eds.), Nonlinear Optimization and Related Topics, Kluwer Academic Publishers, Boston, pp. 445–462.
Yu, Z.S. & Y. Qin, 2011.A cosh-based smoothing Newton method for P0 nonlinear complementarity problem, Real World Applications, 12:875–884.
Zhang, H.-C. et al. 2004 Anonmonotone line search technique and its application to unconstrained optimization, SIAM. J. Optimiz. 14:1043–1056.
Zhang, J. & K.C. Zhang, 2009. A variant smoothing Newton method for P0-NCP based on anew smoothing function, Journal of Computation and Applied Mathematics 225(1): 1–8.

Advances in Materials Science, Energy Technology and Environmental Engineering – Patty & Zhou (Eds)
© 2017 Taylor & Francis Group, London, ISBN 978-1-138-19668-1

Research on measuring method for the groove parameters of the shaft washer of steering bearing

Zhenzhi He, Jinhe Wu & Xiangning Lu
School of Mechanical and Electrical Engineering, Jiangsu Normal University, Xuzhou, China

ABSTRACT: An automatic measurement system for the groove parameters of shaft washer is described in this paper. It consists of two parts which are the mechanical system and the controlling system. Three inductive sensors are used to obtain the displacements of groove profile along the horizontal and vertical directions respectively on both side of groove profiles on the cross section. And the least square curve fitting method is used to determine the arc parameters of groove profile, including arc radius, center distance and the height of center. The controlling system consists of the motion system controlled by SCM and the measurement system to detect and data processing. The experimental results show that the system meets the requirements of measurement accuracy within ±10 μm.

1 INTRODUCTION

Steering bearing, as a standard part, has the advantages of high interchangeability, low friction and simple lubrication, easy to replace and so on. It is an indispensable part of the modern mechanical structure. The performance, fatigue life and reliability of the steering bearing depend on the design, manufacture and testing of the bearing, and the detection is one of most important factors in improving the performance of the bearing (Rowlands et al. 2002). The precision of the steering bearing lies on the machining precision of the groove size, which directly affects the precision of the installation and operation.

Groove parameter Measurement is an important part of the controlling of the shaft washer size, including the circular arc radius, center distance and the height of center. The shaft washer and the housing washer of steering bearing are designed to be two separated parts. In addition, the cross section of the shaft washer is a small circle arc, and the bottom of groove does not exist on the shaft washer, which makes it difficult to measure the ditch radius.

There are some kinds of measuring methods for the groove parameters of shaft washer. And the most popular one in plant is contrastive measurement by using the shape template matching measurement. However this artificial measurement method will cause many errors, which is related to the degree of proficiency of the operator. Especially for high precision measurement of groove parameters, template control method cannot be achieved.

The second method is using profiler to measure the profiles of groove. The bearing should be fixed by the dedicated V-block on the profiler when measuring the groove parameter. However, the positioning precision and clamping position of workpiece are hardly controlled, which can lead to the deviation of measurement results. In addition, the measuring head of profiler needs to go through the entire shaft washer section at a low speed when measuring the center distance. So the measurement efficiency is not high because of the big moving range of measuring head together with the lower moving speed, hence it is suitable for sampling detection.

Another method is to use the coordinate measuring machine, also called CMM, which is a modern measurement technology with high-precision. The spatial position of the probe can be accurately measured and displayed, however this kind of method still takes a long time to measure the groove parameters of shaft washer.

In order to solve the problem of the above measurement methods, an automatic measuring system was described in this paper for the groove parameters of shaft washer of sheering bearing, including the circular arc radius, center distance and the height of center.

2 THE PRINCIPLE OF GROOVE PARAMETER MEASUREMENT

The key of the groove parameter measurement is to measure the circular arc center and its radius. According to the principle of three points could define a circle, the circular arc center and radius can be determined when any of three points on the arc are known. Therefore, two displacement sensors are used to measure the coordinate values of horizontal and vertical directions of points A_1, B_1 and C_1 on the groove surface of shaft washer, which can be seen in Figure 1. Then the groove arc equation can be obtained to determine its center coordinates, radius of curvature and other information.

In fact, the arc obtained according to the three measured points would be different from the actual one because of the deviation of bearing processing, measurement and other factors. Thereby, in order to improve the measurement accuracy, the curve fitting methods was introduced in this paper to get the arc parameters, which is based on the measured multi-point coordinates.

Circle equation could be written as the following:

$$x^2 + y^2 + Ax + By + C = 0 \tag{1}$$

The center coordinates a and b can be provided next:

$$a = -A/2, b = -B/2 \tag{2}$$

And the fitting radius R can be obtained as:

$$R = \sqrt{a^2 + b^2 - C} \tag{3}$$

When measuring more than three points, all of the coordinate of measured point can't satisfy the equation (1). Therefore, there is the residual error that the measured point deviated from the true value, which can be defined as:

$$e_i = x_i^2 + y_i^2 + Ax_i + By_i + C \tag{4}$$

According to the principle of least squares fitting, the accumulated error function can be obtained as follows:

$$\sum_{i=1}^{n} e_i^2 = \sum_{i=1}^{n} (x_i^2 + y_i^2 + Ax_i + By_i + C)^2 = f(A,B,C) \tag{5}$$

The least squares estimated values of A, B and C are the ones which make the equation (5) minimum. The equation of extreme value condition of equation (5) can be provided:

$$\begin{cases} \dfrac{\partial f(A,B,C)}{\partial A} = 0 \\ \dfrac{\partial f(A,B,C)}{\partial B} = 0 \\ \dfrac{\partial f(A,B,C)}{\partial C} = 0 \end{cases} \tag{6}$$

Equation (6) can also be described as the following form:

$$\begin{bmatrix} A \\ B \\ C \end{bmatrix} = \begin{bmatrix} \sum_{i=1}^{n} x_i^2 & \sum_{i=1}^{n} x_i y_i & \sum_{i=1}^{n} x_i \\ \sum_{i=1}^{n} x_i y_i & \sum_{i=1}^{n} y_i^2 & \sum_{i=1}^{n} y_i \\ \sum_{i=1}^{n} x_i & \sum_{i=1}^{n} y_i & n \end{bmatrix}^{-1} \cdot \begin{bmatrix} -\sum_{i=1}^{n} (x_i^3 + y_i^2 x_i) \\ -\sum_{i=1}^{n} (y_i^3 + x_i^2 y_i) \\ -\sum_{i=1}^{n} (x_i^2 + y_i^2) \end{bmatrix} \tag{7}$$

The parameters of A, B and C could be obtained by solving equation (7) when the coordinate x_i and y_i are known. Then the fitting formula of circle could be obtained to meet the conditions that the sum of square of the error between the measurement point and the circle fitted equations is minimum (Guest 2012; Zhu et al. 2009; Shao & Wang 2013; Gavin 2011). The other side of the groove parameters of bearing could be obtained by using the same method which contains the circle center coordinate and radius of curvature. And then center distance and the height of center can be calculated using the above values. The measurement principle of groove parameters of the shaft washer is presented in Figure 2.

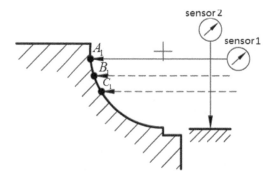

Figure 1. Arc parameters measuring principle.

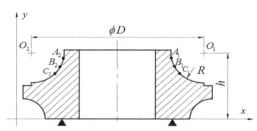

Figure 2. Measurement principle of the groove parameters of the shaft washer.

3 THE STRUCTURE OF MECHANICAL SYSTEM

The schematic diagram of the mechanical system for measuring groove parameters is shown in Figure 3 according to the measurement principle mentioned above. The guide rod and the positioning head are introduced in the mechanical system to position the workpiece. And the sensor system is driving up and down to measure the groove displacement through the stepper motor and the ball screw. The measurement procedure is as follows: first of all, the shaft washer to be measured is put on the test platform through the feeding mechanism and ready to test. At the Beginning of measurement, the stepper motor is controlled by Single Chip Microcomputer (SCM) at the high speed to drives the ball screw rotates and the sensor bracket falling. The positioning head on the guide rod centering the shaft washer by the centering cone when the positioning head is contact with the shaft washer during the falling of the bracket. And then the shaft washer is pressed tightly to ensure the position of the shaft washer unchanged during the measurement.

And then the motor rotates at a lower speed and drives the horizontal direction sensor contacts to the groove, and it starts to record the data of the groove surface. At the same time, the vertical displacement sensors obtain information in the vertical direction. The motor drives the ball screw to start reversing and lift the sensor bracket and makes the sensor out of the workpiece when the motor is moving to the lower limit position. Finally, the collected data could be analyzed and processed in IPC to obtain the groove parameters of the shaft washer.

4 HARDWARE STRUCTURE OF MEASUREMENT SYSTEM

The measurement system described in this paper consists of two parts which are the motion control

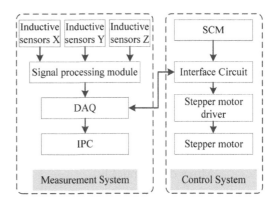

Figure 4. Hardware structure of groove parameters measurement system.

system and the groove parameters measurement system (Yang et al. 2014). The hardware structure of this system is shown in Figure 4 which consists of SCM, IPC, inductive sensors, stepper motor, stepper motor driver and other components. In the motion control system, the stepper motor is mainly controlled by SCM to change the rotation direction and speed, which could control the sensor's movement.

The inductive sensors are adopted in the groove parameters measurement system. When the sensor begin to contact with the surface of groove, the displacement of measuring point will be detected, which can be defined as x. Then, the displacement x is converted to the corresponding electric signal $e(x)$. When calculate processing is completed, the proportional to the displacement of the voltage signal $V(x)$ would be outputted. Then the voltage signals $V(x)$ are converted into the digital ones through the data acquisition card. The IPC will use these data for analysis, storage, and display. Simultaneously data exchange is carried out through the digital outputs of data acquisition card with SCM.

5 SOFTWARE STRUCTURE OF MEASUREMENT SYSTEM

The detection software can be divided into two parts which contain the motion controlling and parameter measurement. The motion control flow of measurement system is shown in Figure 5.

When the workpiece is in place and begin to be measured, the rotation direction and speed of the stepper motor are controlled by SCM to realize the positioning of the workpiece, the movement of sensors, and send the sampling inform signal to the host computer.

The groove parameters measurement system is mainly composed of six parts, as shown in Fig-

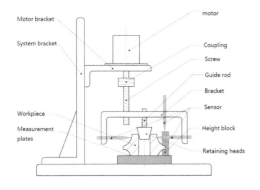

Figure 3. Schematic diagram of the mechanical system.

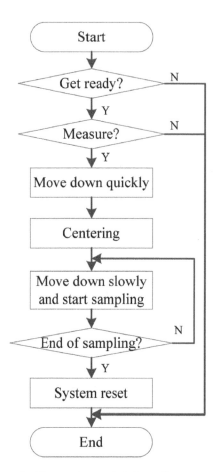

Figure 5. The motion control flow of measurement system.

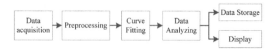

Figure 6. The structure of measurement system.

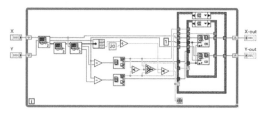

Figure 7. Preprocessing block diagram.

According to Equation (7), we should use the least square fitting method to deal with the measured coordinates, and the value of parameters A, B, C could be obtained. Then the groove parameters including the center coordinates and radius of the circular arc could be calculated according to Equations (2) and (3).

ure 6, which are the data acquisition, preprocessing, least squares curve fitting, data analysis and processing, data storage and display respectively (Ruixia 2002; Chen 2014).

1. Preprocessing. Firstly, through the data fitting method, we can get the groove parameters of center coordinates and radius values according to the collected coordinates, then the measured points could be checked by the obtained groove parameters. Therefore the measurement points with larger error would be removed. And then the remaining measuring points are used to curve fitting by the least square method. So we can get the more accurate fitted curves, which can be used to calculate the best fit value of center coordinates and radius of the groove (Ji-rong 2004; Wen et al. 2012). The block diagram of preprocessing procedure is shown in Figure 7.

2. Least square curve fitting. The arc parametric curve fitting program is shown in Figure 8.

6 EXPERIMENTAL RESULTS

The accuracy of the system is detected by taking the shaft washer of steering bearing in type 91683/22 for example. Firstly, the groove parameters was measured using profiler with the following results, the circular arc radius R is 3.320 μm, the center distance D is 33.410 μm, and the height of center H is 7.010 μm, and they are set as the true values. The shaft washer was repeatedly measured 10 times in the system, and the deviations between tested results and the true values are shown in Figure 9.

The standard deviation of single measurement in measuring list can be calculated from the results as follows:

$$\sigma = \sqrt{\frac{\sum_{i=1}^{n}\left(x_i - \overline{X}\right)^2}{n-1}} \tag{8}$$

By taking a significant degree $\alpha = 0.05$, the limit error of single measurement would be as follows:

$$\delta_l = \pm t_\alpha \sigma = \pm 2.09 \sqrt{\frac{\sum_{i=1}^{n}\left(x_i - \overline{X}\right)^2}{n-1}} \tag{9}$$

According to the results, it shows that the system meets the requirements of measurement accuracy within ±10 μm.

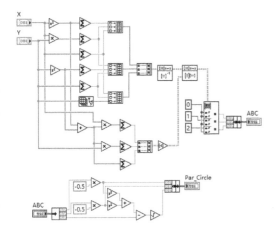

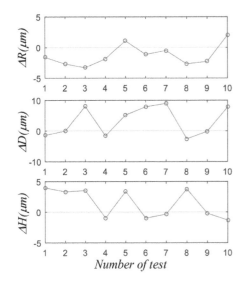

Figure 8.   Arc parametric curve fitting program.

Figure 9.   Deviations between tested results and the true values.

Table 1.   The analysis results of the three parameters according to the above method.

	Average of deviations	Standard deviation	Limit error
	(μm)	(μm)	(μm)
Circular arc radius R	−1.3	1.72	±3.61
Center distance D	−3.2	4.76	±9.96
Height of center H	−1.4	2.31	±4.83

7   CONCLUSION

This article describes a method for measuring the groove parameters of shaft washer of steering bearing. The stepper motor is controlled by SCM, and the sensors are drove up and down to obtain the displacement of multi-point on the circle surface of the groove. The least squares curve fitting method was used to analyze the groove parameters and then the circular arc radius, center distance and height of center can be obtained. The measurement accuracy is within ±10 μm.

ACKNOWLEDGMENTS

This research work is supported by the National Natural Science Foundation of China (Grant No. 51505201), the Jiangsu Normal University Foundation (Grant No.14XLR023).

REFERENCES

Chen, T.P. (2014). Inspection method of outer ring raceway precision of double row angular contact ball bearings for ball screw supporting. *Mechanical & Electrical Engineering Technology*, 43(8), 52–54.

Gavin, H. (2011). The Levenberg-Marquardt method for nonlinear least squares curve-fitting problems. Department of Civil and Environmental Engineering, Duke University,1–15.

Guest, P.G. (2012). Numerical methods of curve fitting. Cambridge University Press, 10, 85–90.

Ji-rong, L.I. (2004). Research on the Application of LabVIEW in Curve Fitting. Journal of Wuyi University (Natural Science Edition), 18(3), 57–63.

Rowlands, D.D., Ray, R.D., Chinn, D.S., & Lemoine, F.G. (2002). Short-arc analysis of intersatellite tracking data in a gravity mapping mission. Journal of Geodesy, 76(6–7), 307–316.

Ruixia, L.W.C. (2002). A New Method of Measuring Inner Radius of Non-full-round Arcs. Tool Engineering, 5(36), 32–33.

Shao, W.G., & Wang, X. (2013). Research on Non-Contact Coordinate Measuring Short-Arc Method. Measurement & Control Technology, 6(32), 140–147.

Wen, H., Ma, L., Zhang, M., & Ma, G. (2012, June). The comparison research of nonlinear curve fitting in Matlab and LabVIEW. In Electrical & Electronics Engineering (EEESYM), 74–77.

Yang, S., Wei, C., Zhang, Y., & Zhang, D. (2014). Measuring scale for ring raceway central diameter of extra large size thrust ball bearings. Bearing, 4, 54–55.

Zhu, J., Li, X.F., Tan, W.B., Xiang, H.B., & Chen, C. (2009). Measurement of short arc based on center constraint least-square circle fitting. Optics and Precision Engineering, 17(10), 2486–2493.

*Advances in Materials Science, Energy Technology
and Environmental Engineering – Patty & Zhou (Eds)
© 2017 Taylor & Francis Group, London, ISBN 978-1-138-19668-1*

A self-adaptive density re-initialization method for SPH techniques

Song Chai
*State Key Laboratory of Ocean Engineering, School of Naval Architecture, Ocean and Civil Engineering,
Shanghai Jiao Tong University, Shanghai, China
Wison (Nantong) Heavy Industry Co. Ltd., Nantong, Jiangsu, China*

Guoping Miao & Yuan Yang
*State Key Laboratory of Ocean Engineering, School of Naval Architecture, Ocean and Civil Engineering,
Shanghai Jiao Tong University, Shanghai, China*

ABSTRACT: In this work, the density re-initialization scheme in Smoothed Particle Hydrodynamics (SPH) techniques is examined. An improved density re-initialization scheme is proposed, in which a filtering threshold for evaluating the oscillation of density field is introduced; and density re-initialization will be performed once the threshold is exceeded. Numerical test on the benchmark case of 2D dam break illustrates that the re-initialization frequency can be self-adjusted according to density field oscillation during simulation, and density oscillation can be reduced. The test result also illustrates that the re-initialization process actually produces additional disturbance to the density field.

1 INTRODUCTION

Originally suggested by (Lucy 1977; Gingold & Monaghan 1977) in an astrophysical context, Smoothed Particle Hydrodynamics (SPH) is a completely mesh-free, fully conservative hydrodynamics method. By now, SPH has spread far beyond its original scope and also found wide range of applications in engineering.

The numerical results from SPH simulations are generally realistic, however large pressure oscillations may happen in the pressure field of the particles, especially for the nonlinear problem in hydrodynamics. Several approaches have been developed to overcome this problem including correcting the kernel (Bonet & Lok 1999) and integrating an incompressible solver. One of the most straight forward and computationally least expensive is to perform a filter over the density of the particles and re-assign a density to each particle (Colagrossi et al. 2003).

In practice such filter is normally performed at fixed step intervals, however, as the oscillation of density field is non-linear, such method may fail to perform density re-initialization at the right time. Till now, no evidence shows that the fix-step scheme can prevent the propagation of the density oscillation in simulation. The major challenge in this respect is to design a filtering scheme which can perform the density re-initialization according to the oscillation of density field.

2 SPH METHODOLOGY

The main features of the SPH method are described in detail in the following reviews (Monaghan 2005; Rosswog 2009; Liu 2010; Springel 2010; Monaghan 2012; Price 2012). At the heart of SPH is the smooth representation of any function by using an interpolating kernel, W. Within the kernel definition, h is the smoothing length, controls the size of the area where the contribution from the rest of the particles must be estimated. In addition, the kernel function should meet several requirements including positivity, compact support and normalization. Detailed discussion about the kernel function can be found in the literature (Benz 1990; Monaghan 2005; Liu 2010). The two-dimensional basic conservation equations can be represented in SPH notation following (Monaghan 2005), while particles were moved by using the so-called XSPH (Monaghan 1989).

2.1 Kernel function

The so-called Wendland functions (Wendland 1995) are used as kernel function here, as these kernels are not prone to the pairing instability which has been discussed in detail in (Dehnen et al. 2012). In the following numerical tests the C2 smooth in 2-dimension is applied

$$W(r,h) = \frac{7}{4\pi h^2}\left(1-\frac{q}{2}\right)^4 (1+2q) \quad 0 \le q \le 2 \qquad (1)$$

where $q = r / h$, r = distance between pairing particles, h = smoothing length.

2.2 Time step criteria

A variable time step scheme is used according to (Monaghan & Kos 1999):

$$\delta t = 0.3 \min \left(\min_a \left(\sqrt{\frac{h}{|f_a|}} \right), \min_a \left(\frac{h}{c_S + \max_b \left| \frac{h \vec{v}_{ab} \vec{r}_{ab}}{r_{ab}^{-2}} \right|} \right) \right) \tag{2}$$

where $|f_a|$ = force per unit mass, c_s = speed of sound.

2.3 Equation of state

By using an equation of state, the incompressibility condition is approximated with the density variation $\Delta\rho < 0.01\rho$. Following (Monaghan & Kos 1999, Batchelor 2000), the relation between pressure and density follows the expression

$$P = B \left[\left(\frac{\rho}{\rho_r} \right)^\gamma - 1 \right]$$

$$B = c_0^2 \rho_r / \gamma \tag{3}$$

$$c_0 = \sqrt{(\partial P / \partial \rho)} \Big|_{\rho_r}$$

where $\gamma = 7$; ρ_r = reference density; c_0 = speed of sound at the reference density.

2.4 Density correction

The Moving Least Squares (MLS) approach, developed by (Dilts 1999) and applied successfully by (Colagrossi et al. 2003; Panizzo 2004), is applied. The density field will be corrected by

$$\rho_a^{new} = \sum_b \rho_b \tilde{W}ab \frac{m_b}{\rho_b} = \sum_b m_b \tilde{W}ab \tag{4}$$

where, the detailed correction kernel W_{ab} can be found in the work of (Colagrossi et al. 2003).

3 SELF-ADAPTIVE DENSITY RE-INITIALIZATION

In incompressible fluid problems, the fluid density will stay unchanged to its theoretical value. The oscillation can be can be measured as the mean value of density deviation of all fluid particles, following

$$\varepsilon = \left(\sum_{i=1}^N |\rho_i - \rho_0| / N \right) \tag{5}$$

where N = number of fluid particles; ρ_i = density of the ith fluid particle; ρ_0 = density of the fluid in research (for water, ρ_0 = 1000 kg/m^3). In practice, ε is calculated in every time step as $\varepsilon(it)$, where it = time step.

The basic concept is to perform the density re-initialization when significant density oscillation appeared. So it is critical to find an appropriate way to evaluate significance of the oscillation. For this purpose, a threshold is introduced, as

$$T(it) = 0.01\rho_0 \varepsilon(it) \tag{6}$$

$T(it)$ will be updated in every time step to obtain the current density field oscillation.

In each time step, each particle's density ρ_i will be compared with the threshold $T(it)$. If $|\rho_i - \rho_0| > T(it)$, the density re-initialization will be performed; if not, the re-initialization will be skipped.

4 BENCHMARK TEST

4.1 2D Dam break on wet bed

A dam break on a wet bed corresponding to the experiment by (Janosi et al. 2004) is considered. Figure 1 shows a schematic sketch of the experimental setup.

For comparison, fix-step scheme, $T(it) = 5‰ \times \rho_0$ and $T(it) = 10‰ \times \rho_0$ will also be tested respectively as Case 1 (fix-step scheme), Case 2 ($T(it) = 5‰ \times \rho_0$) and Case 3 ($T(it) = 10‰ \times \rho_0$), while Case 4 is for self-adaptive scheme. Each numerical experiment is performed for 0.6 seconds.

4.2 Results and discussion

32786 fluid particles in total are generated, however, as the density of particle (ID > 20000) almost stays unchanged during the simulation, only the particle (ID < = 20000) are plotted for better readability. The results of particle density at 0.1 s, 0.3 s and 0.5 s are presented, as shown in Figures 2–4. (Vertical axis presents density, while horizontal axis presents particle ID).

As the particles are generated respectively for the upstream fluid and downstream fluid isolated

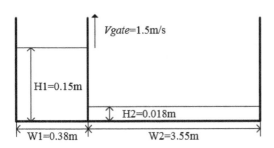

Figure 1. Sketch of experiment setup.

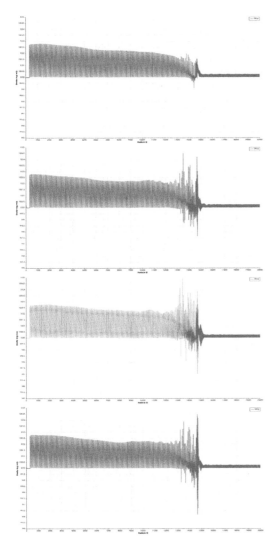

Figure 2. Particle density at $t = 0.1$s (case of fix-step scheme, $T(it) = 5\% \times \rho_0$, $T(it) = 10\% \times \rho_0$, and self-adaptive scheme are shown from top to bottom respectively).

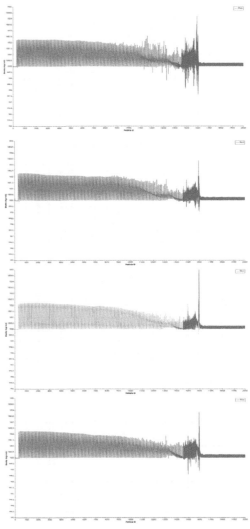

Figure 3. Particle density at $t = 0.3s$ (case of fix-step scheme, $T(it) = 5\% \times \rho_0$, $T(it) = 10\% \times \rho_0$, and self-adaptive scheme are shown from top to bottom respectively).

by gate, the initial particle density oscillation follows the particle configuration to some extent.

At $t = 0.1$ s (see Fig. 2), the fix-step scheme generates the least density oscillation, while the other three Cases show almost the same trend that the most significant density oscillation happens with the particles (13000 < ID < 15000). The density re-initialization can reduce the overall density oscillation, but it is also an disturbance to the density field. In the beginning of the simulation, the frequency of initialization performed by fix-step scheme is the smallest. As a result, the oscillation is much smaller than the other Cases.

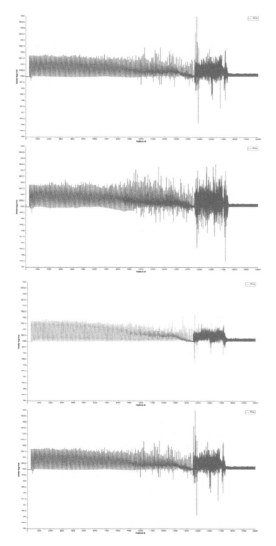

Figure 4. Particle density at $t = 0.5s$ (case of fix-step scheme, $T(it) = 5‰ \times \rho_0$, $T(it) = 10‰ \times \rho_0$, and self-adaptive scheme are shown from top to bottom respectively).

At $t = 0.3$ s (see Fig. 3), the significance of density oscillation generated by fix-step scheme increase compared with which at $t = 0.1$ s, while the other three Cases show a decreasing oscillation. Also, it can be found from Figure 3 that the oscillation of Case 2 and Case 4 is slightlly smaller than which of Case 3. This is in accordance with the general condition that the more re-initialization performed the smaller the oscillation will be.

At $t = 0.5$ s (see Fig. 4), the significance of density oscillation generated by fix-step scheme keep increasing. All the four Cases show almost the same trend. However, compared with others,

the oscillation of Case 3 is much smaller. Further more, the oscillation of Case 4 is smaller than that of Case 1 & 2. Considering the re-initialization frequency performed by Case 3 is the least during this period (see Fig. 5), it can be seen as a clear evidance that the density re-initialization can disturb the density field.

As the re-initialization frequency of fix-step scheme is very predictable, only the other three Cases' frequency are presented in Figure 5. The time steps performed re-initialization are also calculated in Table 1.

From Figure 5 and Table 1, Case 2 & 3 can be seen as a cut off solution, which should be evaluated and defined before the simulation. As the threshold is critical to achieve the best density re-initialization result, it may be a little difficult to decide such value before each simulation.

From the above analysis, the self-adaptive scheme (Case 4) shows a promising result for that the density field oscillation is relatively smaller considering the re-initialization frequency and its own disturbance, and that no parameters need to be pre-defined as this scheme can generate the threshold according to the density oscillation.

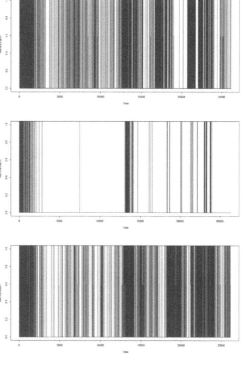

Figure 5. Re-initialization frequency (case of $T(it) = 5‰ \times \rho_0$, $T(it) = 10‰ \times \rho_0$ and self-adaptive scheme are shown from top to bottom respectively).

Table 1. Time steps that re-initialization performed.

	Case 1	Case 2	Case 3	Case 4
Re-initialization	872	1929	292	2239
% by total	2.7%	5.9%	0.9%	6.8%

5 CONCLUSION

A new method to perform re-initialization on density field in SPH simulation is proposed in this paper. At the heart of this method, a threshold is introduced to decide whether the re-initialization should be performed or not. This so-called self-adaptive method is tested with numerical simulation of two-dimension dam break problem. The numerical results proved that: with this new method, the density oscillation can be significantly reduced; the self-adjusting frequently can reflect the density oscillation trend.

In addition, it is proved in this paper that the density re-initialization introduces disturbance to the density field, which however is not taken into consideration in this proposed method. More efficient scheme may be developed in future research.

REFERENCES

Batchelor, G.K. 2000. *An introduction to fluid dynamics.* Cambridge: Cambridge university press.

Benz, W. 1990. Smooth particle hydrodynamics: a review. *The Numerical Modelling of Nonlinear Stellar Pulsations; Volume 302 of the series NATO ASI Series:* 269–288. Dordrecht: Springer Netherlands.

Bonet, J. & Lok, T.S.L. 1999. Variational and momentum preservation aspects of Smooth Particle Hydrodynamic formulations. *Review of. Computer Methods in Applied Mechanics and Engineering* 180 (1–2): 97–115.

Colagrossi, et al. 2003. Numerical simulation of interfacial flows by smoothed particle hydrodynamics. *Review of. Journal of Computational Physics* 191 (2): 448–75.

Dehnen, et al. 2012. Improving convergence in smoothed particle hydrodynamics simulations without pairing instability. *Review of Monthly Notices of the Royal Astronomical Society* 425 (2): 1068–82.

Dilts, G.A. 1999. Moving least squares particle hydrodynamics I. Consistency and stability. *Review of International Journal for Numerical Methods in Engineering* 44 (8): 1115–55.

Gingold, R.A. & Monaghan, J.J. 1977. Smoothed particle hydrodynamics: theory and application to non-spherical stars. *Review of. Monthly Notices of the Royal Astronomical Society* 181 (3): 375–89.

Janosi, I.M., et al. 2004. Turbulent drag reduction in dambreak flows. *Experiments in Fluids* 37: 219–229.

Liu G.R. 2010. *Meshfree methods: moving beyond the finite element method.* Boca Raton: CRC press.

Lucy, L. 1977. A numerical approach to the testing of the fission hypothesis. *Review of. The astronomical journal* 82:1013–24.

Monaghan, J.J. 2005. Smoothed particle hydrodynamics. *Review of. Reports on Progress in Physics* 68 (8): 1703.

Monaghan, J.J. 2012. Smoothed Particle Hydrodynamics and Its Diverse Applications. *Review of. Annual Review of Fluid Mechanics* 44 (1): 323–346.

Monaghan, J.J. 1989. On the problem of penetration in particle methods. *Review of. Journal of Computational Physics* 82 (1): 1–15.

Monaghan, J.J, & Kos, A. 1999. Solitary waves on a Cretan beach. *Review of. Journal of waterway, port, coastal, and ocean engineering* 125 (3): 145–55.

Panizzo, A. 2004. Physical and numerical modelling of subaerial landslide generated waves. PhD Thesis, University of L'Aquila.

Price, D.J. 2012. Smoothed particle hydrodynamics and magnetohydrodynamics. *Review of. Journal of Computational Physics* 231 (3): 759–94.

Rosswog, S. 2009. Astrophysical smooth particle hydrodynamics. *Review of. New Astronomy Reviews* 53 (4–6): 78–104.

Springel, V. 2010. Smoothed Particle Hydrodynamics in Astrophysics. *Review of. Annual review of astronomy and astrophysics* 48 (1): 391–430.

Wendland, H. 1995. Piecewise polynomial, positive definite and compactly supported radial functions of minimal degree. *Review of. Advances in computational Mathematics* 4(1): 389–96.

Advances in Materials Science, Energy Technology
and Environmental Engineering – Patty & Zhou (Eds)
© 2017 Taylor & Francis Group, London, ISBN 978-1-138-19668-1

Design and analysis of an EDM machine tool

Y. Li & S.F. Wang
Beijing Institute of Electro-Machining, Beijing, China
Beijing Key Laboratory of Electrical Discharge Machining Technology, Beijing, China

ABSTRACT: In order to improve the automation level of electrical discharge machining (EDM), an EDM machine tool with a lifting work tank was designed. The 3D model of the EDM machine tool was established using the SOLIDWORKS software. Then, the finite element model was constructed using the ANSYS Workbench software. Static analysis was performed to evaluate the rationality of the structure design. The analysis results show that the deformations of the bed, middle saddle, upper saddle, and worktable are relatively smaller, while the column, head, and connecting shaft are the main deformation parts. The design and analysis results provide an effective basis for the subsequent optimization of the EMD machine tool.

1 INTRODUCTION

EDM is one of the most widely used non-conventional material removal processes. It can be used to generate 3D complex-shaped features and components of difficult-to-machine materials (Rajurkar et al. 2013). With the development of high-precision products, diversified materials, and complicated machining, there is a growing need for an automotive EDM machine tool. The EDM process is based on removing the material from a part by means of a series of repeated electrical discharges between the tool called electrode and workpiece in some medium (Abbas et al. 2007). A dielectric medium for EDM is used to provide suitable electric discharge conditions, cool the electrodes, and take debris away. Comparing with open-door work tanks, lifting work tanks have more advantages, such as convenient workpiece installation and high automation level.

Finite element analysis has been an effective tool for structure design. The deformation of the structure can be obtained by using finite element analysis before completing the prototype. Sun et al. (2015) used finite element analysis to investigate the static and dynamic characteristics of the key vertical parts of a large-scale ultra-precision optical aspherical machine. Sun et al. (2015) used FEM to study the structural seakeeping stability of the marine compound NC machine tool in the ship environment. Liu et al. (2015) analyzed the dynamic and static performance of the turning and milling units with a finite element method to ensure the high accuracy of a high-speed machine tool. Dong et al. (2012) established the finite element analysis model of the three-degrees-of-freedom

parallel machine tool, carried out modal analysis of the whole machine, and found out the weak link of the machine. Cho et al. (2011) carried out finite element analyses and vibration tests to determine the specifications of the composite materials of a tabletop machine tool. Chen et al. (2012) analyzed the dynamic characteristics of the spindle using the ANSYS software, and compared the simulation result with the measurement result. In this study, an EDM machine tool with a lifting work tank is designed. The 3D finite element model is constructed by using the ANSYS Workbench software. Static analysis is performed to estimate deformations of the machine tool. The results will provide valuable design basis for the new EDM machine tool.

2 STRUCTURE DESIGN

The EDM machine tool belongs to C type. The schematic diagram of the EDM machine tool is shown in Figure 1. It is made up of several parts: bed, column, middle saddle, upper saddle, head, etc. The bed supports other parts of the EDM machine tool. The column is fixed on the bed. The head is installed on the front of the column. The head accomplishes the function of moving up and down. The middle saddle can move left and right along the bed. The upper saddle can move forward and back along the middle saddle. The work tank is installed on the upper saddle. The height of the work tank can be adjusted. The worktable is fixed on the upper saddle. There are several pairs of guide rails and sliders, which connect parts of the EDM machine tool. By using the SOLIDWORKS

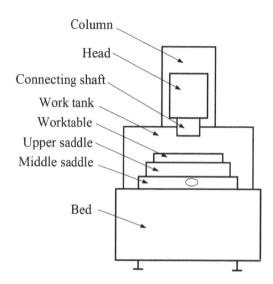

Column

Head

Connecting shaft

Work tank

Worktable

Upper saddle

Middle saddle

Bed

Figure 1. Schematic diagram of the EDM machine tool.

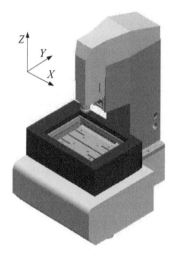

Figure 2. 3D model of the EDM machine tool.

software, a 3D model of the EDM machine tool is set up as shown in Figure 2.

3 CONSTRUCTION OF THE 3D FINITE ELEMENT MODEL

The finite element software is mainly used for analysis of models, but seldom used for building complex models. There is an interface between the finite element software and the CAD software. The 3D models built using the CAD software can be imported into the ANSYS Workbench software. The flowchart of finite element modeling is shown in Figure 3. As the machine tool has many complex parts, the simulation is time consuming. It is necessary to simplify the model before simulation. It is assumed that: (1) the material is isotropic and (2) the deformation is tiny. Meanwhile, the EDM machine tool should be simplified. It is necessary to neglect small chamfers, fillets, and other small pieces. In order to reduce the calculation amount, it is essential to simplify the machine tool accessories, such as the work tank, oil, fixtures, etc. After the machine tool is simplified, it is then imported into the finite element analysis software.

The material properties of key elements need to be set before finite element calculation. According to the design requirement of the EDM machine tool, material parameters of elements are selected as shown in Table 1. The bed, column, middle saddle, and upper saddle are all made of gray cast iron, while the connecting shaft, guide rail, and slider are made of structural steel. In order to reduce the deformation of the workpiece, natural granite is selected as the material of the worktable. Material properties such as density, elastic modulus, and Poisson's ratio should be defined before analysis.

As the parts of the EDM machine tool are connected in turn, the connection type should be set according to the assemble methods. For the connecting elements, the module of "automatic generation of the contact surface", which can quickly identify the contact relationship between the elements, is

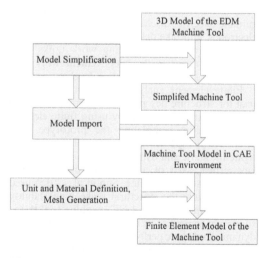

Figure 3. Flowchart of finite element modeling.

Table 1. Material properties of key elements.

Key elements	Density (kg/m³)	Elastic modulus (Pa)	Poisson's ratio
Bed, column, middle saddle, upper saddle	7200	1.1E11	0.28
Worktable	2700	5.5E10	0.3
Connecting shaft, guide rail, slider	7750	1.93E11	0.3

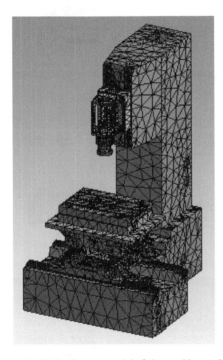

Figure 4. Finite element model of the machine tool.

selected. The bonded type is selected for the fixed elements, while the frictionless type is selected for the moving parts. After defining the connection type, the next step is meshing. Meshing is an important part of finite element analysis, and it influences the accuracy of the sequent analysis results. The automatic meshing function is used during meshing. The finite element model (Figure 4) is set up after meshing.

2 STATIC ANALYSIS

The preprocessing including constraints and loads is required before solving the problem. First, because the bed is bonded to the ground by anchors, the constraint of the bed should be chosen as a fixed support. Second, the gravity is an important affecting factor, which cannot be neglected. Third, it is essential to include the equivalent loads of the accessories, such as the work tank, oil, workpiece, fixtures, etc. The solve command can be conducted after preprocessing. In order to investigate the deformation of the whole machine tool, several representative motion positions are chosen. After solving the deformation of the whole machine tool, the results can be checked through the solution module.

The positions of moving parts influence the deformation of the machine tool, therefore several limit positions of the moving parts are chosen. Static analysis of the finite element model is performed. The deformation nephograms of the whole machine tool are shown in Figure 5. As the electrode is installed under the head, and the workpiece is installed on the worktable, the deformations of the head and the worktable are important factors that influence the machining accuracy. In the case that the head is on the bottom of Z-axis stroke, the maximum deformation value is 0.01981 mm as shown in Figure 5a. In the case that the head is on the top of Z-axis stroke, the maximum deformation value is 0.02157 mm as shown in Figure 5b.

In the case that the middle saddle is on the right of X-axis stroke, and the upper saddle is on the front of the Y-axis stroke, the maximum deformation value of the worktable is 0.00955 mm (Figure 5c). In the case that the middle saddle is on the right of X-axis stroke, and the upper saddle is on the back of Y-axis stroke, the maximum deformation value of the worktable is 0.00738 mm (Figure 5d).

In the case that the middle saddle is on the left of X-axis stroke, and the upper saddle is on the front of the Y-axis stroke, the maximum deformation value of the worktable is 0.00745 mm (Figure 5e). In the case that the middle saddle is on the left of X-axis stroke, and the upper saddle is on the back of the Y-axis stroke, the maximum deformation value of the worktable is 0.00949 mm (Figure 5f).

It can be observed from the above analysis that the deformations of the bed, middle saddle, upper saddle, and worktable are relatively smaller, while the column, head, and connecting shaft are the main deformation parts. With the head moving upward, the deformations increase gradually. The deformation of the worktable is within the permissible range of error.

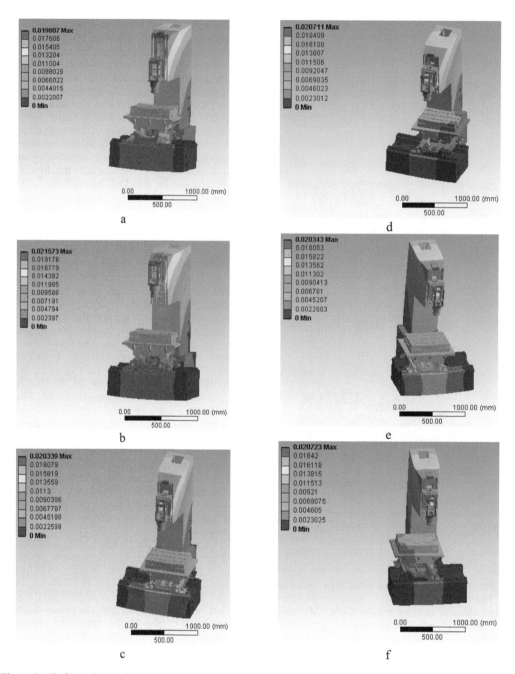

Figure 5.   Deformation nephograms: (a) the head is on the top of Z-axis stroke; (b) the head is on the bottom of Z-axis stroke; (c) the middle saddle is on the right of X-axis stroke, while the upper saddle is on the front of Y-axis stroke; (d) the middle saddle is on the right of X-axis stroke, while the upper saddle is on the back of Y-axis stroke; (e) the middle saddle is on the left of X-axis stroke, while the upper saddle is on the front of Y-axis stroke; and (f) the middle saddle is on the left of X-axis stroke, while the upper saddle is on the back of Y-axis stroke.

3 CONCLUSION

Making use of finite element analysis to solve technical problems is one of the important parts in engineering design. The 3D model of the EDM machine tool with a lifting work tank was established first, and then the finite element analysis was used. The finite element model was established. Static analyses show that the maximum deformation of the worktable is below 0.01 mm, which is within the permissible range of error. The column, head, and connecting shaft are the main deformation elements. The research results can form a valuable basis for the optimization of the EDM machine tool.

ACKNOWLEDGMENTS

This work was supported by the Beijing Academy of Science and Technology Program for Young Backbone Personnel (No. 201426) and the Beijing Natural Science Foundation (No. 3154033). The authors would also like to thank the anonymous reviewers whose comments greatly helped in making this paper better organized and more presentable.

REFERENCES

Abbas N M, Solomon D G, Bahari Md F (2007). A review on current research trends in electrical discharge machining (EDM). *International Journal of Machine Tools and Manufacture* 47(7–8), 1214–1228.

Chen D J, Fan J W, Zhang F H (2012). Dynamic and static characteristics of a hydrostatic spindle for machine tools. *Journal of Manufacturing Systems* 31(1), 26–33.

Cho S K, Kim H J, Chang S H (2011). The application of polymer composites to the table-top machine tool components for higher stiffness and reduced weight. *Composite Structures* 93(2), 492–501.

Dong X, Li R, Yang M (2012). Dynamic finite element analysis for the three degrees of freedom parallel machine tool. *Modular Machine Tool & Automatic Manufacturing Technique* (6), 49–52.

Liu Y, Chen S, Wang Z C, et al (2015). Structural design and analysis of desktop reconfigurable high-speed machine tool. *Mechanical Science and Technology for Aerospace Engineering* 34(8), 1217–1221.

Rajurkar K P, Sundaram M M, Malshe A P (2013). Review of electrochemical and electrodischarge machining. *Procedia CIRP* 6, 13–26.

Sun L, Yang S M, Zhao P, et al (2015). Dynamic and static analysis of the key vertical parts of a large scale ultra-precision optical aspherical machine Tool. *Procedia CIRP* 27, 247–253.

Sun Y W, Zhang Z J, Zhang J Y, et al (2015). Effects of transient slamming and harmonic swings on the marine compound NC machine tool. *Ocean Engineering* 108, 606–619.

*Advances in Materials Science, Energy Technology
and Environmental Engineering – Patty & Zhou (Eds)*
© 2017 Taylor & Francis Group, London, ISBN 978-1-138-19668-1

Plasma etching fabrication and graphene lubrication of IRBRR based on a multi-wafer MEMS

Huan Liu & Yaojin Cheng
Science and Technology on Low-Light-Level Night Vision Laboratory, Xi'an, Shaanxi, China

Shanshan Wang
School of Optoelectronical Engineering, Xi'an Technological University, Xi'an, Shaanxi, China

ABSTRACT: The strengths of micro engines such as high energy density, long sustainability, environmental protection, and economical benefit make the engine have a great prospect of application in fields of micro aerial vehicle, micro robots, and portable electronics. In this research, a multi-wafer Micro-Electro-Mechanical System (MEMS)-based Internal-Rotor Burnt Rotating Ramjet (IRBRR) with a planar structure, high efficiency of compressor, and high utilization of space was proposed. The structural and layout design of rotor, and fabrication of multi-wafer MEMS IRBRR were discussed. The solid lubrication for the bearing with multilayer graphene sheets assembled by Langmuir-Blodgett is also studied. Results indicate that the multilayer graphene sheets assembled by Langmuir-Blodgett possess better friction reduction.

1 INTRODUCTION

The term "PowerMEMS" was first suggested by Epstein and Senturia in 1996 to describe microsystems, which generated power or pumped heat. The promise identified was MEMS heat engines whose power densities equaled or exceeded those of the classical large-scale engines. Thereafter, the Power-MEMS system has evolved into a broader concept (Alan, 2010).

The energy density by weight and volume of different kinds of energy source are different. Several chemical energy sources are also compared in terms of both the gross gravimetric energy in the chemistry and an estimate of the net energy output. Two current practical implementations of lithium battery chemistries, lithium sulfur dioxide and lithium thionyl chloride, have about the same theoretical energy density as TNT. However, both of them are onetime use primary batteries. Rechargeable battery formulations have 30–50% less energy density. Batteries are widely available and inexpensive, and hence, they represent a baseline for all portable power sources. PowerMEMS concepts must compare favorably to be of value. The energy density of both lithium batteries and explosives is low compared with other chemical energy sources, as they contain both fuel and an oxidizer, while the other chemistries use atmospheric oxygen as the oxidizer, which therefore need not be carried by the power source. Heat engines or solid oxide fuel cells that directly burn hydrocarbon fuel are attractive alternatives, if they can be realized in the needed sizes and efficiencies (Alan, 2010; Wang, 2005; Epstein, 2004).

Micro engine based on MEMS technology is a key component of Power MEMS. However, micro engines are faced with some technical problems such as structural design, fabrication, bearing lubrication, etc.

2 DESIGN

2.1 Structure design

The scheme of structure, thermodynamic cycle, and performances of the new conceptive engine has been studied as shown in Figure 1 and 2. The result indicated that the thermodynamic cycle of the rotating ramjet is a Breton cycle including two anti-cycle processes. Owing to the advantage of thermodynamic performances and bladeless structure, this rotating ramjet may promote the study on power MEMS.

2.2 Layout design

A new three-layer rotating ram-rotor has been developed (Figure 4), and the rotating cylinder is internal combustion chamber in the center of rotor in which the revolved stream can be burnt and ram compressed.

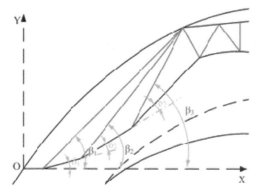

Figure 1. Shock waves in the air inlet of the ram-rotor.

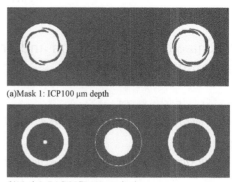

(a)Mask 1: ICP100 μm depth

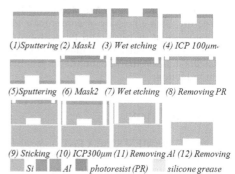

(b)Mask 2: BACK ICP 300 μm depth

Figure 4. Mask layout(a)(b) of the three-layer rotating ram-rotor.

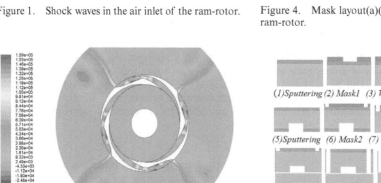

Figure 2. Air pressure field (at 30000 rpm).

(1)Sputtering (2) Mask1 (3) Wet etching (4) ICP 100μm.

(5)Sputtering (6) Mask2 (7) Wet etching (8) Removing PR

(9) Sticking (10) ICP300μm (11) Removing Al (12) Removing
   Si ■ Al ■ photoresist (PR) ■ silicone grease

Figure 5. Schematic diagram of process flow.

Figure 3. Schematic design of a three-layer rotating ram-rotor (thickness: 300 μm).

3 FABRICATION

After the study of structure, principles, and dynamics, we designed the fabrication process flow (Figure 5). Deep Reactive Ion Etching (DRIE) and flip-chip bonding were used to fabricate and package the multi-wafer MEMS IRBRR.

The DRIE process was run on Inductively Coupled Plasma (ICP) systems from Surface Technology Systems (STS) with Bosch process to precisely

Figure 6. Components of Multi-wafer MEMS IRBRR.

etch high-aspect ratio features into Si wafers. Figures 6 and 7 show the component and assembled Multi-wafer MEMS Internal-Rotor Burnt Rotating Ramjet.

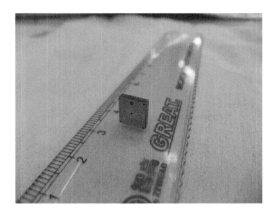

Figure 7. Packaged Multi-wafer MEMS IRBRR.

Figure 8. Homemade Langmuir-Blodgett trough.

4 LUBRICATION

However, one of the problems of MEMS IRBRR is that the rotor must rotate to high speed to form shock wave and ram-compress, and hence, the lubrication of bearing is very important. Liquid lubrication cannot be used in MEMS because of its surface tension. Solid lubrication or gas lubrication or a combination of both can be used to reduce the friction.

We have designed and fabricated journal bearings and thrust bearings for multi-wafer MEMS IRBRR. Specially, we studied the tribological properties of multilayer graphene sheets assembled by Langmuir-Blodgett (LB), and used for rotating MEMS machine bearing lubricating applications.

Graphene has good tribological performance because of its derivation from self-lubricating graphite (Lin, 2011; Geim, 2007; Novoselov, 2004). It is an outstanding solid lubricant at temperatures ranging from $-270°C$ to more than $1,000°C$. Self-Assembled Monolayer (SAM) graphene sheets are revealed outperforming even graphite due to reduced adhesion. However, for rotating MEMS, when the machine is in long-time rotating at a high speed, single and bilayer graphene sheets are not wear-resisting enough. Therefore, it needs to assemble multilayer graphene sheets.

Multilayer graphene sheets can be assembled by LB when the substrate or bearings of MEMS IRBRR were pulled through the 2D air-water interface. Therefore, we designed and fabricated a Langmuir-Blodgett trough as shown in Figure 8.

Multilayer graphene sheets can be transferred to a solid substrate with a density continuously tunable from dilute, closepacked to overpacked of graphene sheets. In addition, the thicknesses of grapheme sheets can also be controlled by the times

of pull. To obtain an accurate measurement and compare the tribological performance between the RGO (Reduced Graphene Oxide) and silicon (Si), we use Si wafers in the experiments.

Prior to assembly, Si wafers were cleaned in a piranha solution (a mixture of 7:3 (v/v) 98% H_2SO_4 and 30% H_2O_2) at 90°C for 30 min. Silicon wafers were then treated with 1:1:5 $NH_4OH:H_2O_2:DI$ solution for 15 min. The Graphene Oxide (GO) solution was sonicated for 24h using a tabletop ultrasonic cleaner. Moreover, the average size of the GO sheets can be controlled by the time and power of sonication.

GO solution was slowly spread onto the water surface dropwise using a glass syringe. The solution was spread with a speed of 300 µL/min up to a total of 6 mL. A GO film with faint brown color could be observed at the end of the compression. The film was compressed by barriers at a speed of 2 cm/min. The dimensions of the trough are 10 cm × 25 cm. The initial and final surface areas we used are 120 and 40 cm², respectively.

The GO multilayers were transferred to substrates at the end of the compression by vertically dipping the substrate into the trough and slowly pulling it up at a speed of 2 mm/min.

Finally, GO was thermally reduced at 200°C for 2h at a heating rate of $5°C \cdot min^{-1}$ under an argon atmosphere (Dikin, 2007; Lin, 2011) in a vacuum Rapid Thermal Processor (RTP-500, China).

Tribological tests were run on a UMT-2 tribometer (CETR; USA) in a ball-on-plate contact configuration. Commercially available steel balls (d = 9 mm, mean roughness = 0.02 µm) were used as the stationary upper counterparts, whereas the lower tested samples were mounted onto the flat base and driven to slide reciprocally at a distance of 0.5 cm. The friction coefficient-versus-time curves were recorded automatically.

The macrotribological behavior of the bare Si assembled with RGO samples was tested, and results in Figure 10 and 11 show that LB-RGO exhibits excellent friction reduction and wear resistance properties under a low applied load, and LB is a simple and feasible route for fabricating solid lubricant material-based graphene.

The friction coefficient-versus-time curves of Si wafer are shown in Figure 9, and the friction coefficient is about 0.6. In Figure 10, the friction coefficient-versus-time curves clearly show that multilayer graphene sheets assembled by Langmuir-Blodgett display good tribological properties under the same testing conditions with bare Si wafer in Figure 9 of 1 N and 1 Hz. Once RGO is assembled, the sample exhibits improved tribological properties. The friction coefficient of the

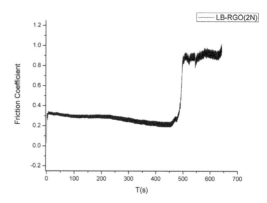

Figure 11. Friction coefficient-versus-time curve of Si assembled with RGO (applied load 2 N).

sample reduced from ~0.6 to ~0.21. The antiwear life under the applied load 1 N is more than 10 min. If in MEMS device, the applied load will be very small, and the antiwear life will be very long.

It can also be observed from Figure 9 and Figure 10 that the fluctuation of the friction coefficient-versus-time curves grows bigger as the time increases, as the applied load 1 N is the boundary value of UMT-2 tribometer measurement range. Therefore, we add the applied load from 1 N to 2 N, and this value is in the measurement range. The fluctuation of the curves is steady-going, and the antiwear life is shortened to 500 s. However, this value is bigger than some graphene solid lubricant. For example, the research by Ou (2010) shows that under the applied load of 0.2 N and 1 Hz, APTES-RGO shows a shorter antiwear life of ~300 s. This may be attributed to the thickness of multilayer graphene sheets assembled by Langmuir-Blodgett, which is bigger than graphene assembled by APTES. If applied load decreases to 0.1 N, the antiwear life may become longer than 10000 s. Therefore, graphene sheets are more effective as solid film lubricant at relatively low applied loads.

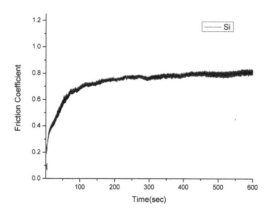

Figure 9. Friction coefficient-versus-time curve of Si (applied load 1 N).

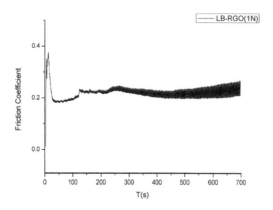

Figure 10. Friction coefficient-versus-time curve of Si assembled with RGO (applied load 1 N).

5 CONCLUSIONS

In this paper, a micro rotating ramjet engine with a planar structure, high efficiency of compressor, and high utilization of space was proposed. The ram compression mechanism, structural design, and graphene solid lubrication of the micro ramjet engine were researched. The layout and technological process of the micro ramjet engine were designed. The deep reactive ion etching and package process of the micro rotating ramjet engine was realized. Macrotribological properties of multilayer graphene sheets assembled by LB are developed.

ACKNOWLEDGMENTS

This work was financially supported by the "Fund of the Key Laboratory of Science and Technology on Low-light-level Night Vision" (No. J20130203) and the "Pre-research fund" (9140C380203140C38176).

REFERENCES

Alan H. Epstein, Multi-Wafer Rotating MEMS Machines: Turbines, Generators, and Engines, Springer, 2010.

Dikin DA, Stankovich S, Zimney EJ, et al. Preparation and Characterization of graphene oxide paper[J]. Nature, 2007, 448:457–60.

Epstein AH. Millimeter-scale, micro-electro-mechanical systems gas turbine engines. Journal of Engineering for Gas Turbines and Power-Transactions of the Asme, 2004, 126 (2): 205–226.

Geim A K, Novoselov K S. The rise of grapheme[J]. Nature material, 2007, 6: 183–191.

Lin J, Wang L, Chen G. Modification of Graphene Platelets and Their Tribological Properties as a Lubricant Additive[J]. Tribol.Lett. 2011, 41, 209–215.

Lin J, Wang L,. Modification of Graphene Platelets and Their Tribological Properties as a Lubricant Additive[J]. Tribol. Lett. 2011, 41, 209–215.

Novoselov K S, Geim A K, Morozov S V, et al. Electric field effect in atomically thin carbon films[J]. Science, 2004, 306: 666–669.

Ou J F, Wang J Q, Liu S, Mu B, Ren J F, Wang H G and Yang S R "Tribology Study of Reduced Graphene Oxide Sheets on Silicon Substrate Synthesized via Covalent Assembly," 2010, Langmuir, 26 15830.

Wang Yun. Primary Research of New Conceptive Internal-Rotor Burnt Rotating Ramjet. Nanjing: Nanjing University, 2005.

Advances in Materials Science, Energy Technology
and Environmental Engineering – Patty & Zhou (Eds)
© 2017 Taylor & Francis Group, London, ISBN 978-1-138-19668-1

Research on the real-time scheduling strategy of a sensor layer node in CPS

Benhai Zhou, Limei Liu & Libo Xu
Teaching Department of Computer Science, Shenyang Institute of Engineering, Liaoning Province, China

ABSTRACT: Cyber-Physical Systems (CPS) combine physical and computing systems tightly. Node Operating Systems (OS) are fundamental units in CPS. There are still many problems unsolved when designing CPS especially CPS node OS in aspects of predictability, reliability, robustness, etc. In this paper, we propose a mixed priority real-time scheduling method to enhance the CPS node OS performance. Generally, the CPS node operating system uses a single method to assign priority. The single priority selection method will lead to the high deadline miss ration, which makes the node OS performance drop sharply. Aiming at the problem, this paper proposes the effective shortest time priority algorithm and the adaptive shortest time priority algorithm. Experimental results indicate that, compared with the traditional FIFO (advanced First in First Out) and LSF (Least Slack First) algorithm, the algorithm proposed in this paper effectively reduces the deadline miss ratio. As a result, the real-time performance of the CPS is effectively improved.

1 INTRODUCTION

Cyber-physical systems are supposed to be a bridge to connect the physical world and virtual (information) world. They realize the interaction of two worlds' information flows. CPS can break the fence between physical world and virtual world using information technology and make the information interaction more comprehensive and thorough. CPS collect information from physical world with underlying physical data sensing units, complete information processing and fusion, and then give feedback to physical processing units using computer, communication and control (3C) technologies. CPS are comprehensive systems to realize information flows interaction at a higher level through finishing physical flows interaction at a lower level of the two worlds.

As CPS have diverse computational and resource characteristics, they are complicated and there are still many research challenges. One example is how to describe the architecture of cyber-physical systems. In this paper, we propose a four-layer architecture for CPS as shown in Figure 1.

1.1 *Physical interaction layer*

This layer is at the bottom of CPS architecture including a large number of CPS nodes, on which node operating systems are running. This layer is directly responsible for the interaction with physical systems. Nodes collect physical data. They not only transmit data up to the upper layer, but also accept feedback data. They control physical operations of reactors. Node operating system controls interaction devices and other hardware facilities of nodes to complete the feedback loop.

1.2 *Data Internet layer*

This layer should receive sensor data and send control data seamlessly to ensure information two-way exchange within horizontal and vertical systems transparently.

1.3 *Information integrating layer*

The data collected from lower layers are different in types and formats. Even some data are

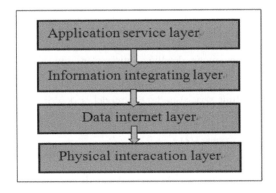

Figure 1. Four-layer architecture of the cyber-physical systems.

uncertain. Therefore, CPS must do the corresponding work (data cleaning, data reorganizing, data fusion, etc) to pick out useful information from raw data. Meanwhile, this layer also transforms the reverse interaction information to the corresponding control data for distribution to the lower layer.

1.4 *Application service layer*

This layer is the interface layer between CPS and upper applications. It could encapsulate all functions from lower layers as various services to provide applications as interface specifications or standards.

Real-time scheduling is an important part of the real-time system and it is also a critical part of the CPS system, which is the key to guarantee the real-time task. It is also a widely studied problem. In the current research, many real-time scheduling methods have been proposed for various task types. However, in the CPS, important parameters of real-time tasks with the release time and deadline will be more dependent on nodes in the CPS location and migration time and other factors. Therefore, traditional real-time scheduling algorithms of the physical influence factors were considered insufficient, easy to cause the system overload operation, which makes the real-time task deadline missed. The system's real-time perceiving performance is seriously affected, causing major production safety hazards as well as economic losses. In view of these problems, this paper proposes a CPS real-time scheduling algorithm considering the physical environment factors.

2 THE REAL-TIME SCHEDULING MODEL OF CPS

In this paper, we propose a real-time scheduling algorithm for the mobile real-time service node (CPS), which has many nodes in the network. This paper proposes a CPS real-time scheduling algorithm, which enables the system to achieve the maximum deadline satisfaction rate. Therefore, based on the real-time scheduling algorithm under the CPS network framework of the mobile real-time service node, the model is described as follows:

l_i: slack time of the task T_i (exponential distribution average $1/\lambda$).

e_i: execution time of the task T_i (evenly distributed on [0, E]).

m_{Ti}: migration time (Uniform distribution on [0, M]).

DMR: rate of meeting the deadline (DMR = tasks of meeting deadline/all of tasks)

In the real-time CPS of the mobile node, in the absence of task conflict, the real-time task DMR is the relaxation time of Ti, which is more than that of the real-time service node. Real-time task DMR Ti is the probability that the relaxation time of Ti is greater than that of real-time service nodes. Given the distribution of l_{Ti} to be $\lambda e^{-\lambda t}$, the DMR of T_i ($DMR_{Ti}(\lambda, m)$) is computed as follows:

$$DMR_{Ti}(\lambda, \text{m}) = \int_m^\infty \lambda e^{-\lambda t} dt = e^{-\lambda m} \qquad (1)$$

As m is assumed to be evenly distributed [0, M], an average DMR is:

$$Mean(DMR_{Ti}(\lambda, M)) = \frac{1}{M} \int_0^M e^{-\lambda m} dm = \frac{1}{\lambda M}(1 - e^{-\lambda M}) \qquad (2)$$

In case of conflict between the two real-time tasks, the paper will give the FIFO (first-in, first-out service), LST (shortest time priority), and ELST (CPS effective shortest time first) algorithm to calculate the average DMR.

3 IMPROVED EFFECTIVE LSF ALGORITHM

The effective LSF algorithm is an optimal algorithm in the real-time scheduling algorithm. However, when the real-time tasks are scheduled in the CPS system, DMR is needed to consider factors of the physical environment. In the real-time service CPS, the real-time service node migration time is one of the physical factors that need to be considered. Therefore, the effective shortest relaxation time of the real-time service node is considered. Therefore, when the real-time task is scheduled, the task with the most short time of slack time is first scheduled. We can use $l_{elsf, Ti} = l_{Ti} - m_{Ti}$, denoting the effective slack time of T_i.

Calculation methods are as follows: As the distribution function of l_{Ti} is $\lambda e^{-\lambda t}$, the distribution function of $l_{elsf, Ti} > 0$ is given as follows:

$$\frac{1}{M} \int_0^M \lambda e^{-\lambda(t+m)} dm = \frac{e^{-\lambda t}}{M}(1 - e^{-\lambda M}) \qquad (3)$$

with the distribution of $l_{elsf, Ti}$ in (−M,0), the function is as follows:

$$\frac{1}{M} \int_{-t}^M \lambda e^{-\lambda(t+m)} dm = \frac{1}{M}(1 - e^{-\lambda(M+t)}) \qquad (4)$$

when T1 and T2 have no conflict, the DMR of T1 should satisfy the probability of $l_{elsf, T1} > 0$, which is described as follows:

$$Mean(DMR_{T1}(\lambda,\mathbf{m})) = \frac{1}{M}\int_0^\infty \frac{e^{-\lambda t}}{M}(1-e^{-\lambda M})dt$$
$$= \frac{1}{\lambda M}(1-e^{-\lambda M}) \qquad (5)$$

The deadline meet ratio of task B following task A is:

$$Mean(DMR_{T2}(\lambda,\mathbf{m}_1,\mathbf{m}_2,\mathbf{e}_1))$$
$$= \int_{m1+m2}^\infty \frac{e^{-\lambda t}}{M}(1-e^{-\lambda M})\,dt$$
$$= \frac{(1-e^{-\lambda M})}{\lambda M}e - \lambda(\mathbf{m}_1+\mathbf{e}_1) \qquad (6)$$

As we assume m1 and e1 as evenly distributed on [0, M] and [0, E], respectively, the mean DMR of m_2 is computed as:

$$Mean(DMR_{T2}(\lambda,m_1,e_1,m_2))$$
$$= \frac{1}{ME}\int_0^E\int_{m+e}^\infty \frac{e^{-\lambda t}}{M}(1-e^{-\lambda M})\times e^{-\lambda(m_1+e_1)}d_{m1}d_{e1}$$
$$= \frac{(1-e^{-\lambda M})^2(1-e^{-\lambda E})}{\lambda^3 M^2 E} \qquad (7)$$

It can be concluded from the results of the ELSF, average deadline meet ratio ($Mean(DMR_{T1}$ (λ,m)) of T_1 equals average deadline meet ratio ($Mean(DMR_{T2}$ $(\lambda,m_1,e_1,m_2))$) of T2. In addition to ELSF, the other parameters are the same as the FIFO algorithm and C algorithm except $l_{elsf,Ti} = l_{Ti} - m_{Ti}$. Also, the FIFO algorithm applies $l_i > m_i$, and the ELSF algorithm is applied by $l_{elsf,Ti} = l_{Ti} - m_{Ti}$. Both of DMR are the same.

Therefore, the FIFO and ELSF algorithms meet the deadline rate similarly. It can be concluded that the ELSF algorithm cannot effectively improve the real-time performance, and hence, the optimization of the ELSF algorithm is studied in this paper.

Let p be the probability of DMR of first scheduled T1. Then, let q be the probability of DMR of the secondly scheduled T2.

$$p = Mean(DMR_{T1}(\lambda,\mathbf{m}))$$
$$= \int_0^\infty \frac{(e^{-\lambda t}1-e^{-\lambda M})}{M}dt = \frac{1-e^{-\lambda M}}{\lambda M} \qquad (8)$$

$$q = Mean(DMR_{T2}(\lambda,m_1,e_1,m_2))$$
$$= \frac{1}{ME}\int_0^E\int_{m+e}^\infty \frac{e^{-\lambda t}}{M}(1-e^{-\lambda M})\times e^{-\lambda(m_1+e_1)}d_{m1}d_{e1}$$
$$= \frac{(1-e^{-\lambda M})^2(1-e^{-\lambda E})}{\lambda^3 M^2 E} \qquad (9)$$

4 SCHEDULABILITY FOR IMPROVED ELSF ALGORITHM

An optimal ELSF algorithm is designed in this paper, considering the real-time service node in the mobile time and relaxation time, in order to improve the CPS nodes task deadline satisfaction rate. When the tasks T1 and T2 are in conflict, the ELSF algorithm will maximize the system to meet the deadline rate. For the $T_1{\rightarrow}T_2$ task sequence, it will be divided into the following four scheduling methods.

1. When tasks T1 and T2 meet the deadline, the probability meets $p \times q$. Under the optimal scheduling ELSF algorithm, it will not change the scheduling strategy, which selects the sequential execution of T1 and T2.
2. When only T1 meets the deadline, the probability meets A. If the optimal ELSF algorithm changes the scheduling order of T1 and T2, it can make the two tasks meet their deadlines. In the scheduling way, the probability of satisfying is $(p-q)/(1-q)*q/p$, which means that when the order is $T_1{\rightarrow}T_2$, under the condition of T2 not meeting the deadline, the probability of T2 under the situation of changed execution order is multiplied by the probability of T1 under the situation of changed execution order. When the order is $T_1{\rightarrow}T_2$, the condition of T2 meets the deadline.
3. The probability of only T2 meeting the deadline is $(1-p) \times q$. In this way, T1 cannot meet the deadline under any scheduling methods.
4. When T1 and T2 are not satisfying the deadlines, the probability is $(1-p) \times (1-q)$. In this case, if the task T2 is able to meet the deadline, the optimal ELSF algorithm will change the scheduling order for $T_1{\rightarrow}T_2$. Meanwhile, the probability of T2 meeting the deadline is $(p-q) \times (1-q)$. It means that, in the execution sequence $T_1{\rightarrow}T_2$, under the T2 not meeting the deadline, there is a probability of T2 after changing scheduling sequence. In this case, T1 cannot meet the deadline in both the sequences.

Similarly, for sequential $T_2{\rightarrow}T_1$, scheduling methods also can also be divided into four ways. The optimal ELSF algorithm will consider the service node travel time and the slack time. It will select the best scheduling approach, which can achieve the maximum DMR.

We can compute the task number of meeting DMR and the probability of the scheduling way for two adjacent tasks. Then, we can compute the probability of the optimal ELSF algorithm. That is the average of DMR of two kinds of task execution ways. Overall, the DMR of optimal ELSF is as follows:

$$DMR_{opti\_elsf} = p(1+q)-(p^2+q^2)/2 \qquad (10)$$

5 EXPERIMENTS AND ANALYSIS

In this paper, we test the performance of the real-time scheduling algorithm through simulation experiments. To verify the performance of the algorithm in the CPS system, the performance of the ELSF algorithm is verified in the case of multi-task conflict. In order to verify the performance of the real-time scheduling algorithm, the following parameters are configured in this paper:

Ei is the execution time at node i with the exponential distribution of average $1/\lambda$. In the experiments, we set $\lambda = 0.02$ and $\lambda = 0.2$. Di is the deadline at the node i with a variable range. S is represented by the moving distance between computing node c and serviced node i. Let $s = \sqrt{(x_i - x_c)^2 - (y_i - y_c)^2}$. The node C and service node I are distributed in a square of 100 by 100. Ri is the release time, and its value is 0. N is the number of simulations, which is 1000 times.

The paper uses DMR to test the performance of the optimal ELSF algorithm, LSF algorithm, and LSF algorithm in task conflict situations. Experimental results are shown in Fig. 2.

Experimental results indicate that the ELSF algorithm and LSF algorithm have different performances in the case of both information and physical factors (ei and mi), regardless of the specific gravity of the information factor, and the traditional LSF algorithm is better than the ELSF algorithm in the special condition. From the simulation results of Figure 3, the ELSF scheduling algorithm can show better performance when the information factor of CPS is relatively large. On the other hand, the traditional LSF algorithm has better performance than the ELSF algorithm when the physical factors are more than the information factor. However, in the above two cases, the optimization of the algorithm ELSF are able to exhibit better performance. The optimal ELSF scheduling algorithm can crease the DMR by up to 20% compared with other ways.

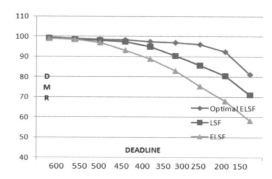

Figure 3. Experimental results when the cyber factor is quite small in comparison to the physical factor.

ACKNOWLEDGMENTS

This research was supported by the Research foundation of Liaoning Provincial Education Department (L2014532, L2014520, L2014523), and the National Social Science Fund (15CGL050).

Corresponding author: 13734610r@qq.com (Benhai Zhou).

REFERENCES

Edward A.L. and Sanjit A.S. (2011). Introduction to Embedded Systems, A Cyber-Physical Systems Approach. NY: Lee & Seshia, 119–158.

Fang-Jing Wua, Yu-Fen Kaob and Yu-Chee Tseng (2011). From wireless sensor network towards cyber physical systems, Pervasive and Mobile Computing, Vol. 7, Issue 4, 397–413.

Gann H. (2012). A Distributed Wireless Sensor Networks Mobile Communication Technology Research[C]. In Proceedings of Multimedia Information Networking and Security (MINES), UK, 203–207.

Hussain D., Illon T. Event (2011). Cyber-Physical Systems Method. In proceedings of Object Component Service Oriented Real-Time Distributed Computing, Beijing, IEEE press, 355–349.

John A. Stankovic (2008). When Sensor and Actuator Networks Cover the World, ETRI Journal, Vol. 30, No. 5, 627–633.

Shah, A. (2012). Cross-Layer Framework for QoS Support in Wireless Multimedia Sensor Networks[J], IEEE Transactions on Multimedia, 14(5): 1442–1455.

Smita P., Prabha K. (2012). A Qos Based Mac Protocol for Wireless Multimedia Sensor Network[J]. Journal of Electronics and Communication Engineering, 1(5): 30–35.

Stankovic, J., I. Lee, A. Mok, R. Rajkumar (2005). Opportunities and obligations for physical computing systems, IEEE Computer, Volume 38, Issue 11, 23–31.

Yong F., Qiang G. (2013). Design of a Wireless Sensor Network Platform for Real-Time Multimedia Communication [C]. In Proceedings of International Conference on Communication, Electronics and Automation Engineering, NY, IEEE press, 1305–1311.

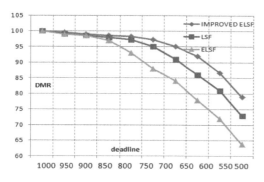

Figure 2. Experimental results when the cyber factor is quite big in comparison to the physical factor.

Advances in Materials Science, Energy Technology and Environmental Engineering – Patty & Zhou (Eds)
© 2017 Taylor & Francis Group, London, ISBN 978-1-138-19668-1

Fault early warning of a wind turbine pitch system based on the least square method

Jinpeng Yang & Wanye Yao
Department of Automation, North China Electric Power University, Hebei Baoding, China

Jianming Wang
ZhongHengBoRui, Beijing, China

ABSTRACT: By studying the structure and typical fault of wind turbine and by mining the intrinsic link of parameters of the variable pitch system, combined with the analysis of large numbers of historical and measured data from SCADA system of Hebei Chigu wind farm, a new method of fault early warning of the wind turbine pitch system is proposed. The historical data curve is fitted by the least square method, and the health model of normal operation condition is formed. Based on the Euclidean distance method, the output residual error is analyzed and the warning threshold value is determined, and then the abnormal rate is calculated. Taking the wind speed and pitch angle as an example, select data between rated wind speed and cut-out wind speed as the research object, and the simulation in Matlab shows that this method can accurately forecast the fault of pitch system of wind turbine.

1 INTRODUCTION

With the continuous development of wind power, the scale of wind farm is increasing rapidly. Wind power is treated as a kind of new energy. In order to realize the goal of long-term stable development, it is necessary to control the cost. The cost of wind farm mainly includes two parts, one is the fan manufacturing, processing, and installation cost, and the other is the operation and maintenance cost. Wind farm is usually built in remote areas, and the running condition is rather harsh. Together with the uncertainty and randomness of wind resources, the running condition of the fan is complex. Under this condition, most wind farms use pitch control technology (Wang, 2011). Therefore, the pitch devices act frequently, it will inevitably lead to failure frequently. To reduce the fan operation and maintenance costs, realizing the online early warning and diagnostic of large parts of fan is of significant meaning.

Wind farm produces a huge amount of data everyday, which include both normal data and fault information. The potential information of these data is of great value to the development and efficient management of wind farm. How to excavate and analyze the data, and apply the analysis result to fault early warning policy has become the focus of wind field. The fault warning algorithm has been deeply studied at home and abroad. Among them, the NEST algorithm is deeply researched and widely used in various scenes; the least square algorithm is becoming more and more popular among scholars with the advantages of simple processing, relatively small calculation amount and suitable for coupling parameters.

2 LEAST SQUARES METHOD

2.1 Least squares curve fitting

Through a set of historical data $\{(x_i, y_i), i = 0, 1, 2\ldots\ldots m\}$, calculate the function between variables x and y based on the principle of least squares. A curve is obtained based on extensive historical data samples to make sure the sample data points are in this curve or not far away. It not only reflects the overall distribution of data, but also changes the trend of original data (Wu, 2000). This method is intended to obtain a function, in which the sum of squared deviations is minimum, given in formula (1):

$$R^2 = \sum_{i=1}^{n} \Delta^2 = \min \sum_{i=1}^{n} [f(x_i) - y_i]^2 \tag{1}$$

In formula (1), R^2 is the minimum sum of squared deviations, and n is the number of sample data.

Determining the form of function is the first thing need to be done, the data with well correlation usually use the polynomial fitting model, i.e.:

$$f(x_i) = a_0 + a_1 x_i + a_2 x^2 + \ldots\ldots + a_k x^k \qquad (2)$$

In the formula (2), $a_i(i = 0,1,2\ldots\ldots k)$ is the model coefficient to be determined, the value of k depends on the actual situation, mostly be 3 or 4.

To get the minimum value of R^2, the coefficients should meet the principle of least squares shown in formula (3):

$$\begin{cases} \dfrac{\partial(R^2)}{\partial a_0} = 0 \\[6pt] \dfrac{\partial(R^2)}{\partial a_1} = 0 \\[4pt] \ldots\ldots \\ \ldots\ldots \\ \dfrac{\partial(R^2)}{\partial a_k} = 0 \end{cases} \qquad (3)$$

Take formula (1) into equation (3), the coefficient matrix A is shown in formula (4):

$$A = (X^T X)^{-1} X^T Y \qquad (4)$$

In formula (4),

$$X = \begin{bmatrix} 1 & x_1 & \ldots & x_1{}^k \\ 1 & x_2 & \ldots & x_2{}^k \\ : & : & \ldots & : \\ : & : & \ldots & : \\ 1 & x_n & \ldots & x_n{}^k \end{bmatrix}, A = \begin{bmatrix} a_0 \\ a_1 \\ : \\ : \\ a_n \end{bmatrix} Y = \begin{bmatrix} y_1 \\ y_2 \\ : \\ : \\ y_n \end{bmatrix}.$$

2.2 *Analysis of deviation based on Euclidean distance method*

The historical data are fitted to a curve based on the least squares method. Within the allowable range, the calculated values can be seen as predicted values (Zhu, 2002). In order to describe the degree of dispersion between fitted values and real values, introduce the concept of degree of deviation, i.e.

$$\varepsilon = \sqrt{\sum_{i=1}^{n}(y_1 - y_0)^2} \qquad (5)$$

In formula (5), y_1 is the theoretical value when the value of x is x_1, y_0 is the actual value.

As the fitting data sample is large enough, an uncertainty of the input model exists. Through the statistical analysis, the degree of deviation ε follows normal distribution and 99.35% of the data are in the range between μ-3δ and μ+3δ. It can cover almost all of the data. Based on field experience, the data can be considered normal when the degree of deviation ε is in the range between μ-3δ and μ+3δ, otherwise it is treated as abnormal information.

Table 1. Basic information of Goldwind 1500 unit.

Parameter index	Index value
Rated power (kW)	1500
Rated speed (rpm)	1800
Pitch range (°)	[0 90]
Cut-in speed (m/s)	3
Rated speed (m/s)	12
Cut-out speed (m/s)	25

3 DATA PREPROCESSING

In this paper, the historical data of Hebei Chigu wind farm are selected as statistical sample. It takes 10 minutes as the acquisition unit. The data are chosen between February 10, 2015 and February 10, 2016 from the SCADA system as the sample library. Selecting the change condition of # 1 blade pitch angle of pitch system as predicted performance, the sample data selection mainly depends on the following principles:

1. Select the data, in which the wind speed is between the rated wind speed and the cut-out wind speed, as the statistical sample.
2. Data obtained when fan does not work should be excluded
3. Exclude the unusual data point of fan because of the causal factor.
4. Exclude the fault data.

The historical data are from Goldwind 1500 kW models unit, and the basic information of this type is shown in Table 1:

4 CURVE FITTING AND RESIDUAL ANALYSIS

Based on the principle of data preprocessing, # 1 blade with the appropriate pitch angle is selected as the processing object. Comparison before and after pretreatment is shown in Figure 1.

From the comparison of two figures, it is known that the most singular point and failure data have little influence on the data analysis after pretreatment. The preprocessed data can nearly represent the normal operating conditions of pitch angle. The function expression of fitting result is shown in formula (6):

$$y = -0.0075x^3 + 0.147x - 0.1x + 7 \qquad (6)$$

To verify whether the curve can accurately represent the normal operating conditions, it is necessary to analyze the calculating residual value (Gao, 2007). The results are shown in Table 2:

142

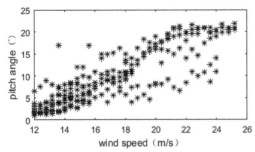

a. Before pretreatment

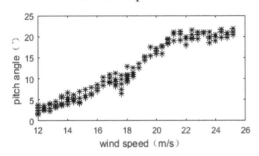

b. After pretreatment

Figure 1.   Preprocessed data.

Table 2.   Residual analysis.

Conditions (m/s)	Number of samples	Number of exceeding limitation	Coverage
12~25	17950	204	98.86%

Coverage is the degree that how accurate this fitted curve can represent the actual normal operating conditions. In Table 2, the coverage of fitting curve is 98.86%. Theoretically, this fitting curve can represent the actual working conditions. In actual operation, within the allowable deviation range, no matter what the real-time wind speed is, the pitch angle should fit the curve function. Otherwise, the fan is in abnormal state. If this abnormal state lasts long, then the fan will alarm (Chen, 2011).

5   VERIFICATION

Select the data of Hebei Chigu wind farm from March 10, 2016 14:00 to April 10, 2016 14:00 as the test sample. Take 10 minutes as acquisition unit and 200 groups of abnormal condition test samples were obtained after pretreatment. Take the 200 groups of test data into the health model, and the result is shown in Figure 2:

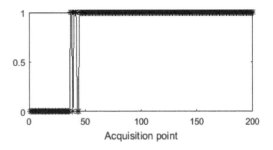

Figure 2.   Detection diagram of abnormal point.

Table 3.   Information record of the SCADA system.

Time	Acquisition point	Residual value	Reason
2016-03-11 15:40	41	5.8756	Pitch devices act frequently as
2016-03-11 16:10	44	8.4431	wind speed varies, leading PLC to
2016-03-11 16:30	46	11.1229	lose control

In Figure 2, 0 represents the normal data and 1 represents the abnormal data. From the graph, it is known that when reaching the fortieth collection point, abnormal data emerge. The abnormal rate increases and reaches 100% at last. According to the principle of early warning, the system alarms when abnormal data emerge and abnormal rate increases continuously (Li, 2015). Querying the information record of the SCADA system, the result is shown in Table 3.

From the analysis of Figure 2 and Table 3, it is known that the period between the acquisition point of 40~50 is interim from normal state to fault and the fiftieth acquisition point is already in fault state. According to the early warning rules, if the system alarms when the abnormal state emerges, then it can predict the fault accurately and carry out the plan ahead of schedule, and it is of practical significance for the economic and efficient operation of wind farm.

6   CONCLUSION

Through the statistical analysis of pitch angle and wind speed data of Hebei Chigu wind farm, the paper reveals a large number of potential valuable information. The paper fits a health model of normal operating conditions based on the least square method and analyzes the residuals of predicted and historical values based on the Euclidean distance method, and then sets the early warning threshold value and rule based on field experience combined with the results of residual analysis. Finally, this method is simulated and verified, and the results show that this method has a high accuracy in the fault early warning and effectively prevent the occurrence of faults. It can eliminate the hidden fault in time through the warning signal ahead of occurrence of faults, improving the safety and economy of fan operation.

REFERENCES

Chen Zisi, Research on fault early warning and forecasting method based on SCADA data of wind turbine; Beijing, Power technology and Application. 2011. 11.

Gao Xiangbao, Dong Lingqing. Mathematical analysis and SPSS application[M]. Beijing: press of Tsinghua University, 2007.

Li Dazhong, Xu Binkun, Chang Cheng. The data preprocessing and model establishment of large scale wind turbine[J]; Shan Dong, Journal of State and Technology College, 2015. 04.

Wang Zhaohua. The main trend of new energy development in the world and thoughts on the development of new energy resources in China [J], Economic Forum, 2011.

Wu Huirong. Statistical principle [M], Shanghai: Press of Shanghai Jiao tong University, 2000.

Zhu Yonghua. Learning Guidance of Application of statistical[M]. Wuhan: Press of wuhan University, 2002.

*Advances in Materials Science, Energy Technology
and Environmental Engineering – Patty & Zhou (Eds)*
© 2017 Taylor & Francis Group, London, ISBN 978-1-138-19668-1

Research on the swirling flow effect of the combustor–turbine interaction on vane film cooling

Shi Liu & Hong Yin
*Electric Power Research Institute of Guangdong Power Grid Corporation, Guangzhou,
Guangdong Province, P.R. China*

ABSTRACT: Lean premixed combustion technology has been adopted widely for heavy-duty gas turbine application. At the combustor inlet section, the basic burner arrangement is multiple-swirl configuration. The multiple-swirl structure creates complicated swirling flow field downstream, which is characterized as a non-uniform flow-field and has impacts on the turbine vane. The issue of combustor-turbine interaction effect has become quite prominent.

This paper introduces a new test rig for the combustor-turbine interaction research, which is designed to investigate the influence of multiple-swirl on the turbine vane system. The test rig consists of a combustor simulator and a first-stage turbine vane with the cooling system. Measurement techniques including the pressure-sensitive paint and five-hole probe are applied.

The experimental test results show that the multiple-swirl combustor flow field has a significant impact on the vane cooling system due to the residual swirl intensity at the combustor outlet. The stagnation line at the vane leading edge is obviously altered compared with uniform inflow. Film cooling effectiveness distribution has distinct characteristics under different conditions. The leading edge is most significantly influenced, while the pressure side film cooling system is affected slightly. Under certain conditions, the suction side film cooling is influenced locally.

1 NOMENCLATURE

Symbols

C	Mass concentration
D	Diameter, mm
I	Light intensity
M	Blowing ratio, $\rho_c V_c / \rho_\infty V_\infty$
p	Pressure, kpa
PSP	Pressure-sensitive paint
y+	y plus

Subscripts

ref	Reference value
aw	Adiabatic
∞	Free stream condition
C	Coolant condition
hub	Swirler hub
sw	Swirler
0	Calibration condition

2 INTRODUCTION

The research and development trend of heavy-duty and industrial gas turbine always seeks higher efficiency, less pollutant emission, more stability, and adaptability. The urgent requirement has led to a number of improvements, which include the lean premixed combustion and the advanced cooling system for the turbine vane. As the combustor becomes quite compact and introduces the strong swirling flow strategy (Davis, 1999), more and more researchers put attention on the interaction of the combustor and turbine. At the combustor/turbine interface area, the non-uniform flow and temperature field are determined by the design of aerodynamics, heat transfer, combustion condition, and the local geometrical arrangement. The turbine first-stage vane usually adopts plenty of film cooling holes and consumes a large quantity of coolant, which even complicated the situation of the interface area.

A comprehensive phenomenon needs to be factored out to deepen insight into the interaction effect. To the knowledge and research of the authors' group, the actual content of interaction can be categorized into four issues: turbulence intensity, hot streak, swirl, and radiation. It is difficult to clarify the individual effect at the engine condition because of the complex reacting flow process. Therefore, basically, the review of literature listed below is concentrated in each factor, respectively. Radiation factor can be referenced in the publication (Wang, 2011), and will not be discussed in this paper.

The topic of "Turbulence Intensity" has been studied extensively both experimentally and numerically. The vane surface heat transfer enhancement and film cooling effectiveness reduction are mostly concerned. Some experts expect the free stream turbulence intensity as high as 20% at the combustor exit. Ames et al. considered several kinds of turbulence generator simulating the combustor component and explored the effect of inlet turbulence and Reynolds number on endwall heat transfer in a turbine vane cascade. Results show the heat transfer coefficient increased by using the low NO_x combustor with a higher turbulence intensity. Chong et al. (2012) conducted turbulence measurements using the CTA system in a novel full annular gas turbine rig and also LES simulation. Besides, the turbulence intensity enhances the mixing of coolant and air, which impairs the cooling effectiveness. However, turbulence can sometimes prevent the lift-off of coolant ejection locally.

The "Hot Streak" has always been a hot topic since the turbine designers need the temperature profile as a basis for the aerodynamic and cooling system design. Hot streaks entering the stage are convected through the NGV and redistributed in rotation stages through the "segregation effect" phenomenon. As the hot fluid migrates through the stage, it tends to accumulate on the pressure side of the rotor blade. Furthermore, Shang showed that the hot fluid moved towards the rotor hub endwall. Analysis confirmed that the segregation effect is the dominant mechanism of redistribution.

Jenkins investigated the interactions of hot streaks from the combustor with the film cooling of the first row of guide vanes of the high-pressure turbine. The hot streak clocking position was also considered relatively to the stator vane. Results show that the hot streak attenuates differently when passing the passage at various clocking positions. The maximum attenuation of the hot streak exists when aimed at the leading edge of the profile because of the cooling ejection. Ong, Qureshi (2008) respectively take further step into the hot streak migration and rotor temperature distributions study comparing homogeneous inlet temperature with hot streak inlet profiles.

The factor "Swirl" can be interpreted as the non-uniform velocity condition, which attracts much attention in recent years.

Barringer pursued the goal to design and build combustor simulators to generate pressure and temperature profiles, which are close to reality under realistic velocity distribution to gain a better understanding of the effect on aerodynamics and heat transfer of the first row of nozzle guide vanes. Colban (2002) described a movement of the passage vortex in the direction of the suction side of the investigated vane and the development of another vortex above the passage vortex.

Qureshi (2011) described a combustor simulation module for a transonic turbine research facility. It is reported that the peak swirl angles at the front plane of the nozzle guide vanes are approximately +/- 40°, which can be characterized as aggressive swirl. Then, the researcher investigated the effects of a swirling inflow on the aerodynamics and heat transfer characteristics of a high-pressure turbine vane. The measurements were carried out in a rotating transonic turbine test rig with an upstream swirl generator. Results show a shifting of the stagnation point as well as a changing vane load over the vane's height.

As computational resources developed, more and more computational achievements research on integrated simulations of combustor chamber and Nozzle Guide Vane (NGV) were accomplished. Turrell (2004) applied a Reynolds stress model to predict the effect of high swirling flow exiting the combustor chamber on the vane aerodynamics, with the hot gas transport visualized. In Roux S. et al. (2003), Large Eddy Simulation (LES) of a combustor chamber with and without vanes were conducted. The flow field in the downstream half of the combustor chamber is affected by the presence of the vanes. Jiang (2009) compared the decoupling computation with the coupling method for a micro-combustor and NGVs case, and concluded that the NGV had limited effect only one span upstream. Medic (2006) built up an integrated simulation system including solver and boundary data exchanging software and also presented an example of aero-engine simulation in Turbulence Research Center of Stanford. Klapdor extended a combustion code by compressibility effects and applies it to a combustor chamber including the NGV.

Coupled simulation appears to be better solution and gives deeper insight into the combustor/vane interface although this method is still under development. From the present research progress, obviously the upstream combustor has significant influence on the vane. However, the vane might have quite limited effect on the combustor flow field. (Roux, 2003; Jiang, 2009)

The research group of Schiffer (Giller, 2012; Schmid, 2012; Pyliouras, 2012) at Technology University Darmstadt conducted an experimental study on the interaction between combustor swirl and high-pressure vane, along with a numerical study on inlet swirl in a turbine cascade. These publications consisted of a systematic description of the flow phenomenon under swirl (orientation, intensity, clocking), also considering the vane heat transfer and film cooling condition. One of the findings is that it presented the adiabatic effectiveness distribution along the vane span. The central part of the leading edge has relatively poor cooling. In the next step, the researchers conducted the numerical simulation (Schmid, 2014) to identify the influence of swirling flow, total pressure, and turbulence level on the aerodynamic and thermal loss and support the corresponding large-scale test rig design process. The research group from Oxford University reported a new test rig (Luque, 2014) to investigate combustor-turbine interactions with multiple can combustors simulating the MHI heavy-duty gas turbine.

The author conducted research (Yin, 2013) on the single swirl effect on the model leading edge film cooling effectiveness distribution by experimental and numerical methods. Both results showed that the model leading edge film cooling is quite sensitive to the inlet flow condition, including the swirling intensity, distance between the swirler and the test model. Extensive research has been carried out to cover many combustor/turbine swirl topics, but the performance of the film cooling system needs more work. As the film cooling system performance is critical for the first-stage vane, this research established an experimental test rig considering the combustor and turbine interaction. The experimental test results are presented considering the impact of swirling flow on the film cooling effectiveness of a vane leading edge and some flow features of swirling flow.

3 EXPERIMENTAL SET-UP

3.1 Wind tunnel and test section

The configuration of the test wind tunnel is demonstrated in Figure 1. The main air flow is filtered and pressurized by a centrifugal blower. Then the air flows through the duct and then into the filter, a converging nozzle, a turbulence generator, and a square duct. The test piece box is the location where the cold flow model combustor is installed.

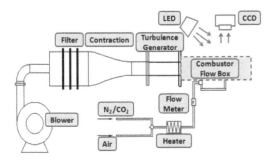

Figure 1. Schematic of the whole experimental set up with pressure-sensitive paint measurement.

In addition, the secondary coolant air system consists of the coolant source tanks, an air compressor, a gas heater and a flow meter, providing the coolant measurement for the test piece. Moreover, the coolant system can be adjusted conveniently in order to cooperate with the different combustor-turbine vane configurations. The whole display of the experimental facility is presented in Figure 2.

To obtain a uniform free stream flow, a small scale turbulence grid is installed in front of the rectangular duct. A turbulence intensity of 6.0% in front of the swirler is measured by a single sensor hot-film anemometer. The distance from the turbulence grid to the inlet of the swirler is approximately 30 cm, which is considered to be long enough to attain uniform flow. The free stream flow velocity reaches 20 m/s, with a static temperature of 310 K. The film cooling flow rate is controlled by mass flow meters, in order to regulate the mass blowing ratio, and coolant-to-gas density ratio can be adjusted by changing the coolant type.

3.2 Cold flow combustor configuration

Multiple-swirl structure is a commonly adopted technology in the advanced gas turbine combustor chamber, where complicated phenomenon of swirl interaction exists. Based on the previous research [23] on interaction of single swirl and the leading edge film cooling system, this paper made

Figure 2. Display of the wind tunnel.

a further step to design and establish an experimental platform with a multiple-swirl combustor chamber and the first-stage vane. To simulate the cold flow under real combustor-turbine configuration, the interaction between multiple-swirl combustor and the first-stage vane was investigated. The experimental section mainly comprises a combustor chamber, the vane passage, and an auxiliary aerodynamic measuring section.

The combustor chamber is the core component in the experimental test rig, and its basic structure is schematically shown in Figure 3. The combustor structure mainly consists of many sections such as the holding plate, swirler segment, combustor chamber, outer square, and transition piece. The swirlers are fixed on the holding plates and reach across the end plate of combustor chamber cylinder. Significantly, the sealing of contact surface of the swirlers and the end plate is imperative. Different arrangements of the swirlers can be realized according to the requirement of the test cases.

The horizontal arrangement of the cold flow combustors is shown in Figure 4. The left-hand side is one big central swirler arrangement, and the right-hand side is a 7-swirler layout, which will be described later. The demonstration of the horizontal flow path is shown in Figure 5 with the blue arrows as symbols. The main air stream flows into the combustor from the back side, crossing the holding plate from void area of the plate. Then through the swirlers and the combustor chamber cylinder, the mainstream air is finally led out from the transition piece, whose traverse section changes from round shape to rectangular shape gradually. At the transition piece outlet in this case, the auxiliary aerodynamic measurement section or the vane passage can be installed for different purposes. As shown in Figure 6, the inclined structure is even more realistic to the gas turbine combustor arrangement. The axis of the combustor chamber center intersects with the horizontal line by an angle of 30 degrees. From the flow path demonstration, the reverse flow pattern is similar to the combustor flow. Compared with the horizontal configuration, the outer square duct is covered by a solid plate. Under this layout, the

Big swirler   Small Swirler

Figure 4. Two schemes of horizontal combustor arrangement.

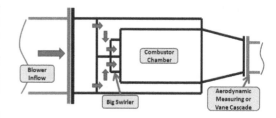

Figure 5. Flow path demonstration of horizontal big swirler arrangement.

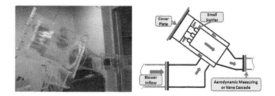

Figure 6. Flow path demonstration of inclined multiple small swirlers arrangement.

mainstream air from the blower enters the square cavity through the rectangular duct, and then flows reversely to the swirlers and the holding plate. The transition piece needs to be able to change the direction from the inclined direction to the horizontal line. In the meantime, the transition piece changes from the circular shape to the rectangular shape while changing the angle gradually.

In the experiment, the first-stage vane from an F-class gas turbine was selected for the experimental test model. Taking the 50% span airfoil shape of the vane as base, the test vane is two dimensional and straight. In the experiment, the clocking relation between the combustor chamber and the first-stage vane model is one combustor chamber corresponding to two vane passages, as shown in Figure7. The two vane passages are formed by installing three vanes on the flat plates. The two vanes at the left and right sides are solid vanes, and the middle vane with a full coverage film cooling system can be inserted into the central area. The

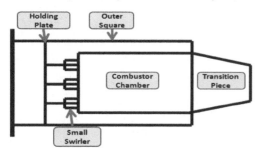

Figure 3. Schematic layout of the combustor.

middle vane faces the upstream combustor central axis directly, providing the cooling performance data similar to the realistic turbine vane geometry.

For better control of coolant distribution during the experiments, the internal structure of the middle vane model is divided into seven cavities, as shown in Figure 8. Specifically, the leading edge area includes two coolant chambers, respectively, supplying the Shower Head (SH) cooling and two rows of cooling holes (SS1, SS2) on the front suction surface. For the pressure side film cooling (PS1, PS2, PS3, PS4), and lateral Suction Surface film cooling (SS3), one independent cooling chamber is designed for each row. Considering the film cooling hole shape issue, the round hole is applied in the leading edge shower head cooling (six rows with one cooling chamber), while the rest of cooling holes adopt the shaped hole.

The auxiliary aerodynamic measurement section can be installed at the outlet of the transition piece section in order to measure the swirling flow field. It should be mentioned that the aerodynamic measurement is conducted without the vane downstream. When conducting the aerodynamic measurement of the swirling flow field at the outlet, the vane passages test section should be replaced with the auxiliary aerodynamic measurement section. To measure the flow field characteristics of transition piece outlet, the five-hole probe can be inserted into the measuring plane through the slot on the bottom plate and move with dynamic sealing. Figure 9 shows the auxiliary aerodynamic measurement section and the aerodynamic measuring installation.

Figure 9. The auxiliary aerodynamic measurement section and the real-time measurement test.

3.3 Multiple swirl configuration and swirler design

There are two different kinds of swirl arrangement in the experiment: one big central swirler and seven small swirlers. For the seven-swirler layout, one swirler lies in the center and the other six swirlers lie around the central axis uniformly. In addition, three kinds of multiple-swirl combination are considered in the experiment, which is schematically shown in Figure 10. The co-swirl combination means that the seven swirls are all towards the clockwise direction (viewing from the holding plate to the downstream combustor and transition piece). The counter-swirl combination means that the central swirler rotates in the anti-clockwise direction while the outer six swirlers rotate clockwise. The stagger-swirl combination means that four swirlers rotate clockwise, while the other three swirlers rotate anti-clockwise.

As the experiment is conducted under cold flow and ambient pressure conditions, the structure and temperature requirements are not so high. The swirler material adopts the ABS plastics, which can be manufactured conveniently and economically. For the multiple-swirl test, small swirlers with both clockwise and anti-clockwise direction are manufactured in order to realize the different swirl combinations. The number of small swirlers is 10 in total, while only one big swirler is made for the single swirler case. The small swirler model and the real product is shown in Figure 11.

The design principle of the swirler is very important, which must consider the aerodynamic and structural considerations. Especially, the swirl number is the key parameter for the design process. According to the technical publication by Lefebvre (Lefebvre, 1998), for the axial swirlers with flat vanes and a constant angle of θ, the swirl number expression is shown in Eq. (1):

$$S_N = \frac{2}{3}\tan\theta\frac{1-(D_{hub}/D_{sw})^3}{1-(D_{hub}/D_{sw})^2} \qquad (1)$$

Based on this swirl number definition, the design swirl number must exceed 0.4 to reach a

Figure 7. The vane passages model and real test rig with film cooled vane in the center.

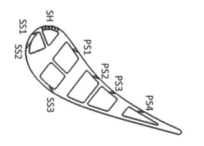

Figure 8. The cooling chamber design.

recirculation zone. Furthermore, only when the swirl number is larger than 0.6, it can be categorized as a strong swirl. Commonly, one would take the swirl number high enough in industry and research practice. According to the definition, the design swirl number of the small swirler is 0.54 by adjusting the swirl angle and diameters.

The outer diameter of the small swirler is 60 mm while the hub diameter is 28 mm. Then the flat vane set angle could be adjusted during the design process to reach the swirl number.

The main purpose of adding the big central swirler in the experimental test is to compare the cooling performance with the previous experimental results (Yin, 2013), which characterizes as the single swirl and leading edge model. The geometrical modeling of the big central swirler and the real product is shown in Figure 12. The design swirl number is also 0.54, while the flow area of the big swirler inlet is 7 times of the small swirler.

It needs to be mentioned that although the swirl number of the big swirler is the same as the small one, the interaction between multiple swirls causes the resulting swirling flow intensity to change. Therefore, the flow field characteristic of the combustor of multiple-swirl configuration is significantly different from the single swirl case.

To investigate the swirling flow effect, the uniform inflow conditions is set up as the reference cases, which is named as No Swirl (NS). The test cases with the big central swirler are named Single Swirl (SS) series. The seven swirlers cases are named Multiple Swirl (MS) series, with three kinds of swirl combinations described above.

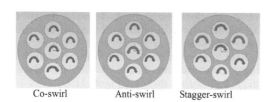

  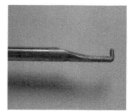

Co-swirl   Anti-swirl   Stagger-swirl

Figure 10.   Three multiple-swirl combinations.

Figure 11.   The small swirler model and plastic product.

3.4   Five-hole probe technique

The five-hole probe experimental system was developed with a probe diameter of 2 mm and a rake diameter of 6 mm. Moreover, three dimensional displacement controllers including two translational and one rotational freedom was also matched, shown in Figure 13. The measurement range and experimentation error were also listed in Table 1.

Five-hole measurements were performed using the rakes at the selected square traverse areas of the outlet. The whole dimension of the auxiliary aerodynamic measurement section is 206 mm X114 mm, which is the size of the transition piece outlet. For the actual measurement movement, the central area with the dimension 100 mm X90 mm is designated to satisfy the geometrical and measuring restrictions. The whole area and measurement area are displayed in Figure 14, which should be paid attention when discussing the results.

Figure 12.   The big swirler model and plastic product.

Five-hole probe   Displacement system

Figure 13.   Display of the five-hole probe measurement system.

Table 1.   Parameters of the five-hole probe system.

	Parameters	Range	Error
Five-Hole probe	Total pressure	−50~400 kPa	~0.2%
	Mach Number	0.01~0.6	~0.005
	Yaw Angle	−30°~30°	~±0.5°
	Pitch Angle	−20°~20°	~±0.5°
Displacement Controller	Translation	300 mm	<0.005 mm
	Rotation	360°	<0.005°

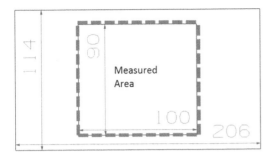

Figure 14.   The measured area in the auxiliary aerodynamic measurement section.

3.5   *Pressure-Sensitive Paint (PSP) technique*

PSP technique is frequently employed for adiabatic film cooling effectiveness measurement, which based on oxygen quenched photoluminescence. Photoluminescence is the property of a compound (the active part of PSP) to emit light after being excited by a suitable light source. This process, however, is interrupted by collisions with oxygen molecules: the excited PSP molecules may relax back to their unexcited state without emitting visible light. Since the intensity of the emitted light depends on the oxygen partial pressure, the emitted light intensity directly relates to the pressure of a surrounding gas, which contains oxygen. A test setup of the PSP application to obtain film cooling effectiveness includes the test section, the CCD camera, and the light sources (Li, 2010). The Ruthenium-based paint used in this study was excited at 450 nm by Light-Emitting Diode (LED) lights. To detect and record the emitted light, which contains pressure and concentration information, a high spectral sensitivity CCD camera, fitted with a 600 nm band pass filter, was used. The measuring facilities are shown in Figure 15. The basic principle of PSP is taking mass/heat transfer analogy, which can be illustrated in the following chart (Figure 16).

The test section included a transition duct located in front of the existing cascade test section, which was made of stainless steel. The swirler and leading edge model is mounted at the middle of the test section. A transparent window was located over the test section. The CCD camera is mounted directly above the window at a distance suitable to view from each side of the leading edge. A LED light is located beside the camera to provide the excitation light. The images of luminescence intensity distribution from the leading edge model, recorded by the CCD camera, are originally 16-bit gray-scale. These images are saved and then processed using an in-house program. Nitrogen/carbon dioxide gas is heated to the temperature of the free stream to eliminate any temperature effects; it was

Figure 15.   The CCD camera and LED light source.

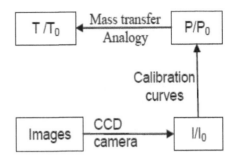

Figure 16.   Basic principle and flow chart of the PSP measurement operation process.

then ejected through the film cooling holes into the main flow. There are rubber coverings with electrical heaters on the outside of the supplying tubes to eliminate heat losses to the environment. The secondary flow temperature difference can be controlled to be within 0.5 K.

The film cooling effectiveness under adiabatic wall temperature hypothesis is defined as Eq. (2), which is suitable for low-speed flow with constant property:

$$\eta = \frac{T_{aw} - T_\infty}{T_c - T_\infty} \tag{2}$$

This research tries to measure the combustor swirling flow field and its comprehensive effect on the first-stage vane film cooling performance. The coolant control parameter such as the blowing ratio and density ratio should be considered. To consider the control parameter in the experiment, the blowing ratio can be adjusted with the value of 0.5, 1.0, and 1.5 by controlling the coolant mass flow rate. Two different density ratios with values 1 and 1.5 are considered by adopting the coolants N_2 and CO_2.

Before conducting the experiments, the PSP measurement should be calibrated by using a specially designed system (Li, 2010). In terms of the 95% confidence interval, the uncertainty of film cooling effectiveness measurement is evaluated about 3% around the value of 0.5, while it will reach 8% when the effectiveness is around 0.05.

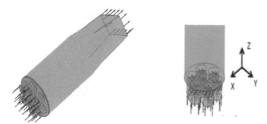

Figure 17. The seven-swirl numerical model.

Table 2. Boundary conditions for the multiple swirl simulations.

	Boundary conditions	Value
U	Inlet velocity, m/s	40
T	Inlet total temperature, K	310
TI	Inlet turbulence intensity	5%
P_o	Outlet static pressure, bar	1

4 NUMERICAL METHOD

This paper also presents some accompanying computational results for the combustor cold flow field. Special attention has been put on the multiple-swirl configurations since the swirl interaction is complicated. The computed results of the flow field can be compared with the experimental results and provide insight into the fluid dynamics mechanism. The numerical model can be set up in the CAD software as demonstrated in Figure 17.

The grid-independence check was performed and the final mesh version is chosen with a total number of about 10 million unstructured meshes. The mesh is locally densified on the wall surface and in the area between swirls. The numerical software ANSYS CFX 12.1 is taken and, for turbulence modeling, the SST model is selected. For the calculation, the boundary condition set up is listed in Table 2.

5 RESULTS AND DISCUSSION

5.1 Numerical results of the multiple-swirl flow

First, the numerical computation results of the combustor flow field are presented. The whole axial length of the combustor and transition is 750 mm. The diameter of the combustor is 230 mm, the same as the test rig configuration. At the traverse section, which is just 100 mm downstream of the swirlers in Figure 18, the velocity distribution is given. For different swirl combinations, the swirl interaction occurs intensively in each case with different characteristics. Basically, in the co-swirl

combination, there exists strong shear between two adjacent swirls. In the anti-swirl combination, the outer six swirl shows a shear effect on each other, while the central swirl keeps well. However, in the stagger-swirl case, three pairs of swirls form in the outer ring and influence the central swirl.

The flow field of the transition piece outlet is also presented corresponding to the experimental measurement. Since the shape of transition piece changes from the circular to rectangular shape, the swirling flow field will be changed due to the geometrical change. The computed flow characteristic of Co-, Anti-, Stagger-swirl are given in Figure 19, 20 and 21 respectively.

Each figure gives the yaw angle and pitch angle distribution with flow vectors superimposed by the CFD post-processing. It is quite obvious that at the combustor chamber outlet, the flow filed is significantly inhomogeneous. The yaw angle value is slightly larger than the pitch angle, with maximum yaw angle reaching 45 degrees and pitch angle of 36 degrees. The value range is limited to ±25 degrees in order to unify the legend and make it convenient for comparison. The maximum yaw angle and pitch angle occurs in the co-swirl combination, while the minimum in stagger-swirl case.

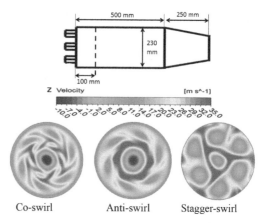

Co-swirl   Anti-swirl   Stagger-swirl

Figure 18. The velocity field plot of the three multiple-swirl combinations.

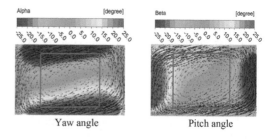

Yaw angle   Pitch angle

Figure 19. Co-swirl combination flow field.

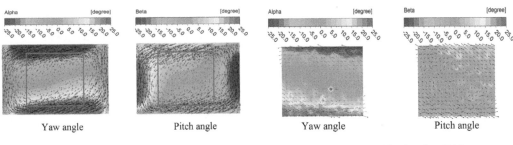

| Yaw angle | Pitch angle | Yaw angle | Pitch angle |

Figure 20. Anti-swirl combination flow field.

Figure 22. Co-swirl combination flow field.

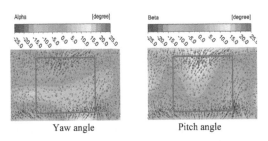

Yaw angle   Pitch angle

Figure 21. Stagger-swirl combination flow field.

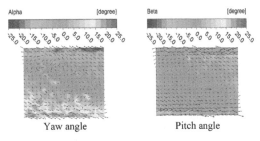

Yaw angle   Pitch angle

Figure 23. Anti-swirl combination flow field.

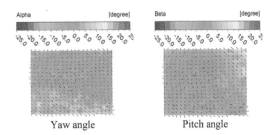

Yaw angle   Pitch angle

Figure 24. Stagger-swirl combination flow field.

The swirl intensity is strongly reduced due to the strong mixing in the stagger-swirl. In the contour plot, the central rectangular region corresponds to the aerodynamic measurement area. Since the measuring limit of the five-hole probe, only the local central area is measured.

5.2 Experimental results of the multiple-swirl flow

The aerodynamic measurement result of the flow field at the transition piece outlet is introduced in this section. As the measurement area is in the central area aligned with the middle vane, it is very effective to provide information for the cooling effectiveness measurement. The measured flow characteristics of co-, anti-, and stagger-swirl are given in Figure 22, 23, and 24 respectively. The flow field is displayed as the yaw and pitch angle distribution with the two-dimensional vector plots. The flow characteristics of the co-swirl and anti-swirl are quite similar. The swirl intensity is stronger as the location is closer to the wall, which coincides with the trend of the numerical results. The swirl intensity of co-swirl combination is slightly larger than the anti-swirl combination. In the measured area of co-swirl case, the yaw angle reaches the maximum value about 25 degrees, while in the anti-swirl case, it is only 20 degrees. After a long process of mixing through the transition piece, the non-uniformity of the stream-wise velocity has been basically eliminated. Compared with the co-swirl and anti-swirl combination, the yaw and pitch angle distribution are quite small in the stagger-swirl condition.

Based on the presented data, the measured data generally matches the CFD data. The distribution of the tangential flow angle (swirl intensity) can reach more than 20 degrees in the central area. The combination of experimental and numerical methods can give a general picture of the swirling flow field especially in the central area at the transition piece outlet.

5.3 Experimental results of swirling flow effect on film cooling performance

The distribution of film cooling under the single swirl and uniform flow are displayed first. The experimental data of uniform inflow is cited from the plane cascade work of the author's group (Han, 2014), as shown in Figure 25. Within the single swirl configuration, the film cooling effectiveness

Effectiveness: 0 0.10.20.30.40.50.60.70.80.9 1

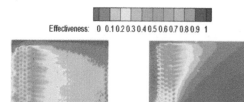

Uniform inflow   Single-swirl inflow

Figure 25. Film cooling effectiveness under uniform and single-swirl inflow.

Co-Swirl   Anti-Swirl   Stagger-Swirl

Figure 26. Leading edge film cooling effectiveness under three multiple-swirl combinations.

test under the single swirl condition is also carried out. The PSP test results are shown in Figure 25.

For the uniform inflow test results, the leading edge is covered well enough by the coolant. As the coolant migrates from the leading edge to the pressure side, the turbulent mixing and heat transfer causes the cooling effectiveness to reduce. Meanwhile, the coolant near the endwall is swept forward by the passage vortices into the mainstream. In the single swirl condition, the distribution of film cooling effectiveness is changed significantly. The upper and lower ends of the leading edge show strong local injection or no ejection of the coolant because of the altered stagnation line. While at the middle position, the blow-off phenomenon occurs locally. The single swirl effect on film cooling effectiveness is basically the same in the model leading edge condition [23] and vane cascade test environment. In general, the film cooling performance under the swirl condition is altered significantly and not uniform at the leading edge. Some local area lacks the coolant air, which would cause the vane material to deteriorate or melt.

5.4 Experimental results of film cooling under multiple-swirl condition

This section describes the film cooling characteristics at different positions of the first stage vane. Experimental test results of the three kinds of multiple-swirl combination conditions are reported. As the characteristic of multiple-swirl flow field is different from the single swirl, the swirl intensity and total pressure distribution are different. Specific analysis of the vane cooling distribution on different locations is given as follows:

1. Leading edge film cooling. As shown in Figure 26, the swirling flow effect of multiple-swirl combination is not so obvious as the single swirl configuration. The leading edge film cooling jets are only slightly turned. The swirl intensity is decreased heavily due to the dissipation of swirl interaction. The multiple swirl distribution determines the outlet pressure distribution to be quite uniform. The coolant air on the vane leading edge is less deflected and the middle section does not show obvious blow off phenomenon. The leading edge film deflection is most obvious under the co-swirl combination. For the stagger-swirl combination, the cooling jet turning is not prominent at all.

2. Pressure side surface film cooling. The four rows of pressure side film cooling are all measured and only the first-row cooling performance is presented here, while the second, third, and fourth rows of holes are not influenced by the swirling flow effect.

As shown in Figure 27, the swirling flow effect on the vane pressure surface is quite limited. The deflection effect caused by residual swirl on first rows of holes (PS1) can be illustrated. To certain extent, the lower end shows slightly worse cooling coverage than the upper end in the co-swirl and anti-swirl cases. While in the stagger-swirl combination, the cooling effectiveness is more uniform from the lower to the upper end and less disrupted by the swirl.

3. Suction side surface film cooling. There are three rows of film cooling holes on the suction side. As the third row of holes located downstream of the throat is not visible to the camera, only the cooling effectiveness of the front two suction side rows is analyzed here. As shown in Figure 28, the swirl intensity has a certain effect on the suction side of the vane coolant distribution.

The front two rows of cooling holes on the suction surface are quite close to the leading edge. Under the three kinds of multiple-swirl combination conditions, the cooling effectiveness at the first row (SS1) cannot inject locally and realize full

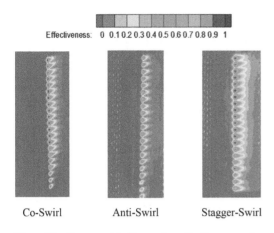

Effectiveness: 0 0.1 0.2 0.3 0.4 0.5 0.6 0.7 0.8 0.9 1

Co-Swirl   Anti-Swirl   Stagger-Swirl

Figure 27.   Pressure side film cooling effectiveness under three multiple-swirl combinations.

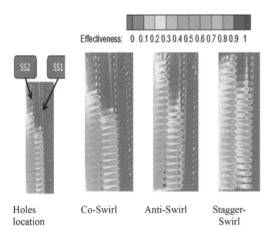

Effectiveness: 0 0.1 0.2 0.3 0.4 0.5 0.6 0.7 0.8 0.9 1

Holes   Co-Swirl   Anti-Swirl   Stagger-
location   Swirl

Figure 28.   Suction side film cooling effectiveness under three multiple-swirl combinations.

coverage. It is indicated that the impact brought by the stagnation line deflection is quite obvious at the SS1 location. With stronger residual swirl intensity in the flow, the stagnation line offset could be larger towards the suction surface. As in the co-swirl combination, more cooling holes show no coolant coverage than the other two schemes. For the second row (SS2) location, the cooling effectiveness shows better coverage under the stagger-swirl combination. The residual swirl intensity is not strong, which does not affect the SS2 row much. While for the co-swirl and anti-swirl combination, the SS2 row film cooling is still affected to some extent. In summary, the swirling flow has obvious effects on the suction surface film cooling, since the SS1 and SS2 rows are

quite close to the leading edge. The film cooling effectiveness distributions basically show the same trend under different combinations. The pressure distribution significantly influences the suction film cooling due to the stagnation line deflection. It is dangerous that some areas are not covered by the coolant.

6 CONCLUSION

This research introduces an experimental facility for combustor-turbine interaction research and presents preliminary test results. Both the aerodynamic results and cooling effectiveness distribution were presented. To help the analysis process, numerical tools were also adopted to get the detailed flow field information.

The core component of the experimental facility is the cold flow combustor, including several parts like the holding plate, swirlers, combustor chamber, and transition piece. Two types of swirlers are selected: the single big swirler configuration is used to verify the swirling flow effect in the vane cascade condition; while the multiple small swirlers are used to simulate the real combustor geometry. The basic layout of the combustor can be installed as the horizontal and 30 degrees inclined form. However, for the current paper, only the horizontal test results are introduced.

Based on the aerodynamic measurement and computations, the flow field of the multiple-swirl combination is investigated. The residual swirl intensity at the transition piece outlet is quite obvious which can have evident impact on the first stage vane. Compared with the stagger-swirl, the flow angle of the co-swirl and anti-swirl can reach 20 degrees or even above. In addition, these two combinations are the frequently chosen design arrangement in heavy-duty gas turbine application.

The PSP test results include two parts. The first part is to verify the swirling flow effect by comparing the single swirl and axial uniform inflow. In the second part, the vane passage cooling performance under the three multiple-swirl combinations is investigated and compared. The residual swirl intensity under multiple-swirl combinations is smaller than the single swirl because of the dissipation by the swirl-swirl interaction. Therefore, the leading edge film cooling is deflected by the swirling flow to certain extent, depending on the swirl intensity. On the pressure side, the first row of cooling holes is affected slightly while the other three rows are not influenced. The film cooling rows on the suction side are close to the leading edge and affected by the shifted stagnation line quite sensitively.

ACKNOWLEDGMENTS

The authors would like to acknowledge the financial support from the International Cooperation and Exchange Project (No.51110105013) supported by the National Natural Science Foundation of China, and the research project on gas turbine combined cycle power plant supported by China Southern Power Grid Company Ltd (No.K-GD20140492). They would also like to thank Dr. Haojie Tang and Experimental Engineer Dechu Lin for their beneficial discussions and experimental assistance.

REFERENCES

Ames F.E., Barbot P.A., Wang, C. "Effects of Catalytic and Dry Low NO$_x$ Combustor Turbulence on Endwall Heat Transfer Distributions". Journal of Heat Transfer, 127(4), pp. 414–424.

Barringer M.D, Thole K.A., Polanka M.D."Effect of Combustor Exit Profiles on Vane Aerodynamic Loading and Heat Transfer in a High Pressure Turbine". ASME Journal of Turbomachinery, 131, p. 021008.

Cha C.M., Ireland P.T., Denman P.A., Savarianandam V. "Turbulence levels are high at the combustor-turbine interface. ASME paper GT2012–69130.

Colban W.F., Lethander A.T., Thole K.A., Zess G. "Combustor Turbine Interface Studies—Part 2: Flow and Thermal Field Measurements". Proceedings of ASME Turbo Expo 2002, GT2002–30527.

Davis L.B, Black S.H. Dry Low NO$_x$ Combustion Systems for GE Heavy-Duty Gas Turbines [R] GE Power Systems_GER-3568G, 1999.

Giller L, Schiffer H P."Interactions between the Combustor Swirl and the High Pressure Stator of a Turbine", ASME paper GT2012–69157.

Han C., Ren J., Jiang H. "Experimental investigations of SYCEE film cooling performance on a plate and a tested vane of an F-class gas turbine". Proceedings of ASME Turbo Expo, 2014, ASME paper GT2014–25774.

Han J.C, Dutta S. "Gas Turbine Heat Transfer and Cooling Technology". New York: Taylor & Francis.

Jenkins S.C., Bogard D.G. "The Effects of the Vane and Mainstream Turbulence Level on Hot Streak Attenuation". Journal of Turbomachinery, 127(1), pp. 215–221.

Jiang L.Y., Carscallen B., Okulov P., Gallien R., Rigaudier G."Effect of Nozzle Guide Vanes on Flow Parameters at the Exit of a Micro Gas Turbine Combustor". ASME Paper GT2009–59694.

Klapdor E.V. "Simulation of Combustor-Turbine Interaction in a Jet Engine". PhD thesis, TU Darmstadt

Lefebvre A.H. "Gas Turbine Combustion, second edition". Taylor & Francis press, 1998.

Li J., Ren J., Jiang H."Film cooling performance of the embedded holes in trenches with compound angles". Proceedings of ASME Turbo Expo, 2010, ASME paper GT2010–22337.

Luque S., Kanjirakkad V., Aslanidou I., Lubbock R., Rosic B. "A New Experimental Facility to Investigate Combustor-Turbine Interactions in Gas Turbines with Multiple Can Combustor". ASME paper GT2014–26987.

Medic G., Kalitzin G., You D., Herrmann M., Ham F., Weide E., Pitsch H., Alonso J.J. "Integrated RANS/LES Computations of Turbulent Flow Through a Turbofan Jet Engine", Center for Turbulence Research, Annual research Briefs 2006, p. 275–285.

Ong J., Miller R.J. "Hot Streak and Vane Coolant Migration in a Downstream Rotor". ASME Conference Proceedings, 2008 (43161), pp. 1749–1760.

Pyliouras S., Schiffer H.P., Janke E., Willer L. "Effects of Non-uniform Combustor Exit Flow on Turbine Aerodynamics". ASME paper GT2012–69327.

Qureshi I., Beretta A., Povey T. "Effect of Simulated Combustor Temperature Non-uniformity on HP Vane and End Wall Heat Transfer: An Experimental and Computational Investigation". ASME J. Eng. Gas Turbines Power, 133, 031901.

Qureshi I., Smith A.D., Povey T. "HP Vane Aerodynamics and Heat Transfer in the Presence of Aggressive Inlet Swirl", ASME paper GT2011–46037.

QureshiI, Povey T. "A Combustor-Representative Swirl Simulator for a Transonic Turbine Research Facility". Proceedings of the Institution of Mechanical Engineers, Part G: Journal of Aerospace Engineering 2011.

Roux S., Cazalens M., Poinsot, T. "Outlet-Boundary-Condition Influence for Large Eddy Simulation of Combustion Instabilities in Gas Turbines". Journal of Propulsion and Power, 24(3), May-June, pp. 541–546.

Schmid G., Krichbaum A., Werschnik H., Schiffer H.P. "The Impact of Realistic Inlet Swirl in a 1 1/2 Stage Axial Turbine". ASME paper GT2014–26716.

Schmid G., Schiffer H.P. "Numerical Investigation of Inlet Swirl in a Turbine Cascade". ASME paper GT2012–69397.

Shang T., Epstein A.H. "Analysis of Hot Streak Effects on Turbine Rotor Heat Load". ASME J. Turbomach., 119, pp. 544–553.

Turell M.D., Stopford P.J., Syed K.J., Buchanan E. "CFD Simulation of the Flow within and downstream of a High-Swirl Lean Premixed Gas Turbine Combustor". ASME Paper GT2004–53112.

Wang W., Sun P., Ren J., Jiang H. Radiative Effectiveness on the Aero—and Thermodynamics in a Highly Thermally Loaded Film Cooling System". ASME Paper GT2011–45592.

Yin H., Qin Y., Ren J., Jiang H. "Effect of Inlet Swirl on the Model Leading Edge of Turbine". ASME paper GT2013–94471.

Zhang L.Z., Baltz M., Padupatty R. "Turbine Nozzle Film Cooling Study Using the Pressure Sensitive Paint (PSP) Technique". ASME paper No. 99-GT-196.

Advances in Materials Science, Energy Technology and Environmental Engineering – Patty & Zhou (Eds)
© 2017 Taylor & Francis Group, London, ISBN 978-1-138-19668-1

Adsorption isotherms of aliphatic-based superplasticizer in CFBC ash-Portland cement paste

Jingxiang Liu, Litao Ma, Haoyu Wang & Jiwei Liu
Anhui Singular Environmental Protection Co. Ltd., Suzhou, Anhui, China

Jingyong Yan & Yuyan Wang
School of Environmental and Material Engineering, Yantai University, Yantai, China

ABSTRACT: This paper studies the adsorption isotherms of aliphatic-based superplasticizer in CFBC ash-cement pastes. UV-visible absorption spectroscopy was used to evaluate the adsorption isotherms. Results show that the adsorption of aliphatic-based superplasticizer in coal ash-Portland cement pastes can be characterized as a monolayer adsorption and described by the Langmuir adsorption model. CFBC ash-cement pastes possess higher adsorption ability than the Pulverized Coal Combustion (PCC) fly ash-cement pastes and the adsorption increases with the increase of ash replacement ratio. It is concluded that the high amount of unburnt carbon along with irregular shape and porous surface morphology of CFBC ashes leads to high adsorption of superplasticizer.

1 INTRODUCTION

Circulating Fluidized Bed Combustion (CFBC) ashes are produced from the combustion of coal with injection of limestone for sulfur capture. With rich content of active SiO_2 and Al_2O_3, CFBC ashes have been reported to exhibit good pozzolanic activity[1–3] and recognized as a potential supplementary cementitious material to partially replace cement for concrete production.

CFBC ashes are produced at a much lower temperature (850–900°C) than Pulverized Coal Combustion (PCC) fly ashes (1200–1400°C). The physical properties and chemical compositions of CFBC ashes are therefore distinct from those of PCC fly ashes. For example, the content of unburnt carbon of CFBC ashes is greater than that of PCC fly ashes and the shape of CFBC particles are irregular with loose and porous surface structure[4].

The adsorption effect of cement or PCC fly ashes on water-reducing agents has been stud-ied extensively. However, no study reports the efficiency of water reducer in CFBC ash-cement system. There is a need to study the adsorption of water-reducing admixtures in the CFBC ash-cement paste. This paper investigates the effects and the underlying mechanisms of CFBC ashes on the adsorption isotherms of aliphatic-based superplasticizer in the ash-cement pastes.

2 EXPERIMENTAL PROGRAM

2.1 Materials

The chemical composition of CFBC fly and bottom ashes, PCC fly ashes and ordinary Portland (PO42.5) cement clinker are summarized in Table 1. The aliphatic-based superplasticizer with a solid content of 27.39% was used in this study.

As shown in Table 1, Loss On Ignition (LOI) of CFBC ashes is higher than that of PCC fly ashes and cement clinker.

Table 1. Chemical composition (by mass) of coal ashes and Portland cement clinker.

Sample	SiO_2	Fe_2O_3	Al_2O_3	CaO	MgO	Na_2O	K_2O	SO_3	Free lime	LOI	Sum
CFBC fly ashes	37.54	5.84	23.10	10.52	1.29	1.17	0.55	4.80	3.03	13.24	98.05
CFBC bottom ashes	56.08	4.91	24.28	3.62	1.11	1.97	0.79	1.48	0.93	4.87	99.05
PCC fly ashes	53.91	4.12	28.81	4.83	2.68	1.20	0.44	0.98	0.95	2.03	99.00
Cement clinker	20.70	3.56	4.94	62.49	3.38	0.11	0.82	0.48	0.82	1.41	97.89

Table 2 summarizes the physical properties of the CFBC ashes and the PCC fly ashes. As can be seen, the water demand of CFBC ashes is nearly twice that of PCC fly ashes. The CFBC bottom ashes have larger particle size with lower specific surface area as compared to the CFBC and the PCC fly ashes.

2.2 Tests

The adsorption of superplasticizer in ash-cement pastes can be calculated based on the following equation.

$$\Gamma = (C_0 - C) \cdot v / 1000\, W \qquad (1)$$

where Γ is the adsorption amount, mg/g; C_0 is the initial concentration of superplasticizer, mg/L; C is the residual concentration of superplasticizer, mg/L; v is the volume of superplasticizer solution, mL; and W is the quality of the mix of coal ashes and cement, g.

2.2.1 Adsorption isotherms of aliphatic superplasticizer in CFBC ash-cement pastes

To evaluate the adsorption isotherms of aliphatic superplasticizer in CFBC ash-cement pastes, 22 ash-cement pastes incorporating different ashes and various superplasticizer concentrations were prepared. The ashes investigated were PCC fly ash (control), CFBC fly ash, and CFBC bottom ash. The ash-to-cement ratio in all mixes was kept at 3-to-7 and the initial concentration of superplasticizer varied from 14 to 377 mg/L.

For each mix, 200 g of binder (cement and ash) was first mixed for 5 minutes. 100 mL of a certain concentration of superplasticizer solution was added to the dry powder and mixed for another 150 minutes. The paste was then filtrated to get the supernatant, followed by the determination of the residual concentration of superplasticizer by the UV-visible spectrophotometry. The adsorption of aliphatic superplasticizer in each ash-cement paste was calculated based on Eqn. (1) and the adsorption isotherm can be obtained by plotting the adsorption against the initial concentration of superplasticizer.

2.2.2 Adsorption of aliphatic superplasticizer in CFBC ash-cement pastes

To evaluate the effect of ash-to-cement ratio on the adsorption of aliphatic superplasticizer in CFBC ash-cement pastes, 12 ash-cement pastes incorporating different ashes at various ash-to-cement ratios were prepared. Again, the ashes investigated were PCC fly ash (control), CFBC fly ash, and CFBC bottom ash. The ash-to-cement ratio investigated ranges from 1-to-9 to 4-to-6 and the mix proportion follows 100: 70: 1.2 (binder: water: superplasticizer).

The binder was dry-mixed for 5 minutes prior to the addition of water and superplasticizer. The fresh paste was then mixed for another 150 minutes, followed by filtration of paste in order to get the supernatant. The residual concentration of superplasticizer in the supernatant was determined by the UV-visible spectrophotometry and the adsorption of superplasticizer in each ash-cement paste was calculated based on Eqn. (1).

3 RESULTS AND DISCUSSION

3.1 Adsorption isotherms of aliphatic superplasticizer in CFBC ash-cement pastes

The adsorption isotherms of aliphatic superplasticizer in coal ash-cement pastes are plotted in Fig. 1. According to the classification of solution adsorption isotherm by Giles [5], all three ash-cement

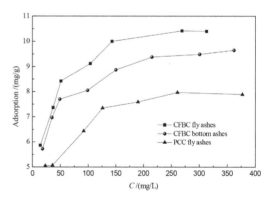

Figure 1. Adsorption isotherms of aliphatic superplasticizer in coal ash-cement pastes.

Table 2. Physical characteristics of coal ashes.

Sample	Water demand (%)	Specific surface area (m²/kg)	Average particle size (μm)
CFBC fly ashes	183	391.4	21.40
CFBC bottom ashes	172	297.6	27.59
PCC fly ashes	95	402.5	20.33

paste isotherm curves may be classified as the type L2 isotherm and can be described by the Langmuir isothermal adsorption equation as follows,

$$C/\Gamma = 1/(\Gamma_m \cdot k) + (1/\Gamma_m) \cdot C \qquad (2)$$

where C is the aqueous concentration, Γ is the amount adsorbed, Γ_m is the maximum amount adsorbed as C increases, and K is the Langmuir equilibrium constant.

Table 3 summarizes the Langmuir adsorption parameters, K and Γ_m, of aliphatic superplasticizer in coal ash-cement pastes through linear regression. The results show the adsorption of aliphatic superplasticizer in coal ash-cement pastes corresponds well with the Langmuir adsorption isotherm model and it can be described by the Langmuir equation with high coefficient of correlation ($R^2 > 0.99$). This suggests the adsorption of aliphatic superplasticizer molecules in coal ash-cement pastes can be characterized as a monolayer (lying flat) adsorption on the surface of coal ashes and cement particles. In general, the adsorption capacity increases with K and Γ_m. As such, the CFBC fly ashes should have higher adsorption capacity than the CFBC bottom ashes, followed by the PCC fly ashes.

3.2 Effect of ash-to-cement ratio on the adsorption of aliphatic superplasticizer

Figure 2 shows the effect of ash-to-cement ratio on the adsorption of aliphatic superplasticizer.

Table 3. Langmuir adsorption parameters of aliphatic superplasticizer in coal ash-cement pastes.

Sample	K (10^{-2} L·mg^{-1})	Γ_m (mg·g^{-1})	R^2
CFBC fly ashes	6.32	10.96	0.9993
CFBC bottom ashes	5.86	10.05	0.9992
PCC fly ashes	4.91	8.40	0.9984

As can be seen, the adsorption of aliphatic superplasticizer in coal ash-cement paste increases with ash-to-cement ratio. This may be attributed to the higher amount of unburnt carbon in coal ashes as compared to that in cement. With increase of ash-to-cement ratio, more unburnt carbon is available in the paste which adsorbs water reducing admixtures. For a given ash-to-cement ratio, the CFBC ash-cement pastes have higher adsorption of the aliphatic superplasticizer than the PCC fly ash-cement paste.

Figure 3 shows the SEM photos of the CFBC and the PCC ashes. As can be seen, the CFBC ashes are irregular with a loose surface structure which is significantly different from that of the PCC fly ashes. However, CFBC is irregular and loose. The loose and porous surface morphology of CFBC ashes may enable the penetration of superplasticizer into the inner surface of ash particles. The surface characteristics of CFBC ashes along with the high unburnt carbon content in the CFBC ashes may contribute to high adsorption of aliphatic superplasticizer in the CFBC ash-cement paste system.

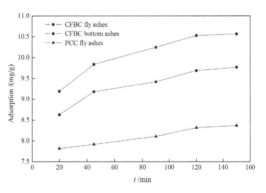

Figure 2. Adsorption of aliphatic superplasticizer versus time in coal ash-cement pastes.

  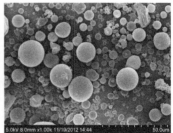

(a) CFBC fly ashes   (b) CFBC bottom ashes   (c) PCC fly ashes

Figure 3. SEM photographs of coal ashes.

159

4 CONCLUSIONS

The adsorption of aliphatic-based superplasticizer in coal ash-Portland cement pastes can be characterized as a monolayer adsorption and described by the Langmuir isothermal adsorption equation. The saturated adsorption capacity Γ_m and the adsorption equilibrium constant K of CFBC ash-Portland cement paste are greater than those of PCC fly ash-Portland cement paste, which suggest CFBC ash-cement pastes possess higher adsorption ability of the aliphatic superplasticizer than the PCC fly ash-cement pastes. In addition, the adsorption of aliphatic superplasticizer increases with the increase of CFBC ash replacement ratio in the paste.

When CFBC ashes are used as supplementary cementitious material; however, potential loss on workability and water-reducer efficiency needs to be considered in mix design.

ACKNOWLEDGMENTS

The authors would like to acknowledge the financial support from the Programs for Science and Technology Development of Shandong Province, China (No. 2011GGX10705).

REFERENCES

[1] Behr-Andres, C.B. & N.J. Hutzler (1994). Characterization and use of fluidized-bed-combustion coal ash. *J. Journal of Environmental Engineering. 120*, 1488–1506.

[2] Chindaprasirt, P. & U. Rattanasak (2010). Utilization of blended Fluidized Bed Combustion (FBC) ash and pulverized coal combustion (PCC) fly ash in geopolymer. *J. Waste Manage. 30(4)*, 667–672.

[3] Li, X.G. Chen, Q.B. & K.Z. Huang. (2012). Cementitious properties and hydration mechanism of circulating fluidized bed combustion (CFBC) desulfurization ashes. Constr Build Mater;36: 182–187.

[4] Qian, J.S. Zheng, H.W. & Y.M. Song. (2008). Special properties of fly ash and slag of fluidized bed coal combustion. *J.J. Chin Ceram Soc. 36*, 1396–1400.

[5] Anthony, E.J. & D.L. Granatstein. (1994). Sulfation phenomena in fluidized bed combustion systems. *J. Prog Energ Combust Sci. 27*, 215–236.

Energy science and environmental engineering

Situation division for the high-temperature zone based on the utility function

Xiaofeng Ma

School of Equipment and Engineering, University of CPFA, Xi'an, China
School of Energy and Resources, Xi'an University of Science and Technology, Xi'an, China

ABSTRACT: This paper reveals the parameter of high-temperature distribution rule and builds the mathematic model with several parameters, using an improved experimental apparatus. A utility function that describes the temperature anomaly was newly defined. This study proposes a quantitative evaluation criterion. The function applying to the field is in good accordance with the actual observations value.

1 INTRODUCTION

The temperature anomaly seriously threatens the mine safety, which causes major fatal accidents of casualties. The spontaneous combustion of float coal is affected by five factors in the goaf, and it is coupled with oxygen, temperature, seepage field, and chemical reaction.

The traditional division of the self-heating zone is based on the heat release rate and the heat rate of coal and oxygen compounds. The method cannot suit conditions of the multi-field coupling. This paper defines a utility function to describe the dangerous degree through the experimental data and absolute parameters obtained by the heat transfer theory and aerodynamics theory.

2 EXPERIMENTAL DETERMINATION

The laboratory apparatus is circular shaped, with a maximum loading height of 200 cm, and an internal diameter of 120 cm. It is mainly composed of an oven shell, a gas circuit, and a control test, gas detection and analysis system. The fire wall which was composed of polyurethane acted as a thermal insulation layer and measured water layer. The electro-thermal tube and the inlet branch in the water were installed a get pipe separately. Inside the oven, there were 140 temperature probes and 24 gas sampling points arranged. The experimenter cramped out the gas sample and analyzed the composition and concentration by gas chromatography. The experimental conditions are given in Table 1.

It can be concluded from the experimental result that (Figure 1) the temperature rises rapidly at the middle to bottom part in the first stage of oxidation of coal body. The points whose rate changes faster move towards the wind inlet eventually at the edges of coal body as time goes by, and finally form an open flame. The coal sample radiating intensity is low when the coal temperature is below 110°C; however, the heat release rate increases considerably when the coal temperature increases more than 110°C gradually. According to the heat transfer theory, aerodynamics theory, and the experimental data, an approximate solution of the maximum intensity of air leakage, the minimum thickness of float coal, the minimum oxygen concentration, and the maximum mean particle size can be obtained by the spontaneous combustion of coal.

Table 1. Experimental conditions.

Volume	Density	Volume weight	Mean grain size	Coal height	Coal weight	Void ratio	wind supply volume	Initial temperature
(cm³)	(g/cm³)	(g/cm³)	d_{50} (mm)	(cm)	(kg)		(m³/h)	(°C)
2147760	1.40	1.093	4	190	2231.95	0.2577	0.15~0.4	26.2

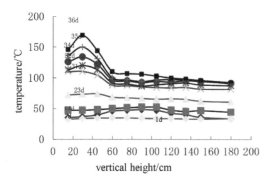

Figure 1. Relationship between the temperature and the height.

3 ABSOLUTE PARAMETER MODEL

3.1 Minimum thickness

When the value is less than the specified thickness, the heat produced by coal is equal to the coal dissipate. The specified thickness is termed the minimum thickness of float coal. According to the experimental data, it can be defined as:

$$\frac{1}{2}q_0(T_{ci}+T_f) - \lambda_c \frac{2\times(T_{ci}-T_f)}{(h/2)^2}$$

$$-\rho_g \cdot c_g \cdot \bar{Q} \cdot \frac{T_{ci}-T_w}{d} > 0 \qquad (1)$$

Abbreviating Equation (1), the model of the minimum thickness of float coal can be defined as

$$h_{\min} = 4\sqrt{\frac{(T_{ci}-T_f)\cdot\lambda_c}{q_0(T_{ci}+T_f)-2\rho_g \cdot c_g \cdot \bar{Q}\cdot(T_{ci}-T_w)/d}} \qquad (2)$$

where $h_{\min}$ is the minimum thickness of float coal (m), h is the thickness of float coal (m), $\bar{Q}$ is the air leakage intensity in the goaf (m/s), T_{ci}, T_f, and T_w are the highest temperature in the coal, the rock temperature, and the wind temperature, respectively (°C), λ_c is the thermal conductivity of loose coal, $q_0(T_{ci})$ is the oxidation thermal intensity (kJ·m^{-3}·s^{-1}), and d is the distance between the goaf and the working face (m).

3.2 Model for minimum oxygen concentration

Oxygen is constantly drained when the wind goes through the coal gap and follows on the surface of the coal. Oxygen can be defined as follows during the move.

$$P_x \frac{\partial^2 C_{O_2}}{\partial x^2} + P_y \frac{\partial^2 C_{O_2}}{\partial y^2} + P_z \frac{\partial C_{O_2}}{\partial z^2} - V_{O_2}(T) \qquad (3)$$

where, P_x, P_y and P_z are the oxygen diffusion coefficients (m^2/s), $V_{O_2}(T)$ is the oxygen concentration (mol/m^3), and $V^0_{O_2}$ is the oxygen consumption rate in the fresh air $(mol/(m^3 \cdot K))$.

According to the hot balance principle and the mechanism of spontaneous combustion of coal, the oxidation of coal body radiating strength $q(T_{ci})$ is proportional to the oxygen concentration, i.e.

$$q(T_{ci}) = \frac{C_{O_2}}{C^0_{O_2}}q_0(T_{ci}) \qquad (4)$$

where $q(T_{ci})$ is the average heating capacity (kJ·m^{-3}·s^{-1}), $C^0_{O_2}$ is the oxygen concentration in the fresh air (mol/m^3), and C_{O_2} is the actual oxygen concentration (mol/m^3).

This implies that the temperature of the coal body cannot rise if the heat produced by coal oxidation is less than that produced when it radiates. C_{min} is likely to change with the parameters of the coal temperature, the air leakage intensity, the distance, and the thickness of the float coal from Eq.4. The minimum oxygen concentration can be obtained under conditions of different thicknesses and coal temperatures, excluding the enthalpy change and thermal control.

3.3 Model of maximum intensity of air leakage, g

Owing to the factors of the shape, the magnitude and the distribution are all random. The influence of the wind can be excluded, and the permeability can be seen as isotropy, namely $k_x = k_y = k_z$.

If other things are equal, the air leakage becomes the main factor to determine the heat dissipating capacity, when the thickness is higher than $h_{\min}$ and the oxygen supply is adequate. The heat coal and oxygen compound to produce will be taken off entirely beyond a special amount of the wind, thus the maximum intensity of air leakage is defined as

$$\bar{Q}_{\max} = 16x\left[\frac{q_0(T_{ci}+T_f)}{32\rho_g \cdot c_g \cdot (T_{ci}-T_w)} - \frac{\lambda_c(T_{ci}-T_f)}{\rho_g \cdot c_g \cdot h^2 \cdot (T_{ci}-T_w)}\right] \qquad (5)$$

Model for maximum mean particle size

The mean grain size of the coal has an inverse relationship with the oxidation of radiating. It can be fitted as follows.

When specific conditions are met, the porosity increases as the particle size is too large. The coal cannot undergo combustion if the heat which is produced by coal and oxygen compounds is taken off entirely. The maximum mean particle size is defined as

$$\bar{d}_{max} =$$

$$d_r \cdot \exp\left\{\frac{div(\rho_g c_g \bar{Q} T) - div[\lambda_e grad(T)]}{n \cdot q_{o_2}(T)} - \zeta\right\} \quad (6)$$

where $\zeta = -m/n$ is dimensionless and $\bar{d}_{max}$ is the maximum mean particle size (m).

3.4 Model for the mechanized speed

The physical conditions of the field change continuously in advance of working face. It still requires enough time to remain unchanged under the conditions in the region. Spontaneous combustion can occur possibly.

$$v < \left(v_{min}\frac{L_{max}}{\tau_{min}}\right) \quad (7)$$

where, $\tau > \tau_{min}$, τ_{min} is the minimum period of combustion (d), $v \le L_{max} = \max\{L\}$, v is the face speed (m), and L_{max} is the maximum distance (m).

The course of heat output of coal-oxygen compounds is relevant to the main parameters above. According to the references, the necessary and sufficient conditions for spontaneous combustion of residual coal in the goaf are defined as:

$$\left(h > h_{min}\right) \cap \left(C_{O_2} > C_{min}\right) \cap \left(\bar{Q} < \bar{Q}_{max}\right)$$
$$\cap (d < \bar{d}_{max}) \cap (v < v_{min}) \quad (8)$$

4 UTILITY FUNCTION

4.1 The definition of utility function

The necessary and sufficient conditions for spontaneous combustion give only a qualitative judgment form in Eq.8, in order to realize the quantitative division of the hazard zone and move forward a single step. The necessary and sufficient conditions for spontaneous combustion can be defined as

$$P\left\{\left(h > h_{min}\right) \cap \left(C_{O_2} > C_{min}\right) \cap \left(\bar{Q} < \bar{Q}_{max}\right)\right.$$
$$\left.\cap \left(d < \bar{d}_{max}\right) \cap \left(v < v_{min}\right)\right\} \quad (9)$$

During the process of the spontaneous combustion of coal, the limit parameters causing the spontaneous combustion of coal belong to the mutually independent random variables. Eq.9 can be further expanded in Eq.10.

$$P(h > h_{min})P(C_{O_2} > C_{min})P(\bar{Q} < \bar{Q}_{max})$$
$$P(d < \bar{d}_{max})P(v < v_{min}) \quad (10)$$

Now, we can define a new probability function of spontaneous combustion of coal, and the function is defined as

$$F_c = \prod_{i=1}^{5} P(x_i) \quad (11)$$

where F_c is the newly defined probability function, x is the actual observed value, and x_{min} or x_{max} is the key parameter.

Integrating the key parameters into Eq.11, and then into Eq.12, the following can be obtained:

$$F_c = (h / h_{min}) \cdot (C_{O_2} / C_{min}) \cdot \left(\bar{Q}_{max} / \bar{Q}\right) \cdot$$
$$\left(\bar{d}_{max} / d\right) \cdot (v_{min} / v) \quad (12)$$

The probability function has the following characteristics: the hazard zone can be divided by the different coupling key parameters, and it can singly increase or decrease, thus the coupled relation among the key parameters can be more quantitative.

If the ratio between the actual value of the parameter and the parameter values is 1/3 (Eq.13), the hazard zone belongs to the fully extinguished zone:

$$F_c = 1/3^5 \quad (13)$$

If the ratio between the actual value of the parameter and the parameter values is 1/2 (Eq.11), the hazard zone belongs to the base extinguished zone:

$$F_c = 1/2^5 \quad (14)$$

In a similar way, if

$$1/2^5 \le F_c \le 1$$

it belongs to the tropical zone.
If

$$F_c \ge 1 \quad (15)$$

it belongs to the oxidation temperature zone.

4.2 Method and procedure

1. Based on laboratory test conditions, the key parameters of spontaneous combustion of coal are determined.
2. According to the field conditions, the actual value is given.
3. Based on the first two steps of the data, the hazard zone by the following block diagram is judged.

165

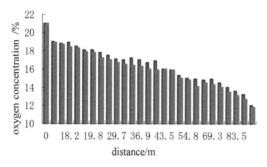

Figure 2. Oxygen concentration curve and the distance.

5 APPLICATION EXAMPLE

According to the actual parameters in the 15109 workface and the procedure shown in Fig. 5, the result can be observed from Fig. 8 that the tropical distribution has a similar law to the distribution of float coal thickness. As the float coal thickness is only 0.57 m, it does not meet the minimum requirement for the float coal thickness. Through the probability function, the value of F_c can range from 1/25 to 1, and therefore, it belongs to the tropical zone. The oxygen concentration is still maintained high at a distance of 40 m, where the float coal thickness reaches 2.48 m, and the value of F_c is higher than 1, and thus the region belongs to the oxidation temperature zone. The value of F_c at a distance of 97 m is less than $1/3^5$, and it belongs to the fully extinguished zone.

Considering all of these factors, the oxidation temperature zone in the goaf on the inlet side can extend to 45 m around, and on the exhaust section, it amounts to 35 m around because of the coal pillar. The hazard zone is most likely in the inlet side. Some measures such as nitrogen injection and grouting must be done to prevent and control the self-ignition. Through the probability function, the value of F_c can range from 1/25 to 1, and hence, it belongs to the tropical zone. The oxygen concentration still remains high at a distance of 40 m, where the float coal thickness reaches 2.48 m, and the value of F_c is higher than 1, and thus the region belongs to the oxidation temperature zone. The value of F_c at a distance of 97 m is less than $1/3^5$, and it belongs to the fully extinguished zone.

6 SUMMARY

1. A probability function is newly defined to quantitatively determine "three belt", and set up the standard of criterion, $1/3^5$, $1/2^5$, 1, and then the procedure and method are carried out to judge the dangerous zone.
2. With the probability function applied to determine the dangerous area in the workface, the oxidation temperature zone can extend to 45 m around on the inlet side and 35 m around on the exhaust section. The results conform to the actual conditions.

REFERENCES

Garcia P. The use of differential scanning calorimetry to identify coal ssusceptible to spontaneous combustion [J]. Thermochimicaacta. 1999.336(1–2):41–46.

Lopez D. Effect of low-temperature oxidation of coal on hydrogen-transfer capability Fuel. 1998, 77(14): 1623–1628.

Luo Ji'an, Wang Lianguo, Tang Furong, He Yan, Zheng Lin. Variation in the temperature field of rocks overlying a high-temperature cavity during underground coal gasification [J]. Mining Science and Technology (China) 21 (2011) 709–713.

McNabb A., Please C.P. & McElwain D.L.S. Spontaneous combustion in coal pillars: Buoyancy and oxygen starvation. Mathematical Engineering in Industry. 1999, 7(3): 283–300.

Nugrohol Y.S., McIntosh A.C. & Gibbs B.M. Low-temperature oxidation of single and blended coals. Fuel, 2000, 79(15): 1951~1961.

Ryzhkov A. Yu, Makarov E. Ya & Bonetskij V.A. About an influence of antipyrogenous processing on the process of coal spontaneous heating. Fiziko~Tekhnicheskie Problemy Razrabotki Poleznykh Iskopaemykh, 1991, 2: 81~84.

Salinger Andrew G et al. Modeling the spontaneous ignition of coal stockpiles. AIChE Journal, 1994, 40, (6): 991~1004.

Schmal Dick et al. Model for the spontaneous heating of coal. Fuel. 1985, 64(7): 963~972.

Sujanti, Zhang Wiwik, Chen Dongke, et al. Low~temperature oxidation of coal studied using wrie~mesh reactors with both steady~state and transient methods[J]. Combustion and Flame, 1999, 117(3): 646~651.

*Advances in Materials Science, Energy Technology
and Environmental Engineering – Patty & Zhou (Eds)*
© *2017 Taylor & Francis Group, London, ISBN 978-1-138-19668-1*

Network security risk assessment model based on temperature

Yaping Jiang, Congcong Cao, Xiao Mei & Hao Guo
*School of Computer and Communication Engineering, Zhengzhou University of Light Industry,
Zhengzhou, China*

ABSTRACT: In view of the traditional network security risk assessment model of real time, accuracy, and characterization, it has certain limitations. Reference to the mechanism of biological immune system, imbalance caused by temperature changes, a Network Security Assessment Model based on Temperature (NSRAM-T) was proposed. The model adopts an r-contiguous bits non-constant matching rate algorithm to improve the detection quality of a detector, reducing the missing rate or the false detection rate. According to activating mature detector and cloning memory detector, it can cause an increase in the antibody concentration by the mechanism. The equation of antibody concentration increase factor is derived. Then, the antibody concentration quantitative calculation model is established. The reference to the mechanism by which the antibody concentration changes can effectively reflect the network risk, which established the temperature assessment model. The simulation results indicated that, according to the temperature value, the proposed model has more effective, real-time to assess the network security risk.

1 INTRODUCTION

With the continuous expansion of the network size, and the increasingly complex network structure and the rapid development of information technology, the research of assessment model has become one of the hot topics in the network security field. Kotenko et al. had proposed a safety assessment framework based on attack graph (Igor, 2013). The method has higher computational complexity. Taking into consideration the impact of time and environmental factors, Masoud proposed a quantitative risk assessment method by using Bayesian attack graphs (Masoud, 2014). Using experiments (Wu Di et al.), it has been proved that the effectiveness of the security threat recognition and analysis method were based on attack graphs (Wu, 2015). The literature (Chen, 2015) proposed a network security situation assessment method based on immune danger theory, but the method cannot perceive more situational factors and complex network security situation.

Although the above literatures can be accurately used to evaluate network security, the result of evaluating the network environment lacks a certain flexibility. This paper puts forward a Network Security Risk Assessment Model based on Temperature (NSRAM-T), and the model makes characterization of the network of the immune system more in line with the biological immune system. It can be used to assess the network security risk.

2 THE MODEL OF BASIC THEORY AND DESIGN IDEA

Biological immune system is a highly distributed self-adaption learning system. It has a sound mechanism to resist the invasion of foreign pathogens. After the body is infected with a pathogen, it can produce a specific antibody and effector T cells to improve the immunity of the pathogen. However, the biological immune system itself before the recovery of adaptive regulation will produce fever and other symptoms; with the increase in virus threat intensity, the biology of temperature can also be increased. Thus, when a computer is attacked by outside illegal attacks or internal network illegal activity, according to the mechanism of the biological immune system, an antibody (detector) in the computer immune system can quickly recognize these antigens (illegal activity). Similar to the increase in the antibody concentration, the computer temperature will also increase with a certain rising trend; at the same time, the network of multiple computers can also evaluate the temperature status of the entire network based on the importance of each computer. According to the temperature, the network risk is evaluated; the temperature value size can be more convenient, can more directly determine risk levels, and can make the corresponding protective measures.

The NSRAM-T model is composed of intrusion detection, antibody concentration, and three parts for temperature assessment. Design ideas are briefly summarized as follows: (1) the detected attacks are classified by the blood; (2) according to the matching process of antigen and antibody, it could calculate the corresponding attack types of the antibody concentration; and (3) temperature is based on the antibody concentration, and the temperature range could also be mapped to a defined temperature area, through the temperature assess network risk.

3 RISK ANALYSIS AND CALCULATION BASED ON THE NSRAM-T MODEL

In order to perform the real-time assessment network security risk more accurately, the model uses the r-contiguous bits non-constant matching rate algorithm to improve the the detection quality of the detector (Feng, 2014). In order to meet the real network environment, self, various detectors, and the corresponding tolerance all are dynamic changes. For obtaining risk assessment more intuitively, the model determines the risk level by changes in temperature.

Let self/non-self in the domain $X \in \{0,1\}^l (l > 0)$, S is the *self* set (the normal behavior of the network), N is the *non-self* set (network illegal behavior or attack),with $S \subset X, N \subset X, S \cup N = X, S \cap N = \varnothing$. Detector set D: $D = \{d_1, d_2, ..., d_n\}, (l_1 \leq l, i \in N)$. Antigen $(Ag \subset X)$ is defined as network intrusion behaviors, identifying the antigen of antibody $(Ab \subset D)$ as a detector.

3.1 *r-contiguous bits non-constant matching algorithm*

Owing to the constant r-contiguous bits matching the algorithm will not be able to detect illegal network behavior more accurately, so the matching antigen and antibody use the r-contiguous bits non-constant matching rate algorithm (Gao, 2013).

The algorithm utilizes the segmentation technology and key position, according to the importance of each section set different matching thresholds. In order to avoid the "black hole" and reduce the missing rate or the false detection rate, while improving the detection quality of the detector. "1" represents "match" and "0" represents "mismatch". The matching calculation method, as shown in equations (1)–(4), is

$$g_{match}(F_1, F_2, r) = \begin{cases} 1, & f \geq r, F_1 \in D, F_2 \in X \\ 0, & otherwise \end{cases}. \quad (1)$$

$$f = \sum_{i=1}^{n} \alpha_i match(i). \quad (2)$$

$$\sum_{i=1}^{n} \alpha_i = 1. \quad (3)$$

$$match(i) = \begin{cases} 1, & iff \exists i, j \left(x_{Key_{i,j}} = x_{Key_{i,j}}, i, j \in N^* \right) \\ \vee \exists i, m, n \left(n - m \geq r, 0 < m, n \leq L/M, r \in N^* \right). \\ 0, & otherswise \end{cases} \quad (4)$$

where the lengths of the match string F_1, F_2 are L and they are respectively divided into M segments, set key position $Key_{i,j}$ in the key field; the matching threshold of each field is set at α_i, $x_{Key_{i,j}} = x_{Key_i}$, which represents the key position of fragment i to be the same as $Key_{i,j}$. f is defined as the sum of each fragment of the matching threshold multiplied by 1 or 0.

3.2 *The antibody concentration quantitative calculation model*

A change in the antibody concentration is due to the illegal intrusion (antigen) computer immune system producing the immune response caused by the imbalance in the immune system; more antigens caused more serious imbalance, i.e. change in the antibody concentration is more obviously increasing, and after the antigen has disappeared (killed), they gradually tend to be normal, but there is a certain duration, if no matching with antigen is done for a long time, the antibody concentration will be attenuated according to certain rules.

Definition 1 The formula of increasing antibody concentration is defined as

$$C_{ab}(t) = C_{ab}(0) + kC_{ab}(t-1), k \in (0,1). \quad (5)$$

where $k = \frac{D_{Ma}^{active}(t) + D_{Mr}^{clone}(t)}{D_{initial}(t) + D_{Ma}(t) + D_{Mr}(t)}$, $D_{Ma}^{active}(t)$ shows that the amount of memory detectors activated by the mature detector, $D_{Mr}^{clone}(t)$, is the number of clone memory detectors, $D_{initial}(t)$ is the number of immature detectors, $D_{Ma}(t)$ is the total number of mature detectors, $D_{Mr}(t)$ is the total number of memory detectors, $C_{ab}(0)$ is the initial antibody concentration, and k is the antibody concentration increase factor.

The above formula can be converted into $C_{ab}(t) = \frac{1-k^t}{1-k} C_{ab}(0)$ $(0 < k < 1 \wedge t > 0)$, when $t \to +\infty$ antibody concentration tends to be $C_{ab}(t) = \frac{1}{1-k} C_{ab}(0), k \in (0,1)$.

Definition 2 Without considering the threat of attack types and the importance of equipment in the network, the host s under q attack of antibody concentration formula is defined as

$$C_{ab}^{sq}(t) = C_{ab}^{sq}(0) + k_{sq}C_{ab}^{sq}(t-1). \quad (6)$$

where $k_{sq} = \frac{D_{Ma}^{active(q)}(t)+D_{Mr}^{clone(q)}}{D_{initial}^{q}(t)+D_{Ma}^{q}(t)+D_{Mr}^{q}(t)}$, k_{sq} is the antibody concentration increase factor under q attack, $D_{Ma}^{active(q)}(t)$ is the number of activated memory detectors under q attack, $D_{Mr}^{clone(q)}$ is the number of clone memory detectors under q attack, $D_{initial}^{q}(t)$ is the number of immature detectors under q attack, $D_{Ma}^{q}(t)$ is the number of mature detectors under q attack, $D_{Mr}^{q}(t)$ is the number of memory detectors under q attack, and $C_{ab}^{sq}(0)$ is the initial antibody concentration before q attack.

Definition 3 The threat of τ attack is σ_{τ}, and $\sigma_{\tau} = 1 - 2e^{-\sqrt{10num_{attack}^{\tau}+10str^{\tau}}}$; the host s under all of attacks of antibody concentration formula is defined as

$$C_{ab}^{s}(t) = \frac{1}{q_n}\sum_{\tau=q}^{q_n}\delta_{\tau}C_{ab}^{s\tau}(t). \qquad (7)$$

where num_{attack}^{τ} is the number of τ attacks, and str^{τ} is the intensity of the τ attacks.

Theorem 1 In the case of threat constant, the antibody concentration of the host s was strengthened with the increase in the number of categories of attacks, i.e. $\delta_q C_{ab}^{sq}(t) < \left(\delta_q C_{ab}^{sq}(t)+\delta_q C_{ab}^{sq_1}(t)\right)/2 < ...$ $< \left(\sum_{\tau=q}^{q_n}\delta_q C_{ab}^{s\tau}(t)\right)/q_n q_n$ shows the types of attacks and $q_n \in N*$.

Proof:

When t is zero,
$\delta_q C_{ab}^{sq}(0) = \left(\delta_q C_{ab}^{sq}(0)+\delta_q C_{ab}^{sq_1}(0)\right)/2 = ... = \left(\sum_{\tau=q}^{q_n}\delta_q C_{ab}^{s\tau}(0)\right)/q_n$

When t is greater than zero and τ is equal to q_1,
$(\delta_q C_{ab}^{sq}(t)+\delta_q C_{ab}^{sq_1}(t))/2 = \left(\sum_{\tau=q}^{q_1}\delta_q C_{ab}^{s\tau}(t)\right)/2 = \left(\sum_{\tau=q}^{q_1}\delta_q \frac{1-k_{s\tau}^t}{1-k_{s\tau}}C_{ab}^{s\tau}(0)\right)/2$

$= \left(C_{ab}^{s}(0)\delta_q \sum_{\tau=q}^{q_1}\frac{1-k_{s\tau}^t}{1-k_{s\tau}}\right)/2$. And so on, when t is greater than zero and τ is equal to q_n, $\left(\sum_{\tau=q}^{q_n}\delta_q C_{ab}^{s\tau}(t)\right)/q_n = \left(C_{ab}^{sq}(0)\delta_q \sum_{\tau=q}^{q_n}\frac{1-k_{s\tau}^t}{1-k_{s\tau}}\right)/q_n$.

Therefore, it is necessary to only prove that
$\frac{1-k_{sq}^t}{1-k_{sq}} < \left(\frac{1-k_{sq}^t}{1-k_{sq}}+\frac{1-k_{sq_1}^t}{1-k_{sq_1}}\right)/2 < ... < \sum_{\tau=q}^{q_n}\frac{1-k_{s\tau}^t}{1-k_{s\tau}}/q_n$.

Owing to the immune system according to the rules of the LRU to weed out all kinds of detectors, the overall size stays the same, and so $k_{sq} = \frac{D_{Ma}^{active(q)}(t)+D_{Mr}^{clone(q)}}{D_{initial}^{q}(t)+D_{Ma}^{q}(t)+D_{Mr}^{q}(t)} = \frac{D_{Ma}^{active(q)}(t)+D_{Mr}^{clone(q)}}{D_{initial}(t)+D_{Ma}(t)+D_{Mr}(t)}$, with the increase in the number of categories of attacks, and from the above formula, molecular increases and the denominator remains the same, then $k_{sq} < k_{sq_1} < ... < k_{sq_n}$.

$\left(\left(\frac{1-k_{sq}^t}{1-k_{sq}}+\frac{1-k_{sq_1}^t}{1-k_{sq_1}}\right)/2\right)/\frac{1-k_{sq}^t}{1-k_{sq}} \approx \left(\left(\frac{1}{1-k_{sq}}+\frac{1}{1-k_{sq_1}}\right)/2\right)$

$*(1-k_{sq}) = \left(1+\frac{1-k_{sq}}{1-k_{sq_1}}\right)/2$

As $k_{sq} < k_{sq_1}$, $\left(1+\frac{1-k_{sq}}{1-k_{sq_1}}\right)/2 > 1$, i.e. $\frac{1-k_{sq}^t}{1-k_{sq}} < \frac{1-k_{sq}^t}{1-k_{sq}}+\frac{1-k_{sq_1}^t}{1-k_{sq_1}}/2$

Similarly, $\left(\sum_{\tau=q}^{q_n}\frac{1-k_{s\tau}^t}{1-k_{s\tau}}\right)/q_n\sum_{\tau=q}^{q_{n-1}}\frac{1-k_{s\tau}^t}{1-k_{s\tau}}/q_{n-1} > 1$.

Thus, with the more kinds of attacks, the antibody concentration is also rising.

Definition 4 When μ_s is the importance of the host s in the network, at the moment t, all hosts s_n under q attack of antibody concentration formula are defined as $C_{ab}^{q}(t) = \frac{1}{s_n}\sum_{s=1}^{s_n}\mu_s C_{ab}^{sq}(t)$, where $\mu_s = 1 - 2e^{-\sqrt{\alpha_s+\beta_s}}$, α_s is the *price* of the host s and β_s is the memory of the host s. Then, all hosts s_n (i.e. entire network) under all attacks of antibody concentration formula are defined as $C_{ab}(t) = \frac{1}{s_n}\sum_{s=1}^{s_n}\mu_s\sum_{\tau=q}^{q_n}\delta_{\tau}C_{ab}^{s\tau}(t)$.

3.3 Temperature assessment model

According to the mechanism of the biological immune system, the temperature is caused by elevation in the face of external viruses and other harmful substances (fever phenomenon), indicating that the invasion of harmful substances alters the physiological regulation of equilibrium. Network subjects to risks caused by external attacks with which they have the same purpose. Therefore, in order to distinguish the network degree of risk more conveniently and intuitively, using the temperature to assess the network risk, the temperature will be divided into different stages and defined different colors; depending on the different colors, the danger zone can be quickly determined.

The host s under q attack of temperature calculation formula, as shown in equation (8), is as follows:

$$T_{sq}(t) = 3\left(1-\ln\left(1+e^{-\sqrt{\delta_q C_{ab}^{sq}(t)}}\right)\right)-1. \qquad (8)$$

The host s under all of attacks of temperature calculation formula, as shown in equation (9), is as follows:

$$T_s(t) = 3\left(1-\ln\left(1+e^{-\sqrt{C_{ab}^{s}(t)}}\right)\right)-1$$
$$= 3\left(1-\ln\left(1+e^{-\sqrt{\frac{1}{q_n}\sum_{\tau=q}^{q_n}\delta_{\tau}C_{ab}^{s\tau}(t)}}\right)\right)-1. \qquad (9)$$

All hosts under q attack of temperature calculation formula, as shown in equation (10):

169

$$T_q(t) = 3\left(1 - \ln\left(1 + e^{-\sqrt{C_{ab}^q(t)}}\right)\right) - 1$$

$$= 3\left(1 - \ln\left(1 + e^{-\sqrt{\frac{1}{s_n}\sum_{s=1}^{s_n}\mu_s C_{ab}^{sq}(t)}}\right)\right) - 1. \qquad (10)$$

All hosts under all of attacks of temperature calculation formula, as shown in equation (11):

$$T(t) = 3\left(1 - \ln\left(1 + e^{-\sqrt{C_{ab}(t)}}\right)\right) - 1$$

$$= 3\left(1 - \ln\left(1 + e^{-\sqrt{\frac{1}{q_n s_n}\sum_{s=1}^{s_n}\mu_s \sum_{\tau=q}^{q_n}\delta_\tau C_{ab}^{s\tau}(t)}}\right)\right) - 1. \qquad (11)$$

Owing to the temperature range of 0 to 1 and as the defined temperature range is different, the temperature T needs to adopt deviation standardization of the inverse function to T^* (Naderpour, 2013), i.e. $T^* = T(\max - \min) + \min = 5T + 1$. The standardized temperature range is 1 to 6. The function of network temperature is defined as: $T' = 34 + T^*$.

4 SIMULATION EXPERIMENTS AND ANALYSIS

This model uses the r-contiguous bits non-constant matching rate algorithm in the stage of invasion, selects Artificial Immune Algorithm (AIA), and $r \in [2,10]$. It has been proved that the matching algorithm can improve the detection rate of the non-self and reduce the false detection rate of self, as shown in Figure 1, 2.

In order to verify the feasibility and effectiveness of the method described in this paper, this paper used the typical types of attacks (such as SYN Flood, Land, Smurf attacks, etc.) of simulation experiment carried out. The experimental network is composed of twenty hosts, and the hosts s_1, s_2,

etc., are monitored. In this experiment, the selected parameters are: initial antibody concentration is 0.015, the host s_1, s_2 price is, respectively, 0.3, 0.6 a thousand yuan, hosts' memory is respectively 2G, 4G, the intensities of the attacks of SYN Flood, Land, and Smurf attack are, respectively, 0.5, 0.8, and 0.1, and the number of attacks are 0.2, 0.1, and 0.15, respectively.

From Figure 3, it can be observed that the antibody concentration and temperature change trend has good consistency. The temperature evaluation of the entire network is illustrated in Figure 4. As we can see from Figure 4, in the moment of 40, the temperature significantly increased, and the temperature value was at a low-risk stage. In the moment of 50 to 60, the temperature was slowly decreasing, and it was still at a low-risk stage,

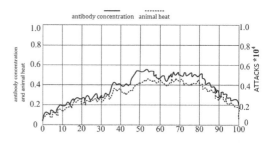

Figure 2. The self flase detection rate of two matching algorithms.

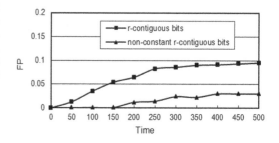

Figure 3. Host s_1 of antibody concentration and temperature evaluation factor.

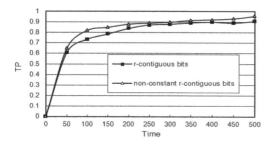

Figure 1. The non-self detection rate of two matching algorithms.

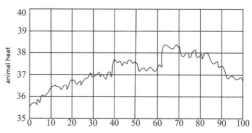

Figure 4. The temperature evaluation of the entire network.

because the system does not take measures. The temperature decreased because the system did not suffer new attacks in a certain period of time, and a part of the mature detector results in death. In the moment of 70, the temperature increased, and the temperature value was at a moderate risk stage. However, in the moment of 80, the temperature value was decreased to a low-risk stage, during this period, indicating that the system takes the corresponding measures.

5 CONCLUSION

This paper references the mechanism of temperature change caused by biological immune system imbalance, and analyzes the change in the antibody concentrations caused by the change process of various types of detectors in the computer immune system. This paper proposed a Network Security Assessment Model based on Temperature (NSRAM-T). The model established the evaluation equation of antibody concentration and temperature in this paper, and temperature values mapped to more easily convenient intuitive judgment dangerous levels of body temperature range, make it more in line with the mechanism of biological immune system, more practical significance. Simulation results indicate that the model can be based on the temperature value and the corresponding color, relatively more effective, real-time and intuitive to assess network security risk.

ACKNOWLEDGMENTS

This work was financially supported by the National Natural Science Foundation (No. 61272038), Henan Science and Technology Agency-funded science and technology research projects (No. 0624220084), and Henan Science and Technology Department of Basic and cutting-edge technology projects (No. 122300410255).

REFERENCES

Di Wu, Yi-feng Lian, Kai Chen, Yu-ling Liu. A security threats identification and analysis method based on attack graph [J]. Chinese Journal of Computers, Vol. 35(2012), p. 1938–1950.

Igor Kotenko, Andrey Chechulin. A cyber attack modeling and impact assessment framework [C]. The 5th International Conference on Cyber Conflict, (2013), p. 1–24.

Masoud Khosravi-Farmad, Razieh Rezaee, Abbas Ghaemi Bafghi. Considering temporal and environmental characteristics of vulnerabilities in network security risk assessment [C]. The 11th International ISC Conference on Information Security and Cryptology (ISCISC), (2014), p. 186–191.

Naderpour, M., J. Lu, G. Zhang. A fuzzy dynamic bayesian network-based situation assessment approach [C], 2013 IEEE International Conference on Fuzzy Systems (FUZZ). (2013), p. 1–8.

Xiang Feng, Mei-yi Ma, Tian-ling Zhao, Hui-qun Yu. Intrusion detection system based on hybrid immune algorithm [J]. Journal of Computer Science, Vol. 41(2014), p. 43–47.

Yan-ling Chen, Guang-ming Tang, Yi-feng Sun. Assessment of network security situation based on immune danger theory[J]. Journal of Computer Science, Vol. 42(2015), p. 167–170.

Zhi-qiang Gao, Xiao-qin Hu. Design and implementation of real-time network risk control system based on antibody concentration [J]. Journal of Computer Applications, Vol. 33(2013), p. 2842–2845.

*Advances in Materials Science, Energy Technology
and Environmental Engineering – Patty & Zhou (Eds)*
© 2017 Taylor & Francis Group, London, ISBN 978-1-138-19668-1

Influencing factors analysis of the testing system for the fire detection and fire alarm system

Tingting Wang
Shanghai Institute of Quality Inspection and Technical Research, Shanghai, China

ABSTRACT: A widely used testing system for fire detection and fire alarm systems is introduced. This system includes a smoke tunnel, a Measuring Ionization Chamber (MIC), an obscuration meter, and an aerosol generator. The system testing repeatability is analyzed here. In addition, it concludes that the rate of increase in the aerosol density, the flow rate of aerosol, and the position will affect the testing result of the system. All these parameters should be closed properly. This work may be beneficial to testing laboratory, system manufactory, and standard revision personnel.

1 INTRODUCTION

The fire detection and fire alarm system is widely used in public places for safety purposes. The false alarm especially the lag of alarm will cause adverse consequences. Therefore, the testing of the fire detection and fire system is very important. Many standards specify the detail testing method, such as BS EN54-7:2001 "Fire detection and fire alarm systems", etc.

Smoke density can be measured by different techniques (Chen et al. 2007; Jackson & Robins 1994; Shibata et al. 2010; Heskeatad & Newman 1992). As we know, there are two kinds of widely used smoke measuring equipment (Aggarwal & Motevalf 1997), which are obscuration meter and Measuring Ionization Chamber (MIC) (Bernigau & Luck 1986; Scheidweiler 1976; Helsper et al. 1983). They can be integrated in a testing system to satisfy different testing requirement. A typical testing system for fire detection and fire alarm system includes a smoke tunnel, a Measuring Ionization Chamber (MIC), an obscuration meter, and an aerosol generator.

In this paper, the factors influencing the overall measuring accuracy of the testing system are analyzed. This will be a beneficial reference for testing laboratories and manufactories that will use and set up the testing system. It is also a reference for testing standard revision.

2 TESTING SYSTEM BASED ON A SMOKE TUNNEL

The testing system is based on a smoke tunnel. Integrate the Measuring Ionization Chamber

(MIC), the obscuration meter, and the aerosol generator into this system. Use software to control the rate of increase in aerosol density and record the response threshold of detectors such as y, X, and m. The set-up principle which comes from the instruction manual for smoke measuring equipment MIC type EC-912 of Delta company is shown in Figure 1. The appearance of the system is shown in Figure 2.

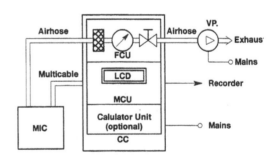

Figure 1.　Schematic diagram of the set-up principle.

Figure 2.　The appearance figure of the system.

3 ANALYSIS OF THE INFLUENCING FACTORS

Testing environment: air temperature is 20.6°C and air humidity is 45.4%. Air velocity is 0.2 m/s. Test equipment used in this study are as follows: Measuring Ionization Chamber (MIC): Delta MIC EC-912; smoke tunnel: AWT 2800.

In the flowing, the testing results of Measuring Ionization Chamber (MIC) indicate that the value X or y is used to verify the system repeatability and analyze the influencing factors.

The measured smoke density is expressed in terms of the dimensionless quantity defined by the equation:

$$X = (I_0 - I)/ I_0, \tag{1}$$

where I_0 is the quiescent ionization chamber current in clean air and I is the ionization chamber current in the presence of smoke.

Moreover, there is a relationship between the values X and y, which is

$$y = X(2 - X)/(1 - X). \tag{2}$$

3.1 System testing repeatability

Here, a given Measuring Ionization Chamber (MIC) is used to verify the system repeatability. For the given testing parameters, the test is repeated six times, the value y of MIC is obtained and the system repeatability is verified by comparing the value y denoting the smoke density in the smoke tunnel in different testing cycles. The testing result is shown in Figure 3.

In the testing standards, a scope of the rate of increase in the aerosol density is given. For detectors using scattered or transmitted light, it is

$$0.015 \text{ dB}^{-1} \text{min}^{-1} \leq \Delta m / \Delta t \leq 0.1 \text{ dB}^{-1} \text{min}^{-1}.$$

For detectors using ionization, it is

$$0.05 \text{min}^{-1} \leq \Delta y / \Delta t \leq 0.3 \text{min}^{-1}.$$

The chamber voltage (VDC) is 19.3V. The rate of increase in the aerosol density is $0.020 \text{ db}^{-1}\text{min}^{-1}$. The flow rate of aerosol is 14 l min⁻¹. The flow rate of the vacuum pump is 30 l min⁻¹. Preheat the smoke tunnel for 1 hour and the control unit of the Measuring Ionization Chamber (MIC) for 15 minutes. The smoke tunnel must be emptied before each testing.

From Figure 3, it can be seen that the value y of the MIC fits well, but still has a tiny difference. It has been predicted that this tiny difference comes from many factors. The smoke tunnel is not completely emptied, which is shown in curve 8. The actual rate of increase in the aerosol density cannot perfectly fit with the set-up value. This may come from the inappropriate setting of the flow rate of aerosol.

3.2 Analysis of influencing factors

In chapter 3.1, the system repeatability is discussed. It is found that there are some tiny differences between difference testing circles. Here, some influencing factors will be analyzed.

Two similar Measuring Ionization Chambers (MIC) are required to perform the test. The error between them is less than 0.1%. The samples were labeled as No. 1 and No. 2, respectively.

3.2.1 The rate of increase in the aerosol density

The samples No.1 and No. 2 are placed in the vertical direction of the smoke flow. The schematic diagram is shown in Figure 4.

Adjust the aerosol generator. The Measuring Ionization Chambers (MICs) (sample 1 and 2) are exposed to a slowly increasing smoke density

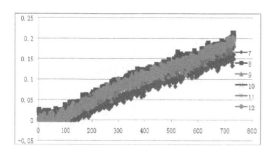

Figure 3. The result of testing of the system repeatability.

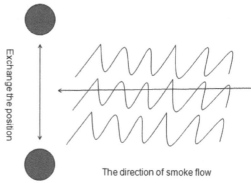

The direction of smoke flow

Figure 4. Schematic diagram of the position of the samples.

produced by a paraffin oil aerosol generator. The flow rate of aerosol is 20 1 min⁻¹. The rates of increase in the aerosol density are 0.05 min⁻¹ and 0.20 min⁻¹. The values X of the samples No. 1 and No. 2 are recorded at different rates of increase in the aerosol density. The values are recorded as X^{1-1}, X^{2-1}, X^{1-2}, X^{2-2}. X^{1-1} and X^{2-1} are denoted as abscissa and ordinate and X^{1-2} and X^{2-2} are denoted as abscissa and ordinate, respectively. Carry on the linear numerical fitting. Compare the testing difference between the two samples at different rates of increase in the aerosol density. Table 1 shows the testing result of different rates of increase in the aerosol density. Figure 5 shows the numerical fitting.

From Figure 5, it can be concluded that there is a smaller difference when the testing is performed in a high rate of increase in the aerosol density. In addition, the difference between these two situations is very small. The error is 0.036%.

Table 1. The testing result of different rates of increase in the aerosol density.

The rate of increase in the aerosol density 0.05 min⁻¹		The rate of increase in the aerosol density 0.20 min⁻¹	
X_{1-1}	X_{2-1}	X_{1-2}	X_{2-2}
0.00	0.00	0.00	0.00
0.04	0.05	0.03	0.05
0.09	0.10	0.07	0.10
0.14	0.15	0.12	0.15
0.19	0.20	0.17	0.20
0.24	0.25	0.23	0.25
0.29	0.30	0.27	0.30
0.33	0.35	0.33	0.35
0.39	0.40	0.38	0.40
0.43	0.45	0.43	0.45
0.49	0.50	0.47	0.50

3.2.2 The flow rate of aerosol

The samples No.1 and No. 2 are placed in the vertical direction of the smoke flow. The schematic diagram is shown in Figure 4.

Adjust the aerosol generator. The Measuring Ionization Chambers (MICs) (sample No. 1 and No. 2) are exposed in a slowly increasing smoke density produced by a paraffin oil aerosol generator. The rate of increase in the aerosol density is 0.05 min⁻¹. The flow rates of aerosol are 20 1 min⁻¹ and 30 1 min⁻¹, respectively. The values X of the samples No. 1 and No. 2 are recorded at different flow rates of aerosol. Record as X^{1-1}, X^{2-1}, X^{1-2}, X^{2-2}. Denote X^{1-1} and X^{2-1} abscissa and ordinate and denote X^{1-2} and X^{2-2} abscissa and ordinate, respectively. Carry on the linear numerical fitting. Compare the testing difference between the two samples at different flow rates of aerosol. The testing result at different flow rates of aerosol are given in Table 2. The numerical fitting is shown in Figure 6.

From Figure 6, it can be concluded that there is a small difference when the testing is performed at different flow rates of aerosol. The error is 1.06%.

Table 2. The testing result at different flow rates of aerosol.

flow rate of aerosol 20 1 min⁻¹		flow rate of aerosol 30 1 min⁻¹	
X_{1-1}	X_{2-1}	X_{1-2}	X_{2-2}
0.00	0.00	0.00	0.00
0.05	0.03	0.05	0.03
0.10	0.09	0.10	0.08
0.15	0.14	0.15	0.13
0.20	0.19	0.20	0.18
0.25	0.24	0.25	0.23
0.30	0.30	0.30	0.29
0.35	0.35	0.35	0.35
0.40	0.40	0.40	0.39
0.45	0.45	0.45	0.46
0.50	0.50	0.50	0.50

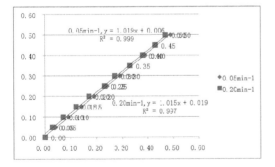

Figure 5. Numerical fitting of testing result at different rates of increase in the aerosol density.

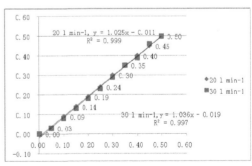

Figure 6. Numerical fitting of testing result at different flow rates of aerosol.

When in high flow rate of aerosol, the generation of the smoke is uneven. This may cause a larger error between the same samples.

3.2.3 *Position of the samples*

The samples No. 1 and No. 2 are placed in the vertical direction of the smoke flow. The schematic diagram is shown in Figure 4. Then, the positions are swapped and the same test repeated.

Adjust the aerosol generator. The Measuring Ionization Chambers (MICs) (sample No. 1 and No. 2) are exposed to a slowly increasing smoke density produced by a paraffin oil aerosol generator. The rate of increase in the aerosol density is 0.05 min⁻¹. The flow rate of aerosol is 30 l min-1. Record the values X of the samples 1 and 2 at different flow rates of aerosol. Record as X^{1-1}, X^{2-1}, X^{1-2}, X^{2-2}. Denote X^{1-1} and X^{2-1} abscissa and ordinate and denote X^{1-2} and X^{2-2} abscissa and ordinate, respectively. Carry on the linear numerical fitting. Compare the testing difference between the two samples at different positions. The testing results at different positions are given in Table 3. The numerical fitting is shown in Figure 7.

Table 3. The testing results at different positions.

Position 1		Position 2	
X_{1-1}	X_{2-1}	X_{1-2}	X_{2-2}
0.00	0.00	0.00	0.00
0.03	0.05	0.05	0.05
0.06	0.10	0.11	0.10
0.12	0.15	0.16	0.15
0.17	0.20	0.20	0.20
0.22	0.25	0.26	0.25
0.27	0.30	0.32	0.30
0.33	0.35	0.36	0.35
0.38	0.40	0.41	0.40
0.43	0.45	0.46	0.45
0.48	0.50	0.50	0.50

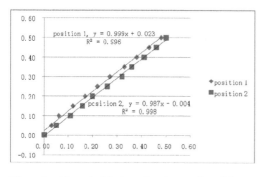

Figure 7. Numerical fitting of testing result at different positions.

From Figure 7, it can be concluded that there is a small difference when the testing is performed at difference positions. The error is 1.2%. It can be seen that the position has an effect on the testing result.

4 CONCLUSION

In this paper, a widely used testing system for fire detection and fire alarm systems is introduced. This system includes a smoke tunnel, a Measuring Ionization Chamber (MIC), an obscuration meter, and an aerosol generator. The system testing repeatability is analyzed here. In addition, it concludes that the rate of increase in the aerosol density, the flow rate of aerosol, and the position will affect the testing result of the system. All these parameters should be chosen properly. However, in the current valid standards, there is only a provision of the scope of the rate of increase in the aerosol density. For other parameters, there is no clear requirement. This paper may be beneficial to testing laboratories, system manufactories, and standard revision personnel.

This paper was funded by the project "14DZ2294100 The platform construction project of professional testing technical service for digital electronic product" by the Shanghai Science and Technology Committee.

REFERENCES

Aggarwal S., & V. Motevalf (1997). Investigation of an approach to fuel identification for non-flaming sources using light-scattering and ionization smoke detector response. *Fire safety J.* 29, 99–112.

Bernigau, N.G. & H.O. Luck (1986). The principle of the ionization chamber in aerosol measurement techniques—A review. *J. Aerosol Sci.* 17, 511–515.

Chen, S.J., D.C. Hovde, K.A. Peterson, & A.W. Marshall (2007). Fire detection using smoke and gas sensors. *Fire safety J.* 42, 507–515.

Helsper, C. H. Fissan, J. Muggli, & A. Scheidweiler (1983). Verification of ionization chamber theory. *Fire Tech.* 19, 14–21.

Heskeatad, G. & J.S. Newman (1992). Fire detection using cross-correlations of sensor signals. *Fire safety J.* 18, 355–374.

Jackson, M.A. & I. Robins (1994). Gas sensing for fire detection: Measurements of CO, CO_2, H_2, O_2, and smoke density in European standard fire tests. *Fire safety J.* 22, 181–205.

Scheidweiler, A. (1976). The ionization chamber as smoke dependent resistance. *Fire Tech.* 12, 113–123.

Shibata, S., T. Higashino, A. Sawada, T. Oyabu, Y. Takei, H. Nabto, & K. Toko (2010). Fire Detection Using tin Oxide Gas Sensors Installed in an Indoor Space. *IEEJ Trans. on Sensors and Micromachines* 130, 38–43.

*Advances in Materials Science, Energy Technology
and Environmental Engineering – Patty & Zhou (Eds)
© 2017 Taylor & Francis Group, London, ISBN 978-1-138-19668-1*

Study on differential games for discrete-time stochastic control systems with Markov jump and multiplicative noise

Y.L. Liu, M. Li & S.M. Ge

The College of Electrical Engineering and Automation, Shandong University of Science and Technology, Qingdao, Shandong, China

ABSTRACT: This paper defines the observability and detectability and the corresponding Popov–Belevitch–Hautus (PBH) criterion for discrete-time stochastic control systems with Markov jump and multiplicative noise. First, we derive three important theorems on two-player differential games. Then, we provide the optimal strategies (Nash equilibrium strategies) and the optimal cost values for discrete-time stochastic control systems. It can be seen that stochastic differential games are associated with four coupled Generalized Algebraic Riccati Equations (GAREs). Finally, we propose an iterative inversion algorithm to solve the four coupled equations, and verify the effectiveness and fast convergence of the algorithm by performing a numerical simulation.

1 INTRODUCTION

A common phenomenon, namely the randomness and interference of noise, often appears in most of the control systems in engineering practice. So this requires adjustment by the controller to realize the stability of a stochastic control system, which is likely to be influenced by both internal and external factors. For systems with time delay, interference may be nonlinear, uncertain parameters or Markov switch, etc. For example, the Markov jump system (Sheng, 2014) is a special class of stochastic control system. Thus, the main aim of this paper is to analyze and comprehensively research discrete-time stochastic control systems with multiplicative noise and Markov jump.

Differential game belongs to the intelligent control theory. It refers to the countermeasure process in which decision makers apply control behavior simultaneously to the stochastic control system, to realize each other's optimal and suboptimal targets. The Linear Quadratic (LQ) differential game in the stochastic control system plays an extraordinary role in the field of economic, military and even intelligent robots. This article focuses on the non-zero cooperative differential game problems for infinite-time horizon time-invariant LQ discrete stochastic control systems with multiplicative noise and Markov jump.

The Markov jump system is one of the common stochastic control systems. A robust fuzzy control problem for a nonlinear Markov jump control system has been studied in a paper (Wu, 2007). Random countermeasures in the jump system have

been adopted in another article (Zhu, 2014); The H^∞ robust control problem has been transformed into a two-person zero-sum differential game model in the literature (Zhu, 2013). Although many achievements have been made in differential games for these different systems, the study about the LQ differential game of discrete-time MJSCS (Markov jump stochastic control system) is rare. In this paper, we study in detail the problem of a LQ two-player differential game for discrete-time MJSCS with (x, u, v) -dependent multiplicative noise, in order to obtain the LQ two-player differential games related to GARE solutions and provide the best strategies and performance indicators.

The rest of this paper is organized as follows. Section 2 describes the discrete-time MJSCS and gives the PBH criterion for observability and detectability, considering many important lemmas. Section 3 presents important results for the LQ two-player differential games. Section 4 presents GARE solutions that solve the differential game through the reverse iterative algorithm, and the systems are simulated by numerical examples to verify the effectiveness and convergence of the algorithm. Finally, Section 5 concludes this paper.

For convenience, we adopt the following notations throughout the paper: $\mathbb{R}^n$: the n-dimensional real linear vector space with the corresponding 2-norm $\|\cdot\|$; $\mathbb{R}_{m \times n}$: the vector space of all $m \times n$ matrices with entries in $\mathbb{R}$. $\mathbb{S}_{n+}^N$: the linear space in the composition of all $n \times n$-dimensional real symmetric matrices. U': the transpose of a matrix U, with $U > 0(U \geq 0)$, where U is a positive definite (positive semi-definite) symmetric matrix. $L^2(\infty,\mathbb{R}^k)$: the space of $\mathbb{R}^k$-valued, square-summable random vectors. Finally, $\mathbb{N}=\{0,1,2,\cdots\}$, $\mathbb{N}_t=\{0,1,2,\cdots,t\}$; $\mathbb{C}$: the complex plane; $E(\cdot)$: the mathematical expectation.

2 PRELIMINARIES

Given a basic probability space $(\Omega,F,P,\{\mathcal{F}_k\})$, it contains a real random sequence $h(k)$ (i.e., multiplicative noise) and a discrete Markov chain r_k. $\mathcal{F}_k$ represents the σ-algebra generated by $h(k)$ and r_k, i.e., $\mathcal{F}_k=\sigma\{r_s,h(s)\,|\,s=0,1,2,...,k\}\subset F$.

Consider the following MJSCS defined on the basic probability space:

$$\begin{cases} x(k+1)=A_{r_k}x(k)+B_{r_k}u(k)+C_{r_k}v(k) \\ \qquad +\big[\bar{A}_{r_k}x(k)+\bar{B}_{r_k}u(k)+\bar{C}_{r_k}v(k)\big]h(k) \\ x(0)=x_0\in\mathbb{R}^n,\ k\in\mathbb{N} \\ y^\tau(k)=Q_{r_k}^\tau x(k),\ \tau=1,2 \end{cases} \quad (1)$$

where $\{x(k),r_k;k\in\mathbb{N}\}$ are the system state on the set $\mathbb{R}^n\times\mathcal{X}$.$\{r_k;k\in\mathbb{N}\}$ are the value of discrete Markov chains on a finite set $\mathcal{X}=\{1,2,...,N\}$. $(u(k),v(k))\in\mathbb{R}^r\times\mathbb{R}^r$ and $y^\tau(k)$ $\in\mathbb{R}^m$ are, respectively, the system-controlled input and output measurements. The initial distribution of the Markov jump is μ, and the transition probability matrix is $\mathbb{P}=\big[p_{ij}\big]$, $p_{ij}:=\mathbb{P}(r_{k+1}=j|r_k=i),\forall i,j\in\mathcal{X},k\in\mathbb{N}$.

$\mathcal{X}$ contains the various operating modes of system (1). For each $r_k=i\in\mathcal{X}$, $(A_{r_k},\bar{A}_{r_k},B_{r_k},\bar{B}_{r_k},C_{r_k},\bar{C}_{r_k},Q_{r_k}^\tau)\in\mathbb{R}_{n\times n}\times\mathbb{R}_{n\times n}\times\mathbb{R}_{n\times r}$ $\times\mathbb{R}_{n\times r}\times\mathbb{R}_{n\times r}\times\mathbb{R}_{n\times r}\times\mathbb{R}_{m\times n}$ will hence forth be referred to as $(A_i,\bar{A}_i,B_i,\bar{B}_i,C_i,\bar{C}_i,Q_i^\tau)$.

They are related to the i-th operation modes of the system as the real constant matrix with proper dimensions. $\{h(k),k\in\mathbb{N}\}$ is a sequence of random variables defined on a complete probability space $(\Omega,F,P;\{\mathcal{F}_k\})$, which is a second-order process, with $E(r(k))=0$ and $E(r(i)r(j))=\delta_{ij},i,j\in\mathbb{N}$, and δ_{ij} being a Kronecker function.

Assume that $h(k)$ and r_k are independent of each other. For convenience, they are denoted by $x(k)=x_k$. When system (1) is stable, we can also say that $(A_i,\bar{A}_i)$ is stable.

The quadratic cost function associated with each player is:

$$\begin{aligned} J^\tau(u,v)=\sum_{k=0}^{\infty}E[&x'(k)(Q_{r_k}^\tau)'Q_{r_k}^\tau x(k) \\ &+u'(k)R_{r_k}^\tau u(k)+v'(k)S_{r_k}^\tau v(k)],\ \tau=1,2 \end{aligned}$$
$$(2)$$

where $Q_{r_k}^\tau=Q_i^\tau\ge 0$, $0<R_{r_k}^\tau=R_i^\tau\in\mathbb{R}_{r\times r}$, $0<S_{r_k}^\tau=S_i^\tau$ $\in\mathbb{R}_{r\times r}$.

Therefore, our ultimate aim is to seek the optimal control to satisfy the following equation:

$$J^1(u^*,v^*)\le J^1(u^*,v),\quad J^2(u^*,v^*)\le J^2(u,v^*) \quad (3)$$

where $(u^*(k),v^*(k))\in L^2(\infty,\mathbb{R}^{r_u})\times L^2(\infty,\mathbb{R}^{r_v})$.

To ensure the convergence of the infinite-time quadratic cost function, we restrain the admissible control set to the constant linear feedback strategies, i.e., $u(k)=K_{r_k}^1x(k)$, $v(k)=K_{r_k}^2x(k)$, where $K_{r_k}^1\in\mathbb{R}_{r\times n}$, $K_{r_k}^2\in\mathbb{R}_{r\times n}$, $(K_{r_k}^1,K_{r_k}^2)\in\mathbb{K}$, and $\mathbb{K}:=\{(K_{r_k}^1,K_{r_k}^2)\,|$ system (1) can be stabilized with $u(k)=K_{r_k}^1x(k)$, $v(k)=K_{r_k}^2x(k)$.

The optimal differential games u^* and v^* determined by (3) are also called the Nash equilibrium strategy. Here, only two participants can use the constant feedback control, which can ensure the global Nash equilibrium solution.

When $v(k)=0$, system (1) is as follows:

$$\begin{cases} x(k+1)=A_ix(k)+B_iu(k) \\ \qquad +[\bar{A}_ix(k)+\bar{B}_iu(k)]h(k) \\ x(0)=x_0\in\mathbb{R}^n,\ k\in\mathbb{N} \end{cases} \quad (4)$$

For the moment $k\in\mathbb{N}$, we have:

$$X(k)=E[x(k)x'(k)]\quad Y(k)=E[y(k)y'(k)].$$

Thus, we can obtain the following equation:

$$\begin{aligned} X(k+1)&=E[x(k+1)x'(k+1)] \\ &=\sum_{j=1}^{N}h_{ji}(A_j+B_jK_j)X_j(k)(A_j+B_jK_j)' \\ &+\sum_{j=1}^{N}h_{ji}(\bar{A}_j+\bar{B}_jK_j)X_j(k)(\bar{A}_j+\bar{B}_jK_j)' \quad (5) \end{aligned}$$

Then, according to formula (5), we obtain a linear operator and its spectrum as follows.

Definition 1 (Zhang, 2013). For any given feedback gain matrix K_j related to system (4), linear operators from the set $\mathbb{S}_{n+}^N$ to $\mathbb{S}_{n+}^N$ are defined as follows:

$$\begin{aligned} \mathcal{L}_i(Z)=&\sum_{j=1}^{N}p_{ji}(A_j+B_jK_j)Z_j(k)(A_j+B_jK_j)' \\ &+\sum_{j=1}^{N}p_{ji}(\bar{A}_j+\bar{B}_jK_j)Z_j(k)(\bar{A}_j+\bar{B}_jK_j)' \\ &\forall Z_i\in\mathbb{S}_{n+}^N,i\in\mathcal{X} \end{aligned}$$

where $\sigma(\mathcal{L}_i)=\{\lambda\in\mathbb{C}:\mathcal{L}_i(Z)=\lambda Z_i,Z_i\in\mathbb{S}_{n+}^N,Z_i\ne 0$, $i\in\mathcal{X}\}$ is a spectrum set of linear operator $\mathcal{L}_i(Z)$. The matrix group Z_i with eigenvalues λ is corresponding eigenvectors.

Definition 2. In system (4), let the control input $u(k)\equiv 0$. For any $x_0\in\mathbb{R}^n$ and the initial distribution μ, if $\lim_{k\to\infty} E\left[x(k)x'(k)\right]=0$, then the MJSCS is asymptotically mean square stable. We can also say that $\left(A_i, \bar{A}_i\right)$ is stable.

Definition 3. If there is a state feedback control $u(k)=K_{r_k}x(k)\left(K_{r_k}=K_i, r_k=i\in\mathcal{X}\right)$ making system (5) asymptotically mean square stable, i.e., $\lim_{k\to\infty} E\left[x(k)x'(k)\right]=0$, then system (5) is called stabilizable in the mean square sense, and $\left(A_i+B_iK_i, \bar{A}_i+\bar{B}_iK_i\right)$ is also stabilizable.

Definition 4. For any initial distribution μ and $k_0\in\mathbb{N}, k_0\neq 0$, if the corresponding output response of x_0 is always equal to zero, i.e., for almost any $k\in\mathbb{N}_{k_0}$, then $y(k)\equiv 0$. Thus, we can say that $x_0\in\mathbb{R}^n$ is an unobservable state. If system (1) is no longer the unobservable state except the zero initial condition, i.e., for almost $0\leq k\leq T, \forall T>0$, then $y(k)\equiv 0\Rightarrow x_0=0$. Thus, we can say that system (1) is accurately observable.

Lemma 1. For the stochastic control system $[A_i, \bar{A}_i\,|\,Q_i]$, the following two statements are equivalent to each other:

a. In the sense of definition 4, the system $[A_i, \bar{A}_i\,|\,Q_i]$ is accurately observable;
b. Random PBH criteria: there is no non-zero $Z_i\in\mathbb{S}_{n+}^N\,(i\in\mathcal{X})$ making $\mathcal{L}_i(Z)=\lambda Z_i, (Q_1Z_1, Q_2Z_2,...,Q_NZ_N)=0, \lambda\in\mathbb{C}$

Definition 5. For any $x_0\in\mathbb{R}^n$ and the initial distribution μ, if almost all the $0\leq k\leq T, \forall T>0, y(k)= =0\Rightarrow\lim_{k\to\infty} E\left[x(k)x'(k)\right]=0$, we can then say that the system $[A_i, \bar{A}_i\,|\,Q_i]$ is accurately detectable.

Lemma 2. For the stochastic control system $[A_i, \bar{A}_i\,|\,Q_i]$, the following two statements are equivalent to each other:

a. In the sense of definition 5, the system $[A_i, \bar{A}_i\,|\,Q_i]$ is accurately detectable;
b. Random PBH criteria: there is no non-zero $Z_i\in\mathbb{S}_{n+}^N\,(i\in\mathcal{X})$ making $\mathcal{L}_i(Z)=\lambda Z_i, (Q_1Z_1, Q_2Z_2,..., Q_NZ_N)=0, |\lambda|\geq 1$

Lemma 3. If $\left[A_i, \bar{A}_i|Q_i\right]$ is observable, and if and only if the discrete Lyapunov equation

$$-L_i+A_i'\,\mathcal{E}_i(L)A_i+\bar{A}_i'\,\mathcal{E}_i(L)\bar{A}_i+Q_i'\,Q_i=0 \qquad (6)$$

has a unique positive definite solution $L_i>0$, then the system $\left(A_i, \bar{A}_i\right)$ is asymptotically mean square stable, where $\mathcal{E}_i(L)=\sum_{j=1}^N p_{ij}L_j, i\in\mathcal{X}$.

Lemma 3 is the key to explore the two-player stochastic differential game. Through the PBH criterion for observability and detectability, we obtain the following theorem.

Lemma 4. If the system $[A_i, \bar{A}_i\,|\,Q_i]$ is observed/detected, then $[A_i+B_iK_i^1+C_iK_i^2, \bar{A}_i+\bar{B}_iK_i^1+\bar{C}_iK_i^2\,|\,\sum]$ is observable/detectable, where $\sum=Q_i'Q_i+K_i^{1'}R_iK_i^1+K_i^{2'}S_iK_i^2$.

Proof. It proves the important lemma by using reduction to absurdity. First,

assume that the stochastic control system $[A_i+B_iK_i^1+C_iK_i^2, \bar{A}_i+\bar{B}_iK_i^1+\bar{C}_iK_i^2\,|\,\sum]$ is unobservable. Then, by Lemma 1, there is a non-zero $Z_i\in\mathbb{S}_{n+}^N$, making the following formula hold:

$$\begin{cases} \sum_{j=1}^N p_{ji}\left(A_j+B_jK_j^1+C_jK_j^2\right)Z_j\left(A_j+B_jK_j^1+C_jK_j^2\right)' \\ +\sum_{j=1}^N p_{ji}\left(\bar{A}_j+\bar{B}_jK_j^1+\bar{C}_jK_j^2\right)Z_j\left(\bar{A}_j+\bar{B}_jK_j^1+\bar{C}_jK_j^2\right)' \\ =\lambda Z_i\left(Q_i'Q_i+K_i^{1'}R_iK_i^1+K_i^{2'}S_iK_i^2\right)Z_i=0, i\in\mathcal{X} \end{cases} \qquad (7)$$

By rearranging the first equation of equation (7), we obtain:

$$\sum_{j=1}^N p_{ji}A_jZ_jA_j'+\sum_{j=1}^N p_{ji}\bar{A}_jZ_j\bar{A}_j'+\sum_{j=1}^N p_{ji}(A_jZ_jK_j^{1'}B_j'$$
$$+A_jZ_jK_j^{2'}C_j'+B_jK_j^1Z_jA_j'+B_jK_j^1Z_jK_j^{1'}B_j'+C_jK_j^2Z_jA_j'$$
$$+B_jK_j^1Z_jK_j^{2'}C_j'+C_jK_j^2Z_jK_j^{1'}B_j'+C_jK_j^2Z_jK_j^{2'}C_j')$$
$$+\sum_{j=1}^N p_{ji}(\bar{A}_jZ_jK_j^{1'}\bar{B}_j'+\bar{A}_jZ_jK_j^{2'}\bar{C}_j'+\bar{B}_jK_j^1Z_j\bar{A}_j'$$
$$+\bar{B}_jK_j^1Z_jK_j^{1'}\bar{B}_j'+\bar{C}_jK_j^2Z_j\bar{A}_j'+\bar{B}_jK_j^1Z_jK_j^{2'}\bar{C}_j'$$
$$+\bar{C}_jK_j^2Z_jK_j^{1'}\bar{B}_j'+\bar{C}_jK_j^2Z_jK_j^{2'}\bar{C}_j')=\lambda Z_i \qquad (8)$$

Premultiplying Z_i' on the second equation of equation (7), we have:

$$Z_i'Q_i'Q_iZ_i+Z_i'K_i^{1'}R_iK_i^1Z_i+Z_i'K_i^{2'}S_iK_i^2Z_i=0,$$

where $Q_i'Q_i\geq 0, R_i>0, S_i>0$ and $Z_i\neq 0$, and it is possible to obtain

$$K_i^1Z_i=Z_iK_i^{1'}=0, K_i^2Z_i=Z_iK_i^{2'}=0, Q_iZ_i=Z_iQ_i'=0.$$

Therefore, equation (7) transforms to the following form:

$$\begin{cases} \sum_{j=1}^N p_{ji}A_jZ_jA_j'+\sum_{j=1}^N p_{ji}\bar{A}_jZ_j\bar{A}_j'=\lambda Z_i Q_iZ_i=0, i\in\mathcal{X} \end{cases}$$

So the system $[A_i, \bar{A}_i|Q_i]$ is unobservable. It contradicts with each other according to the assumptions in Lemma 1, where $[A_i, \bar{A}_i\,|\,Q_i]$ is observable. Thus, the proof of observability is completed. Similarly, detectability can also be proved, which is omitted in this paper.

3 TWO-PLAYER DIFFERENTIAL GAME

By defining observability, we obtain the following important theorem. Then, the finite horizon

Theorem 1 in (Sun, 2012) is generalized to a discrete infinite horizon.

Theorem 1. For the control system (1), assume that the coupling equations (9)-(12) have a solution $\left(L_i^1, L_i^2; K_i^1, K_i^2\right)$, with $L_i^1 > 0, L_i^2 > 0$:

$$
\begin{cases}
-L_i^1 + \left(A_i + B_i K_i^1\right)' \mathcal{E}_i\left(L^1\right)\left(A_i + B_i K_i^1\right) + \left(\bar{A}_i + \bar{B}_i K_i^1\right)' \\
\quad \times \mathcal{E}_i\left(L^1\right)\left(\bar{A}_i + \bar{B}_i K_i^1\right) + Q_i^{1'} Q_i^1 + K_i^{1'} R_i^1 K_i^1 \\
\qquad -K_i^{3'} H_i^1\left(L^1\right)^{-1} K_i^3 = 0 \\
\qquad\qquad H_i^1\left(L^1\right) > 0
\end{cases}
$$

$$\tag{9}$$

$$K_i^1 = -H_i^2\left(L^2\right)^{-1} K_i^4 \tag{10}$$

$$
\begin{cases}
-L_i^2 + \left(A_i + C_i K_i^2\right)' \mathcal{E}_i\left(L^2\right)\left(A_i + C_i K_i^2\right) + \left(\bar{A}_i + \bar{C}_i K_i^2\right)' \\
\quad \times \mathcal{E}_i\left(L^2\right) \times \left(\bar{A}_i + \bar{C}_i K_i^2\right) + Q_i^{2'} Q_i^2 + K_i^{2'} S_i^2 K_i^2 \\
\qquad -K_i^{4'} H_i^2\left(L^2\right)^{-1} K_i^4 = 0 \\
\qquad\qquad H_i^2\left(L^2\right) > 0
\end{cases}
$$

$$\tag{11}$$

$$K_i^2 = -H_i^1\left(L^1\right)^{-1} K_i^3 \tag{12}$$

where

$$H_i^1\left(L^1\right) = S_i^1 + C_i' \mathcal{E}_i\left(L^1\right) C_i + \bar{C}_i' \mathcal{E}_i\left(L^1\right) \bar{C}_i$$

$$K_i^3 = C_i' \mathcal{E}_i\left(L^1\right)\left(A_i + B_i K_i^1\right) + \bar{C}_i' \mathcal{E}_i\left(L^1\right)\left(\bar{A}_i + \bar{B}_i K_i^1\right)$$

$$H_i^2\left(L^2\right) = R_i^2 + B_i' \mathcal{E}_i\left(L^2\right) B_i + \bar{B}_i' \mathcal{E}_i\left(L^2\right) \bar{B}_i$$

$$K_i^4 = B_i' \mathcal{E}_i\left(L^2\right)\left(A_i + C_i K_i^2\right) + \bar{B}_i' \mathcal{E}_i\left(L^2\right)\left(\bar{A}_i + \bar{C}_i K_i^2\right)$$

$$\mathcal{E}_i\left(L^\tau\right) = \sum_{j=1}^N p_{ij} L_j^\tau, \tau = 1, 2$$

If the random control system $[A_i, \bar{A}_i | Q_i^\tau](\tau = 1, 2)$ is observable, then:

1. $\left(K_i^1, K_i^2\right) \in \mathbb{K}$;
2. Infinite horizon discrete stochastic differential game has a pair of optimal control solution $\left(u^*(k), v^*(k)\right)$ and $u^*(k) = K_i^1 x(k), v^*(k) = K_i^2 x(k)$;
3. By using the optimal differential game $\left(u^*(k), v^*(k)\right)$, the optimal quadratic cost function of the system is $J_i\left(u^*(k), v^*(k)\right) = x_0' L_i^\tau x_0, \tau = 1, 2$.

Proof. In Lemma 3, $Q_i' Q_i$ is a symmetric matrix. For system (1), according to Lemma 3, we can obtain the following equation:

$$
\left(A_i + B_i K_i^1 + C_i K_i^2\right)' \mathcal{E}_i\left(L^1\right)\left(A_i + B_i K_i^1 + C_i K_i^2\right)
$$
$$
+ \left(\bar{A}_i + \bar{B}_i K_i^1 + \bar{C}_i K_i^2\right)' \mathcal{E}_i\left(L^1\right)\left(\bar{A}_i + \bar{B}_i K_i^1 + \bar{C}_i K_i^2\right)
$$
$$
+ Q_i^{1'} Q_i^1 + K_i^{1'} R_i^1 K_i^1 + K_i^{2'} S_i^1 K_i^2 = L_i^1
$$

$$\tag{13}$$

and

$$
\left(A_i + B_i K_i^1 + C_i K_i^2\right)' \mathcal{E}_i\left(L^2\right)\left(A_i + B_i K_i^1 + C_i K_i^2\right)
$$
$$
+ \left(\bar{A}_i + \bar{B}_i K_i^1 + \bar{C}_i K_i^2\right)' \mathcal{E}_i\left(L^2\right)\left(\bar{A}_i + \bar{B}_i K_i^1 + \bar{C}_i K_i^2\right)
$$
$$
+ Q_i^{2'Q_i^2} + K_i^{1'R_i^2 K_i^1} + K_i^{2'S_i^2 K_i^2} = L_i^2
$$

$$\tag{14}$$

By rearranging (13) and (14), we can respectively obtain the coupled equations (9) and (10) in Theorem 1.

If the discrete stochastic control system $[A_i, \bar{A}_i | Q_i^\tau](\tau = 1, 2)$ is observable, then by using Lemma 4, we can obtain the random control systems as follows:

$$[A_i + B_i K_i^1 + C_i K_i^2, \bar{A}_i + \bar{B}_i K_i^1 + \bar{C}_i K_i^2 | Q_i^{1'}$$
$$Q_i^1 + K_i^{1'} \ R_i^1 K_i^1 + K_i^{2'} S_i^1 K_i^2]$$
$$[A_i + B_i K_i^1 + C_i K_i^2, \bar{A}_i + \bar{B}_i K_i^1 + \bar{C}_i K_i^2 | Q_i^{2'}$$
$$Q_i^2 + K_i^{1'} \ R_i^2 K_i^1 + K_i^{2'} S_i^2 K_i^2]$$

which are observable. Equation (9) has a solution $L_i^1 > 0$, equation (10) has a solution $L_i^2 > 0$, i.e., (13) and (14) have positive definite solutions $L_i^1 > 0, L_i^2 > 0$. According to Lemma 3, the stochastic control system is asymptotically mean square stable, i.e., there is a state feedback control $u(k) = K_i^1 x(k), v(k) = K_i^2 x(k)$, making the random control system stabilizable. Thus, the proof of Theorem 1 (1) is completed.

Then we prove (2) and (3) in Theorem 1. Note that $u^*(k) = K_{r_k}^1 x(k), r_k = i$, and by substituting $u^*(k)$ into (1), we can obtain the following system:

$$
\begin{cases}
x(k+1) = \left(A_{r_k} + B_{r_k} K_{r_k}^1\right) x(k) + C_{r_k} v(k) \\
\qquad\quad + [(\bar{A}_{r_k} + \bar{B}_{r_k} K_{r_k}^1) x(k) + \bar{C}_{r_k} v(k)] h(k) \\
x(0) = x_0 \in \mathbb{R}^n, \quad k \in \mathbb{N}
\end{cases}
$$

For convenience, let $x(k) = X_k$. Using the Lyapunov function $Z(x_k) = x_k' L_{r_k}^1 x_k$, we have:

$$
E\left[\Delta Z\left(x_k\right)\right] = E\left[Z\left(x_{k+1}\right) - Z\left(x_k\right)\right]
$$
$$
= E\{-x_k' L_{r_k}^1 x_k + [\left(A_{r_k} + B_{r_k} K_{r_k}^1\right) x_k
$$
$$
+ C_{r_k} v_k]' \mathcal{E}_i\left(L^1\right)\left[\left(A_{r_k} + B_{r_k} K_{r_k}^1\right) x_k
$$

$+C_{r_k}v_k] + [(\overline{A}_{r_k} + \overline{B}_{r_k}K^1_{r_k})x_k$
$+\overline{C}_{(r_k)}v_k]' \; \mathcal{E}_i(L^1)[(\overline{A}_{(r_k)} + \overline{B}_{(r_k)}K^1_{(r_k)})x_k$
$+\overline{C}_{(r_k)}v_k]\}$, due to

$$\sum_{k=0}^{\infty}E[\Delta Z(x_k)] = E\left[\sum_{k=0}^{\infty}\Delta Z(x_k)\right] = E[Z(x_\infty) - Z(x_0)]$$
$$= -x_0'L^1_i x_0$$

By the first equation (9) of Theorem 1, and using the square technology, we have:

$$J^1(u^*,v) = \sum_{k=0}^{\infty}E\left[x_k'(Q^{1'}_{r_k}Q^1_{r_k} + K^1_{r_k}{}'R^1_{r_k}K^1_{r_k})x_k\right]$$

$$+v_k'S^1_{r_k}v_k] + \sum_{k=0}^{\infty}E[\Delta Z(x_k)] + x_0'L^1_i x_0$$

$$= x_0'L^1_i x_0 + \sum_{k=0}^{\infty}E[x_k'K^{3'}_iH^1_i(L^1)^{-1}K^3_i x_k$$

$$+ x_k'K^{3'}_i v_k + v_k'K^3_i x_k + v_k'H^1_i(L^1)v_k]$$

$$= x_0'L^1_i x_0 + \sum_{k=0}^{\infty}E\{[v(k)-K^2_i x(k)]'\; H^1_i(L^1)[v(k)$$
$$-K^2_i x(k)]\}, \tau = 1$$

then by (3), we obtain the optimal control and the optimal LQ cost function, respectively, $v^*(k) = K^2_i x(k)$, $J^1(u^*,v^*) = x_0'L^1_i x_0 \le J^1(u^*,v)$, Finally, substituting $v^*(k)$ into system (1), we similarly obtain:

$$u^*(k) = K^1_i x(k), J^2(u^*,v^*) = x_0'L^2_i x_0 \le J^2(u,v^*), \tau = 2.$$

Theorem 2. If the stochastic control system $[A_i,\overline{A}_i|Q^\tau_i](\tau=1,2)$ is observable for system (1), assuming that Theorem 1 has four coupled equation solutions $(L^1_i,L^2_i;K^1_i,K^2_i)$ and satisfies $(K^1_i,K^2_i) \in \mathbb{K}$, then:

1. $L^1_i > 0, L^2_i > 0$;
2. Infinite horizon discrete stochastic differential game has a pair of optimal control solution $(u^*(k),v^*(k))$ and $u^*(k) = K^1_i x(k), v^*(k) = K^2_i x(k)$;
3. By using the optimal differential game $(u^*(k),v^*(k))$, the optimal quadratic cost function of the system is $J_i(u^*(k),v^*(k)) = x_0'L^\tau_i x_0, \tau = 1,2.$

Theorem 3. If the stochastic control system $[A_i,\overline{A}_i|Q^\tau_i](\tau=1,2)$ is detectable for system (1), assuming that Theorem 1 has four coupled equation solutions $(L^1_i,L^2_i;K^1_i,K^2_i)$ and satisfies $L^1_i \ge 0, L^2_i \ge 0$, then:

1. $(K^1_i,K^2_i) \in \mathbb{K}$;
2. Infinite horizon discrete stochastic differential has a pair of optimal control solution (u*(k), v*(k)) and $u^*(k) = K^1_i x(k), v^*(k) = K^2_i x(k)$;

3. By using the optimal differential game $(u^*(k),v^*(k))$, the optimal quadratic cost function of the system is $J_i(u^*(k),v^*(k)) = x_0'L^\tau_i x_0, \tau = 1,2.$

4 ITERATIVE ALGORITHM AND SIMULATION EXAMPLE

This section presents a reverse iterative algorithm to solve four coupling equations (9)–(12) in Theorem 1. As the infinite horizon equation is very difficult to solve, it is converted into a finite horizon equation, i.e., to obtain the result of an infinite iteration through a limited number of iterations, where n represents the number of iterations:

$$\left\{\begin{array}{l} L^{1n}_i(k) = (A_i + B_i K^{1n}_i(k))' \mathcal{E}_i(L^1(k+1))^n \\[4pt] \times (A_i + B_i K^{1n}_i(k)) + (\overline{A}_i + \overline{B}_i K^{1n}_i(k))' \\[4pt] \times \mathcal{E}_i(L^1(k+1))^{n'}(\overline{A}_i + \overline{B}_i K^{1n}_i(k)) + Q^{1'}_i Q^1_i \\[4pt] + K^{1n}_i(k)' R^1_i K^{1n}_i(k) - K^{3n}_i(k)' H^1_i \\[4pt] \times (L^{1n}(k+1))^{-1} K^{3n}_i(k) \\[4pt] L^{1n}_i(k+1) = 0 \\[4pt] H^1_i(L^{1n}(k+1)) > 0 \end{array}\right. \quad (15)$$

$$K^{1n}_i(k) = -H^2_i(L^{2n}(k+1))^{-1}K^{4n}_i(k) \quad (16)$$

$$\left\{\begin{array}{l} L^{2n}_i(k) = (A_i + C_i K^{2n}_i(k))' \mathcal{E}_i(L^2(k+1))^n \\[4pt] \times (A_i + C_i K^{2n}_i(k)) + (\overline{A}_i + \overline{C}_i K^{2n}_i(k))' \\[4pt] \times \mathcal{E}_i(L^2(k+1))^n (\overline{A}_i + \overline{C}_i K^{2n}_i(k)) + Q^{2'}_i Q^2_i \\[4pt] + K^{2n}_i(k)' S^2_i K^{2n}_i(k) - K^{4n}_i(k)' H^2_i \\[4pt] \times (L^{2n}(k+1))^{-1} K^{4n}_i(k) \\[4pt] L^{2n}_i(k+1) = 0 \\[4pt] H^2_i(L^{2n}(k+1)) > 0 \end{array}\right. \quad (17)$$

$$K^{2n}_i(k) = -H^1_i(L^{1n}(k+1))^{-1}K^{3n}_i(k) \quad (18)$$

where $H^1_i(L^{1n}(k+1)) = S^1_i + C_i'\mathcal{E}_i(L^1(k+1))^n C_i$
$$+ \overline{C}_i'\mathcal{E}_i(L^1(k+1))^n \overline{C}_i \quad (19)$$

$$K^{3n}_i(k) = C_i' \; \mathcal{E}_i(L^{1n}(k+1))(A_i + B_i K^{1n}_i(k))$$
$$+ \overline{C}_i' \; \mathcal{E}_i(L^{1n}(k+1))(\overline{A}_i + \overline{B}_i K^{1n}_i(k+1)) \quad (20)$$

$$H^2_i(L^{2n}(k+1)) = R^2_i + B_i' \; \mathcal{E}_i(L^{2n}(k+1))B_i$$
$$+ \overline{B}_i' \; \mathcal{E}_i(L^{2n}(k+1))\overline{B}_i \quad (21)$$

181

$$K_i^{4n}(k) = B_i' \ \mathcal{E}_i\left(L^{2n}(k+1)\right)\left(A_i + C_iK_i^{2n}(k+1)\right)$$
$$+\bar{B}_i' \ \mathcal{E}_i\left(L^{2n}(k+1)\right)\left(\bar{A}_i + \bar{C}_iK_i^{2n}(k+1)\right)$$
$$\mathcal{E}_i(L^\tau) = \sum_{j=1}^{N}p_{ij}L_j^\tau, \ \tau = 1,2 \qquad (22)$$

We obtain solution (9)–(12) through the above iterative equation. The iterative process is described as follows:

a. Given the appropriate constants n (generally about 100), and given the zero initial value:

$$L_i^{1n}(n+1) = 0, \ L_i^{2n}(n+1) = 0, \ K_i^{1n}(n+1) = 0,$$
$$K_i^{2n}(n+1) = 0$$

b. According to the values of $L_i^{1n}(n+1)$, $L_i^{2n}(n+1)$, $K_i^{1n}(n+1)$, $K_i^{2n}(n+1)$, and coupled algebraic equations (19)–(22), we obtain $H_i^1(L_i^{1n}(n+1))$, $H_i^2(L_i^{2n}(n+1))$, $K_i^{3n}(n)$, $K_i^{4n}(n)$.

c. By equations (16) and (18), we obtain, respectively, $K_i^{1n}(n)$ and $K_i^{2n}(n)$, and by equations (15) and (17), respectively, we obtain $L_i^{1n}(n)$ and $L_i^{2n}(n)$.

d. Let

$$L_i^{1n}(n+1) = L_i^{1n}(n), \ L_i^{2n}(n+1) = L_i^{2n}(n),$$
$$K_i^{1n}(n+1) = K_i^{1n}(n), \ K_i^{2n}(n+1) = K_i^{2n}(n)$$

e. Then, let $n = n-1$, repeating steps (b)–(d), until iteration $n+1$ times, we eventually obtain $L_i^{1n}(0)$, $L_i^{2n}(0)$, $K_i^{1n}(0)$ and $K_i^{2n}(0)$.

Based on the literature (Zhang, 2007), we conclude the results of the GARE asymptotic analysis of the continuous time stochastic LQ optimal control. For any $x_0 \in \mathbb{R}^n$, under the premise stabilization and observability of the random control system, we can obtain

$$\begin{cases} \lim_{n\to\infty} x_0'L_i^{1n}(0)x_0 = \lim_{n\to\infty}\min J^{1n}\left(u_n^*,v\right) \\ \qquad = \min J^{1\infty}\left(u^*,v\right) = x_0'L_i^1x_0 \\ \lim_{n\to\infty} x_0'L_i^{2n}(0)x_0 = \lim_{n\to\infty}\min J^{2n}\left(u_n^*,v\right) \\ \qquad = \min J^{2\infty}\left(u^*,v\right) = x_0'L_i^2x_0 \end{cases}$$

$$\lim_{n\to\infty} K_i^{1n}(0) = K_i^1, \lim_{n\to\infty} K_i^{2n}(0) = K_i^2 \text{ so}$$

$$\lim_{n\to\infty}\left(L_i^{1n}(0), L_i^{2n}(0); K_i^{1n}(0), K_i^{2n}(0)\right) = \left(L_i^1, L_i^2; K_i^1, K_i^2\right)$$
$$(23)$$

where $\left(L_i^1, L_i^2; K_i^1, K_i^2\right)$ is the optimal solution of coupled equations (9)–(12).

The reverse iterative algorithm for solving differential game solutions is effective and reliable. In order to verify its validity, we give the following examples of a two-dimensional digital simulation. Assume that the number of iterations is $n = 100$ in the stochastic control system (1).

Let the initial state of this stochastic control system be:

$$x_0 = \begin{bmatrix} 2 & 0 \\ 0 & 3 \end{bmatrix}$$

1. when $r_k = 1$, we set

$$A_1 = \begin{bmatrix} 0.85 & 0 \\ 0 & 0.8 \end{bmatrix}, \ \bar{A}_1 = \begin{bmatrix} 0.4 & 0 \\ 0 & 0.45 \end{bmatrix},$$

$$B_1 = \begin{bmatrix} 0.6 & 0.55 \\ 0.55 & 0.7 \end{bmatrix}, \ \bar{B}_1 = \begin{bmatrix} 0.65 & 0.75 \\ 0.75 & 0.35 \end{bmatrix}$$

$$C_1 = \begin{bmatrix} 0.75 & 0.6 \\ 0.6 & 0.85 \end{bmatrix}, \ \bar{C}_1 = \begin{bmatrix} 0.5 & 0.4 \\ 0.4 & 0.25 \end{bmatrix}$$

$$Q_1^1 = \begin{bmatrix} 0.55 & 0 \\ 0 & 0.65 \end{bmatrix}, Q_1^2 = \begin{bmatrix} 0.75 & 0 \\ 0 & 0.25 \end{bmatrix},$$

$$R_1^1 = R_1^2 = S_1^1 = S_1^2 = \begin{bmatrix} 1 & 0 \\ 0 & 1 \end{bmatrix}$$

We obtain the optimal solution of coupled algebraic equations (15)–(18) by using the above reverse iterative algorithm as follows:

$$\begin{cases} L_1^{1n}(0) = \begin{bmatrix} L_1^1(1,1) & L_1^1(1,2) \\ L_1^1(2,1) & L_1^1(2,2) \end{bmatrix} = \begin{bmatrix} 1.5988 & -1.1819 \\ -1.1819 & 1.7887 \end{bmatrix} \\ L_1^{2n}(0) = \begin{bmatrix} L_1^2(1,1) & L_1^2(1,2) \\ L_1^2(2,1) & L_1^2(2,2) \end{bmatrix} = \begin{bmatrix} 3.3513 & -2.5465 \\ -2.5467 & 2.7158 \end{bmatrix} \\ K_1^{1n}(0) = \begin{bmatrix} K_1^1(1,1) & K_1^1(1,2) \\ K_1^1(2,1) & K_1^1(2,2) \end{bmatrix} = \begin{bmatrix} -0.2388 & -0.0047 \\ -0.2538 & 0.1933 \end{bmatrix} \\ K_1^{2n}(0) = \begin{bmatrix} K_1^2(1,1) & K_1^2(1,2) \\ K_1^2(2,1) & K_1^2(2,2) \end{bmatrix} = \begin{bmatrix} -0.1960 & -0.0442 \\ 0.2070 & -0.4240 \end{bmatrix} \end{cases}$$

According to (23), $(L_1^{1n}(0), L_1^{2n}(0); K_1^{1n}(0), K_1^{2n}(0))$ are optimal solutions of the four coupling equations. The results indicate that L_1^1 and L_1^2 are positive definite. We obtain the iterative process of $L_1^{1n}(0)$, $L_1^{2n}(0)$; $K_1^{1n}(0)$, $K_1^{2n}(0)$, as shown in Figure 1 and Figure 2, in which each curve represents the value of each iteration. Moreover, the figures clearly show the convergence and effectiveness of the inverse iterative algorithm.

The optimal Nash equilibrium is given by:

$$u^*(k) = K_1^1 x(k) = K_1^{1n}(0)x(k)$$
$$= \begin{bmatrix} -0.2388 & -0.0047 \\ -0.2538 & 0.1933 \end{bmatrix} x(k)$$

$$v^*(k) = K_1^2 x(k) = K_1^{2n}(0)x(k)$$
$$= \begin{bmatrix} -0.1960 & -0.0442 \\ 0.2070 & -0.4240 \end{bmatrix} x(k)$$

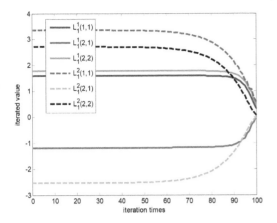

Figure 1. Iterative process values of $L_1^{1n}(k)$ and $L_1^{2n}(k)$.

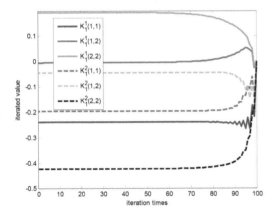

Figure 2. Iterative process values of $K_1^{1n}(k)$ and $K_1^{2n}(k)$.

Optimal performance index:

$$J_1\left(u^*(k), v^*(k)\right) = x_0' L_1^1 x_0 = x_0' L_1^{1n}(0) x_0$$
$$= \begin{bmatrix} 6.3952 & -7.0914 \\ -7.0914 & 16.0983 \end{bmatrix}$$
$$J_2\left(u^*(k), v^*(k)\right) = x_0' L_1^2 x_0 = x_0' L_1^{2n}(0) x_0$$
$$= \begin{bmatrix} 13.4044 & -15.2790 \\ -15.2790 & 24.4422 \end{bmatrix}$$

The iterative process values are shown in Figure 1 and Figure 2.

When $r_k = 2$, $(L_2^{1n}(0), L_2^{2n}(0); K_2^{1n}(0), K_2^{2n}(0))$ can be easily obtained as the solutions of (15)–(18), with $L1 > 1$ and $L2 > 1$. Because it is the same as the above process ($r_k = 1$), we do not repeat it again.

5 CONCLUSIONS

This paper defines the exact observability and detectability of the discrete-time MJSCS with control and state-dependent noise using the corresponding PBH criterion, and studies the infinite horizon LQ differential games of the stochastic control system. Based on the Lyapunov equation and the mean square stability, we obtained three useful theorems for solving two-player differential games; and proposed a reverse iterative algorithm to compute four coupled GAREs. Finally, we verified the effectiveness and fast convergence of the iterative algorithm by performing a numerical simulation.

ACKNOWLEDGMENTS

The author acknowledges the financial support from the Shandong University of Science and Technology (Grant No.YC150346).

REFERENCES

Sheng, L., W. Zhang, and M. Gao, "Relationship Between Nash Equilibrium Strategies and Control of Stochastic Markov Jump Systems With Multiplicative Noise," *IEEE Transactions on Automatic Control,* vol. 59, no. 9, pp. 2592–2597, 2014.

Sun, H. Y., L. Y. Jiang, and W. H. Zhang. Feedback control on Nash equilibrium for discrete-time stochastic systems with Markovian jumps: finite-horizon case, *International Journal of Control, Automation and Systems,* vol. 10, no. 5, pp. 940–946, 2012.

Wu, H. N. and K. Y. Cai, Robust fuzzy control for uncertain discrete-time nonlinear Markovian jump systems without mode observations, *Information Sciences,* vol. 177, no. 6, pp. 1509–1522, 2007.

Zhang, W. H., C. Tan. On detectability and observability of discrete-time stochastic Markov jump systems with state-dependent noise, *Asian Journal of Control,* vol. 16, no. 1, pp. 1–10, 2013.

Zhang, W. H., Y. L. Huang, and H. S. Zhang. Stochastic H2/H∞ control for discrete-time systems with state and disturbance dependent noise, *Automatica,* vol. 43, no. 3, pp. 513–521, 2007.

Zhu, H. N., C. K. Zhang, and B. Ning, Infinite horizon linear quadratic stochastic Nash differential games of Markov jump linear systems with its application, *International Journal of Science Systems,* vol. 45, no. 5, pp. 1196–1201, 2014.

Zhu, H., C. Zhang, and S. O. Management, H∞ Control Based on Game Theory Approach for Markov Jump Linear Systems, *Information and Control,* vol. 42, no. 4, pp. 423–429, 2013.

Advances in Materials Science, Energy Technology
and Environmental Engineering – Patty & Zhou (Eds)
© 2017 Taylor & Francis Group, London, ISBN 978-1-138-19668-1

A testifying method for the instructing efficiency of information security courses with statistical probability models

X.M. Jia

Information Security College, Yunnan Police College, Kunming, China

ABSTRACT: In this paper, a novel method based on the statistical probability models has been proposed to testify the instructing efficiency of these courses for the information security major. A new measure, as well as the increment in the description length, is employed to improve the previous measure used for testifying these courses. When this similarity measure is given, a scale of clustering algorithms can be used to help the analysis for instructing efficiency. As the improvement, the increment in description length satisfies the symmetrical feature. The experiment results indicate that the proposed similarity measure can lead to better clustering results than some other previous similarity measures and the testifying results can hold higher accuracy.

1 INTRODUCTION

The courses are the basis of the instruction architecture. The result, actually the instructing efficiency, is one of the normal standards to illustrate the ability of these courses to improve the corresponding major. In predecessors' work, some testifying methods were proposed to sketch the probability that the data can indicate the efficiency when some parameters are obtained in the form of data, which actually change the texture information into some numeric values. Based on these changes, some algorithms for analysis can be used to help the testifying process. Most of these previous works suggested that the statistical model could be considered to illustrate the efficiency of the instructions of those courses. Meanwhile, some clustering algorithms are employed to implement the analysis. As well, the clustering operation can indeed reduce the complexity of analysis by reducing the scale of data-set or reduce the feature patterns. However, it is different from data clustering, and the merging operation for probability distributions is similar to vector clustering, which implies that the traditional similarity measure, such as Euclidean Distance, is not suitable for vector clustering. Actually, the relative entropy between two probability distributions can be used to measure the similarity between these two distributions. However, the relative entropy is asymmetric, which does not satisfy the law that one similarity measure should hold. In practice, the similarity among probability distributions should be relative to their estimation process, i.e. these probability distributions are not known in advance, and they need to be estimated by using corresponding count vec-

tors. This estimation process actually has influence on the similarity between each two of these distributions. Rissanen (1983; 2001) proposed a new parameter named description length to describe the complexity of the estimation process. In Forchhammer (2004; 2007), the description length was used to help the intelligence algorithms to achieve optimal context quantization. However, there are two problems in using the description length as the similarity measure. One is that the description length is not a similarity measure, and another is that the description length should be calculated with a higher computing complexity. In Chen (2013), the rapid calculation algorithm for the description length is proposed. Based on this approximation, the calculation of description length can be accelerated. However, it is also not suitable for the similarity measure, which reasons to that the description length is just related to only one count vector. In order to tackle this problem, in this paper, we give a novel similarity measure, the increment in description length. It comes from the theory of description length, but more efficient than those previous similarity measures. Its details will be discussed in section 2.

On the other hand, the data analysis is widely used to evaluate the efficiency of education. In 2012, the educational data mining is discussed. In Bapler (2013), some pattern recognition algorithms are suggested to mine education data to make the course establishment. Furthermore, the information security instruction is very important to train the people who will protect the security of the networks. In this case, the efficiency of the instruction will directly influence the training process. Based on the statistical model, with the clustering algorithm, the purpose can be achieved. In this paper, we try to use the clustering algorithm to analyze the police training data and to influence the corresponding course establishment using our similarity measure proposed. The K-means algorithm is employed to implement our application. The details of our algorithm will be given in section 3.

2 INCREMENT IN THE DESCRIPTION LENGTH

In instruction efficiency analysis, the data is obtained first by using some mapping methods which can put those texture information mapping into numeric values. Then, the probability distribution is constructed to describe the statistical feature of these observing data. Based on this observation, the prediction of the future event is made. The clustering operation is suggested to reduce the scale of the prediction space. Namely, the number of possible distributions which are used to describe the feature of one event tailored by clustering. In this case, the data fusion process will come from less distributions with reasonable computing complexity. However, to achieve this objective, the clustering operation should be executed first.

For probability distribution clustering, the first problem needed to be considered is the similarity measure. In predecessors' works, the relative entropy (K-L distance) between two probability distributions is used to describe the distance between these two distributions and this "distance" is considered as their similarity measure. However, the relative entropy is asymmetric, i.e. it does not satisfy the properties which one distance measure should hold. When this similarity measure is used as the criterion in probability distribution clustering, the results may be different, when the clustering operation goes from different sides (from distributions A to B, or from distributions B to A). In order to tackle this problem, in this paper, we give a novel similarity measure between two distributions to obtain the reasonable clustering results.

In practice, especially in probability distribution clustering for data big in size, each probability distribution is estimated by using its corresponding count vector. It means that the counting number of the observing data is used to calculate the probability using the classical probability model. Considering two count vectors on 3-ary case, they are described by (1).

$$
\begin{array}{cccc}
 & 0 & 1 & 2 \\
\mathbf{CV}_1: & n_0 & n_1 & n_2 \\
\mathbf{CV}_2: & m_0 & m_1 & m_2
\end{array} \tag{1}
$$

In Chen (2013), we give the conclusion that each of this count vector holds a parameter named "description length". The description length implies that description complexity, which actually denotes the code length when these counting symbols are coded. For the count vector $\mathbf{cv}_1$, its corresponding description length L_1 can be calculated by (2)

$$
L_1 = \log(V_1 - 1)! - \sum_{i=0}^{2} \log n_i! - \log(3 - 1)! \tag{2}
$$

When Stirling's formula (3)

$$
\log n! \approx (n + \frac{1}{2}) \log n - \log \sqrt{2\pi} - n \tag{3}
$$

is used to approximate the logarithm operation in (2), the description length can be represented by (4).

$$
\begin{aligned}
L_1 = & V_1 \log V_1 - n_0 \log n_0 - n_1 \log n_1 - n_2 \log n_2 \\
& - \frac{1}{2} \log \frac{V_1}{n_0 n_1 n_2} + \sigma
\end{aligned} \tag{4}
$$

where $\sigma = -\log 3! - 3\log\sqrt{2\pi}$. From (4), it is obvious that the description length is related to the number of training data with different values. Meanwhile, it is also related to representation (5).

$$
\zeta = \log \frac{n_0 n_1 n_2}{V_1} \tag{5}
$$

Let us consider the relative entropy between uniform distributions with the count vector that holds V_1 training data and each probability in this distribution can be calculated by V_1 / I. Then, the relative entropy between the probability distribution with count vector cv_1 and the uniform distribution can be described as:

$$
D = \mu - \zeta \tag{6}
$$

where μ denotes a constant value and ξ is obtained from representation (5). Therefore, the relative entropy is correlated with representation (5). Meanwhile, in count vector $\mathbf{CV}_1$, if the number of data with value 0 is close to the total number of training data which this count vector obtains, the value of representation (5) will be near to value 0. It implies that the probability distribution performs well, and the value of representation (5) will become smaller. Based on this discussion, representation (5) in our work is referred to as the amazing measure.

On the other hand, let L denote the description length when $\mathbf{CV}_1$ and $\mathbf{CV}_2$ are merged into one, and L_1 and L_2 denote the description length of $\mathbf{CV}_1$ and $\mathbf{CV}_2$, respectively. Considering the increment in the description length ΔL between two count vectors $\mathbf{CV}_1$ and $\mathbf{CV}_2$, ΔL can be described as:

$$\Delta L = L - (L_1 + L_2) \tag{7}$$

After derivation, using (3), ΔL can also be transformed to (8):

$$
\begin{aligned}
\Delta L_{mk} &= n_m \sum_{i=1}^{I} \left(\frac{n_i^{(m)}}{n_m} \right) \log\left[\left(\frac{n_i^{(m)}}{n_m} \right) \bigg/ \left(\frac{n_i^{(mk)}}{n_{mk}} \right) \right] \\
&\quad + n_k \sum_{i=1}^{I} \left(\frac{n_i^{(k)}}{n_k} \right) \log\left[\left(\frac{n_i^{(k)}}{n_k} \right) \bigg/ \left(\frac{n_i^{(mk)}}{n_{mk}} \right) \right] \\
&\quad - \frac{I-1}{2} \log \frac{n_m n_k}{n_m + n_k} \\
&= n_m D\big(p(x \mid c_m) \parallel p(x \mid c_{mk})\big) + \\
&\quad n_k D\big(p(x \mid c_k) \parallel p(x \mid c_{mk})\big) - \frac{I-1}{2} \log \frac{n_m n_k}{n_m + n_k}
\end{aligned}
\tag{8}
$$

where n_m and n_k are the number of data in $\mathbf{CV}_1$ and $\mathbf{CV}_1$. Apparently, ΔL is equivalent to the weighting of two relative entropies. It implies that the increment in the description length can be considered as the similarity measure between two count vectors. Meanwhile, the probability distribution is obtained by using its corresponding count vector; therefore, the increment in the description length can also be considered as the similarity measure between two probability distributions.

From (3), some properties of ΔL can be obtained as follows:

i. ΔL is symmetric. This property is one necessary condition for the similarity measure, which concurs the flaw of the relative entropy.
ii. ΔL contains the information about the similarity measure, which was described as the relative entropy. Although triangle inequities are not satisfied by ΔL.

Above all, the increment in the description length can be considered as the similarity measure, when each two probabilities are merged.

When the similarity measure is given, some clustering algorithms can be employed to implement the merging operation for big data analysis. In this paper, the simplest clustering algorithm, K-means, is used to help the clustering. The steps of the proposed algorithm are listed as follows:

Step 1: Constructing some count vectors for estimating their corresponding probability distributions.
Step 2: Using training data to fill these count vectors.
Step 3: Giving the number of centers and K-means is executed. For the calculation of the distance, the increment in the description length is used to testify the similarity between two count vectors instead of the relative entropy.
Step 4: After iterations, the clustering results are obtained.

3 INFORMATION SECURITY COURSES ESTABLISHMENT BASED ON CLUSTERING OPERATION

The proposed similarity measure is suggested to analyze the efficiency of the setting of the police officer training course which comes from some pure courses.

They are:

Information Theory
Cybergraphy
Network security
Communication system
Digital signal processing

Meanwhile, due to the function of Yunnan Police College, the students who need to accept the corresponding security study come from various areas of Yunnan Province, such as DaLi, ChuXiong, and DeHong.

If one course has to be established, its utilization should be testified first by using the statistical method. A large size of investigation data consists of the training data. There are four results for evaluating this course: emergency (E), needed (N), Normal (O), and no need (NO). The investigation table is given in Table 1.

For every area where we give this investigation table, the respective statistical number of persons who choose one answer from E, N, O, and No, respectively, are filled into the table as the similar table to Table 1. Then, this filled investigation table becomes the count vector, which can be used to estimate the corresponding probability distribution that describes the request of one course for its respective area. For example, the count vector for the course "Information theory" on Kunming is given in Table 2.

Meanwhile, for different areas, such as DaLi, ChuXiong, etc., their count vectors may be different. It implies that the course "Information security" may not be needed for those areas. When our training courses are establishing, this course should not be considered for those areas where this training course "Information security" is not needed. In this case, the request for one course should be obtained first.

Table 1. The format of the investigation table (number of persons Investigated: XX).

Item	E	N	O	NO
Number	XX	XXX	XXX	XX

Table 2. Count vector for Kunming at the course "Information security" (2000 persons).

Item	E	N	O	NO
Number	300	1200	323	177

To tackle this problem, in this paper, we suggest the clustering algorithm to help obtain the course request. For one course, there are many count vectors as Table 2 for many areas. Then, these count vectors are clustered. Those count vectors which locate into the same class give the request information for the current course. When every course is determined by a similar clustering algorithm. All courses for one special area are established. Therefore, clustering operation is a key problem for our application and the similarity measure is also a key problem for the clustering operation.

4 EXPERIMENTS AND RESULTS

To test our proposed similarity measures, some experiments are employed. Fifty count vectors for those courses listed above which come from the key course of the "Information security" major from 29 areas are used as the test data. First, in experiment 1, we testify the efficiency of the increment in description length. It is easy to understand that if the clustering results are reasonable, the total description length of these count vectors should be shorter. In Table 3, the total description length based on the proposed similarity measure is listed. For comparison, the description lengths based on the relative entropy are also listed in Table 3.

From Table 3, it is easy to find that the similarity measure proposed is better than the relative entropy, since the description length is shorter based on our proposed measure.

In experiment 2, the proposed clustering algorithm is used to establish the police training course ("Information security") for different areas. There are four levels to describe the request of one course; therefore, the number of class is set to 4. Twenty-

nine count vectors are joined in clustering. In Table 4, the numbers of areas which are located into their corresponding centers respectively are listed.

From Table 4, using the clustering algorithm, the establishment of training courses can be made with a reasonable distribution. Meanwhile, based on the similarity measure proposed, the clustering algorithm can be used to help the implementation of our applications. After this analysis, according to the results from the proposed algorithm, the results indicate that the instruction efficiency can be testified more accurately than the other previous algorithms, which come from the advantage that the novel similarity measure is suggested.

5 CONCLUSION

The increment in the description length is suggested as the similarity measure between two count vectors, which are corresponding to their probability distributions based on discussion and experiment results. This measure can be employed to help the implementation of police training course establishment and the reasonable results can be achieved by using the proposed algorithm.

ACKNOWLEDGMENTS

This work was supported by the Educational Science Foundation of Yunnan Province (under grant 2014YESK-03) and the Natural Science Foundation of Yunnan Province (under grant 2013FD042).

Table 3. The comparison of description lengths based on two similarity measures.

Count vectors	Description length (bit)	
	Proposed measure	Relative entropy
Total these 50 count vectors	27,488,399	28,022,323

Table 4. Results of our clustering algorithm.

Levels	Number of areas
E	10
N	31
O	7
NO	2
Total	50 counting vectors

REFERENCES

Bapler. P & Murdoch, Academic Analytics on Data Mining in Higher Education. *International Journal for the Scholarship of Teaching and Learning*, Vol.4(2), pp.1926–1933, 2013.
Enhancing Teaching and Learning through Education Data Mining and Learning Analytics[J], *Education Department of America*, pp.336–339, 2012.
Forchhammer, S., Wu, X. Context quantization by minimum adaptive code length, in: *Proc. of IEEE Inter. Symposium on Information Theory*, Nice, France, pp.246–250, June 2007.
Forchhammer, S., Wu, X., Andersen, J.D. Optimal context quantization in lossless compression of image data sequences, *IEEE Transactions on Image Processing* 13(4), pp.509–517, Apr. 2004.
Min Chen, Jianhua Chen, Affinity propagation for the Context quantization, *Advanced Materials Research*, Vols. 791, pp.1533–1536, 2013.
Rissanen, J. A universal data compression system, *IEEE Trans. Inform. Theory*, vol. 29, pp. 656–664, Sept. 1983.
Rissanen, J. Strong optimality of the normalized ML models as universal codes and information in data, *IEEE Trans. on Information Theory*, vol. IT-47, No. 5, pp.1712–1717, 2001.

*Advances in Materials Science, Energy Technology
and Environmental Engineering – Patty & Zhou (Eds)*
© 2017 Taylor & Francis Group, London, ISBN 978-1-138-19668-1

Influence of the shear retention factor on the fracture behavior of fiber reinforced concrete

Xiaoyue Zhang
Department of Hydraulic Engineering, Zhejiang University of Water Resources and Electric Power, Hangzhou, China

Dong Wang
The Qiantang River Administration of Zhejiang Province, Hangzhou, China

ABSTRACT: The shear transferred across concrete cracks is usually calculated by a shear retention factor. It is the most important factor that related to mode II of concrete fracture model and shows the extent of the aggregate interlock and the crack shear sliding. By using the smeared crack rotating theory, various physical models of shear retention factor are analyzed in this paper. According to a four-point bending beam test of the fiber reinforced concrete, the relationship between shear retention factor and crack normal strain and that between shear stiffness and crack normal strain are obtained. The result indicates that it is reasonable to use the physical models of shear retention factor in strain hardening stage and use a constant value in the strain softening stage for the fiber reinforced concrete, and the range of the shear retention factor is also given.

1   INTRODUCTION

The finite numerical simulation of the fracture process of fiber reinforced concrete can be verified by the smeared crack theory (Anderson, 2005). The material model depends on some parameters, including three modes that prescribe the mechanism of fracture: with mode I defined as the opening mode, which is caused by direct and indirect tensile load, with mode II represented by a plane or crack shear, which is caused by a sliding action between fracture surfaces, with mode III caused by tearing action (Alecci, 2013).

Sophisticated tensile softening formulations have been developed to simulate the mode I cracking behavior of concrete. However, less attention has been paid to the shear transfer of concrete crack. The mode II corresponds to a relation between shear stress and shear strain across the crack. In the past, mode II was often assigned as a constant value (Ramani, 2010), and then an improvement was obtained by making the shear stiffness after cracking a decreasing function of the crack normal strain.

Shear Retention Factor β (SRF) is the most important factor that related to mode II of concrete fracture model and shows the extent of the aggregate interlock and the crack shear sliding. Researchers have focused on the value of SRF since the 1970s and β is usually treated as a constant value, which is a rather coarse method. While β is assigned a constant, the shear stress increment and the shear strain increment across the concrete crack will have a linear ascending relation.

For simulating the smeared crack model of the fiber reinforced concrete, both tension (mode I) and shear (mode II) should be considered. Thus, the sensitivity of SRF needs to be studied for its impact on mode II. In this paper, the mode II and SRF of the fiber reinforced concrete is researched by simulating several mathematical models, by using a four-point bending beam test.

2   MODELS OF SHEAR RETENTION FACTOR

2.1   *Constitutive relation for fiber reinforced concrete*

There are two basic methods that can simulate the crack fundamental alternatives of concrete, and the first is the discrete-crack approach, while the second is the smeared-crack approach. In this paper, the smeared crack approach was used for modeling the cracked elements. The stress-strain relation for the cracked concrete can be written as Porter (2001):

$$\Delta\sigma = [D^{co} - D^{co}\hat{N}[\hat{D}^{cr} + \hat{N}^T D^{co}\hat{N}]^{-1}\hat{N}^T D^{co}]\Delta\varepsilon \quad (1)$$

where $\Delta\sigma$ is the increment of global stress, D^{co} is the elastic modulus matrix, $\Delta\varepsilon$ is the increment of global strain, $\hat{N}$ is the transformation matrix reflecting the orientation of the crack, and $\hat{D}^{cr}$ is a 2×2 matrix incorporating the stiffness of the material in fracture mode I D^I and mode II D^{II}.

After cracking, the constitutive matrix in the principle directions $\hat{D}^{cr}$ is as follows:

$$D^{cr} = \begin{bmatrix} \dfrac{\mu}{1-\mu}E & 0 \\ 0 & \dfrac{\beta G}{1-\beta} \end{bmatrix} \qquad (2)$$

where G is the shear stiffness, E is Young's modulus, and μ is the reduction factor for the mode I stiffness.

From Equation (2), the smeared crack theory represents the overall shear stiffness of concrete after cracking was reduced from G to $\beta G/(1-\beta)$ by the shear retention factor. $\beta G/(1-\beta)$ is associated with the total strain increment while the crack shear modulus D^{II} is associated with the crack strain increment.

2.2 Sena-Cruz's model

Theoretically, SRF can be a value between 0.0 and 1.0, and in fact, by Eqn. (2), while the value of SRF is between 0.5 and 1.0, it is physically unrealistic, because the shear stiffness would increase after cracking. The shear retention factor is always treated as a constant varying from 0.1 to 0.5, and most literatures keep SRF constant during the entire response of concrete cracking. In fact, it is more correct to assume β as a function of the crack history.

To simplify the calculation of β, it is demonstrated by normal stress and normal strain instead of shear stress and shear strain, shown as Sena-Cruz's model. In this model, an assumed relationship between β and ε^{cr}_{nn} is given (Amit Rana, 2013):

$$\beta = \left[1 - \frac{\varepsilon^{cr}_{nn}}{\varepsilon^{cr}_u}\right]^q \qquad (3)$$

where ε^{cr}_u is the stress-free crack normal strain, ε^{cr}_{nn} is the crack normal strain at the beginning of the load increment, and q is a constant and here q defines the decrease level of β with the increase in the crack normal strain.

2.3 Rots' model

Rots gave a simple equation to demonstrate the effect of SRF as following (Luca, 2013):

$$\beta = 1.0/(1 + 4447\varepsilon^{cr}_{nn}) \qquad (4)$$

In this model, SRF is reduced by the cracking strain ε^{cr}_{nn}. This model is the simplest one of all

the mathematical models that have been proposed, with SRF only related with the cracking strain.

2.4 Al-Mahaidi model

Use variable SRF suggested by Al-Mahaidi as following (Beddoe, 2005):

$$\beta = 0.4/(1 - \varepsilon^{cr}_{nn}/\varepsilon_e) \qquad (5)$$

Here, ε_e is the total cracked concrete strain in the direction perpendicular to the crack direction, and $\varepsilon_e = \varepsilon_{t0} + \varepsilon^{cr}_{nn}$, ε_{t0} is the cracking tensile strain of concrete. This equation gives an initial value of 0.40 for β at the onset of cracking, and while the cracks open, the shear stiffness of cracked concrete reduces gradually.

2.5 Kolmar's model

Shear stiffness of the cracked concrete can be reduced by using Kolmar's formula, which relates the SRF to the crack strain (Benavent, 2010):

$$\beta = c_3 \frac{-\ln\left(\dfrac{1000\varepsilon^{cr}_{nn}}{c_1}\right)}{c_2} \qquad (6)$$

where $c_1 = 7 + 333\,(p - 0.005)$ and $c_2 = 10 - 167\,(p - 0.005)$, p is the transformed reinforcing ratio, $0 \le p \le 0.02$; c_3 is the user's scaling factor, which can be treated as 1.0.

3 EXAMPLE

The models described in the previous section are implemented in this section. Here, we employ a four-point bending beam test (Farnam Y, 2010). The flexural test setup is shown in Fig. 1, with prism of 51 mm by 51 mm cross-section and length of

Figure 1. Flexural test setup.

432 mm, and formed by fiber reinforced concrete. The parameters of the test are given in Table 1.

Applying the models of SRF mentioned earlier to the bending test, the relationship of the shear stiffness and the crack normal strain is obtained, as shown in Fig. 2. From Fig. 2, as the value of SRF should between 0.0 and 0.5, Sena-Cruz's

Table 1. Parameters of the test.

Parameter	Unit	Value
Modulus of elasticity	[Gpa]	61
Poisson's ratio		0.2
Stress-free crack normal strain ε_u^{cr}		0.009
Density	[kg/m³]	2570
Transformed reinforcing ratio p		0.0
Sena-Cruz's model parameter q		2.0
Cracking tensile strain ε_{t0}		0.002

model can be used while the crack normal strain ε_{nn}^{cr} is larger than 0.3%, Kolmar's model can be used while the crack normal strain ε_{nn}^{cr} is smaller than 0.5%, Rots's model can be used while the crack normal strain ε_{nn}^{cr} is larger than 0.03%, and Al-Mahaidi model can be used in the global area of the crack normal strain ε_{nn}^{cr}.

From Fig. 3, it is known that the shear stiffness decreases with the decrease in the shear retention factor. In addition, if SRF is chosen as a constant value from 0.1 to 0.5, the shear stiffness will be a constant from 3.0 to 27.0 in this example. From Fig. 3, it can be observed that while the crack normal strain is larger than 0.5%, the change in shear stiffness is not very obvious. Considering the strain hardening stage and strain softening stage of the fiber reinforced concrete, it is suitable to use the models studied above for shear retention factor in the early stage of cracking and use a constant value

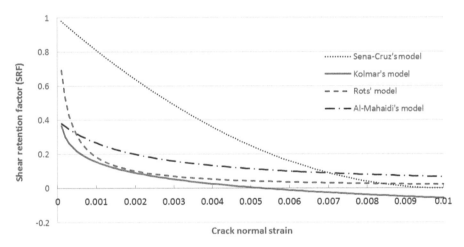

Figure 2. Relationship between the shear retention factor and the crack normal strain.

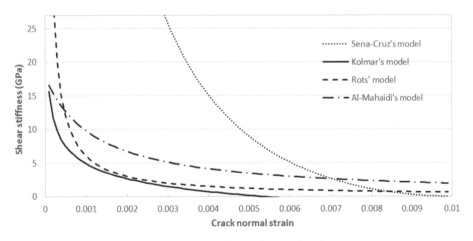

Figure 3. Relationship between the shear stiffness and the crack normal strain.

in large crack strain condition. While the value of SRF is very small, it will lead to ill-conditioning stiffness matrix of cracked concrete, and SRF should not be too large, otherwise each step of the load will lead to cracking. Considering the value of shear stiffness and Fig. 3, it has been suggested that the suitable value of shear retention factor for strain softening stage of fiber reinforced concrete is from 0.1 to 0.2.

4 CONCLUSION

In this paper, the influence of mode II and shear retention factor on cracking of the fiber reinforced concrete was discussed. Different models and constant values from 0.1 to 0.5 for shear retention factor are researched. The relationship between shear retention factor and crack normal strain and the relationship between shear stiffness and crack normal strain are obtained. According to this paper, it is reasonable to use the shear retention factor of the physical models in strain hardening stage and use a constant value in the strain softening stage for the fiber reinforced concrete, and the constant value can be from 0.1 to 0.2.

ACKNOWLEDGMENTS

This work was supported by Zhejiang Provincial Natural Science Foundation (Grant No. LQ15E090003).

REFERENCES

Anderson, T.L. 2005. *Fracture Mechanics Fundamentals and Applications.* Florida: Boca Raton CRC Press.

Alecci, V. Fagone, M. Rotunno, T. & Stefano, M. 2013. Shear strength of brick masonry walls assembled with different types of mortar. *Construction and Building Materials* 40: 1038–1045.

Amit Rana. 2013. Some studies on steel fiber reinforced concrete. *International Journal of Emerging Technology and Advanced Engineering* 3(1): 120–127.

Beddoe, R.E. & Dorner, H.W. 2005. Modelling acid attack on concrete: Part I. The essential mechanisms. *Cement and Concrete Research* 35(12): 2333–2339.

Benavent, C.A. & Zahran R. 2010. An energy-based procedure for the assessment of seismic capacity of existing frames: application to RC wide beam systems in Spain. *Soil Dynamics and Earthquake Engineering* 30: 354–367.

Farnam Y., Mohammadi S. & Shekarchi M. 2010. Experimental and numerical investigations of low velocity impact behavior of high-performance fiber-reinforced cement based composite. *International Journal of Impact Engineering* 37(2): 220–229.

Luca, F. Vamvatsikos, D. Iervolino, I. 2013. Near-optimal piecewise linear fits of static pushover capacity curves for equivalent SDOF analysis. *Earthquake Engineering and Structural Dynamics* 42(4): 523–543.

Porter, K.A. & Kiremidjian, A.S. 2001. Assembly-based vulnerability of buildings and it use in performance evaluation. *Earthquake Spectra* 17(2): 291–312.

Ramani, A. 2010. A pseudo-sensitivity based discrete-variable approach to structural topology optimization with multiple materials. *Structural and multidisciplinary optimization* 41(4): 913–934.

*Advances in Materials Science, Energy Technology
and Environmental Engineering – Patty & Zhou (Eds)*
© *2017 Taylor & Francis Group, London, ISBN 978-1-138-19668-1*

Study on primary recrystallisation annealing of 3% Si low temperature nitriding HiB steel

Jianwen Gao, Gang Zhao, Gang Tang, Yong Xu & Tao Xiong
*Key Laboratory for Ferrous Metallurgy and Resources Utilization of Ministry of Education,
Wuhan University of Science and Technology, Wuhan, Hubei, China*

ABSTRACT: In this paper, the optimal annealing temperature and time of primary recrystallisation annealing of 3% Si low temperature nitriding HiB steel were investigated. The production process referred to the actual production of cold-rolled silicon steel. The temperature range is from 750°C to 880°C. The time range is from 3 min to 9 min. After analyzing the results of Electron Back Scattered Diffraction (EBSD), it was found that the optimum temperature for primary recrystallisation annealing was 820°C, and the optimum annealing time was 5 minutes.

1 INTRODUCTION

Oriented silicon steel manufactured by using rcrystallization phenomenon is mainly used as the transformer core material. In order to obtain the ideal magnetic, it is important to control the generation of appropriate texture in the process of production (Zhou, 2009; Heejong, 2014; Huang, 2000). Oriented silicon steel recrystallization annealing includes primary and secondary recrystallization annealing. The eventually texture and organization of the silicon steel are formed through secondary recrystallization, but the primary recrystallization texture has important effect on the microstructure and texture of the secondary recrystallization (Carolina, 2014; Vladimír, 2010; Wu, 2010).

The purpose of oriented silicon steel primary recrystallisation annealing is to get sufficient {110} <001> grains which are main crystal nuclei for secondary recrystallization and microstructure beneficial for the growth of primary recrystallisation grains (Gao, 2006; Zu, 2015; Fang, 2015).Therefore, primary recrystallisation annealing plays a significant role in oriented silicon steel production and is also key to controlling the success of subsequent production therewith (Samajdar, 1998; Takeshi, 2000; Chang, 2007).

In the current research, laboratory-made oriented silicon steels were taken as the raw materials, with reference to the technology used in practice to produce oriented silicon steels. In order to provide a theoretical foundation for industrial production, the optimal temperature and time of primary recrystallistion annealing were researched.

2 EXPERIMENTAL WORK

2.1 *The determination of initial recrystallisation annealing temperature*

Specimens are taken from the the as-fabricated plate. Their chemical composition is as listed in Table 1.

The original thickness was 2.6 mm which was reduced to 0.3 mm after normalising, cold-rolling, intermediate annealing, and double cold-rolling. At the same time, specimens were cut into eight pieces measuring 10 mm × 10 mm × 0.3 mm. If the primary recrystallisation temperature were too low, incomplete recrystallisation will occur. On the other hand, if the primary recrystallisation temperature were too high, the inhibitors will be inefficient. Therefore, the range of primary recrystallisation annealing optimal temperatures is usually regarded as 750°C to 880°C. We initially determined the best primary recrystallization annealing temperature by conducting annealing experiments on five specimens as follows: Y1 (760°C, 7 min); Y2 (790°C, 7 min); Y3(820°C, 7 min); Y4 (850°C, 7 min); and Y5 (880°C, 7 min).

The protective gas used during annealing was nitrogen. After five specimens had been annealed on the basis of the aforementioned annealing

Table 1. Chemical composition of the steel (wt.%).

C	Si	Mn	P	S	Al	Ti	O	N
0.046	3.07	0.09	0.029	0.0044	0.005	0.0012	0.0014	0.0069

parameters, they were electrolytically polished on an EBSD observation plane and subjected to EBSD imaging.

2.2 *Determining the primary recrystallisation annealing time*

The remaining three specimens were annealed at a constant temperature of 820°C for different annealing times and three different regimes as follows: Y6 (820°C, 3 min); Y7 (820°C, 5 min); Y8 (820°C, 9 min).

3 RESULTS AND DISCUSSION

3.1 *Initial recrystallisation annealing temperature*

1. The size of primary recrystallisation grains
The primary recrystallisation grain size was inversely proportional to the driving force of the secondary recrystallisation grain growth: the smaller the primary recrystallisation grain size, the greater the driving force of secondary recrystallisation grain growth, the more conducive it was to developing secondary recrystallisation grains, and the better the magnetic properties of the finished product (Etter, 2002).

Microstructures of samples at different temperature are shown in Fig. 1.

It was concluded that, while being annealed, the grain size increased with increasing annealing temperature. The sample Y1 grain size was the smallest in the metallographic examination, but also showed cold-rolled grain effects, indicative of incomplete recrystallisation. The grain size of sample Y5 was the largest among the five Specimens tested, cold-rolled grains were rare, and some grains were extremely large. This was due to the degradation of the inhibitor volume fraction once the annealing temperature exceeded 850°C (Sha, 2004). The EBSD experimental average grain size is listed in Table 2.

The average grain size of oriented silicon steel should not exceed 15 μm: Y4 and Y5 were eliminated on the grounds of their excessive average grain size. This conclusion agreed with observed metallographic results.

2. Recrystallisation percentage
The recrystallisation percentage of all speximens is shown in Table 3.

Table 3 show that sample Y1 could be eliminated due to its low recrystallisation ratio, which was also consistent with the large number of cold-rolled grains observed.

3. Percentage of texture
From now on, only parameters of Y2 and Y3 could be evaluated, because Y1, Y4, and Y5 had been

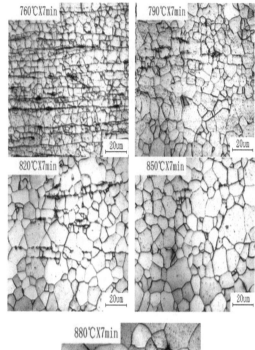

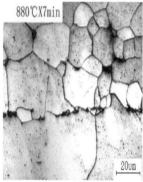

Figure 1. Metallography of the specimens at different temperatures.

Table 2. Average grain sizes.

Sample no.	Average grain sizes (μm)
Y1	9.93534
Y2	13.89955
Y3	14.52091
Y4	16.12994
Y5	25.56571

excluded. Recent studies show that the orientation {111} $<11\overline{2}>$ grains and Goss grains satisfied the orientation relationship for high mobility of grain boundaries. The Goss oriented grains swallowed a large amount of {111} $<11\overline{2}>$ grains because of their faster grain boundary migration; however,

Table 3. Recrystallisation percentages (all specimens).

Sample no.	Recrystallisation (%)	Sub-structured (%)	Deformed (%)
Y1	76.72	13.21	10.06
Y2	92.07	3.49	4.42
Y3	95.87	1.96	2.15
Y4	95.90	1.44	2.66
Y5	96.85	2.36	0.77

Table 4. Percentage texture content of Specimens Y2 and Y3.

Sample no.	$\{111\} <11\bar{2}>$ texture	$\{111\} <1\bar{1}0>$ texture	$\{012\}<001>$ texture	Goss texture
Y2	8.02	12.20	5.54	0.41
Y3	11.40	8.43	2.38	1.90

the $\{111\} <1\bar{1}0>$ texture may have weakened the intensity of secondary recrystallisation Goss texture. Besides, the $\{012\} <001>$ texture can promote the development of secondary recrystallisation and assist secondary recrystallisation magnetic induction and reduce iron loss. The percentages of texture content of Specimens of Y2 and Y3 are shown in Table 4.

From Table 4, it was concluded that the content of detrimental texture $\{111\} <1\bar{1}0>$ of Y2 was greater than that of Y3, while the content of beneficial texture $\{111\} <11\bar{2}>$ of Y3 was greater than that of Y2. The content of $\{012\} <001>$ texture of Y2 was less than that of Y3. What matters most was that the Goss texture content of Y3 was much larger than that of Y2. Sample Y2 was not capable of providing enough Goss grain nuclei for secondary recrystallisation. So the parameters of Y3 were better than those of Y2. Accordingly, the appropriate temperature for primary recrystallisation was 820°C.

3.2 Primary recrystallisation annealing time

1. The primary recrystallised grain size and recrystallisation percentage
At 820°C, the metallography of the Specimens after 3, 5, 7, and 9 minutes are shown in Figure 2.

Fig. 2 shows that the grain sizes of four Specimens were same at a constant annealing temperature, and the grains were substantially equiaxial. There were individual large grains in sample Y8, indicating grain grows therein. The average grain size and percentage of recrystallisation at 860°C is shown in Table 5.

Table 5 shows that the average grain size of these four Specimens differed little, this was because the

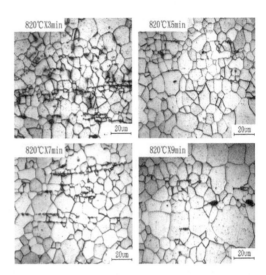

Figure 2. Metallography of specimens in different times.

Table 5. The average grain size and percentage of recrystallisation at 860 °C.

Sample no.	Grain size (μm)	Recrystallised	Sub-structured	Deformed
Y6	13.94681	88.10%	7.54%	4.34%
Y7	14.20258	92.00%	5.31%	2.67%
Y3	14.52091	95.87%	1.96%	2.15%
Y8	15.1035	96.38%	1.76%	1.85%

inhibitor functioned properly at annealing temperatures below 850°C. Although the annealing time differed, the effects of inhibited grain growth were significant. Meanwhile, Table 5 shows that sample Y6 underwent insufficient recrystallisation while the average grain size of sample Y8 was slightly larger than 15 μm, so it may be concluded that the parameters of Y3 and Y7 were superior to that of Y6 and Y8.

2. Percentage of texture
Table 6 lists the texture from EBSD.

Table 6 indicates that the lowest content of harmful texture $\{111\} <1\bar{1}0>$ was found in sample Y7 while Y8 contained the most thereof. Meanwhile, among the beneficial textures, the lowest content of texture $\{111\} <11\bar{2}>$ was in Y8 and the maximum in Y3. Texture $\{012\} <001>$ contents in Y8 were the lowest while those in Y6 were maximised. Moreover, the highest content of the most important Goss texture was found in Y7 and Y8 had the least. Therefore, the texture content of the sample was most unsatisfactory after an annealing time of 9 minutes, which may hinder secondary

195

Table 6. Texture content at different times.

Sample no.	{111}⟨1$\bar{1}\bar{2}$⟩ texture	{111}⟨1$\bar{1}$0⟩ texture	{012}⟨001⟩ texture	Goss texture
Y6	9.63%	7.78%	5.98%	1.03%
Y7	0.40%	3.16%	4.73%	2.46%
Y3	11.40%	8.43%	2.38%	1.90%
Y8	0.10%	17%	1.63%	0.51%

recrystallisation. Referring to judgements about the parameters of Y3, Y6, Y7, and Y8, the parameters of Y7 were regarded as optimal.

4 CONCLUSIONS

In the study of the oriented silicon steel primary recrystallization annealing, a series of heat treatment experiments had been conducted under different temperatures and times. After initial recrystallisation annealing, these eight Specimens were analysed by EBSD which indicated that when the annealing temperature was 820°C and the annealing time was 5 minutes, the optimal average recrystallised grain size, percentage recrystallisation, and variety of textures were obtained.

ACKNOWLEDGMENTS

This work was financially supported by the National Natural Science Foundation of China (No. 51274155).

REFERENCES

Carolina C.S. & Marco A.C. (2014).The influence of internal oxidation during decarburization of a grain oriented silicon steel on the morphology of the glass film formed at high temperature annealing. Journal of Magnetism and Magnetic Materials. 359, 65–69.

Chang S.K. (2007). Texture change from primary to secondary recrystallization by hot-band normalizing in grain-oriented silicon steels. Materials Science and Engineering A. 452–453, 93–98.

Etter A.L., Baudin T. & Penelle R. (2002). Influence of the Goss grain environment during secondary recrystallisationg of conventional grain oriented Fe-3% Si steels. Scipta mater, 47, 725–730.

Fang F. (2015). Evolution of recrystallization microstructure and texture during rapid annealing in strip-cast non-oriented electrical steels. Journal of Magnetism and Magnetic Materials 381, 433–439.

Gao X.H. (2006) New process for production of ultra-thin grain oriented silicon steel. Rare Metals. 25, 454–460.

Heejong J. (2014). Jongryoul Kim, Influence of cooling rateon iron loss behavior in 6.5 wt% grain-oriented siliconsteel. J. Magn Magn Mater. 353, 76–81.

Huang B.Y. & Yamamoto K. (2000). Effect of cold-rolling on magnetic properties of non-oriented silicon steel sheets. J. Magn Magn Mater. 209, 197–200.

Samajdar I. (1998). Primary recrystallization in a grain oriented silicon steel: on the origin of Goss {110} <001> grains. Scripta Materialia. 39, 1083–1088.

Sha Y.H. (2004). Effects of primary annealing condition on recrystallization texture in a grain oriented silicon steel. Journal of Materials Science and Technology. 20, 253–256.

Takeshi K. (2000). Recent progress and future trend on grain-oriented silicon steel. J. Magnetism and Magnetic Materials. 216, 69–73.

Vladimír S. & František K. (2010).Texture evolution in Fe-3% Si steel treated under unconventional annealing conditions. Materials Characterization. 61, 1066–1073.

Wu X.L. (2010). Study on microstructure and texture of primary recrystallization in grainoriented silicon steels. Shanghai Metals. 32, 28–33.

Xia Z. & Kang Y. (2008). Developments in the production of grain-oriented electrical steel. Journal of Magnetism and Magnetic Materials. 320, 3229–3233.

Zhou, Y.J. (2009). EBSD analyses on grain orientation and grain boundary structure of grain-oriented silicon steel. J. Chinese Electron Microscopy Society. 28, 15–19.

Zu G.Q. (2015). Analysis of the microstructure, texture and magnetic properties of strip casting 4.5 wt.% Si non-oriented electrical steel. J. matdes. 07.

Advances in Materials Science, Energy Technology
and Environmental Engineering – Patty & Zhou (Eds)
© 2017 Taylor & Francis Group, London, ISBN 978-1-138-19668-1

Research and application of e-waste resourcing LCSA model

Qinglan Han & Chunxiao Zhu
School of Business, Central South University, Changsha, Hunan

ABSTRACT: A new problem hindering the sustainable development of China is the increasing number of electronic wastes. Resource recovery processes of electronic wastes including collecting, dismantling, recycling and disposing of scrapped parts were discussed in this work and a Life Cycle Cost (LCC) model was developed. Together with eco-indicator scores using Eco-Indicator 99 (EI99) and social impact scores drived from labor times, a Life Cycle Sustainability Assessment (LCSA) was performed for recycling discarded PCBs. The results showed that the model could provide a practical decision-making method for decision makers.

1 INTRODUCTION

At present, China has entered the peak period of electronics scrap. It was estimated that by 2014 the amount of wasted electronic products (e-waste) reached to 113.78 million. The large number of e-waste is becoming a new problem which hinders the sustainable development of China. How to choose the most sustainable resourcing activities is becoming an attractive research area.

It is well accepted that the environmental, economic and social aspects are essential to achieve sustainability. For the resourcing of e-waste, Ruan, Zheng, Noon, Rubin and Bigum carried out environmental impact assessment. Li, Yin, Veerakamolmal and Geyer performed a cost-benefit analysis. Bin, Zheng, Bhuie and Dowdell conducted an environmental-economic integrated evaluation. Lv Bin took the labor time as the intermediate variable to carry on the social impact assessment of the waste desktop. He measured reusability of typical electrical, electronic products and components in of environmental, economic, social aspects; Umair evaluated social life cycle impact of electronic ICT wastes in Pakistan.

According to above literature review, several points of view could be drawn. Firstly, the methods of evaluating environmental impact on the resourcing of e-waste are relatively mature, generally by the LCA method. Secondly, cost-benefit analysis focusing on the demolition stage of each stage of cost lacks detailed analysis. Thirdly, the research on comprehensive evaluation from the perspective of sustainability is less, and most of them are based on case studies. An intact, quantitative and operational sustainability assessment model is in great need. To ameliorate the existing LCC calculation model, this paper constructed the LCSA

model of e-waste resourcing from three dimensions of economy, environmental and society. The evaluation result is a numerical value. Finally, the model is applied to the selection of discarded PCBs resourcing programs for verifying the model. The model aims to provide a practical decision-making method for decision makers, and a reference for the sustainable evaluation of product life cycle.

2 LIFE CYCLE SUSTAINABILITY ASSESSMENT MODEL

LCSA is proposed by UNEP and SETAC on the base of life cycle theory including techniques that assess the three pillars of sustainability. It can be formally expressed in the symbolic equation (1):

$$LCSA = LCA + LCC + SLCA \qquad (1)$$

where LCSA = Life Cycle Sustainability Assessment; LCA = Life Cycle Assessment; LCC = Life Cycle Cost; SLCA = Social Life Cycle Assessment.

As LCC and SLCA have not yet established a set of standard methods, the framework is barely of guiding significance, which indicates little practical guidance. In order to apply the theory into practice, this paper built a LCSA evaluation model as shown in Figure 1.

Initially, to determine the functional unit and system boundaries of research are essential to ensure the comparability of results. Secondly, the input and output—in each stage of the life cycle are analyzed in detail, and then the calculation of LCC. Thirdly, based on the analysis of LCC, the Life Cycle Inventory (LCI) and Social Life Cycle Inventory (SLCI) are established respectively. Fourthly, LCI is used by the EI99 to assess environmental

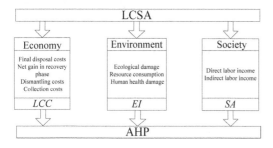

Figure 1. Life cycle sustainability assessment model of e-waste.

effects on human health, ecosystems, and natural resources. Direct and indirect labor incomes of workers made in recycle process are calculated by the SLCI and hourly wages. Besides the local minimum living security expenditure is used to carry on the characteristic, the computation, which captures the social influence value (SA). Finally, the evaluation results of the three aspects are integrated and the value of sustainability index is calculated, while AHP method is applied in the study.

2.1 LCC

In this paper, the process of e-waste resourcing is divided into four stages, that is, collecting, dismantling, recovering, final disposing. Through detailed analysis of the input and output in each stage, constructed costing model could be shown as follows:

i. Collection costs. E-waste collects from consumers, transported and stored in the dismantling center. Thus, costs paid to consumers, transportation costs and storage costs are in need to be included. The formula for calculation is as follows:

$$TC = C_{paid} + C_{tran} + C_{store}$$
$$= P - R_{allow} + T_d \cdot F + (C_l + C_f) \div S \times S' \quad (2)$$

C_{paid} equals to purchasing price (P) minus the environmental subsidies (R_{allow}). T_d is the transport distance and F is the unit cost of transportation, which is affected by weight, labor costs, etc. C_{store} is mainly sensitive to the land rent cost (C_l), building cost (C_f), plant effective use area (S) and stacking area (S').

ii. Dismantling Costs (DC). Dismantling line in disassembling center commonly uses assembly-line, and conveyed through the intelligent transportation equipment, labor and machine dismantling combined. It includes labor costs and equipment costs:

$$DC = C_d \cdot T_d + \sum C_{appi} \div Q \cdot Q' \quad (3)$$

C_d is the unit time artificial dismantling costs and T_d is the total dismantling time. Because the capital equipment investment and life are different, C_d is assessed according to the cost of equipment depreciation per day (C_{appi}) and daily processing capacity of equipment (Q) to make it reasonable. Q' represents the actual amount of processing.

iii. Net revenue in recycling phase. The revenue of this stage is the deduction of the cost of the whole resourcing system. The Revenue (R), Cost (C), Net Revenue (NR) are:

$$R = R_m + R_u + R_s,$$
$$= \sum \omega_{mi} \cdot P_{mi} + \sum \omega_{ui} \cdot P_{ui}$$
$$+ \sum \sum (\omega_{si} \cdot R_{wir}) \cdot R_r \cdot P_r \quad (4)$$

$$C = C_{material} + C_{machine} + C_{labor}$$
$$= \sum \omega_k \times P_k + \sum C_{eql} \div W_{dl} \times W_{pl}$$
$$+ \sum h_m \cdot w_m \quad (5)$$

$$NR = R - C \quad (6)$$

R mainly includes three parts: revenue of remanufacturing (R_m), revenue of reuse (R_u), and revenue of material recycling (R_s). ω_{mi} and P_{mi} are the quantity and the unit price of finished products after remanufacturing of the ith component respectively. ω_{ui} and P_{ui} are the quantity and the unit price of the ith component reuse respectively. ω_{si} is the mass of the ith component of material recycling. R_{wir} is the mass ratio of the ith materials for parts. Different materials are mixed together during recovery. Complete recovery cannot be achieved, thus the recovery of the material (R_r) should be used. The price of the material is P_r.

Cost comprises direct materials ($C_{material}$), direct labor (C_{labor}) and manufacturing costs ($C_{machine}$). ω_k and P_k indicate the consumption and unit price of the ith material respectively. C_{eql}, W_{dl}, W_{pl}, respectively, said day depreciation, daily processing capacity and actual capacity of the ith machine. h_m, w_m, respectively, said the ith process required for labor hours and hours of wages.

iv. Final waste disposal costs (WC). The toxic, harmful components/materials cannot be recovered and need specific disposal. It can be calculated as:

$$WC = \sum C_{ti} \cdot w_{ti} + \sum \sum (\omega_{si} \cdot R_{wir}) \cdot (1 - R_r) \cdot C_{sr} \quad (7)$$

w_{ti} is the weight of toxic and hazardous components. C_{ti} and C_{sr} are the unit disposal cost. Combining the previous equations, the LCC can be calculated as:

$$LCC = TC + DC - NR + WC \quad (8)$$

2.2 LCA

The main environmental impact of e-waste resourcing is its damage to the ecosystem, resources and the human body. This paper selects the EI99 method to evaluate the environmental impact and the steps are shown as follow:

i. Characterization. The characterization factor (α_{ij}) changes the different environmental factors (C_{ij}) into the same unit size, and classifies to ten ecological problems, as shown in equation (9):

$$IC_i = \sum C_{ij} \times \alpha_{ij} \qquad (9)$$

ii. Normalization. The characteristic results are normalized by the normalization factor (γ_i), which makes it comparable, which is shown in equation (10):

$$D_i = IC_i \div \gamma_i \qquad (10)$$

iii. Weighting. Normalized results were classified into three categories and added to value of impact damage, as shown in equation (11). According to the importance of one of three injury categories, N_k was given the weight ω_k. By the formula (12) EI can be drawn:

$$N_k = \sum D_i \qquad (11)$$

$$EI = \sum \omega_k \times N_k \qquad (12)$$

2.3 SLCA

SLCA developed late and is still at the exploration stage. In 2009, the UNEP/SETAC Life Cycle Initiative published Guidelines for Social Life Cycle Assessment of Products that built up a assessment framework based on 31 types of social impact of five different stakeholders, including workers, consumers, local community, society and value chain actors. However, this guide only provides way to various social impact indices and data collection categories. It does not specify how to quantify the research data. At present, the most widely used method is to set up the system of social impact assessment index, and acquire the data through the questionnaire way (e.g. Dong, Hosseinijou and Foolmaun). Obviously, this method is difficult to quantify and is subjective.

According to current situation in our country, e-waste resourcing is the main factor of influence on workers, community and society. It not only affects the health of workers and community residents, but also helps to reduce the unemployment rate. Since LCA has been evaluated on human health, here is the social impact evaluation at aspect of employment. Reference herein Hunkler's method, basing on the LCC analysis, builds a labor time list of e-waste recycling in the whole life, and calculates income workers made in the resourcing process combining per-hour wage. Because of differences in the regional income levels, the minimum living security expenditure (COL_{ai}) where the ith phase's activities took place is the characteristic factor. The income is characterized, and the social impact value (SA) is obtained. The calculation method is shown in the formula (13). Obviously, the indicator is the positive index. The greater the value is, the higher the degree of support for income is, the more favorable to the society is.

$$SA = \sum \frac{h_i \cdot w_i + h_i' \cdot w_i'}{COL_{ai}} \qquad (13)$$

h_i represents the direct labor hours consumption of the ith stage. Resourcing process will have indirect impact on society. For example, during the disposal of consumed electricity, it indirectly increases the power of workers labor time, thereby increasing their income. Therefore, h_i represents the indirect labor hours consumption of the ith stage. w_i, w_i', respectively, said the direct and indirect labor workers' hourly wages for the ith stage.

2.4 LCSA

LCC, EI, SA have different properties, therefore, it is necessary to standardize them. LCC_k', EI_k', SA_k' were normalized index values. Then, according to the actual situation, the AHP is used to determine the weight (w_i) of the three aspects, which is influenced by the state, the region, the evaluation purpose, etc. Finally, the sustainability index values (Score) of different schemes are obtained, shown in formula (14).

$$Score = w_1 \times LCC_k' + w_2 \times EI_k' + w_3 \times SA_k' \qquad (14)$$

3 MODEL APPLICATION

Printed Circuit Board (PCB) is an important part of computer, television, telephone and other electronic equipment. Therefore, the sustainability of resourcing of discarded PCBs research has practical significance. This paper compares the sustainability of two different schemes with 1 tons of discarded PCBs as a functional unit. *Scheme A*: Discarded PCBs were treated first through a hammer mill, and then pyrolysis furnace pyrolysis, and finally air shaker sorted metallic and nonmetallic materials. *Scheme B*: Discarded PCBs were broken twice first in a shear crusher, then fined in a vertical impact crusher, and at last sorted metallic and nonmetallic materials in an air shaker. Since the two programs in the collection, dismantling stage use the same approach, with the same sustainability performance, this paper analyzes only the recycling phase and final disposal stage.

3.1 LCC of discarded PCBs

i. Net revenue in recovery phase. Discarded PCBs at this stage major cost benefits include material recovery revenue, manufacturing costs, direct labor.

① Revenue. In both scenarios, the discarded PCBs are only implemented materials recycling activities. ω_{si} = 1t. In this paper, material composition of discarded PCBs and the mass ratio (R_{wir}) of them used Park's.

Assuming the recovery (R_r) of metal and glass fiber in scheme A was 98%, the metal recovery in scheme B was 95% and the recovery of glass fiber was 78.57%. The price (P_r) of recycled materials in the two schemes can be found form relative website.

According to R_{wir}, R_r and P_r, using the formula (4) to calculate the revenues of the two schemes, the results are shown in Table 1.

② Manufacturing costs. The machines used in the two schemes includes hammer mill, sheared double-toothed roll, etc. Their price, power (kw) and production capacity (t/h) can be found from website. All machines are assumed to have a 6-year depreciation period. Thus, day deprciation (C_{eql}) could be calculated according to the price obtained. According to the machine production capacity and 8 hours per day standard, daily processing capacity (W_{dl}) of the machine can be estimated. Actual processing capacity of the machine is one ton. W_{dl} = 1t. Calculated by formula (5), the cost of manufacturing is shown in Table 1.

③ Direct labor. The working hours of each working procedure (h_m), can be obtained by the production capacity and processing capacity of the machine (ω_{si}). According to the wage level in Hunan Province, it is assumed that the wages of workers in different stages are the same. w_m = $2.30/h. By formula (6) the labor costs are shown in Table 2.

Table 1. LCC of two schemes.

	R	$C_{machine}$	C_{labor}	NR	WC	LCC($)
Scheme A	6990	13	24	6960	8	−6950
Scheme B	6790	9	39	6740	9	−6730

Table 2. The LCI of two schemes.

		Scheme A	Scheme B
Energy (MJ)	Raw coal	3.56E+06	1.11E+05
	Crude oil	1.37E+05	−2.20E+04
	Natural gas	2.74E+04	8.95E+02
Material (kg)	Copper	−1.57E+02	−1.52E+02
	Aluminum	−4.90E+01	−4.75E+01
Emissions	N_2O	3.06E−02	1.03E−03
to air (kg)	SO_2	−1.02E+02	−1.02E+02
	CO	3.75E−03	−8.72E−03
	NO_X	1.57E+00	4.90E−02
	CH_4	1.30E+00	8.20E+00
	CO_2	−1.67E+03	−2.00E+034
	NH_3	4.12E−03	3.21E−02

ii. Final disposal cost. It is assumed that the part of the discarded PCBs resource which cannot be recovered in the two schemes can be disposed of with normal household landfill waste. C_{si} = $0.058/kg. According to R_{wir} and R_r, the quality of the waste of the two schemes are 23.41 kg and 160.36 kg respectively. The final disposal cost of the two schemes can be calculated by the formula (7) as shown in Table 1.

iii. LCC. LCC is calculated by formula (8) of two schemes as shown in Table 1.

3.2 LCA of discarded PCBs

The power consumption of the two schemes are 306.46 kwh and 10.32 kwh, respectively, according to the power output of the machine and processing capacity of the machine. In accordance with primary energy and gas emissions of the consumption of 1 kwh power, the LCI of the recycling phase in two schemes can be calculated. The two schemes mainly recovered copper and aluminum two materials, according to the 1 kg copper and 1 kg aluminum LCI, deductible items of recycling phase can be calculated. Depending on the quality of the final disposal of waste, combined with 1 kg landfill disposal life cycle inventory, the inventory data of the final disposal phase in two schemes can be calculated. Consolidate the lists of all stages, and ultimately get the LCI for the two schemes, as shown in Table 2.

Using characteristic factor and per capita basis based on the basic data of our country, according to the formula (9)–(11), the choice of Goedkoop undifferentiated weights, EI in two schemes was −358 and −546. Negative number indicates that the environmental impact of the implementation of the program to avoid the environmental impact is greater than the program itself, and shows a good performance of the project at aspect of environmental impact.

3.3 SLCA of discarded PCBs

The direct working hours of the recycling phase have been calculated in Section 3.1. The daily treatment capacity of the landfill site is about 4000t, and the numerical value of the working time of the final disposal stage is minimal, thus, it can be ignored. In addition, in the resourcing process, energy is consumed. SO_2, CO_2 and other pollutants are discharged, which indirectly increases the demand for energy and environmental management jobs in the industry. Since the availability of data, this paper only considered the power consumption. The power consumption of the two schemes was obtained in section 3.2. According to the basic data, 1kwh power generation requires the power workers to work for 0.0015 hours. Direct and indirect working hours of the two schemes are shown in Table 3.

It is assumed that that the two schemes have all occurred in Hunan province. The urban minimum living standard in Hunan province is the characteristic factor. $COL_{ai} = \$55.26/month$. $w_i = w_i' = 230/h$. The SA calculated by the formula (13) are 0.49 and 0.76, respectively, as is shown in Table 4.

3.4 LCSA of discarded PCBs

Use the AHP method to determine the weights of the three aspects of economic, environment, society, respectively 0.28, 0.64, 0.07, the test $CR = 0.083 < 0.1$, with a consistency. The results of the three aspects are converted into proper dime, combined with the weight, the value of sustainable indicators can be obtained. The results are shown in Table 4.

Table 3. Total working hours of two schemes.

	Direct hours	Indirect hours	Total hours (h)
Scheme A	11.27	0.46	11.73
Scheme B	18.15	0.02	18.17

Table 4. Results of life cycle sustainability assessment.

	Scheme A	Scheme B
LCC	−6950	−6730
Std. LCC	1	0.97
Wt.		0.28
EI	−358	−546
Std. EI	0.66	1
Wt.		0.64
SLCA	0.49	0.76
Std. SLCA	0.64	1
Wt.		0.07
Score	0.7472	0.9816

From the results, conclusion can be drawn that the evaluation results of scheme B is better than those of scheme A. Therefore, scheme B is more conducive to sustainable development.

4 CONCLUSIONS

In this paper, firstly a detailed input-output analysis of e-waste resourcing in full life cycle was carried out. Secondly, the LCC calculation model was established. Thirdly, the EI99 index was used to quantify the environmental impact. Fourthly, the social impact was evaluated with the labor time as the intermediate variable. Finally, a quantitative operational LCSA model was constructed. It is proved that the model is feasible through case analysis of the discarded PCBs resourcing. Since the data are difficult to obtain, domestic studies are less. Therefore, part of the data needs to be meliorated and revised. Though the comprehensive evaluation of sustainability is still in the initial stage of exploration, it still has a wide application range and a promising future.

ACKNOWLEDGEMENT

This paper has been supported by the National Natural Science Foundation of China under Grant No. 71172101.

REFERENCES

Bhuie A.K., O.A. Ogunseitan, Saphores J.D.M., et al. Environmental and economic trade-offs in consumer electronic products recycling: a case study of cell phones and computers, Electronics and the Environment, International Symposium on IEEE, (2004) 74–79.

Bigum M., L. Brogaard & T.H. Christensen, Metal recovery from high-grade WEEE: A life cycle assessment, J. Hazard. Mater. 207–208 (2012) 8–14.

Bin L., et al., Reusability based on Life Cycle Sustainability Assessment: Case Study on WEEE, Procedia CIRP. 15 (2014) 473–478.

Chen Ailun, Chen Chan, Tao Xiaqiu, et al. Life cycle assessment of sanitary landfill of domestic garbage in Changsha, Environ. Sci. & Technol. 36 (2013) 90–395.

Dong Y.H., S.T. Ng, A social life cycle assessment model for building construction in Hong Kong, Int. J. Life. Cycle. Assess. 20 (2015) 1166–1180.

Dowdell D.C., et al. An integrated life cycle assessment and cost analysis of the implications of implementing the proposed waste from electrical and electronic equipment (WEEE) directive, Proceedings of the 2000 IEEE International Symposium on (2000).

Foolmaun R.K., T. Ramjeeawon, Comparative life cycle assessment and social life cycle assessment of used polyethylene terephthalate (PET) bottles in Mauritius, Int. J. Life. Cycle. Assess. 18 (2013) 155–171.

Geyer R., V.D. Blass, The economics of cell phone reuse and recycling, Int. J. Adv. Manuf. Tech. 47 (2010) 515–525.

Goedkoop M., R. Spriensma, The Eco-indicator'99: A damage oriented method for Life Cycle Impact Assessment, Zoetermeer, (2000).

Havlik T., et al., Leaching of copper and tin from used printed circuit boards after thermal treatment, J. Hazard. Mater. 183 (2010) 866–73.

Hosseinijou S.A., S. Mansour, M.A. Shirazi, Social life cycle assessment for material selection: a case study of building materials, Int. J. Life. Cycle. Assess. 19 (2014) 620–645.

Hunkeler D., Societal LCA Methodology and Case Study (12 pp), Int. J. Life. Cycle. Assess. 2006. 11 (2006) 371–382.

Li Jian, Zhang Jihui, Reverse Logistic Game About Obsolete Computer and the Cost-Benefit Recycling Model, China Resour. Compr. Util. 04 (2008) 10–14.

Lü Bin, Yang Jianxin, Eco-efficiency analysis of recycling strategies of WEEE in China, Chin. J. Environ. Eng. 01 (2010) 183–188.

Lü Bin, Yang Jianxin, Song Xiaolong, Social impact assessment of life cycle and its application in electronic waste management, China Sustainable Development Forum, Jinan, (2010).

Noon M.S. & S. Lee, A life cycle assessment of end-of-life computer monitor management in the Seattle metropolitan region, Resour. Conserv. Recy. 57 (2011) 22–29.

Ruan Jiuli, Guo Yuwen, Qiao Qi, Life cycle assessment of recycling and disposal of waste printers, J. Environ. Eng. Techno. 04 (2015) 323–327.

Rubin R.S., et al., Utilization of LCA methodology to compare two strategies for recovery of copper from printed circuit board scrap, J. Clean. Prod. 64 (2014) 297–305.

Song Guojun, Du Qianqian, Ma Ben, Social cost accounting for solid waste landfill disposal in Beijing, Journal of Arid Land Rsources and Environment. 08 (2015) 57–63.

Umair S., A. Björklund and E.E. Petersen, Social life cycle inventory and impact assessment of informal recycling of electronic ICT waste in Pakistan, Hilty L, Aebischer E, Andersson G, Lohmann W, Proceedings of the First International Conference on Information and Communication Technologies for Sustainability ETH Zurich, (2013).

Veerakamolmal P., S.M. Gupta, A combinatorial cost-benefit analysis methodology for designing modular electronic products for the environment. in Electronics and the Environment, Proceedings of the 1999 IEEE International Symposium on. (1999).

Wen Xuefeng, et al., Review Of The Physical Treatment Of Waste Printed Wiring Boards In China. Min. Metall. 14 (2005) 58–63.

Yin Limeng, Liu Liangqi, Study on Recycle, Reuse and the Cost-Benefit Model of Electronic Waste, Mater. Rev. 09 (2010) 105–107+112.

Zheng Xiujun, Assessment of Waste Mobile Phone Recycling: A Comprehensive Analytical Framework: Study on Life Cycle Assessment and Interdisciplinary Construction of Economics, Ecol. Econ. 09 (2015) 101–104.

Zheng Xiujun, Wang Jingwei, Used Mobile Phones' Resourcing Model in the Concept of Cyclic Economy, Ecol. Econ. 02 (2014) 30–36+142.

*Advances in Materials Science, Energy Technology
and Environmental Engineering – Patty & Zhou (Eds)*
© *2017 Taylor & Francis Group, London, ISBN 978-1-138-19668-1*

Influence of particle size on seepage of water and sediment in fractured rock

Yu Liu
*Kewen College and School of Mechanical and Electrical Engineering, Jiangsu Normal University,
Xuzhou, Jiangsu, China
School of Mines, China University of Mining and Technology, China*

Xiao Zhang, Jiang Bai, Xiaoli Cai & Zhenhua Si
*Kewen College and School of Mechanical and Electrical Engineering, Jiangsu Normal University,
Xuzhou, Jiangsu, China*

ABSTRACT: Coal excavations performed in areas of shallow, thick sand can overburden the strata layers of the excavation site, resulting in fractures in the strata layers. Surface water, groundwater and sand layers can flow into mine goaf through the fractured rock and thus often lead to inrush of water and collapsing of sand. It is important to study the mechanical properties of water and sand in excavations sites under different conditions and the influencing factors of the water and sand seepage system. The viscosity of water-sand mixtures of various particle sizes were tested based on the relationship between the shear strain rate and the surface viscosity. A seepage system test was designed to analyze the permeability of water and sand in fractured rock. According to the seepage test, effective fluidity is in $10^{-8} \sim 10^{-5} m^{n+2} \cdot s^{2-n}/kg$, while the non-Darcy coefficient ranges from 10^5 to 10^8 m^{-1} with the change of particle size of sand. Effective fluidity decreased as the particle size of sand increased. The non-Darcy coefficient ranged from 10^5 to 10^8 m^{-1} depending on particle size and showed contrary results.

1 INTRODUCTION

China relies on coal for approximately 70 percent of its energy. Therefore, coal production is of crucial importance for China's economy and development. Coal reserves occupy 80 percent of the reserves in northwest China (Du, 2014, p1–2). The coal reserves are located at shallow depths and the thin bedrock and thick sand overburdens the strata layers, inducing connected cracks. Surface water, groundwater and sand layers can flow into the mine goaf through the fractured rock and thus lead to inrush of water and collapsing of sand.

From the mechanical perspective, the result of water and sand erupting and permeating fractured rock reflects the instability of the strata layers. Therefore, studying the seepage properties of fractured rocks plays an important role in coal mining engineering. The inrush of water and sand compromises mine safety by causing instability in stress block beams, which creates surface subsidence and water resource run off.

Field tests that are conducted in order to replicate water and sand inrush are difficult; therefore, many scholars suggest conducting experimental simulations of inrushing water and sand. Yang (2009), Yang, Sui, Ji & Zhao (2012) and Sui, Cai & Dong (2007) analyzed the angle of fluid using cemented sand to analyze the mechanisms supporting the inrushing of water and sand. The flow law was examined during various conditions and critical hydraulic gradients of sand inrush currents. Sui, Dong, Cai & Yang (2008) and Xu, Wang & You (2012) analyzed the initial position of inrushing sand based on the structure of water inrush.

Based on underground water dynamic theories, Zhang, Kang & Liu created the critical condition and forecasting formula for the prevention of sand inrush by calculating the hydraulic head (2006, p429–432). Wu designed a mechanical model of sand inrush pseudo structures, and discussed the force during sand inrush and described the theory of expression of sand inrush (2004, p57–59). Zhang, Zhang, Wang & Wu used a case study to discuss drills resulting in sand inrush based on the funnel model (2014, p1–11). Zhang et al studied the relationship between backfill and water through conducting crack zone (2014, p1439–1448). Moreover, river sediment engineering, the theory of sediment transmission and sediment transport mechanics, are excellent subject matters to aid in studying the start and movement of sand in mines. Furthermore, the study of sediment engineering, sediment transport theory and practice, and sediment kinematics can aid in understanding the commencement, flow and inrushing sand problems.

In this work, fluid attributes of water-sand mixtures were obtained through testing by replicating the design system of water-sand seepage in fractures. The influence of mass concentration and particle size of sand on the seepage parameters were tested using specially designed instruments.

2 VISCOSITY TEST OF WATER AND SEDIMENT

Viscous parameters of water-sand mixture in stress-strain relationships were tested using a NDJ-8S viscosimeter.

Shear strain rate γ of water-sand is defined as

$$\gamma = \frac{\pi n_{rot}}{30} \times \frac{d}{(D-d)} \qquad (1)$$

Apparent viscosity μ_a of water-sand mixture were obtained from the NDJ-8S viscosimeter and the shear stress was calculated as follows

$$\tau = \mu_a \gamma \qquad (2)$$

By changing the rotational speed of the NDJ-8S viscosimeter, several values of shear strain rate γ and shear stress were obtained and plotted on a $\gamma - \tau$ scatter diagram. According to the shape of the $\gamma - \tau$ diagram, the water-sand mixture was identified as non-Newton fluid, and the viscous parameter of water-sand was obtained through linear regression.

Sand particles measuring 0.061~0.080 mm, with a mass concentration of 20 kg/m³ and a temperature of 20°C, were measured for the shear strain rate γ, apparent viscosity μ_a and stress τ of the water-sand mixture at various rotation rates (Table 1).

A scatter diagram of $\gamma - \tau$ of sand particles measuring 0.061~0.080 mm, with mass concentration of 20 kg/m³ and a temperature of 20°C, are displayed for the water-sand mixture in Figure 1. It can be seen in Figure 3 that the shear strain rate γ increases monotonously along with the shear strain of the water-sand mixture. Therefore, we assume that the water-sand mixture is a power law fluid as follows

$$\tau = C\gamma^n \qquad (3)$$

where C is the consistency coefficient, n is the power exponent.

Combining Eq. 2 and Eq. 3 yields expression of apparent viscosity as follows

$$\mu_a = C\gamma^{n-1} \qquad (4)$$

Through linear regression, viscous parameters of water-sand (consistency coefficient C and power exponent n) were obtained. It was deduced that the water-sand mixture was a pseudo-plastic fluid, whose viscous parameters changed with sand particle d_s and mass concentration ρ_s. Consistency coefficient C increases with mass concentration as exponential relationship, and decreases along with the increase of sand particle; power exponent n increases along with the increase of mass concentration, and decreases along with the increase of sand particle.

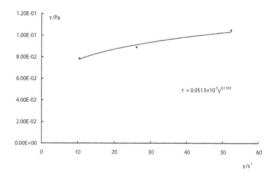

Figure 1. Scatter plot of angle strain rate—shear stress.

Table 1. Angle strain rate, apparent viscosity and shear stresses at different rotating speed.

Rotational speed (rpm)	Shear strain rate (S⁻¹)	Apparent viscosity (Pa.s)	Shear stress (Pa)	Consistency coefficient (Pa.sⁿ)	Power exponent
12	10.47	0.0075	0.0785	0.0513	0.1765
30	26.17	0.0034	0.0889		
60	52.33	0.002	0.1046		

204

3 SEEPAGE TEST OF WATER AND SAND IN FRACTURE

3.1 Test principle

Figure 2 demonstrates a model of seepage in a fracture.

According to Figure 2, we can get Eq. 5.

$$V = \frac{Q}{bh} \qquad (5)$$

where V is the velocity of seepage, Q is the flow of seepage, b is the width of the fracture, and h is height of the fracture.

For the momentum equation, research literature was used to deduce the formula of non-Darcy law for porous media (Li, Miao, Chen & Mao, 2008, pp. 85–92). The relationship of Forchheimer is

$$mc_a \frac{\partial V}{\partial t} = -\frac{\partial p}{\partial X} - \frac{\mu_e}{k_e} V^n - m\beta V^2 \qquad (6)$$

where μ is liquid viscosity, β is non-Darcy factor, the pressure is p, the effective viscosity is μ_e, and the effective permeability is k_e.

The water-sand mixture was a non-Newton fluid, whose liquid viscosity and permeability were related to fluid properties and fracture aperture. Therefore, liquid viscosity and permeability were obtained separately, and the effective fluidity I_e was introduced to simplify the expression.

$$I_e = \frac{k_e}{\mu_e} \qquad (7)$$

The Eq. 6 can be changed into

$$mc_a \frac{\partial V}{\partial t} = -\frac{\partial p}{\partial X} - \frac{1}{I_e} V^n - \beta m V^2 \qquad (8)$$

Eq. 8 calculated the momentum conservation of water-sand seepage in the fracture. For the seepage in Figure 3, the steady-flow method was selected to measure water-sand seepage in the fracture, Eq. 5 can be deduced into Eq. 8.

$$\frac{1}{I_e} V^n + m\beta V^2 = -\frac{dp}{dL} \qquad (9)$$

Substituting Eq. 5 into Eq. 9 yields Eq. 10 and Eq. 11

$$\frac{\mu_e}{k_e} \frac{Q}{bh} + m\beta \left(\frac{Q}{bh}\right)^2 = -\frac{dp}{dL} \qquad (10)$$

or

$$-dp = \frac{\mu_e}{k_e} \frac{Q}{bh} dL + m\beta \left(\frac{Q}{bh}\right)^2 dL \qquad (11)$$

The inner and outer diameter of rock specimen respectively were $2a_1$ and $2a_2$; the pressure of water and sand at the entrance wall were:

$$\begin{cases} p\big|_{r=a_1} = p_0 \\ p\big|_{r=a_2} = 0 \end{cases} \qquad (12)$$

The definite integral of Eq. 11 on the interval [a, b] was

$$p_0 = \frac{\mu_e}{k_e} \frac{Q}{hb} + m\beta \left(\frac{Q}{bh}\right)^2 \qquad (13)$$

Introducing the sign $\lambda_1 = \frac{\mu_e}{hbk_e}$, $\lambda_2 = \frac{m\beta}{(hb)^2}$. Therefore, Eq. 13 was obtained using

$$\lambda_1 Q + \lambda_2 Q^2 - p_0 = 0 \qquad (14)$$

In the test, 5 flows are set as $Q_i, i = 1, 2, \ldots, 5$. Steady state values of inlet pressures were tested, and coefficients λ_1 and λ_2 were fitted. The specific process was as follows:

Eq. 14 was obtained using

$$\Pi = \sum_{i=1}^{5} \left(\lambda_1 Q_i + \lambda_2 Q_i^2 - p_0^i\right)^2 = 0 \qquad (15)$$

In order to get the least value of the flow Q, Eq. 15 can be set as Eq. 16.

$$\begin{cases} \left(\sum_{i=1}^{5} Q_i^2\right)\lambda_1 + \left(\sum_{i=1}^{5} Q_i^3\right)\lambda_2 = \left(\sum_{i=1}^{5} Q_i p_0^i\right) \\ \left(\sum_{i=1}^{5} Q_i^3\right)\lambda_1 + \left(\sum_{i=1}^{5} Q_i^4\right)\lambda_2 = \left(\sum_{i=1}^{5} Q_i^2 p_0^i\right) \end{cases} \qquad (16)$$

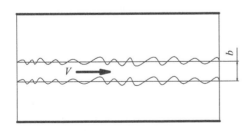

Figure 2. Parallel seepage in fracture.

λ_1 and λ_2 are solved by Eq. 15, effective mobility I_e and non-Darcy β were obtained.

$$I_e = \frac{k_e}{\mu_e} = \frac{1}{hb\lambda_1} \qquad (17)$$

$$\beta = (hb)^2 \lambda_2 \qquad (18)$$

3.2 Experimental equipment and steps

Based on testing principles, a set of experimental systems were designed and manufactured as shown in Figure 3. Radial seepage in the fracture was $JRC4{\sim}6$ and from the five specimen, the velocity of seepage was obtained. Sand comes from the surface of the mine in Northwest of China. Stone is the sandstone from the mine under −265 m.

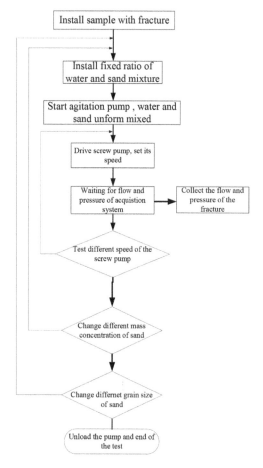

Figure 3.   Flow chart of the test.

Figure 3 illustrates the entire experimental procedure. The test steps were as follows:

1. The test system was assembled according to Figure 3 and the sample was loaded. The leakage of the experiment system was tested.
2. The sand grain with a diameter of 0.038~0.044 mm was placed into the mixing pool and the initial concentration was between 20 kg/m³ and 80 kg/m³. In order to facilitate subsequent analysis, the intermediate value scope of the grain size was obtained.

   To control the motor speed, flow and pressure under different rotational speeds were recorded and the opening of the fracture was controlled at 0.75 mm; the motor speeds, 200 r/min, 400 r/min, 600 r/min, 800 r/min, 1000 r/min were changed separately. Different pressures and seepage velocities of the fracture were obtained using a paperless recorder. The concentration of water and sand were 40 kg/m³, 60 kg/m³, 80 kg/m³.
3. The grain diameters were changed from 0.071 mm to 0.100 mm and 0.150 mm and the flow and pressure under different rotational speeds were recorded.
4. Using Eq. 15 and 16, I_e and β were calculated.

4   RESULTS AND ANALYSIS

Keeping the fracture width 0.75 mm, the permeability parameters of water and sand seepage in the fracture under particle sizes of 0.038~0.044 mm, 0.061~0.080 mm, 0.090~0.109 mm and 0.120~0.180 mm were tested at a concentration of 20 kg/m³, 40 kg/m³ and 60 kg/m³, as shown in Figure 4.

The exponential function was used to fit the relationship between effective fluidity and particle sizes of sand, the same way with the non-Darcy coefficient. The polynomial equations are used to fit the relationship between effective fluidity I_e, the non-Darcy factor β and JRC.

From Figure 4, the following results were obtained:

1. The seepage of water sand in a fracture was nonlinear.
2. Along with the change of grain size of sediment, the relationship between effective fluidity I_e and mass concentration of sediment ρ_s was the negative exponential relationship; the absolute value of the exponent increased along with an increase of sand concentration.
3. Non-Darcy factor β and sand concentration had a positive exponential relationship; the absolute value of the exponent increased along with a decrease of sand concentration.

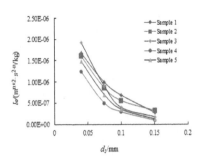

(a) Curve of $I_e - \rho_s$ under 20kg/m³

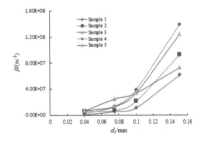

(e) Curve of $I_e - \rho_s$ under 60kg/m3

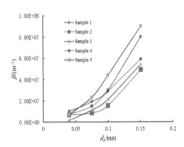

(b) Curve of $\beta - \rho_s$ under 20kg/m³

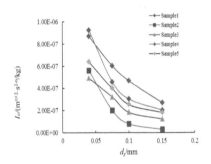

(f) Curve of $\beta - \rho_s$ under 60kg/m3

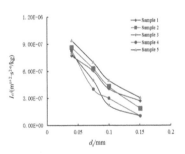

(c) Curve of $I_e - \rho_s$ under 40kg/m3

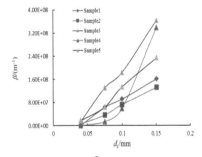

(g) Curve of $I_e - \rho_s$ under 80kg/m3

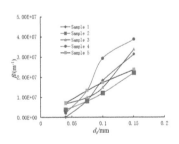

(d) Curve of $\beta - \rho_s$ under 40kg/m3

(h) Curve of $\beta - \rho_s$ under 80kg/m3

Figure 4.   Curves of permeability parameters changing with d_s.

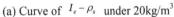

5 CONCLUSION

In this paper, the viscosity of water and sand mixture was discussed and the seepage of water and sand mixture in rude fracture was analyzed.

1. The seepage velocity of water and sand in a fracture increases along with the pressure of the fracture, but the relationship between them is nonlinear.
2. Consistency coefficient ρ_s becomes larger in conjunction with the mass concentration, but decreases along with the particle size of sand. The lower exponent n becomes enlarger along with mass concentration, but decreases along with particle size of sand.

ACKNOWLEDGEMENT

This research was supported by Natural science fund for colleges and universities in Jiangsu Province (14KJB440001), Jiangsu Normal University PhD Start Fund (14XLR032), Fund of Jiangsu Normal University (13XLA12) and Jiangsu Planned Projects for Postdoctoral Research Funds (1402055B), Project supported by the National Natural Science Foundation of China (51574228), All the supports are gratefully acknowledged.

REFERENCES

Du Feng. (2014) Experimental Investigation of Two Phase Water-sand Flow Characteristics in Crushed Rock Mass: China university of mining & technology.

Fan Niannian, Wu Baosheng. (2015) Anomalous diffusion of non-uniform bed load particles based on a stochastic-mechanic model [J]. Journal of Zhejiang university (engineeting science), 49(2): 246–250.

Lu Shouqian, Lu Yongjun, Zuo Liqin, Huang Weihao, Lu Yan. (2014) Incipient motion of sediment in wave and combined wave-current boundary layers. Advances in water science. 25(1): 106–114.

Li Shuncai, Miao Xiexing, Chen Zhanqing, Mao xianbiao. (2008) Experimental Study on Seepage properties of non-Darcy Flow in Confined Broken Rocks [J]. Engineering Mechanics. 25(4): 85–92.

Sui Wanghua, Cai Guangtao, Dong Qinghong. (2007) Experimental research on critical percolation gradient of quicksand across overburden fissures due to coal mining near unconsolidated soil layers, Chinese Journal of Rock Mechanicas and Engineering. 26(10): 2084–2091.

Sui Wanghua, Dong Qinghong, Cai guangtao, Yang Weifeng. (2008) Mechanism and prevention of sand inrush during mining: Geological house.

Wu Yongping. Condition analysis on sand inrush in shallow stope, (2004) Mine pressure and support. 4: 57–59.

Xu Yan-chun, Wang Bo-sheng, You Shen-wu. (2012) Mechanism and criteria of crushing sand near loosening sand stone aquifer. Journal of Xi'an University of Science and Technology. 32(1): 63–69.

Yang Weifeng. (2009) Overburden failure in thin bedrock and characteristics of mixed water and sand flow induced by mining: China University of Mining Technology.

Yang Weifeng, Sui Wanghua, Ji Yubing, Zhao Guorong. (2012) Experimental research on the movement process of mixed water and sand flow across overburden fissures in thin bedrock by mining. 37(1): 141–146.

Zhang Guimin, Zhang Kai, Wang Lijuan, Wu Yu. (2014) Mechanism of water inrush and quicksand movement induced by a borehole and measures for prevention and remediation. Bulletion of Engineering Geology Environment. 1–11.

Zhang Jixiong, Zhang Qiang, Sun Qiang, Gao Rui, Deon Germain, Sami Abro. (2015) Surface subsidence control theory and application to backfill coal mining technology. Environmental Earth Sciences. 74(2): 1439–1448.

Zhang Yujun, Kang Yong-hua, Liu Xiue. (2006) Predicting on inrush of sand of mining under loosening sandstone aquifer [J]. Journal of China coal society. 31(4): 429–432.

Advances in Materials Science, Energy Technology
and Environmental Engineering – Patty & Zhou (Eds)
© 2017 Taylor & Francis Group, London, ISBN 978-1-138-19668-1

Establishment of an evaluation system of "green level" of scale inhibitors for water treatment

Li Zhu & Ning Wang
School of Municipal and Environmental Engineering, Shandong Jianzhu University, P.R. China
Shandong Co-Innovation Center of Green Building, P.R. China

Limi Sun
School of Thermal Energy Engineering, Shandong Jianzhu University, P.R. China

ABSTRACT: Based on the concept of green chemistry, an innovative five-range scoring system was investigated and suggested for the evaluation of "green level" of scale inhibitors used in water treatment. The concept and theory focused on the assessment of raw materials, treatment processes, and products, which was expected as an efficient way to evaluate the environment-friendly properties.

1 INTRODUCTION

1.1 Scale inhibitor for water treatment

Since "green chemistry" was first mentioned in 1995, the concept has become the guideline and goal for the industries, including chemistry and chemical engineering (Fan et al. 2012). Scale inhibitors of a membrane could efficiently prevent reverse osmosis, nanofiltration, and ultrafiltration from the problems scaling and fouling. The development of these chemicals should also comply with the purpose of green chemistry and "green scale inhibitor" has already been mentioned as the future, which should be biodegradable and caused no negative impact on environment and human health during the manufacture and utilization (Ketrane et al. 2009, Tang et al. 2010, Sun et al. 2012).

The evaluation method for the performance of scale inhibitors has received more attention, and research on environmental impact assessment methods of scale inhibitor production, transportation, storage, and use process is lacking. However, current evaluation systems of green scale inhibitor were either misleading or limiting. For instance, only the index of "low phosphorus" or "no phosphorus" was used to grade the chemicals, which was far from enough to describe "green". Then, in this research, the "green level" was used as the index to provide an operable evaluation system for green scale inhibitors in this industry.

1.2 "Green level"

"Green level" in this research is related to the extent to human health and environment-friendship of a type of substance during the whole life history, which could be quantified as "green index" for evaluation.

2 "GREEN LEVEL" EVALUATION SYSTEM FOR SCALE INHIBITORS IN WATER TREATMENT

2.1 Index evaluation system

Based on the life process investigation of production, utilization, and degradation of scale inhibitors for membranes, the "green level" evaluation system was established and is shown in Table 1.

2.1.1 Index and explanation

1. Difficulty level of transport and storage, A_{11}: meaning the temperature and pressure conditions for the transport and storage of raw materials, reflecting the cost, an index, as one of the characteristics presenting the material economy and environmental risk.
2. Hazards and risk, A_{12}: meaning the hazardous properties including flammable, explosive, or reactive raw materials, identified by the national standards "List of dangerous goods (GB12268–2012)" of China, PR, and any item in this list considered as risk.

3. Toxicity, A_{13}: classifying hazardous materials into five categories (0–4) using the US academy method, according to median lethal dose (LD_{50}) (Dixon 1979):

0-no toxicity, $LD_{50} > 15$ g · kg^{-1};
1-little toxicity, 5 g · kg^{-1};
2-light toxicity, 0.5 g · kg^{-1};
3-medium toxicity, 50 mg · kg^{-1};
4-toxicity, LD50 < 50 mg · kg^{-1}.

4. Wastewater, A_{21}: relating to the type, quantity, toxicity, and treating difficulty of the wastewater generated as the production of scale inhibitors, an index reflecting the environment-friendship of the manufacturing process.
5. Waste gas, A_{22}: relating to the type, quantity, toxicity, and treating difficulty of the waste gas generated as the production of scale inhibitors, an index reflecting the environment-friendship of the manufacturing process.
6. Solid waste, A_{23}: relating to the type, quantity, toxicity, and treating difficulty of the solid waste generated as the production of scale inhibitors, one index reflecting the environment-friendship of the manufacture process.
7. Clean of production, A_{24}: relating to fact whether the inhibitor production process belongs to cleaner production or encouraged process by government.
8. Health hazards, A_{31}: relating to the items identified by "Hazardous Chemical List (2015)" of China PR, which also complies with "Globally Harmonized System of Classification and Labelling of Chemicals" of UN (GHS) and to the classification of chemicals into ten categories, namely acute toxicity, skin corrosion/irritation, serious eye damage/eye irritation, respiratory or skin sensitization, germ cell mutagenicity, carcinogenicity, reproductive toxicity, specific target organ toxicity (single exposure), specific target organ toxicity (repeated exposure), and aspiration hazard.
9. Environmental hazards, A_{32}: relating to the items identified by "Hazardous Chemical List (2015)" of China PR, which also complies with "Globally Harmonized System of Classification and Labeling of Chemicals" of UN (GHS) and to the classification of chemicals into two categories, namely hazardous to the aquatic environment and hazardous to the ozone layer.

2.1.2 Principles of index evaluation

A five-range scoring system is built up to evaluate the indexes, meaning that classes I-V corresponded to 5–1 score. The golden rule of scoring is the principle of "the higher first", which means higher class is preferred if the criteria is same. For example, evaluating the hazards and risk (A_{12}), when class I

and class II are both "not belonging to GB12268-2012", it will be defined as class I, scoring 5.

2.2 Criteria for "green level" classification of scale inhibitors

The "green level" classification of scale inhibitors was characterized with the green index, which combined raw material index, process index, and product index, and weighed as follows: 0.4 for product index, and 0.3 for both raw material index and process index, since the product contributes mostly to the green chemistry, followed by both the raw material and the process which are equally important. Raw material green index is equal to the score sum of three indicators of A_{11}, A_{12}, and A_{13}. Process green index and product green index can be calculated by the same method. According to the evaluation standards given in Table 1, the classification of "green level" is given in Table 2.

Based on Table 2, the green levels of scale inhibitor would be classified into five categories, I and II belonging to the green scale inhibitor while others belonging to the non-green scale inhibitor. The green index was descendent along the categories I to V.

According to this classification standard, the "green level" of raw materials, process, and product are evaluated easily, and a scale inhibitor can be determined ultimately its "green level" based on the comprehensive green index.

3 CASE STUDY

Taking three kinds of commercial scale inhibitor of reverse osmosis membrane as an example, the method of using this set of index system is explained. According to Table 1, first of all, the raw materials and processes of three kinds of scale inhibitors were understood clearly by reading product specification, consulting manufacturer and experts, reviewing the literature, and so on. Then, each index was itemized points, and then their raw material green index, process green index, and product green index were calculated, respectively. Finally, green index of them were determined, recorded as I_g, as shown in Table 3.

According to Table 3, GY-319 and TH-613 are evaluated as green scale inhibitors, but the former belongs to the first class (I) and the latter belongs to the second class (II). The others belong to non-green scale inhibitor. The green level of ZG-PRP-009 is the worst, and this shows that it is the least environmentally friendly. As can be seen, with the different products of the green scale inhibitor, differences between them can be distinguished by the

Table 1. "Green level" evaluation system for scale inhibitors in water treatment.

Green index	Evaluation index	Criteria of classification				
		I (5 score)	II (4 score)	III (3 score)	IV (2 score)	V (1 score)
A_1: Raw material index	A_{11}: Difficulty level of transport and storage	Very easy transport, Normal Pressure and Temperature (NPT) storage	Easy transport, NPT storage	Hard to transport, NPT storage	Hard to transport, more than NPT storage	Very hard to transport, more than NPT storage
	A_{12}: Hazards and Risk	Not belonging to GB12268-2012	Not belonging to GB12268-2012	Belonging to GB12268-2012	Belonging to GB12268-2012	Belonging to GB12268-2012
	A_{13}: Toxicity	No toxicity	Little toxicity	Light toxicity	Medium toxicity	Toxicity
A_2: Process index	A_{21}: Wastewater	(1) Pollutants less than two (2) Easy to treat (3) Low quantity (4) No toxicity or light toxicity	(1) Three or four pollutants (2) Easy to treat (3) Low quantity (4) No toxicity or light toxicity	(1) Five or six pollutants (2) Hard to treat (3) Medium quantity (4) Normal toxicity	(1) Seven or eight pollutants (2) Hard to treat (3) Large quantity (4) High toxicity	(1) Pollutants more than nine (2) Very hard to treat (3) Very large quantity (4) High toxicity
	A_{22}: Waste gas	(1) Items less than two (2) Easy to treatment (3) Low quantity (4) No toxicity or light toxicity	(1) Three or four pollutants (2) Easy to treat (3) Low quantity (4) No toxicity or light toxicity	(1) Five or six pollutants (2) Hard to treat (3) Medium quantity (4) Normal toxicity	(1) Seven or eight pollutants (2) Hard to treat (3) Large quantity (4) High toxicity	(1) Pollutants more than nine (2) Very hard to treat (3) Very large quantity (4) High toxicity
	A_{23}: Solid waste	(1) Quantity of solid waste less than 5% of the raw material (2) No toxicity (3) Easy to recycle or reuse	(1) Quantity of solid waste between 5 and 10% of the raw material (2) light toxicity (3) easy to recycle or reuse	(1) Quantity of solid waste between 10–15% of the raw material (2) medium toxicity (3) hard to recycle or reuse	(1) Quantity of solid waste more than 15% of the raw material (2) high toxicity (3) hard to recycle or reuse	(1) Quantity of solid waste more than 15% of the raw material (2) very high toxicity (3) hard to recycle or reuse
	A_{24}: Clean of production	Belonging to cleaner production or encouraged process by government	Belonging to cleaner production or encouraged process by government	Belonging to permitted process by government	Belonging to obsolete process by government	Belonging to banned process by government
A_3: Product index	A_{31}: Health hazards	0 hazardous item	0 hazardous item	2–3 hazardous items	4–5 hazardous items	More than 5 hazardous items
	A_{32}: Environmental hazards	0 hazardous item	0 hazardous item	1 hazardous item	2 hazardous items	2 hazardous items

value of their green level calculated. This evaluation system can promote not only the improvement of raw materials, but also producers to keep the process cleaner. If the product's green level can be included in the product identification content, such practice will help consumers identify and choose a more environment-friendly scale inhibitor products.

Table 2. "Green level" evaluation system for scale inhibitors in water treatment.

Type	Green scale inhibitor		Non-green scale inhibitor		
Green level	I	II	III	IV	V
Raw material index, I_1 $I_1 = I_{11} + I_{12} + I_{13}$	$13.5 < I_1 \leq 15$	$10.5 < I_1 \leq 13.5$	$7.5 < I_1 \leq 10.5$	$4.5 < I_1 \leq 7.5$	≤ 4.5
Process index, I_2 $I_2 = I_{21} + I_{22} + I_{23} + I_{24}$	$18 < I_2 \leq 20$	$14 < I_2 \leq 18$	$10 < I_2 \leq 14$	$6 < I_2 \leq 10$	≤ 6
Product index, I_3 $I_3 = I_{31} + I_{32}$	$9 < I_3 \leq 10$	$7 < I_3 \leq 9$	$5 < I_3 \leq 7$	$3 < I_3 \leq 5$	≤ 3
Green index, I_S $I_S = 0.3I_1 + 0.3I_2 + 0.4I_3$	$13.0 < I_s \leq 14.5$	$13.0 \leq I_s \leq 11.6$	$11.6 < I_s \leq 7.4$	$7.4 < I_s \leq 4.3$	≤ 4.3

Table 3. A case study.

Number	Product code	Principal component	Place of production	I_1	I_2	I_3	I_s
1	GY-319	Sodium of Polyepoxysuccinic Acid	China	14	18	9	13.2
2	TH-613	Acrylic-acrylate-sulfosalt copolymers	China	11	17	8	11.6
3	ZG-PRP-009	Phosphorus containing organic molecules	China	12	14	6	10.8

4 CONCLUSIONS

As environment-friendly scale inhibitor in water treatment is the trend driving the future of marketing, the evaluation system established is beneficial to the development of green scale inhibitor. Based on the properties of raw materials, processes, and products of inhibitor manufacture, the "green level" of scale inhibitor can be classified into five categories by calculating synthetic green index. The higher two classes belong to green scale inhibitor, while the lower three do not. In this way, the green level of the scale inhibitor is simply and efficiently evaluated, enriching the theory system of green scale inhibitor and benefiting the application in practice.

REFERENCES

China National Standardization Committee & General Administration of quality supervision, inspection and Quarantine of the people's Republic of China. List of dangerous goods (GB12268-2012)[S]. Beijing. China Standard Press, 2012.

Chunfang Fan, Amy T. Kan, Ping Zhang et al. (2012). Scale Prediction and Inhibition for Oil and Gas Production at High Temperature/High Pressure. J. SPE journal. 17(2):379–392.

Ketrane R., B. Saidani, O. Gil et al. (2009). Efficiency of five scale inhibitors on calcium carbonate precipitation from hard water: Effect of temperature and concentration. J. Desalination: The International Journal on the Science and Technology of Desalting and Water Purification. 249(3):1397–1404.

Robert L. Dixon (1979). United States governmental efforts to improve the regulation of toxic substances. J. Archives of Toxicology. 43(1):37–38.

State Administration of production safety supervision and Administration, Ministry of industry and information technology of the people's Republic of China, Ministry of Public Security of the People's Republic of China, et al. Hazardous chemicals catalog (2015):2.

Sun Qunfeng, Xia Zhi, Li Pengfei et al. (2012). Research progress of environmental friendly water treatment agent. J. Chemical Industry Times. 10:42–45.

Tang Fei, Zhen Aiping, Song Zhaozheng. (2010). Application and research progress of scale inhibitor for industrial circulating cooling water. J. Chemical science and technology. 18(3):70–74.

Advances in Materials Science, Energy Technology
and Environmental Engineering – Patty & Zhou (Eds)
© 2017 Taylor & Francis Group, London, ISBN 978-1-138-19668-1

Effects of leaf area and biomass on different seasons for individual eucalyptus trees

Xiping Cheng & Liqin Dong
Southwest Forestry University, Kunming, Yunnan, China

Sihai Wang
*Key Laboratory of the State Forestry Administration on Conservation of Rare, Endangered and Endemic
Forest Plants, Yunnan Academy of Forestry, Kunming, Yunnan, China*

ABSTRACT: Leaf is an important organ for photosynthesis and respiration in plants. Leaf area and biomass are important parameters in the physiological processes of trees, which affect the biological structure of trees directly. This research focused on sampling and measurement of wet and dry season in the different eucalyptus age leaves in Kunming area, Yunnan province of China. Image J software was used to analyze and compare leaf area and biomass in dry and wet seasons. The results showed the significant difference between the effects of wet and dry seasons on eucalyptus leaf area. It has great influence on the leaf area of annual eucalyptus, especially in wet season. Response of biomass on dry and wet seasons also showed significant difference. This research can provide theoretical reference for the physiological characteristics and the law of the eucalyptus tree.

1 INTRODUCTION

1.1 *Type area*

We obtained data on single leaf of *Eucalyptus* in the Kunming city of Yunnan Province, eastern P.R. China (25°04' N, 102°46' E; Fig. 1). This area is within the subtropical zone and is characterized by a plateau monsoon climate. As the landform is complex diversiform, the average annual temperature is 16.5°C and the average annual precipitation is 1450 mm. Annual sunshine is for 2250 hours, the frost-free period is 230 days. The maximum and minimum temperatures are 22°C and 6°C, respectively. Dry and wet seasons are trenchant. The wet season is from May to October, temperature is 19°C to 22°C, and rainfall is about 1200 mm; the dry season is from November to next-year April, temperature is 6°C to 8°C, and rainfall is within 300 mm. The soil is acidic red soil and latosol. The main vegetation types are subtropical evergreen broadleaved forest and some coniferous trees such as *Pinus yunnanensis* Franch, *Pinus Massoniana* Lamb, the broad-leaved species mainly including *Machilus yunnanensis* Lec., *Eucalyptus globulus* Labill., *Platanus Orientalis* L., *Cinnamomum camphora* (L.) Presl, *Cinnamomum glanduliferum* (Wall.) Nees, and *Populus yunnanensis* Dode.

Figure 1. Location of study site.

1.2 *Data collection*

In 2015, the sampled leaves were randomly selected from healthy eucalyptus (*Eucalyptus globulus* and *Eucalyptus drepanophylla*) trees in March and October, which are dry and wet seasons, respectively. A total of 400 leaves were sampled from branches at different orders from 40 trees, which 185 were perennial leaves and 215 were annual leaves. Perennial leaves were mainly in the stem or

Table 1. Statistical summary of sampled leaves.

Season	Leaf age	Parameter	Dry weight/g	Fresh weight/g	Area/cm^2	Number
Dry season	Annual leaves	Mean	0.17 ± 0.09	0.57 ± 0.27	6.59 ± 3.14	
		Min	0.02	0.15	1.65	105
		Max	0.59	1.16	19.95	
	Perennial leaves	Mean	0.62 ± 0.29	1.39 ± 0.58	9.60 ± 3.35	
		Min	0.27	0.64	4.69	95
		Max	0.63	3.30	21.50	
Wet season	Annual leaves	Mean	0.51 ± 0.29	1.15 ± 0.62	44.11 ± 23.52	
		Min	0.02	0.12	9.27	110
		Max	1.12	2.59	116.41	
	Perennial leaves	Mean	0.62 ± 0.34	1.25 ± 0.64	36.76 ± 20.13	
		Min	0.22	0.47	10.9	90
		Max	1.34	2.65	57.09	

the middle and base of branch, and their color was dark, and perennial leaves had stopped expanding. By contrast, annual leaves were mainly located on the top of branch and their color was light. In addition, each 200 leaves were collected in the dry and wet season. All the leaves were weighed on the site with an electronic balance with a precision of 0.001 g. After the leaves were taken to the laboratory, a scanner was used to transform the leaves to the images, and digitalize them to attain the leaf area using the MapInfo software. Table 1 shows the statistical summary on sampled leaves.

1.3 *Data analysis*

We adopted model selection as our statistical approach in this study. Using Akaike's Information Criterion to assess the appropriateness of models, we selected the best model from candidate models to identify important patterns in our data-sets. We used mixed-effect models with plots and saplings as random effects. This approach accounted for random fluctuations caused by differences among plots and among saplings. We used the software package R (version 2.8.1; R Development Core Team) for all analyses. To identify major patterns in biomass, we estimated the annual and perennial leaves in different seasons.

2 FORMULATION OF THE PROBLEM

Plant photosynthesis and respiration occurs in leaves, and the size of the leaf area is directly related to the strength of the function; therefore, a study on the differences between the leaf area and biomass in different seasons can provide a theoretical reference for the plant growth strategy (Austin and Margules 1990, Hopkins et al. 2013,

Cheng et al. 2015). In the systematic research on the relationship between leaf area and biomass, the Modified logistic model, the log-normal model and the Gaussian modified model have been applied to the modeling of leaf area and biomass (Giardina and Ryan 2002, Kasuya et al. 2010, Zhang et al. 2013). In addition, the leaf area and biomass of related species are also discussed in detail, such as *Dioscorea nipponica*, and underground biomass and leaf number between the plant heights have an extremely significant linear relationship (Yao et al. 2008). Moreover, the study on the relationship between leaf area index and biomass of *Pseudosasa amabilis* also showed that the leaf area index was closely related to the biomass (Zheng et al. 2001). Guo et al. (2013) targeted the plant leaf area and biomass of five major advantages of Xiangshan in Anhui province, and discussed the dynamic relationship between the relative magnitude of biomass accumulation and the amount of leaf area and biomass, and revealed the differences in the utilization and adaptability of different species to environmental resources. At present, the research on the relationship between leaf area and biomass is mainly focused on the correlation between the leaf area and the biomass, but rarely studies focused on the response of leaf area and biomass to seasonal variation. Just Cheng et al. (2015) and Diao et al. (2010) discussed the relevant issues; however, related to the content of the seasonal change was not furthest.

Therefore, this study focused on the relationship between the leaf area and biomass and the change in wet and dry seasons, and defined the mechanism of growth and accumulation of leaf area and biomass in different seasons based on the same sampling in different seasons, in order to provide the theoretical reference for further study and practical production.

3 RESULTS AND ANALYSIS

3.1 *Effects of annual leaves on different seasons*

The relationship between leaf area and biomass of annual eucalyptus leaves in the two seasons is studied using the collected annual eucalyptus leaf. The biomass with leaf area increased both in dry and wet seasons, and the increasing trend in the dry season was significantly faster than the wet season. In addition, in the same level of leaf area, the biomass of eucalyptus leaf was high in the dry season more than in the wet season, which showed that small area eucalyptus leaf in the dry season can maintain high growth rate (Fig. 2). From the dry and wet season analysis results of annual leaves, it has been demonstrated that the leaf area and biomass are widely distributed in wet season, the maximum leaf area was as high as 116.41 cm^2, and the maximum biomass was as high as 1.12 g. In the dry season, leaf area was below 20 cm^2, and maximum biomass was only 0.60 g (Table 1; Fig. 2); the results show that annual trees of eucalyptus in the wet season have a large number of big leaves, and, in the dry season, they have a large number of small leaves.

Because of the dry season, light is of low intensity, temperature is low, and relative humidity is smaller, transpiration efficiency is very low, and the eucalyptus leaf photosynthetic rate is not high. Therefore, the annual growth of eucalyptus leaves was slow, resulting in a common small leaf, and the change of area and biomass were small. In the wet season, especially from August to September, the precipitation is big, the climate is warm and humid, the soil moisture content is high, the eucalyptus leaf transpiration and the photosynthesis are higher, the transpiration efficiency is also higher, and the photosynthetic rate reaches the maximum value. Therefore, the annual growth of eucalyptus leaves is in the state of rapid growth or steady growth, the growth situation of leaves is good, and the leaf area and biomass are large.

3.2 *Effects of perennial leaves on different seasons*

Dry and wet season biomass with leaf area increased, and the dry and wet season growth rate was basically the same for the perennial leaves. In the comparison of the relationship between leaf area and biomass of perennial eucalyptus, the difference in the slope of wet and dry season was very close to that in the two seasons, which indicated that the trend of the growth and biomass accumulation was similar in different seasons (Fig. 3). From the perennial eucalyptus leaves dry and wet season analysis results, it has been demonstrated that the leaf area and biomass are widely distributed, the maximum leaf area is 57.09 cm^2, the leaf area is 10.9 cm^2, and mean value is 36.76 cm^2 in the wet season. In addition, the distribution of perennial eucalyptus leaf area and biomass was small, and biomass is more than 0.60 g without distribution in dry season (Table 1; Fig. 3), indicating that the difference between the distribution in dry and wet season was significant for the perennial eucalyptus.

Owing to the long-term adaptation and growth for perennial eucalyptus leaves, the stable growth state was in the dry and wet seasons in biology, and hence, the growth environment had little influence on the growth of eucalyptus, the results were characterized by growth trend of perennial eucalyptus leaves in the dry and wet seasons, which was nearly the same as the two seasons. In addition, due to the concentration of precipitation as well as the impact of the intensity of light, the rapid response environment of eucalyptus leaves in the wet season is more than that in the dry season to show stronger growth rapidly developing characteristics. Therefore, the leaf area and biomass of perennial eucalyptus in wet season were significantly higher than those in dry season.

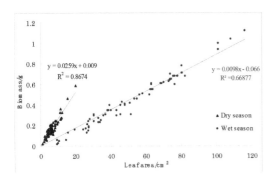

Figure 2. Relationship between leaf area and biomass for annual leaves in different seasons.

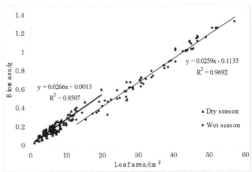

Figure 3. Relationship between leaf area and biomass of perennial leaves in different seasons.

4 DISCUSSION AND CONCLUSIONS

For the different ages of eucalyptus leaf in wet and dry seasons, there were significant differences in leaf area growth and biomass accumulation. The research compares leaf area, wet weight, and biomass of eucalyptus leaf in dry season and found that the leaves of perennial trees were larger than those of annual trees in the value of mean, maximum, and minimum (Table 1).

Wang et al. (1999) made researches on the transpiration and photosynthetic efficiency of trees, and showed that effects of different plants in response to atmospheric and soil drought factors, and plant transpiration physiological regulation mechanism will begin to operate and reduce the role of transpiration to save water. In the dry season, eucalyptus leaves were influenced by the atmospheric and the soil drought factors. The perennial eucalyptus leaves will remain in a stagnant growth state in order to maintain the leaf moisture, although the annual eucalyptus leaves also keep the corresponding mechanism in the dry season, but the short growth period and smaller leaf morphology; therefore, the leaf area and biomass of the perennial eucalyptus trees were greater than those of the annual trees in the dry season.

Analysis on eucalyptus leaf area and biomass of dry and wet seasons in Kunming area found that the biomass and leaf area of annual and perennial eucalyptus leaves were significantly superior in the wet season. Owing to warm climates, high light intensity, big relative humidity, and the transpiration rate and photosynthesis rate are high. Therefore, in the wet season, the response of leaf area for annual eucalyptus leaves was very obvious, and the biomass was similar to that of the perennial eucalyptus leaves.

Some scholars using the model found that the leaf biomass and leaf area was related in different ages of eucalyptus (Abelho and Graça 1996, Souza et al. 2015). The study also indicated that there is a high correlation between biomass and leaf area in wet and dry seasons at different age levels, the coefficient of determination for the regression analysis was from 0.67 to 0.97 (Figs. 1; 2), and the above results were further certified. The results of this study can provide ideas and reference for the physiological growth mechanism of eucalyptus under different seasonal conditions. In order to confirm the effect of eucalyptus leaf growth mechanism, we should also consider the different parameters like light intensity, light quality, soil nutrients, rainfall, and its physiological characteristics, and such issues will need to be further strengthened in the follow-up study to improve.

ACKNOWLEDGMENTS

This study was supported by the National Natural Science Foundation of China (31360164; 31560092), the National Scientific and Technological Basic Work of China (No. 2012FY110300), and Science Fund of China's Yunnan Government (No. 2015BB018).

REFERENCES

Abelho M, Graça M A S. (1996) Effects of eucalyptus afforestation on leaf litter dynamics and macroinvertebrate community structure of streams in Central Portugal. *Hydrobiologia*. 324(3):195–204.

Austin M P, Margules C R. (1990) Measurement of the realised qualitative niche: environmental niches of five Eucalypt species. Ecological Monographs. *Ecological Monographs*. 60(60):161–177.

Cheng Xiping, Wang Yanfang, Ma Yuewei. (2015) Single leaf-level measurement and estimation of eucalyptus based on functional-structural model. *Advances in Energy Science and Equipment Engineering*, 1:51–54.

Diao Jun, Lei Xiangdong, Hong Lingxia, *et al.* (2010) Single leaf area estimation models based on leaf weight of eucalyptus in southern China [J]. *Journal of Forestry Research*. 21(1):73–76.

Giardina C P, Ryan M G. (2002) Total Belowground Carbon Allocation in a Fast-growing Eucalyptus, Plantation Estimated Using a Carbon Balance Approach. *Ecosystems*. 5(5):487–499.

Guo Zhiyuan, He Junjie, Zhou Jiyuan, *et al.* (2013) Preliminary Study on the Leaf Area and Biomass of the Dominant Plant Species in Xiangshan Mountain of Huaibei City. *Anhui Forestry Science and Technology*. 39(3):14–16.

Hopkins M S, Ash J, Graham A W, et al. (2013) Charcoal evidence of the spatial extent of the Eucalyptus woodland expansions and rainforest contractions in North Queensland during the late Pleistocene. *Journal of Biogeography*. 20(20):357–372.

Kasuya M C M, Coelho I D S, Campos D T D S, et al. (2010) Morphological and molecular characterization of Pisolithus in soil under eucalyptus plantations in Brazil. *Revista De Saúde Pública*. 34(6):1891–1898.

Souza R M S, Almeida A Q D, Ribeiro A, et al. (2015) Evaluation of the spatial dependence of dendrometric characteristics for an Eucalyptus plantation. *Acta Scientiarum Agronomy*. 37(4):483–488.

Wang Mengben, Li Hongjian, Chai Baofeng, *et al.* (1999) A comparison of transpiration, photosynthesis and transpiration efficiency in four tree species in the Loess region. *Acta Phytoecologica Sinica*. 23(5):401–410.

Yao Yunsheng, Sun Guangwei, Huang Xiaodan. (2008) Correlation between leaf area and biomass in dioscorea *Nipponica Makino*. *Crops*. 6:38–40.

Zheng Jinshuang, Cao yonghui, Dai Quanlin, *et al.* (2001) A study on relationship between biomass and leaf area index of *Pseudosasa amabilis*. *Journal of Bamboo Research*. 20(1):53–57.

Zhang Yonghe, Chen Wenhui, Guo Qiaoying, *et al.* (2013) Hyperspectral estimation models for photosynthetic pigment contents in leaves of Eucalyptus. *Acta Ecologica Sinica*. 33(3):876–887.

Advances in Materials Science, Energy Technology and Environmental Engineering – Patty & Zhou (Eds)
© 2017 Taylor & Francis Group, London, ISBN 978-1-138-19668-1

Study on the coal seam gas migration law with numerical simulation under the condition of extraction

Gang Xu & Chao Liu
College of Safety Science and Engineering, Xi'an University of Science and Technology, Xi'an, Shaanxi, China

ABSTRACT: Based on the coal seam gas migration theory under multi-field coupling effect, the fluid-solid coupling model of coal seam gas was developed using elastic mechanics, fluid mechanics in porous medium, and effective stress principle. Gas seepage behavior under different original gas pressures was simulated. Results indicated that residual gas pressure, gas pressure gradient, and gas low were bigger when original gas pressure was higher. Coal permeability distribution decreased exponentially when original gas pressure was lower than critical pressure. Coal permeability decreased rapidly first and then increased slowly when the original pressure was higher than the critical pressure.

1 INTRODUCTION

Coal seam gas seepage under multi-field coupling effect is one of the hot topics of gas migration researches. It has great significance to mine gas disaster prevention and improving gas extraction efficiency. At present, many researchers have studied fluid-solid coupling of coal seam gas seepage based on temperature and crustal stress (Guo, 2012; Hu, 2011; Ling, 2000; Wang, 2000). However, these researches did not take enough influencing factors into consideration. Especially, the influence of crustal stress, gas pressure, and adsorption swelling stress on gas seepage needs to be studied further. Therefore, based on the influence of crustal stress, gas pressure, and adsorption swelling stress on gas seepage, the fluid-solid coupling model of coal seam gas was developed and gas migration behavior under gas extraction was obtained using numerical simulation, in the hope of providing guidance to gas extraction and gas disaster prevention in deep coal seam.

In order to establish the model to make the following basic assumptions: (1) Gas containing coal is isotropic linear elastic body and elastic small deformation obeys generalized Hooke's law. (2) Fluid in the coal seam is a single-phase fluid and is saturated, and the gas is ideal gas and gas adsorption obeys the modified Langmuir adsorption equation. (3) Gas flow in coal seam obeys the Darcy law.

2 DYNAMIC PERMEABILITY EQUATION

2.1 Porosity equation of coal

According to the definition of porosity φ (Ran, 1997), the porosity equation expressed by bulk strain can be obtained.

$$\varphi = 1 - \frac{1-\varphi_0}{1+\varepsilon_V}(1+\frac{\Delta V_S}{V_{S0}}) \tag{1}$$

where φ is the current porosity, φ_0 is the original porosity, ΔV_S is the volume change quantity of coal skeleton, V_{S0} is the original volume of coal skeleton, and ε_V is the volumetric strain (volume expansion is positive and volume compression is negative).

Change in matrix block volume strain $\Delta V_S / V_{S0}$ is formed by three parts.

$$\frac{\Delta V_S}{V_{S0}} = -C_S \Delta p + \frac{\Delta \varepsilon_P}{1-\varphi_0} \tag{2}$$

where C_S is the compression coefficient of coal skeleton, Δp is the gas pressure variation, $\Delta p = p - p_0$, p_0 is the original gas pressure, p is the current gas pressure, and $\Delta \varepsilon_P$ is the adsorption expansion dependent variable of unit volume.

Expansion-dependent variable caused by gas pressure can be expressed as follows (Li, 2005):

$$\Delta \varepsilon_P = \frac{2a\rho RT(1-2v)}{EV_m} \ln(1+bp) \tag{3}$$

where E is the elastic modulus, v is the Poisson ratio, ρ is the apparent density, V_m is the molar volume, a, b are the adsorption constant, and R is the gas content.

2.2 Effective stress equation

Relation between effective stress and volume strain of gas-containing coal obeys generalized Hooke's law. As coal body is always affected by stress and gas pressure, the volume strain can be expressed when crustal stress and gas pressure changed:

$$\varepsilon_V = \frac{\sigma'_m}{K} = \frac{2a\rho RT(1-2v)^2}{EV_m}\ln(1+bp)$$
$$-\frac{1-2v}{E}(\sigma-\sigma_0)+\frac{3(1-2v)\alpha_p}{E}\Delta p \qquad (4)$$

where σ is the current stress, σ_0 is the original stress, α_p is the pore pressure coefficient, σ'_m is the average effective stress, and K is the bulk modulus, $K = E / [3(1-2v)]$.

2.3 Dynamic permeability equation

According to the relation between coal permeability and porosity in Kozeny-Carman equation (Kong, 1999), coal permeability equation under multi-field coupling effect can be obtained using equations (1), (2), (3), and (4),

$$k = k_0(\frac{1}{\varphi_0})^3\left[1-\frac{(1-\varphi_0)(1-C_S\Delta p)+\Delta\varepsilon_p}{1+\varepsilon_V}\right]^3 \qquad (5)$$

where k_0 is the original permeability, and k is the permeability after change.

3 FLUID-SOLID COUPLING MATHEMATICAL MODEL OF GAS-CONTAINING COAL

3.1 Deformation field equation of coal skeleton

3.1.1 Equilibrium equation
According to the effective stress theory and ignoring free gas seepage, the inertia force of coal deformation, and body force of gas, differential equation of mechanical equilibrium of gas-containing coal can be expressed as follows:

$$\sigma'_{ij,j}+(\alpha p\delta_{ij})_{,j}+F_i=0 \qquad (6)$$

3.1.2 Geometric equation
Coal deformation obeys generalized Hooke's law and the following equation can be obtained:

$$\varepsilon_{ij}=\frac{1}{2}\left(u_{i,j}+u_{j,i}\right) \quad (i,j=1,2,3) \qquad (7)$$

3.1.3 Constitutive equation
Total strain of gas-containing coal is composed of strain caused by crustal stress, strain caused by gas pressing coal body, and strain caused by coal adsorption expansion, i.e.

$$\varepsilon=\frac{1}{2G}\left(\sigma'-\frac{v}{1+v}\Theta'\right)$$
$$+\frac{C_s}{3}p-\frac{2a\rho RT(1-2v)}{3EV_m}\ln(1+bp) \qquad (8)$$

Using the Lame constant in equation (8), the equation is reorganized and expressed by tensor,

$$\sigma_{ij}=\lambda\varepsilon_V\delta_{ij}+2G\varepsilon_{ij}+\alpha_{PY}p\delta_{ij}$$
$$+\alpha_{PX}aT\ln(1+bp)\delta_{ij} \qquad (9)$$

where λ, G is the Lame constant, and α_{PY} and α_{PX} are stress coefficients caused by gas pressure and gas adsorbing, $\lambda=\frac{2Gv}{(1-2v)}$, $\alpha_{PY}=1-\frac{(3\lambda+2G)C_s}{3}$, $G=\frac{E}{2(1+v)}$, $\alpha_{PX}=\frac{2\rho R(1-2v)(3\lambda+2G)}{3EV_m}$. Equation (9) is a constitutive equation of gas-containing coal expressed by strain.

3.1.4 Stress field equation
Putting constitutive equation (9) expressed by strain into equilibrium equation , equation (6) can be reorganized and expressed by tensor,

$$Gu_{i,jj}+\frac{G}{1-2v}u_{j,ji}+\alpha_{PY}p_{,i}$$
$$+\alpha_{PX}[aT\ln(1+bp)]_{,i}+F_i=0 \qquad (10)$$

Equation (10) is fluid-solid coupling stress field equation of gas-containing coal.

3.2 Seepage field equation

Gas flow obeys the Darcy law and coupling seepage field equation considering terms of source and sink can be obtained by ignoring body force of gas,

$$\nabla\cdot(\rho_g v_g)+\frac{\partial X}{\partial t}=q \qquad (11)$$

where ρ_g is the gas density, v_g is the absolute speed of gas flow, and X is the gas content in the coal seam.

Gas content in coal can be calculated by following equation:

$$X=(\frac{abp}{1+bp}e^{n(t_0-t)}B+\varphi\frac{p}{p_n})\cdot\rho_n \qquad (12)$$

where e is the base of natural logarithm, $e = 2.718$, T_0 is the temperature of adsorption constant testing lab, T is the coal temperature, n is the coefficient which can be calculated by $n = 0.02/(0.993+0.07p)$, B is the coefficient, $B=\frac{\gamma_m}{1+0.31W}\cdot\frac{100-A-W}{100}$, A, W are ash and water contents,%, γ_m is the coal bulk density, p_n is the atmospheric pressure, and ρ_n is the density under atmospheric pressure.

Putting equation (12) into equation (11), and abbreviating equation (11) we get

$$\left[2\varphi + \frac{2abBp_n}{(1+bp)^2}\right]\frac{\partial p}{\partial t} + 2p\frac{\partial \varphi}{\partial t} - \nabla \cdot (\frac{k}{\mu}\nabla p^2) = q \quad (13)$$

Porosity change $\frac{\partial \varphi}{\partial t}$ in gas fluid-solid coupling problem can be expressed by the following equation (Zhou, 1998):

$$\frac{\partial \varphi}{\partial t} = (1 - \frac{C_s}{C_f})\frac{\partial \varepsilon_V}{\partial t} + C_s(1-\varphi)\frac{\partial p}{\partial t} \quad (14)$$

where C_f is the whole compression coefficient of coal body and $C_f = \frac{3(1-2v)}{E}$.

Putting equation (14) into equation (13), the fluid-solid coupling field equation of gas-containing coal can be expressed as

$$\left[2\varphi + \frac{2abBp_n}{(1+bp)^2}\right]\frac{\partial p}{\partial t} + 2p(1 - \frac{C_s}{C_f})\frac{\partial \varepsilon_V}{\partial t} \\ +2pC_s(1-\varphi)\frac{\partial p}{\partial t} - \nabla \cdot (\frac{k}{\mu}\nabla p^2) = q \quad (15)$$

3.3 Definite condition

3.3.1 Definite condition of coal skeleton deformation field
Under boundary conditions, surface force is known and $\sigma = F_i$. F_i is the distribution function of coal surface force. Internal boundary namely borehole wall is free boundary.

Initial condition is $u|_{t=0} = \bar{u}$ or $\frac{\partial u}{\partial x_i}|_{t=0} = F_i$.

3.3.2 Definite condition of seepage field
Boundary condition: external boundary is $p = p_0$ and internal boundary is $p = P$, where P is the extraction pressure in borehole. Original condition is $p = p_0$, when $t = 0$.

4 NUMERICAL SIMULATION OF COAL GAS SEEPAGE UNDER EXTRACTION

4.1 Numerical simulation model

The software in this simulation is an automatic generated system (FEPG) by the finite element program. Based on gas borehole extraction characteristics, the numerical simulation model is developed, as shown in Figure 1. Radius of borehole is 0.1 m, and gas seepage area radius is 3 m. Unite type in computational domain is quadrangle and unit around borehole is concentrated. There are 1578 units in this model (shown in Figure 1).

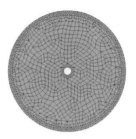

Figure 1. Numerical simulation model and meshing.

Calculation time is 10 days that is $t = 864000$ s and time step is $\Delta t = 8640$ s.

Basic parameters in this simulation are from 3# coal seam in Wuyang Coal mine Lu'an Group. These parameters are shown in Table 1.

In this simulation, the external boundary stress in the coal deformation field is $\sigma_0 = 10$ MPa. The internal boundary is free boundary. Original displacement is $u|_{t=0} = 0$. The external boundary of gas seepage field is defined as gas pressure boundary, which is the same with original gas pressure. In the internal boundary, gas pressure in borehole is $P = 0.1$ MPa. In order to have gas seepage behavior under different gas pressures, the original gas pressure of gas seepage field is set at 0.5 MPa, 1 MPa, 1.5 MPa, 2 MPa, and 2.5 MPa.

4.2 Analysis of numerical simulation results

1. Gas pressure distribution
Gas pressure distribution nephogram under different original gas pressures is shown in Figure 2. Gas pressure distribution along the radius is shown in Figure 3. These figures showed that the residual gas pressure is bigger with bigger original gas pressure when distance from extraction borehole is the same. The gas pressure gradient is bigger with longer distance. The bigger the original gas pressure is, the easier the gas pressure gradient is going to form. The gas pressure gradient is bigger with nearer distance from exposed coal wall at face. These explain coal and rock outburst tendency around face and provide a direction for outburst prevention. That is stress relief slot, deep hole standing shot, and gas extraction through borehole. These measures can decrease the gas pressure gradient and increase the width of pressure relief zone, so that outburst can be prevented.

2. Gas flow distribution
Gas flow distribution along the radius is shown in Figure 4. This figure showed that gas flow is bigger with higher original gas pressure. When gas flow is bigger, gas is more easily to gather as exposed coal wall, higher gas concentration is more easy to

Table 1. Coal mechanical parameters and gas seepage parameters of No.15 coal seam in Ping Ding-shan.

Parameters	ρ/t·m⁻³	E/Mpa	σ_0/MPa	a/m³·t⁻¹	b/MPa⁻¹	φ_0	k_0/10⁻³μm²	p_0/MPa
Value	1.24	150	10	20	1.08	0.06	4.8	0.5
Parameters	v	W/%	A/%	p_n/MPa	R/J·mol⁻¹·K⁻¹	ρ_n/kg·m⁻³	T_0/°C	C_s/MPa⁻¹
Value	0.3	1	10	0.1	8.314	0.059	30	0.001

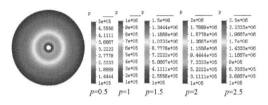

Figure 2. Gas pressure distribution nephogram under different original gas pressures.

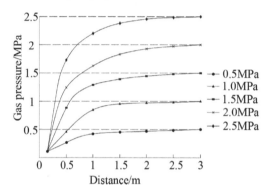

Figure 3. Gas pressure distribution along the radius.

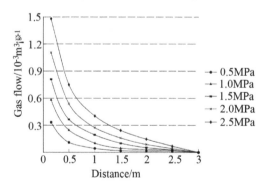

Figure 4. Gas flow distribution along the radius.

appear and gas concentration over-run accidents are easy to happen.

3. Permeability distribution

Permeability distribution along the radius is shown in Figure 5. This figure showed that permeability decreases exponentially with the distance from borehole increasing, when the original

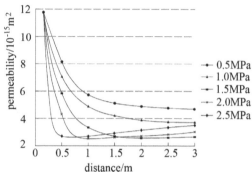

Figure 5. Permeability distribution along the radius.

gas pressures are 0.5 MPa, 1 MPa, and 1.5 MPa. Permeability decrease rapidly first and then increase slowly and finally remain stable with distance from borehole increasing when the original gas pressures are 2 MPa and 2.5 MPa. The author considers that there is a critical gas pressure, and shrinkage of coal matrix is the main effect on permeability when the gas pressure is less than the critical gas pressure. Pore gas pressure is the main effect on permeability when the gas pressure is bigger than the critical gas pressure. According to the calculation, the critical gas pressure in this paper is 1.8 MPa. In order to get a better extraction effect, during extraction, coal seam permeability should be increased through decreasing gas pressure.

5 CONCLUSIONS

1. Based on the porosity definition, the dynamic seepage equation concludes multi-influence factors. Using elasticity and fluid mechanics in porous medium, the fluid-solid coupling mathematical model of gas-containing coal was developed.
2. Residual gas pressure, pressure gradient, and gas flow get bigger and coal and gas outburst hazard get greater when the gas pressure is higher.
3. There is a critical gas pressure. Permeability decreases exponentially when the gas pressure is less than the critical gas pressure. Permeability decreases rapidly first and then increases slowly

when the gas pressure is bigger than the critical gas pressure. Decreasing gas pressure is one of the ways to improve extraction efficiency and prevent coal and gas outburst effectively.

ACKNOWLEDGMENTS

This work was financially supported by the National Natural Science Foundation of China (Program No. 51404189) and the Natural Science Basic Research Plan in Shaanxi Province of China (Program No. 2015 JQ5191).

REFERENCES

Guo Ping, Cao Shu-gang, Zhang Zun-guo, et al. Analysis of solid-gas coupling model and simulation of coal containing gas. *Journal of China Coal Society*, 2012, 37(S2): 330–335.

Hu Guo-zhong, Xu Jia-lin, Wang Hong-tu, et al. Research on a dynamically coupled deformation and gas flow model applied to low-permeability coal. *Journal of China University of Mining & Technology*, 2011, 40(1): 1–6.

Ling Jun, Liu Jian-jun, Fan Hou-bin, et al. The mathematical model and its numerical solution of gas flow under unequal temperatures. *Chinese Journal of Rock Mechanics and Engineering*, 2000, 19(1): 1–5.

Wang Tu-hong, Du Yun-gui, Xian Xue-fu, et al. Coalbed gas seepage equation affected by in-situ stress, geothermal temperature and geo-electric effect. *Journal of Chongqing University* (Natural Science Edition), 2000, 23 (S1): 47–50.

Ran Qi-quan, Li Shi-lun. Study on dynamic models of reservoir parameters in the coupled simulation of multiphase flow and reservoir deformation. *Petroleum Exploration and Development*, 1997, 24(3): 61–65.

Li Xiang-chun, Guo Yong-yi, Wu Shi-yue. Analysis of the relation of porosity, permeability and swelling deformation of coal. *Joutnal of Taiyuan University of Technology*, 2005, 36(3): 264–266.

Kong Xiang-yan. Mechanics of porous media flow. He fei: *Press of University of Science and Technology of China*, 1999.

Zhou Y, Rajapakse R, Graham J.A coupled thermoporoelastic model with thermo-osmosis and thermalfiltration. *International Journal of Solids and Structures*, 1998, 35(34): 4659–4683.

*Advances in Materials Science, Energy Technology
and Environmental Engineering – Patty & Zhou (Eds)
© 2017 Taylor & Francis Group, London, ISBN 978-1-138-19668-1*

Effect of rotational excitation on the reaction Li + DF($v = 0, j = 0 – 5$)

Xianfang Yue

*Department of Physics and Information Engineering, Jining University, Qufu, Shandong Province, China
Beijing Computational Science Research Center, Beijing, China*

ABSTRACT: Effect of reagent rotational excitation on product rotational polarization in the reactions Li + DF($v = 0, j = 0-5$) → LiF + D is investigated by employing quasiclassical trajectory calculation. It is found that differential cross sections display a strongly forward scattering for the Li + DF($v = 0, j = 0$) → LiF + D reaction, but sideways scatterings for the Li + DF($v = 0, j = 1-5$) → LiF + D reactions. This means that the reaction mechanism of title reactions changes from the direct to indirect behavior with the reagent rotational excitation. The product rotational angular momentum *j'* is not only aligned along the direction perpendicular to the reagent relative velocity *k*, but also oriented along the negative direction of the *y*-axis for the reactions. As the rotational quantum number increasing, the product alignment and orientation is changing slightly but not monotonically. The generalized polarization-dependent differential cross-sections distribution demonstrates that the product angular distributions are anisotropic.

1 INTRODUCTION

The Li + HF reaction is one of the simplest collision systems involving three different atoms. It is a benchmark for the study of molecular reaction dynamics. Many groups have carried out lots of experimental and theoretical studies on the title reaction and its isotopes. With the Crossed Molecular Beam (CMB) method, Becker (1980) et al. studied the product angular distributions of the Li + HF → LiF + H reaction in the Center-of-Mass (CM) frame. They found a forward-backward symmetric distribution of the differential cross section (DCS) of the reaction at the collision energy (E_c) of 0.13 eV, while a strongly forward peaked distribution at $E_c = 0.377$ eV. Loesch's(1993) group investigated the influences of translational energy, vibrational excitation, rotational excitation and reagent alignment on the reactions Li + HF(v, j) → LiF(v', j') + H. They observed significant effects of the varying *j*-state populations on the shape of the product angular distributions. The angular distributions of LiF in the CM frame changes from nearly forward-backward symmetric to preferred forward scattering with the rising collision energies ranging from 0.088 to 0.378 eV, which is caused by the significant dependence of the DCSs on the rotational energy of HF rather than by the transition from a long lived complex to a direct mechanism. Steric effects markedly influence the product angular distributions, the partition of available energy, and the integral reaction cross sections of the Li + HF($v = 1, j = 1, m = 0$) → LiF(v', j') + H reaction at a translational collision energy of 0.42 eV.

Theoretically, a lot of time-dependent wave packet, time-independent quantum dynamics, and Quasiclassical Trajectory (QCT) calculations have been carried out. Zanchet et al. performed the state-to-state differential cross sections calculations on the Li + HF(v = 0, j = 0) reaction. They found that the rotational alignment of the LiF(v', j') products indicates that *j'* is perpendicular to the *k-k'* plane for the forward/backward peaks, while it seems to be nearly isotropically distributed for sideways distributions. Krasilnikov(2013) et al. presented the vector correlations to the elementary chemical reaction Li + HF($v_r = 0, j_r = 0-5$) → LiF(v, j) + H at the collision energy of 0.317 eV.

It can be seen from previous studies that the rotational excitation of the reagents plays an import role in the tile reaction. In the present work, effect of reagent rotational excitation on the title reactions are carried out by employing the QCT method with a most popular and accurate ground $1^2 A'$ state PES developed by Aguado et al.

2 THEORY

In the calculation, batches of 100 000 trajectories are run for each reaction and the integration step size is chosen to be 0.1 femtosecond (fs). The trajectories start at an initial distance of 20 Å between the Li atom and the Center of Mass (CM) of the DF molecules. The collision energy is chosen to be 0.10 eV for all reactions. The vibrational and rotational levels of the reagent molecule DF are taken to be $v = 0$ and $j = 0, 1, 2, 3, 4, 5$, respectively.

The CM frame was used as the reference frame in the present study. The reagent relative velocity vector $\boldsymbol{k}$ is parallel to the z-axis. The x-z plane is the scattering plane which contains the initial and final relative velocity vectors, $\boldsymbol{k}$ and $\boldsymbol{k'}$. θ_t is the angle between the reagent relative velocity and product relative velocity (so-called scattering angle). θ_r and ϕ_r are the polar and azimuthal angles of the final rotational angular momentum $\boldsymbol{j'}$. Expressions and descriptions of the product rotational angular momentum distributions, $P(\theta_r)$, $P(\phi_r)$, $P(\theta_r,\phi_r)$ and generalized Polarization-Dependent Differential Cross-Sections (PDDCSs) in the CM frame can be found in our previous work.

3 RESULTS AND DISCUSSION

The product $P(\theta_r)$ distribution describes the $\boldsymbol{k}$–$\boldsymbol{j'}$ vector correlation with $\boldsymbol{k} \cdot \boldsymbol{j'} = cos\theta_r$. Figure 1 shows the calculated product $P(\theta_r)$ distribution of the reactions Li + DF($v = 0, j = 0$–5) → LiF + D. Obviously, each $P(\theta_r)$ distribution is symmetric about $\theta_r = 90°$, and illustrates a distinct peak at $\theta_r = 90°$. This indicates that the product rotational angular momentum vector $\boldsymbol{j'}$ is aligned perpendicular to the reagent relative velocity direction $\boldsymbol{k}$. The product rotational alignment of the reactions Li + DF($v = 0$, $j = 1$–5) → LiF + D is stronger than that of the ground state reaction Li + DF($v = 0, j = 0$) → LiF + D. As increasing of the rotational quantum number, variation of the product rotational alignment is not monotonic. The product rotational alignment is strongest in the Li + DF($v = 0, j = 1$) → LiF + D reaction.

The $P(\phi_r)$ distribution describes the $\boldsymbol{k}$–$\boldsymbol{k'}$–$\boldsymbol{j'}$ vector correlation and can provide both product alignment and orientation information. Figure 2 illustrates the $P(\phi_r)$ distributions for the

Li + DF($v = 0, j = 0$–5) → LiF + D reactions. The $P(\phi_r)$ distributions appear a distinct peak at about $\phi_r = 270°$ and no peak at about $\phi_r = 90°$ for the Li + DF($v = 0, j = 0$) → LiF + D reaction, which means that the orientation of the product rotational angular momentum tends to point to the negative direction of y-axis. That is to say, the product molecules prefer a clockwise rotation (see from the positive direction of y-axis, hereafter) in the plane parallel to the scattering plane. Similar behavior can be found for the Li + DF($v = 0$, $j = 1$–5) → LiF + D reactions, nevertheless, peaks of the $P(\phi_r)$ distributions at about $\phi_r = 270°$ becomes small and broad with the increase of the reagent rotational quantum number.

Alternative representation can provide complementary pictures of the scattering event which can lead to an increased understanding of the mechanism governing molecular collision processes. The use of these alternative representations can help elucidate the product rotational polarization information. As an alternative representation, the joint $P(\theta_r,\phi_r)$ distributions are displayed in figure 3a, b, c, d, e and f corresponding to the rotational quantum number of j = 0, 1, 2, 3, 4 and 5, respectively, for the title reactions.

The generalized PDDCSs describe the $\boldsymbol{k}$-$\boldsymbol{k'}$-$\boldsymbol{j'}$ correlation and the scattering direction of the product molecule. Figure 4 shows the calculated results of the PDDCSs for the title reactions. The PDDCS$_{00}$ is simply proportional to the Differential Cross-Section (DCS), and only describes the $\boldsymbol{k}$-$\boldsymbol{k'}$ correlation or the product angular distributions. Figure 5a plots the PDDCS$_{00}$ as a function of scattering angle θ_t for the above reactions. As clearly seen in Fig. 4a, the PDDCS$_{00}$ distribution shows a strong backward scattering for the Li + DF($v = 0$, $j = 0$) → LiF + D reaction. This discloses a direct dynamical process. The PDDCS$_{00}$ exhibit weakly

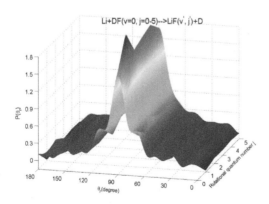

Figure 1. Distributions of the $P(\theta_r)$ as a function of the polar angle θ_r for reactions Li + DF($v = 0, j = 0$–5) → LiF + D.

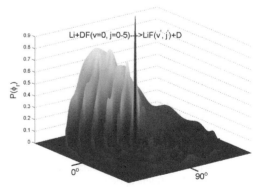

Figure 2. Distributions of the $P(\phi_r)$ as a function of the dihedral angle ϕ_r for reactions Li + DF($v = 0, j = 0$–5) → LiF + D.

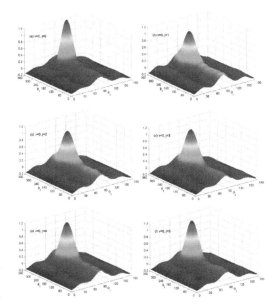

Figure 3. $P(\theta_r, \phi_r)$ distributions as a function of both the polar angle θ_r and dihedral angle ϕ_r for the reactions Li + DF($v = 0, j = 0$–5) → LiF + D.

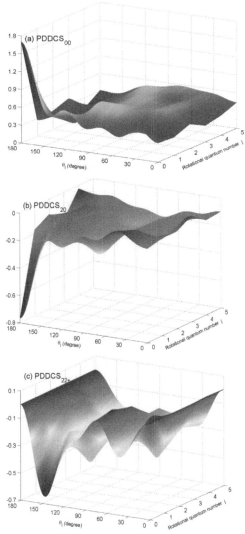

Figure 4. Panel (a) and (b) show the PDDCSs with $(k, q) = (0, 0)$ and $(2, 0)$, respectively. Panel (c) and (d) show the PDDCSs with $(k, q\pm) = (2, 2+)$ and $(2, 1-)$, respectively.

sideways scattering with the rotational excitation of the reactions Li + DF($v = 0, j = 1$–5) → LiF + D, which features a indirect reaction mechanism.

For title reactions, all of the PDDCS$_{20}$ values are negative for all scattering angles, which suggests that the product rotational angular momentum j' polarizes preferentially along the direction perpendicular to k. This is consistent with the product alignment prediction of the $P(\theta_r)$ distribution depicted in Fig. 1. Obviously, the PDDCS$_{20}$ values are more negative for the Li + DF($v = 0, j = 0$) → LiF + D reaction than that of the Li + DF($v = 0, j = 1$–5) → LiF + D reactions.

Figure 4c and d depict the PDDCSs distributions with $q \neq 0$. All of the PDDCSs with $q \neq 0$ are equal to zero at the extremities of forward and backward scatterings. At these limiting scattering angles, the k-k' scattering plane is not determined and the value of these PDDCSs with $q \neq 0$ must be zero. The behavior of PDDCSs with $q \neq 0$ at scattering angles away from extreme forward and backward direction is more interesting, which can provide detailed information about the product rotational alignment and orientation. The PDDCS$_{22+}$ is related to $\langle \sin^2 \theta_r \cos 2\phi_r \rangle$. The negative values of the PDDCS$_{22+}$ correspond with the product rotational alignment along the y-axis, while the positive values with the rotational alignment along the x-axis. The larger the absolute value, the stronger the product rotational align-

ment is along the corresponding axis. As shown in Fig. 4c, the PDDCS$_{22+}$ distribution for each reaction show the negative values for all the scattering angles. This demonstrates that the product j' alignment is along the y axis, which is consistent with the $P(\phi_r)$ distributions described in Fig. 2. The PDDCS$_{21-}$ is related to $\langle -\sin 2\theta_r \cos \phi_r \rangle$, and its behavior is similar with that of the PDDCS$_{22+}$. The PDDCS$_{21-}$ is positive or negative, corresponding to the product rotational angular momentum

j' along the directions of vector $x - z$ or $x + z$. As shown in Fig. 4d, the PDDCS$_{21-}$ values vary with the different scattering angles, which imply that the product angular momentum distributions are anisotropic.

4 CONCLUSIONS

With the accurate ground 1^2 A' state PES and the collision energy of 0.11 eV, we performed QCT calculations on the Li + DF($v = 0, j = 0$–5) $\rightarrow$ LiF + D reactions. Three product rotational angular momentum distributions, $P(\theta_r)$, $P(\phi_r)$, and $P(\theta_r, \phi_r)$ were explored. The $P(\theta_r)$ distribution shows a prominent peak at $\theta_r = 90°$, while the $P(\phi_r)$ distribution displays a large peak at $\phi_r = 270°$ and no peak at $\phi_r = 90°$ for the title reactions. This indicates that the product rotational angular momentum j' is not only aligned along the direction perpendicular to the reagent relative velocity k, but also oriented along the negative direction of the y-axis. As an alternative representation, the joint distribution of the $P(\theta_r, \phi_r)$ is consistent with the $P(\theta_r)$ and $P(\phi_r)$ distributions. Four commonly used PDDCSs were also computed. Differential cross sections were revealed via the PDDCS$_{00}$ distributions, and illustrate strongly forward scatterings for the Li + DF($v = 0, j = 0$) $\rightarrow$ LiF + D reaction, whereas the sideways scatterings for the Li + DF($v = 0$, $j = 1$–5) $\rightarrow$ LiF + D reactions. The PDDCS$_{20}$ and PDDCS$_{22+}$ distributions are consistent with angular momentum distributions. The PDDCS$_{21-}$ distribution demonstrates that the product angular distributions are anisotropic.

ACKNOWLEDGEMENT

The author gratefully acknowledges the financial support provided by the National Natural Science Foundation of China (Grant No. 11447014), the Project of Shandong Province Higher Educational Science and Technology Program (Grant No. J14 LJ09) and the China Postdoctoral Science Foundation (Grant No. 2014M550595).

REFERENCES

Becker, C.H., Casavecchia, P., Tiedemann, P.W., Valentini J.J. & Lee, Y.T. 1980. Study of the reaction dynamics of Li+HF, HCl by the crossed molecular beams method. *J. Chem. Phys.* 73(6): 2833–2850.

Bobbenkamp, R., Paladini, A., Russo, A., Loesch, H.J., Menéndez, M., Verdasco, E., Aoiz, F.J. & Werner H.J. 2005. Effect of rotational energy on the reaction Li + HF ($v = 0, j$) $\rightarrow$ LiF + H: An experimental and computational study. *J. Chem. Phys.* 122: 244304-1–18.

Chen, M.D., Han, K.L. & Lou N.Q. 2003. Theoretical study of stereodynamcis for the reactions Cl + H$_2$/HD/D$_2$. *J. Chem. Phys.* 118(10): 4463–4470.

Höbel, O., Bobbenkamp, R., Paladini, A., Russo, A. & Loesch, H.J. 2004. Effect of translational energy on the reaction Li+HF ($v = 0$)$\rightarrow$LiF + H. *Phys. Chem. Chem. Phys.* 6: 2198–12204.

Krasilnikov, M.B., Popov, R.S., Roncero, O., Fazio, D.D., Cavalli, S., Aquilanti, V., & Vasyutinskii O.S. 2013. Polarization of molecular momentum in the chemical reactions Li + HF and F + HD. *J. Chem. Phys.* 138: 244302.

Liu, S.L. & Shi, Y. 2011. Theoretical study of stereodynamics for the O(^{3}P) + H$_2$ $\rightarrow$ OH + H reaction. *Chin. Phys. B* 20(1): 013404.

Loesch, H.J. & Stienkemeier F. 1993. Steric effects in the state specific reaction Li+HF ($v = 1, j = 1, m = 0$)$\rightarrow$LiF + H. *J. Chem. Phys.* 98(12): 9570–9584.

Yue, X.F. & Feng, P., 2012. A quasi-classical trajectory study on stereodynamics of the F+HCl ($v = 0, j = 0$)$\rightarrow$HF + Cl reactions. *J. Theor. Comput. Chem.* 11(3): 663–674.

Yue, X.F. & Miao, X., 2011. A quasi-classical trajectory analysis of stereodynamics of the H+FCl ($v = 0$–3, $j = 0$–3)$\rightarrow$HCl + F reaction. *J. Chem. Sci.* 123(1): 21–27.

Yue, X.F. 2013. Product polarization and mechanism of Li+HF ($v = 0, j = 0$)$\rightarrow$LiF(v', j')+H collision reaction. *Chin. Phys. B* 22(11): 113401.

Yue, X.F., Cheng, J., Feng, H.R., Li, H. & Wu E.L., 2010. Theoretical study on stereodynamics of reactions of N(^{2}D)+H$_2$ $\rightarrow$NH + H and N(^{2}D)+D$_2$ $\rightarrow$ND + D. *Chin. J. Chem. Phys.* 23(4): 381–386.

Zanchet, A., Roncero, O., González-Lezana, T., Rodríguez-López, A., Aguado, A., Sanz-Sanz, C. & Gómez-Carrasco, S. 2009. Differential cross sections and product rotational polarization in A + BC reactions using wave packet methods: H$^+$ + D$_2$ and Li + HF examples. *J. Phys. Chem. A* 113: 14488–14501.

Zhai, H.S. & Yin S.H. 2012. Stereo-dynamics of the exchange reaction H$_a$+LiH$_b$$\rightarrow$LiH$_a$ + H$_b$ and its isotopic variants. *Chin. Phys. B* 21(12): 128201.

*Advances in Materials Science, Energy Technology
and Environmental Engineering – Patty & Zhou (Eds)*
© 2017 Taylor & Francis Group, London, ISBN 978-1-138-19668-1

Thermal simulation and analysis of high-power white LED light

Y.Z. Zhang, C.X. Li, Y. Wang & L.L. Ran
Shenzhen Institute of Information Technology. Shenzhen City, Guangdong Province, China

ABSTRACT: Based on the finite element software, a thermal model of high-power white LED street light was developed, and the temperature field distribution of the LED light was simulated. The simulation results fitted well with the real temperature data, which showed that the thermal model is accurate. Based on the simulation result, the distribution of the LED light chips is optimized.

1 INTRODUCTION

1.1 Problem description

Compared with traditional high pressure sodium light, LED street light has a series of advantages with high brightness, long life, fast response, a directional light, small light pollution, and low power consumption. However, the energy conversion efficiency of the LED street light is not high. Only 15%–25% of the energy is converted into light energy, the rest is almost converted into heat energy.

The accumulation of heat will cause temperature rise of LED street light chip, shorten the life of the chip, reduce brightness and stability and cause a series of problems (Fu 2011, Hu 2012, Tang 2010). It has been reported that LED luminous flux was reduced by approximately 1% if the chip junction temperature rise every 1 0°C; moreover, when the temperature increases above a certain level, the reliability of LED light will be reduced 10% every 2°C (Yang 2010, Liu 2008). Furthermore, high-power LED street light is usually composed of a few numbers of LED chips; the higher the arrangement density, the easier the accumulation heating. Then, it will affect the performance and reliability of the chip. In conclusion, heat dissipation design is one of the research emphases of LED street light products. LED street light cooling technology is divided into active and passive cooling technology (Gou 2012). It is widely used for the thermal convection technology (passive cooling technology) in engineering practice.

1.2 Research direction

However, the optimization design of LED street light heat sink is still not entirely solved. The weight of street light heat sink is more than 90% of the total weight of LED light for example. Therefore, it is also in urgent for the LED street light heat sink to carry out research on the optimization design.

For multichip LED street light products, this paper simulates and analyzes thermal distribution of LED street light at work and proposes a LED chip installation configuration optimization design based on the research.

2 MODEL BUILDING

2.1 Preparing the sample

Based on the principle of heat conduction and convection in the LED street light, this paper uses the thermal design software ICEPAK, simulates, and analyzes the thermal characteristics of LED light at work. The model used in this paper is mainly composed of LED chips, the substrate and the heat sink.

Manufacturers provide street light front view as shown in Figure 1. The light has 196 LED chips, for which the power of each LED chip is 1 w. The chips are arranged by a 7 × 28 matrix with size parameters: 2.2 cm (width), 3.5 cm (height).

2.2 Building the LED light model

The LED street light works in the open atmosphere environment. The environment temperature is kept at about 20°C, air convection equation is laminar, the number of iterations is 100 times, and error limits are 10^{-8}.

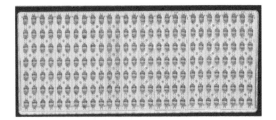

Figure 1. The street light front view. 1.

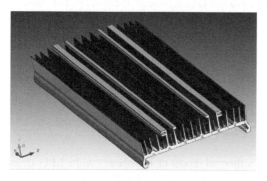

Figure 2. The stereogram of the heating sink.

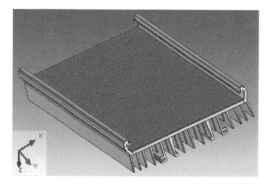

Figure 3. The LED street light model.

Figure 2 is a stereogram for radiator. Finally, the the LED street light model was developed, as shown in Figure 3.

3 SIMULATION RESULT

3.1 *LED substrate temperature distribution*

Software simulation result is shown in Figure 4, and it is obvious that the temperature is higher in the center of the substrate.

The simulation temperature distribution is in accordance with the Gaussian function, and the specific formula is shown in formula (1):

$$T = T_0 + Ae^{-\frac{1}{2}\left(\frac{(x-x_c)^2}{w_1^2} + \frac{(y-y_c)^2}{w_2^2}\right)} \quad (1)$$

The point (x_c, y_c) = (341.5, 141.5) is the center coordinate of aluminum substrate. The curve fits to get T_0 = −6573.4, A = 6626.6, w_1 = 18646.3, w_2 = 6904.8. According to the temperature Gaussian function, the temperature for LED substrate is simulated as shown in Figure 5.

Figure 4. LED substrate simulation temperature distribution.

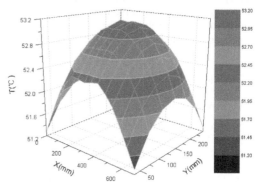

Figure 5. Three-dimensional temperature distribution of LED substrate simulation.

The measuring temperature is shown in Figure 6. Compared with Fig. 5 and Fig. 6, it can be seen clearly that the result of software computing is consistent with the test result and the numerical difference is small which proves that the model is consistent with the real LED street light. Simulation result shows that the LED highest junction temperature is 53.2°C (the highest measured temperature is 55°C), which means the heat sink can reach the requirement of street light. However, the whole LED chips' temperature distribution is uneven. The temperature of the heat sink shows the radial distribution of center high and edge low. Temperature difference is above 2°C (more than the measured up to 5°C). This will create some certain effects on the whole LED light's work as follows:

First, the ambient temperature of each LED chip is different, and the life and working state are also different. Therefore, in some instances, the center LED chip cannot meet the luminous intensity of lighting requirement when the ambient chip is still working regularly. Second, if the heat sink works for long time in the uneven temperature environment, inevitable changes will appear in shape. These also have influence on the unnecessary using of the product. If the two cases mentioned above happened, the entire LED light has to be replace. These have effect on the usage, cause the waste of resources, and furthermore, potentially increase the cost of the product.

3.2 LED substrate thermal homogenization

In order to make the substrate of the LED temperature uniform, the arrangement of LED chips should be changed. The opposite performance should show about the heat flux density and temperature on the base, which means that the center is low and the edge is high. Therefore, it will offset the temperature collection in the central area to uniform the temperature. In order to get the heat flow density function, make the auxiliary function, together with the temperature of the Figure 6: The actual temperature distribution of LED substrate formula (1) makes the substrate temperature of every point to be $T_{ave} = 52.2°C$, the auxiliary function should have the same e index with the original temperature function, and only the coefficient is different.

$$T' = T_0' + A'e^{-\frac{1}{2}\left(\frac{(x-x_c)^2}{w_1^2} + \frac{(y-y_c)^2}{w_2^2}\right)}$$

$$T_{ave} = \frac{T+T'}{2}$$

$$= \frac{1}{2}\left(T_0 + Ae^{-\frac{1}{2}\left(\frac{(x-x_c)^2}{w_1^2} + \frac{(y-y_c)^2}{w_2^2}\right)}\right. \tag{2}$$

$$\left. +T_0' + A'e^{-\frac{1}{2}\left(\frac{(x-x_c)^2}{w_1^2} + \frac{(y-y_c)^2}{w_2^2}\right)}\right)$$

$$= 52.2$$

The above formula is true eternally, so

$$\frac{T_0 + T_0'}{2} = 52.2$$

$$A + A' = 0$$

Among

$$T_0 = -6573.4$$

$$A = 6626.6$$

So

$$T_0' = 6677.8$$

$$A' = -6626.6$$

The e-exponential is the same for heat flow density function and the auxiliary function, and only the coefficient is different:

$$q = q_0 + Be^{-\frac{1}{2}\left(\frac{(x-x_c)^2}{w_1^2} + \frac{(y-y_c)^2}{w_2^2}\right)} \tag{3}$$

The average heat flux density is 0.1 W/cm², the average temperature of 52.2°C, and the proportion coefficient is 522, hence

$$q_0 = \frac{T_0'}{522} \approx 12.7927$$

$$B = \frac{A'}{522} \approx -12.6946$$

Formula (3) heat flow density function is shown in Figure 7. Figure 8 is the LED heat flux density distribution. It can be seen that Figure 9 is the temperature distribution at the arrangement.

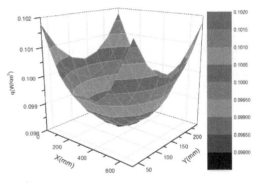

Figure 7. The heat flow density function after LED array is optimized.

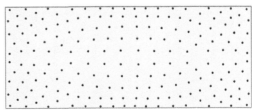

Figure 8. Optimized LED array.

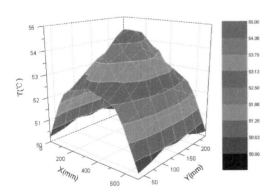

Figure 7. The heat flow density function after LED array is optimized.

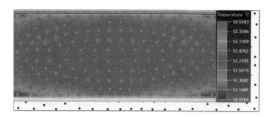

Figure 9. Temperature distribution after the LED array is optimized.

The temperature distribution is uniform after the LED array optimization when compared with that before optimization. Heat will not be gathered in the center of the radiator again and heat dissipation characteristics have been improved greatly. At the same time, it also reduces the reliability problems caused by heat accumulation.

4 CONCLUSION

According to the thermal design software ICE-PAK applied, multichip LED street light studied and the thermal characteristics simulated, the simulation result and test data match well. Moreover, a LED chips' arrangement scheme is designed in this paper to simulate that the uneven temperature distribution of the LED chips causes many losses of the light in daily work. Therefore, a new method is suggested in this paper to improve today's heat sink.

This research focuses on solving the problem of uneven temperature distribution. The light unit temperature difference is controlled between 1 and 3°C, which can make full use of the ability of heating dispersion. The conclusions of this research are as follows:

1. The LED heat sink is simulated for thermal and tested in practical, and the highest temperature is 55°C. The heat sink meets the requirements of street light heat dissipation and makes the whole temperature far below the normal working temperature upper limit of LED. It will not turn up the problem due to high temperature of LED light.
2. Under the condition of no change in the structure of heating sink, the heat flux is distributed reasonably due to optimized LED chip array, decreasing the LED amount in the high temperature region, reducing the heat dissipation pressure and increasing the LED amount in edge region. The temperature difference is in below 3°C after the heat sink is optimized. The temperature distribution homogenization comes true.

REFERENCES

Fu, X.Z. (2011). Research on the Heat Dissipation Problem of LED Lamp. *J. China illuminating engineering journal.* 22[3]: 73–78.
Gou, Y.J. (2012). New research frontier of high power LED lamp cooling system. *J. China light & lighting.* 2: 1–7.
Liu, Y.B. (2008). Research of heat release of technology of base on power-LED. *J. China illuminating engineering journal.* 19[1]: 69–73.
Tang, X.J. (2010). Heat dissipation analysis of high-power LED lamps. *J. China light & lighting.* 2: 17–20.
Yang, G. (2010). Research on schemes about heat dissipation of high-power LED lamp in road lighting. *J. China illuminating engineering journal.* 11: 18–21.
Zhu, X.Z. (2012). Optimization analysis of thermal performance of LED radiators. *J. China illuminating engineering journal.* 23[1]: 62–65.

Advances in Materials Science, Energy Technology and Environmental Engineering – Patty & Zhou (Eds)
© 2017 Taylor & Francis Group, London, ISBN 978-1-138-19668-1

Optimization of the blending process for ABS/PMMA/EMA via response surface methodology

Jin Ding

School of Mechanical, Electrical and Information Engineering, Shandong University, Weihai, China
Department of Mechanical and Electrical Engineering, Weihai Vocational College, Weihai, China

Zhenming Yue, Jiao Sun & Jun Gao

School of Mechanical, Electrical and Information Engineering, Shandong University, Weihai, China

ABSTRACT: In this work, an experimental investigation on impact strength and surface glossiness of Acrylonitrile-Butadienestyrene (ABS)/Poly (methylmethacrylate) (PMMA)/Ethylene methacrylate (EMA) blend has been conducted. The effect of blending parameters such as screw rotation speed, hopper temperature and EMA content on impact strength and surface glossiness is optimized Via Response Surface Methodology (RSM). The mathematical models for impact strength and surface glossiness are established between the parameters and the responses. The adequacy of the models is tested using the sequential F-test, lack-of-fit test and the Analysis of Variance (ANOVA) technique. According to the desired optimization criteria, the optimal blending parameters are obtained, and the results present an acceptable agreement between predicted and experimental values.

1 INTRODUCTION

Nowadays, highlight Acrylonitrile-Butadienestyrene (ABS) resin has attracted considerable attentions due to the increasing environmental concerns and esthetic product appearance (Zhang 2010). The binary blend of ABS with Poly (Methylmethacrylate) (PMMA) shows great potential to replace polymerization-based highlight ABS because of the excellent surface glossiness, easy processability, and high mechanical strength (Zhang 2015, Kim 1996, Wang 2014). Nevertheless, although blenders with good toughness are highly desirable in practical applications, PMMA always exhibits inherent brittlement, which makes ABS/PMMA blend application far from as widespread as expected.

Literature review on modification of ABS/PMMA blend reveals that, the influence of the content of modifier and the extrusion process parameters (such as screw rotation speed, feeding speed, barrel temperature et al.) on the mechanical and surface performance of alloy. A number of experimental programs have been carried out to study the effect of process parameters on ABS/PMMA properties, with various modifiers (Nazari 2012, Zhang 2010, among others). However, no comprehensive research work has been reported on modification ABS/PMMA with Ethylene methacrylate (EMA), so an extensive research is needed to analyze and optimize the ternary blend of ABS/PMMA/EMA. To get the desired impact strength and surface glossiness, the combination of the process parameters should be selected carefully. Response Surface Methodology (RSM) is one the modeling and optimization techniques currently in widespread use in describing the blend performance and finding the optimum of the responses of interest (Acherjee 2012, Agarwal 2012, Taghavian 2015).

The objective of the present study is to predict the impact strength and surface glossiness of ABS/PMMA/EMA blend under different content of EMA and variations of extrusion process parameters using response surface methodology. This approach enables statistical investigation of the individual process parameters and the interactions of the parameters simultaneously. The developed mathematical models are tested for their adequacy using analysis of variance and other adequacy measures. The effect of process parameters on the responses is studied on the basis of the developed models. The mathematical models are further used to find optimum melt blending conditions to achieve the desired properties.

2 EXPERIMENT

2.1 *Materials*

The ABS matrix polymer (757 K) was obtained from Zhenjiang Chimei Chemical Co. Ltd., with the Vicat softening temperature of 105°C and the melt flow index of 1.8 g/10 min. PMMA (CM-211) with the melt flow index of 16 g/10 min and the Vicat softening temperature of 102°C was obtained from Taiwan Chimei Chemical Co., Ltd. EMA was provided by Dongguan AnChenPlastic Technology Co. Ltd., with the Vicat softening temperature of 59°C and the melt flow index of 8 g/10 min.

2.2 Sample preparation

Prior to use, ABS and PMMA were dried at 80°C for 12h in an air circulating oven, as well as EMA, which was dried at 50°C for 12 hours. Melt blending ABS/PMMA/EMA were prepared using a two-screw extruder (the L/D ratio of screw is 40 and the diameter is 21.7 mm, SJSL-20, China) under different process parameters. After pelletized, standard testing specimens were processed with a 68 tons injection molding machine (XL680, China). The injection temperature was set at 225°C–215°C from the hopper to the nozzle, and the holding pressure was 50 MPa.

2.3 Characterization

The notched Izod impact strength was evaluated using an impact tester (XC–5.5D, China) with a

pendulum hammer of 2.75 J. The testing was carried out at room temperature (23°C). The average value reported was obtained from at least five specimens. The surface glossiness was measured by a JFL-BZ60 s (China) gloss meter. For optical test, the average of five values was used in the analysis.

2.4 Experimental design

For the design and analysis of these experiments, RSM was adopted in this study. It is used to approximate numerical or physical experimental data by an expression that is usually a low-order polynomial. The experimental data were analyzed by means of RSM to fit the second-order polynomial equation. The three key steps of the methodology are the following: Response was first fitted to the factors by multiple regressions (Jayle1996, Boyaci San2013). The quality of fit of the model was evaluated by the coefficients of determination (R^2) and the Analysis of Variances (ANOVA). The Design Expert V 8.0.6 software was used to code the variables and to establish the design matrix.

In order to investigating the toughness and surface glossiness of ABS/PMMA/EMA alloy, three independently blending process parameters, namely: rotation speed (S, r/min), temperature (T,°C) and EMA content (C, wt%) are considered as input parameters to carry out the experiments. The selected process parameters and their limits, units and notations are given in Table 1.

The selected design matrix, shown in Table 2, is a three factors five levels central composite rotat-

Table 1. Process control parameters and their limits.

| Parameters | Units | Notations | Limits | | | | |
			−2	−1	0	1	2
Rotation speed	r/min	S		100	120.	150	180	200
Temperature	°C	T		180	190	205	220	230
EMA content	wt%	C		2	4	6	8	10

Table 2. Experimental design and resulted responses.

Experiment no.	S (r/min)	T (°C)	C (wt%)	Impact Strength (kJ/m2)	Surface Glossiness (Gs)
1	0	0	0	71.2	64.8
2	0	0	0	70.6	64.9
3	−1	1	1	75.3	60.3
4	0	0	−2	43.8	66.4
5	0	0	2	76.8	57.7
6	0	2	0.	72.1	64.6
7	1	−1	1	73.8	59.8
8	0	0	0	70.9	64.2
9	0	0	0	73.2	66.8
10	−1	−1	−1	53.1	66.9
11	−2	0	0	71.4	63.6
12	1	1	−1	54.9	68.1
13	−1	1	−1	55.3	67.2
14	1	−1	−1	51.2	67.5
15	2	0	0	72.4	63.1
16	0	0	0	73.2	63.7
17	1	1	1	75.2	60.1
18	0	0	0	73.4	64.3
19	−1	−1	1	73.5	58.2
20	0	−2	0	69.1	62.7

able design with 20 sets of coded conditions and comprising a full replication of 2^3 factorial design plus six center points and six star points. The 20 experimental runs allowed the estimation of the quadratic and two-way interactive effects of the process parameters on the response parameters.

3 RESULTS AND DISCUSSION

The adequacy of the model is tested using the sequential F-test, Lack of fit and the Analysis of Variance (ANOVA) technique using Design Expert V 8.0.6 software to obtain the best-fit model.

3.1 *Analysis of impact strength*

The fit summary for impact strength suggests the quadratic model are significant and the model is not aliased. The ANOVA table of the quadratic model is given in Table 3.

The associated p-value of less than 0.05 for the model indicates that the model terms are satistically significant. In this model, R^2, adjusted R^2 and predicted R^2 are in reasonable agreement and are close to 1, which indicate adequacy of the model. The adequate precison compares the signal to noise ratio and a ratio greater than 4 is disirable. The value of adequate precision ratio of 23.245 indicates adequate model discrimination. The lack-of-fit F-value of 4.35 implies that the lack-of-fit is not sig.nciant relative to the pure error, as this is desirable.

The final mathematical models for impact strength (Y_1), which can be used for prediction within same design space, are as follows:

a. In terms of coded factors:

$$Y_1 = 72.19 - 0.031\,S + 1.04T + 10.16\,C + 0.14ST + 0.31SC - 0.34TC - 0.76S^2 - 1.22T^2 - 4.86C^2 \tag{1}$$

b. In terms of actual factors:

$$Y_1 = -247.83289 + 0.16635S + 2.34174T + 15.87767C + 0.000311ST + 0.0044SC - 0.0095TC - 0.0086S^2 - 0.0055T^2 - 0.85921C^2 \tag{2}$$

3.2 *Analysis of surface glossiness*

For surface glossiness, the fit summary recommends the quadratic model where the additional terms are significant and the model is not aliased. Table 4 presents the ANOVA table of the quadratic model. The ANOVA result shows the significant model terms associated with surface glossiness. The other adequacy measures R^2, adjusted R^2 and predicted R^2 are in reasonable agreement and are close to 1, which indicate adequate model. The value of adequate precision ratio of 13.374 indicates adequate model discrimination. The lack-of-fit F-value of 1.46 implies that the lack-of-fit is not sig0nciant relative to the pure error, as this is desirable.

The final mathematical models for surface glossiness (Y_2), which can be used for prediction within same design space, are as follows:

a. In terms of coded factors:

$$Y2 = 64.76 - 0.15S + 0.48T - 3.36\,C - 0.19ST - 0.012SC + 0.19TC - 0.38S^2 \, 0.27T^2 - 0.84C^2 \tag{3}$$

b. In terms of actual factors

$$Y_2 = -7.7596 + 0.2219S + 0.5712T - 0.6947C - 0.000424ST - 0.000177SC + 0.0053TC - 0.00043S^2 - 0.00124T^2 - 0.1483C^2 \tag{4}$$

Table 3. ANOVA results for analysis of the impact strength.

Sources	Sum of squares	Degrees of freedom	Mean squares	F-value	p-value
Model	1776.15	9	197.35	42.83	0.0001
S	0.013	1	0.013	2.78E-003	0.9590
T	14.65	1	14.65	3.18	0.1049
C	1410.66	1	1410.66	306.15	0.0001
ST	0.15	1	0.15	0.033	0.8598
SC	0.78	1	0.78	0.17	0.6892
TC	0.91	1	0.91	0.20	0.6660
S2	8.31	1	8.31	1.80	0.2091
T2	21.41	1	21.41	4.65	0.0565
C2	340.45	1	340.45	73.89	0.0001
Residual	46.08	10	4.61		
Lack of Fit	37.47	5	7.49	4.35	0.0662
Pure Error	8.61	5	1.72		
Cor Total	1822.23	19			

$R^2 = 0.9747$, adjusted $R^2 = 0.9520$, predicted $R^2 = 0.9376$, adequate Precision = 23.245

Table 4. ANOVA results for analysis of the surface glossiness.

Sources	Sum of squares	Degrees of freedom	Mean squares	F-value	p-value
Model	170.26	9	18.92	13.22	0.0002
S	0.31	1	0.31	0.22	0.6513
T	3.09	1	3.09	2.16	0.1725
C	154.48	1	154.48	107.95	0.0001
ST	0.28	1	0.28	0.20	0.6670
SC	1.250E-003	1	1.250E-003	8.735E-004	0.9770
TC	0.28	1	0.28	0.20	0.6670
S2	2.08	1	2.08	1.45	0.2562
T2	1.08	1	1.08	0.75	0.4059
C2	10.15	1	10.15	7.09	0.0238
Residual	14.31	10	1.43		
Lack of Fit	8.48	5	1.70	1.46	0.3453
Pure Error	5.83	5	1.17		
Cor Total	184.57	19			

$R^2 = 0.9525$, adjusted $R^2 = 0.9127$, predicted $R^2 = 0.9011$, adequate Precision = 13.374

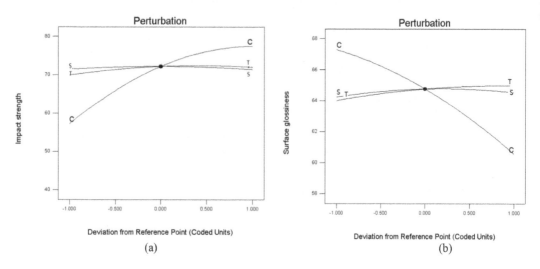

Figure 1. The activities of impact strength (a) and surface glossiness (b) as responses to three factors in perturbation plot A-rotation speed B-temperature and C-EMA content.

3.3 Effect of blending process parameters on responses

The response function and the affected variables could be graphically evaluated by the perturbation plot, the interaction plot and three dimensional (3D) diagrams. The perturbation plot represents the effect of changing one variable while holding the rest constant. The perturbation plot as shown in Figure 1 represents the effect of rotation speed, temperature and EMA content separately from each other on the activity of impact strength and surface glossiness. The impact strength shows an increase as the EMA content is enlarged. A slight increment is observed under different rotation speed and temperature, which is in accordance with the

prediction model that C and C^2 presenting significant. However, the surface glossiness shows an decrease as the EMA content is enlarged. A slight increment is observed under different rotation speed and temperature, which is also in accordance with the prediction model that C and C^2 presenting significant.

Factors interaction plots are used to present the results in graphical form. Figure 2 shows all the significant interaction terms related to impact strength. It is observed from Figure 2 (a) that rotation speed and temperature have a little impact on impact strength. When the EMA content was 6wt%, rotation speed was 150 r/min, and the temperature was 205°C, impact strength was up to the

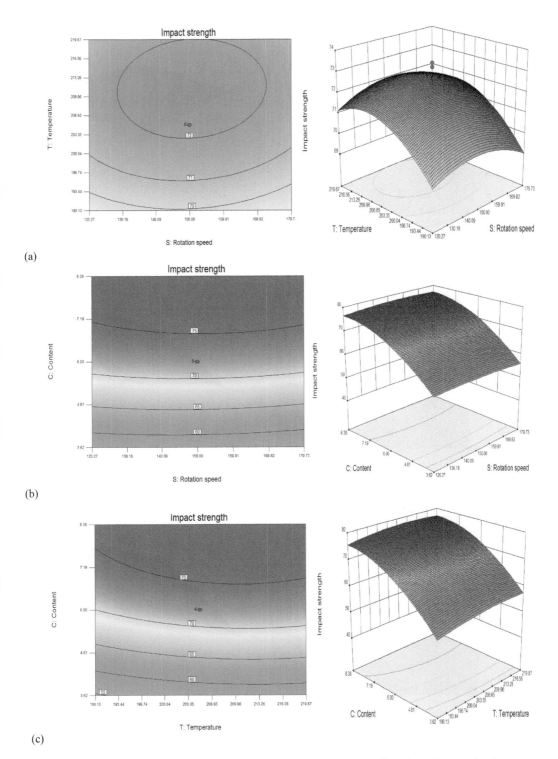

(a)

(b)

(c)

Figure 2.   Interaction effect of parameters on impact strength.(a) interaction effect of rotation speed and temperature; (b) interaction effect of rotation speed and content; (c) interaction effect of temperature and content.

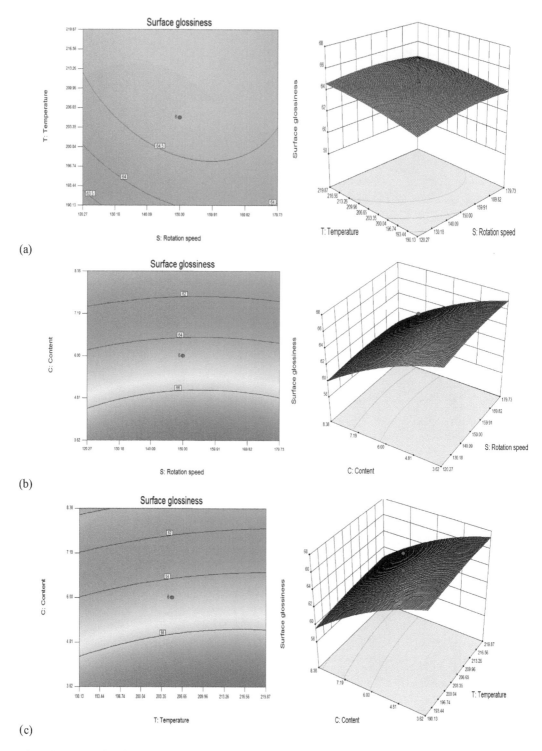

(a)

(b)

(c)

Figure 3. Interaction effect of parameters on surface glossiness (a) interaction effect of rotation speed and temperature; (b) interaction effect of rotation speed and content; (c) interaction effect of temperature and content.

high level. It is evident from Figure 2 (b) that rotation speed and EMA content have a strong interaction effect on impact strength, especially with the change of EMA content. The results from contour map and 3D response surface plot indicated that the impact strength increased with increasing EMA content, and rotation speed influenced impact strength slightly. When EMA content was 6wt% and rotation speed was 150 r/min impact strength was obtained the maximum value. The results from Figure 2 (c) presented that temperature and EMA content have also a strong interaction effect on impact strength, especially with the change of EMA content. When EMA content was 6wt% and temperature was 205°C impact strength was obtained the maximum value.

It is observed from Figure 3, that EMA content has also a significant effect on surface glossiness. In terms of the interaction effect between the rotation speed and temperature (Figure 3 (a)), it was evident that when rotation speed was 150 r/min and temperature was 205 °C, the surface glossiness was obtained high level. Compared with Figure 3 (a), the results of Figure 3 (b) and Figure 3 (c) indicated that the interaction among EMA content, temperature and rotation speed influenced surface glossiness greatly. It is seen from this figure that the surface glossiness has a decline trend with increasing EMA content. EMA is the dominant factor among the interaction effect of EMA, temperature and rotation speed. When EMA content was 6wt%, temperature was 205°C and rotation speed was 150 r/min, surface glossiness was obtained the maximum value.

3.4 Optimization of process parameters

The optimization module in Design Expert searches for a combination of factor levels that simultaneously satisfy the requirements placed on each of the responses and factors. Numerical optimization searches the design space, using the developed regression model to find the factor settings that optimize any combination of one or more goals. The goals are combined into an overall desirability function. The numerical optimization finds a point that maximized this desirability function. The criteria in this numerical optimization is to reach maximum impact strength and surface glossiness with limitation scope of the extruder. Table 5 summarizes these optimization criteria.

Table 6 presents the optimal blending conditions according to the criterion that would lead to maximum impact strength of 71.197 kJ/m² and maximum surface glossiness of 65.305, which is better than that achieved with experimental trials. The table presents three set of the pareto-optimal solutions. In view of the fact that none of the solutions in the pareto-optimal outcome is definitely better than others, each of them is an acceptable solution. According to the criterion, the optimum parametric range for rotation speed has to be 150.98–151.18 r/min, temperature has to be 214.03–214.16°C and EMA content within the range of 5.73–5.74 wt%.

Finally, according to the criterion, three additional experiments are conducted at the parameters setting and average of those three is used for verification. Table 7 present the results of

Table 5. The criterion of numerical optimization.

Parameters	Goal	Lower limit	Upper limit	Importance
S, r/min	Is in range	80	200	3
T,°C	Is in range	180	250	3
C, wt%	Is in range	2	12	3
Impact strength, kJ/m²	Maximize	43.8	100	3
Surface glossiness, Gs	Maximize	57.7	100	3

Table 6. Optimal blending parameters based on the criterion.

Sol. No.	S (r/min)	T (°C)	C (wt%)	Impact strength (kJ/m²)	Surface glossiness (Gs)	Desirability
1	151.04	214.06	5.73	71.1971	65.3049	0.097 selected
2	150.98	214.14	5.74	71.1975	65.3047	0.097
3	151.18	214.03	5.74	71.2238	65.2974	0.097

Table 7. Results of verification experiments.

Results	Predicted	Experimental	|Error(%)|
Impact strength (kJ/m²)	71.1971	72.3	1.55
Surface glossiness (Gs)	65.3049	64.8	0.78

verification tests. A fair agreement between the predicted and experimental results is observed.

4 CONCLUSION

In this study, the blending parameters of screw rotation speed, hopper temperature and EMA content on the impact strength and surface glossiness of ABS/PMMA/EMA blend are analyzed by RSM. The developed response surface models can predict the responses adequately within the limits of blending parameters being used. The analysis of ANOVA presents that the EMA content has the significant effect on impact strength and surface glossiness. When EMA content increases, the impact strength increases, but the surface glossiness decreases. However the rotation speed and temperature influence impact strength and surface glossiness slightly. The optimal blending parameters can be obtained by using the numerical optimization technique, and results obtained represent an acceptable agreement between predicted and experimental values.

ACKNOWLEDGEMENTS

This work was financially supported by the National Science and Technology Support Project (2011BAF16B01) and Weihai city Science and Technology Plan Project (2015GGX030).

REFERENCES

Acherjee, B. (2012). Experimental investigation on laser transmission welding of PMMA to ABS via response surface modeling *Opt. Laser. Technol.* 44, 1372–1383.

Agarwal, P. (2012). Statistical optimization of the electrospinning process for chitosan/polylactide nanofabrication using response surface methodology. *J. Mater. Sci.* 47, 4262–4269.

Boyaci, San. (2013). Analysis of the polymer composite bipolar plate properties on the performance of PEMFC (polymer electrolyte membrane fuel cells) by RSM (response surface methodology). *Energy.* 55, 1067–1075.

Kim, B.K. (1996). ABS Ternary Blends: Morphology, Surface Gloss, and Mechanical Properties. *J. Polym. Sci, Part B: Polym. Phys.* 35, 829–840.

Kim, B.K. (1997). Effects of Annealing in ABS Ternary Blends. *J. Appl. Polym. Sci.* 66, 1531–1542.

Properties of Nanoparticle-Filled Polymer in Rapid Heat Cycle Molding. *J. Appl. Polym. Sci.* 132, 1–9.

Rybnicek, J. (2012). Ternary PC/ABS/PMMA blends—morphology and mechanical properties under quasi-static loading conditions. *J. Polym.* 57(2), 87–94

Taghavian, H. (2015). Optimizing the Activity of Immobilized Phytase on Starch Blended Polyacrylamide Nanofibers-Nanomembranes by Response Surface Methodology. *J. Fiber Polym.* 16, 1048–1056.

Tayebe, N. (2012). Effect of Clay Modification on the Morphology and the Mechanical/Physical Properties of ABS/PMMA Blends. *J. Appl. Polym. Sci.* 126, 1637–1649.

Wang, S.H. (2014). Notch Sensitivity of Poly (Acrylonitrile-Butadiene-Styrene)/Poly (Methyl Methacrylate)/Ethylene Methacrylate Co-polymer (EMA) Composites. *Appl. Mech. Mater.* 496–500, 322–326.

Wang, S.H. (2014). Study on the Mechanical Properties and Surface Gloss of Acrylonitrile–butadiene-styrene / Poly (methyl methacrylate)/Ethylene Methacrylate Copolymer (EMA) composites *Mater. Sci. Eng.* 62, 1–6.

Wang, S.H. (2014). Tensile deformation mechanisms of ABS/PMMA/EMA blends. *Appl. Mech. Mater.* 496–500, 327–330.

Zhang, A.M. (2010). Effect of Acrylonitrile-Butadiene-Styrene High-Rubber Powder and Strain Rate on the Morphology and Mechanical Properties of Acrylonitrile-Butadiene-Styrene/Poly (Methyl Methacrylate) Blends. *Polym. Compos.* 31, 1593–1602.

Zhang, A.M. (2010). Study on Mechanical and Flow Properties of Acrylonitrile-butadiene-styrene/Poly (methylmethacrylate)/Nano-Calcium Carbonate Composites. *Polym. Compos.* 31,1593–1602.

Zhang, A.M. (2015). Effects of Mold Cavity Temperature on Surface Quality and Mechanical

Zhou, X. (2013). Optimization and Characteristics of Preparing Chitosan Microspheres Using Response Surface Methodology. *J. Appl. Polym. Sci.* 127, 4433–4439.

Advances in Materials Science, Energy Technology
and Environmental Engineering – Patty & Zhou (Eds)
© 2017 Taylor & Francis Group, London, ISBN 978-1-138-19668-1

Enhanced synchronous treatment of wastewater and solid pollution by hydrolysis acidification technology

Yangyang Zhang & Jinghui Cui
School of Water Resources and Environment, China University of Geosciences, Beijing, China

ABSTRACT: Many experiments were performed in the laboratory, and the results obtained are discussed in this paper. the mass and heat transfer efficiency of hydrolytic acidification was improved significantly by using a composite anaerobic reactor. The wastewater treatment time and sludge hydrolization time were reduced to 4 hours and 24 hours, respectively. The maximum hydraulic load was 2.4 kg CODCr/ $(m^2 \cdot d)$. The maximum excess sludge load was 9 kg CODCr/$(m^3 \cdot d)$. The SS and CODCr removal efficiency were 94% and 44% for actual domestic wastewater, respectively. The SS and CODCr removal efficiency were 94% and 44% for excess sludge, respectively. The process could be used as pretreatment to reduce load for post treatment. The SS and CODCr removal efficiency were 96% and 75% for mixed liquid of actual domestic wastewater and sludge, respectively. It could achieve synchronous controlling of wastewater & sludge very well.

1 INTRODUCTION

At present, high energy costs in aeration and sludge disposal are serious problems for common municipal wastewater treatment technologies. These problems significantly affect the construction and operation of urban sewage treatment plants in China. The cleaner production of urban wastewater treatment plant has not been given enough attention. Excess sludge, stink and standard attainment wastewater are also discharged from the sewage treatment plants. The living environment of mankind is damaged seriously by all these problems. To solve these problems, this study focused on treating wastewater through synchronous controlling of wastewater and sludge of high-efficiency hydrolytic acidification process, which costs less energy.

On the other hand, anaerobic digestion can be used as a treatment option for sludge from WWTPs in order to transform organic matter into biogas and reduce the amount of sludge. Anaerobic digestion thus optimizes WWTP costs, and its environmental footprint is considered as a major and essential part of a modern WWTP (Appels et al., 2008). In addition, anaerobic digestion is a widely used method for the treatment of sewage sludge (Dereix et al., 2006; Morgan-Sagastume et al., 2011). The optimization of anaerobic digestion process strongly depends on the increase in the hydrolysis efficiency, since the organic matter of sewage sludge exists in particulate form and hydrolysis is the rate-limiting step of the whole process (Miron et al., 2000; Appels et al., 2008; Lv et al., 2010).

The hydrolysis acidification anaerobic reactor controlled wastewater treatment at two stages of hydrolysis and acidification, generally used for pretreatment of wastewater. Hydrolysis acidification can degrade organic matter, intercept suspended solids, and improve sewage biodegradability, reducing the load and difficulty on the subsequent operation process. Hydrolytic acidification method is widely used in a variety of sewage, including urban sewage, pharmaceutical wastewater, meat or beans processing wastewater, sugar water, polyether polyester, and other chemical wastes.

Around the concept of clean production, saving energy and reducing consumption, the high-efficiency and energy-saving hydrolysis acidification process both for sludge reduction and for the pretreatment of the wastewater is quite attractive. In this paper, in the degradation of sewage, sludge treatment is the starting point, and it focuses on sewage sludge by hydrolysis acidification, to obtain detailed research data for complex water quality hydrolysis acidification and it is convenient to be used in future production.

2 MATERIALS AND METHODS

2.1 Wastewater and sludge source

The wastewater used for hydrolysis acidification was obtained from the sewage pump plant of China University of Geosciences in Beijing, China, for which the indicators of the sewage were close to general city sewage. The main characteristics of sewage are as follows: TCOD (Total Chemical Oxygen Demand) 158.12~317.57 mg/L, NH_3^-–N 28.88~48.21 mg/L, TN 45.67~71.50 mg/L, TP 3.2~8.9 mg/L, SS 170~200 mg/L, and pH 7.0~7.5.

Secondary sludge used in this study was obtained from the secondary sedimentation tank of municipal WWTP located in Beijing, China, which was operated with a traditional activated sludge process. The main characteristics of sludge are as follows: TSS (Total Suspended Solids) 21440~24900 mg/L, VSS (Volatile Suspended Solids) 13520~15850 mg/L, TCOD (Total Chemical Oxygen Demand) 18500~21500 mg/L, SCOD (Soluble Chemical Oxygen Demand) 458~680 mg/L, VSS/SS 69~71%, NH_3^--N 84~105 mg/L, TN 1135~1268 mg/L, and pH 6.58~7.04.

2.2 *Experimental set-up*

The compound anaerobic sludge bioreactor is made of organic glass plate, the bioreactor length is 140 cm (each hydrolysis acidification pool length, 25 cm; plate sedimentation tank length, 40 cm), 12 cm wide, and 50 cm high, with effective protection of 12 cm high and the effective volume of 80 L. Increased elastic solid fillers are present in the reactor's upper part, and the lower part is anaerobic sludge (Figure 1).

2.3 *Analytic methods*

TKN (Total Kjeldahl nitrogen), TSS (Total Suspended Solids), VSS (Volatile Suspended Solids), and Suspended Solids (SS) were analyzed according to standard methods (APHA, 1992). Ammonium (NH_4^+) was analyzed on total wastewater by steam distillation using MgO followed by back titration of the boric acid distillates using sulfuric acid (0.1 M). Total and soluble COD was analyzed by the potassium dichromate method.

3 RESULTS AND DISCUSSION

If the compound anaerobic sludge bioreactor is set up and is running continuously for 50 d, then it is said to be successfully started.

3.1 *COD_{Cr} removal rate in sewage*

In a previous investigation, we reported that the best HRT in the compound anaerobic sludge bioreactor is 4 hours, effluent pH range in 4.8 ~ 6.2, test temperature is 25 degrees Celsius, and adjust inflow water COD_{Cr} concentration are 300, 500, 700, 900 and 1100 mg/L, respectively. In order to get stable data, the bioreactor has to run at least one week before getting the test data. The COD_{Cr} removal rate under different concentrations in the reactor is shown in Figure 2.

Figure 2 shows the removal rate of COD_{Cr} at different concentrations. COD_{Cr} concentration is in the range of 300–900 mg/L, with the increase in feed concentration, the removal rate of COD_{Cr} is gently lower, and the concentration has been reduced by more than 40%. In addition, we can conclude that our bioreactor has a certain ability to resist organic load impact. But when the COD_{Cr} concentration is greater than 900 mg/L, the bioreactor COD_{Cr} removal rate is sharply reduced, it also illustrates the volume of the reactor under the condition of load reaching the limit, and the degradation ability of microorganisms limit the COD_{Cr} further removed.

3.2 *$CODCr$ and SS removal rates in Sludge*

In a previous investigation, we reported that when feed for excess sludge, the best HRT range is obtained in 2~8 days. In order to get the fastest processing speed and effect, set HRT for 2 days in this test, and feed for excess sludge. Of course, the test temperature is 25 degrees Celsius. In order to get stable data, the bioreactor has to run at

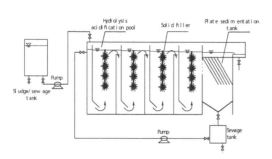

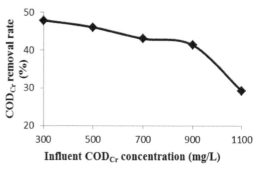

Figure 1. Schematic showing the structure of a compound anaerobic sludge bioreactor.

Figure 2. Showing the COD_{Cr} removal rate under different concentrations.

least one week before getting the test data. The COD_{Cr} and SS removal rate under different times in the reactor are shown in Figure 3 and figure 4, respectively.

From Figure 3, it is easy to see that the reactor is more sensitive to water quality fluctuation. The bioreactor's volume load reached 8.5 $kgCOD_{Cr}/(m^3 \cdot d)$, and effluent COD_{Cr} concentration in the vicinity of 3000 mg/L. This bioreactor, with COD_{Cr} removal rate of about 84%, can be more efficient in the removal of excess sludge COD_{Cr} than the traditional anaerobic reactor.

Structurally, compound anaerobic sludge bioreactor with the folded plate structure has a higher removal rate of organic matter. Folded plate can make the whole overall fluid in the form of pushing flow, so that the local organic load is descending, eventually making microbial population distribution along different. Look from the ability to resist the impact load, as reactors are packed with elastic solid fillers, microbial species are more rich, and their resistance to impact load ability is stronger.

The excess sludge SS was set to 22000 mg/L, as shown in Figure 4, the removal rate is as much as 98%, with the effluent concentration of 260 mg/L. This is mainly due to the reactor in the final set up of a set of efficient inclined plate sedimentation tank, that it is withholding from compound anaerobic baffle out most of the SS, and did not find out the water particles obviously, and only those dots of sludge. That is good for subsequent unit.

3.3 COD_{Cr} removal rate in sewage sludge mixture

To test the domestic sewage and excess sludge by hydrolysis acidification, respectively, the following research synchronous treatment domestic sewage and excess sludge was carried out. Synchronous processing is to make sludge and sewage HRT different. According to the above test, excess sludge HRT at 2 days and sewage HRT at 4 hours obtained good results. Therefore, the test reference residence times of the proportion of the above test set the remaining sludge and sewage water ratio to 1:12. This experiment HRT was set to 4 hours. Test half month, the experimental data are as follows.

Figure 5 shows sewage and sludge mixed by 1:12 ratio and COD_{Cr} is about 1750 mg/L. After a compound anaerobic sludge bioreactor effluent was added, COD_{Cr} is about 440 mg/L, and the removal rate is up to 75%. The feasibility of hydrolysis acidification after mix excess sludge and sewage is very strong.

Figure 6 shows sewage and sludge mixed by 1:12 ratio and SS is about 1850 mg/L. After a compound anaerobic sludge bioreactor effluent was added, SS is about 70 mg/L, and the removal rate is up to 96%. As the flow increases, the sludge funnel gathered a lot of sludge in a plate sedimentation tank, the sludge reflux. No sludge carried out during the test.

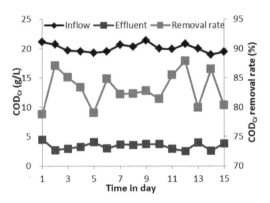

Figure 3. COD_{Cr} removal rate at different operation times.

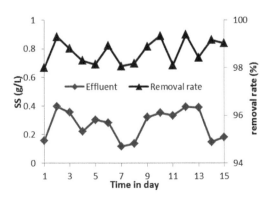

Figure 4. SS removal rate at different operation times.

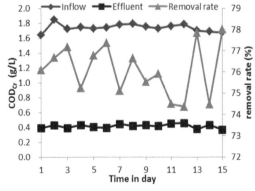

Figure 5. COD_{Cr} removal rate at different operation times.

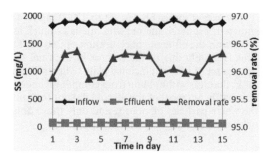

Figure 6. SS removal rate at different operation times.

4 CONCLUSIONS

Through laboratory tests, initially explored a set of synchronization sewage and solid pollutants by efficient hydrolysis acidification technology. The main conclusions are as follows:

1. In this paper, the design of an efficient hydrolysis acidification equipment—compound anaerobic sludge bioreactor, and higher efficiency compared with conventional anaerobic reactors. Owing to the design used mass transfer technology, the bioreactor can shorten the hydraulic retention time of hydrolysis acidification process to 4 hours and can withstand the maximum load of the influent 2.4 kg COD_{Cr}/$(m^3 \cdot d)$, and the COD_{Cr} removal rate can be closer to 45%.
2. The reactor is also very good for excess sludge hydrolysis acidification, COD_{Cr} removal rate of up to 75%, load into the sludge reached 9 kg COD_{Cr}/$(m^3 \cdot d)$, and the SS removal rate was of up to 98% to achieve the reduction of excess sludge hydrolysis efficiently.
3. Feed is a mixture of sewage and sludge, the COD_{Cr} removal rate can reach 75%, and the SS removal rate is 96%, which can achieve good synchronization and efficient sewage and solid-phase pollutants hydrolysis acidification.

ACKNOWLEDGMENTS

The authors would like to thank "The research and demonstration of ecological low-energy-consuming treatment technology and key equipment for processing rural domestic pollution from water resource region" and "Hubei Hanjiang watershed water security key technology research and demonstration (No. 2012ZX07205002-02-02)" project, which provided financial support for this study. They also thank Wang for patiently guiding them.

REFERENCES

APHA, 1992. Standard Methods for the Examination of Water and Wastewater, 18th ed. American Public Health Association, Washington, DC, USA.

Appels, L., Baeyens, J., Degrève, J., Dewil, R., 2008. Principles and potential of the anaerobic digestion of waste-activated sludge. Prog. Energy Combust. Sci. 34, 755–781.

Dereix, M., Parker, W., Kennedy, K., 2006. Steam-explosion pretreatment for enhancing anaerobic digestion of municipal wastewater sludge. Water Environ. Res. 78 (5), 474–485.

Lv, W., Schanbacher, F.L., Yu, Z., 2010. Putting microbes to work in sequence: recent advances in temperature-phased anaerobic digestion processes. Bioresour. Technol. 101 (24), 9409–9414.

Miron, Y., Zeeman, G., van Lier, J.B., Lettinga, G., 2000. The role of sludge retention time in the hydrolysis and acidification of lipids, carbohydrates and proteins during digestion of primary sludge in CSTR systems. Water Res. 34 (5), 1705–1713.

Morgan-Sagastume, F., Pratt, S., Karlsson, A., Cirne, D., Lant, P., Werker, A., 2011. Production of volatile fatty acids by fermentation of waste activated sludge pretreated in full-scale thermal hydrolysis plants. Bioresour. Technol. 102 (3), 3089–3097.

Advances in Materials Science, Energy Technology
and Environmental Engineering – Patty & Zhou (Eds)
© 2017 Taylor & Francis Group, London, ISBN 978-1-138-19668-1

An evaluation model of soft foundation treatment based on the analytic hierarchy process and gray theory

Yiliang Liu, Xu Fu, Xiaoli Liu & Ligai Bai
Department of Architectural Engineering, North China Institute of Aerospace Engineering, Langfang, China

ABSTRACT: There are many decision problems in the comprehensive evaluation of the highway soft foundation treatment. The factors, such as environmental impact, geological condition, and personnel condition, are difficult to make a quantitative analysis, and increase the uncertainty of foundation treatment effect. How to change the decision from the experience to the objective intelligent decision and how to select a reasonable soft foundation treatment plan are the main research directions of soft foundation treatment evaluation. First, we established a hierarchy structure model of the evaluation index system, and then used the AHP method to determine the weight of each index for soft ground treatment. According to the gray theory, an evaluation model of soft foundation treatment is created and applied to an engineering sample. The results show that the model has certain theoretical significance and application value.

1 INTRODUCTION

Since the 1980s, the highway has a rapid development in China, and the total mileage is more than 104,000 kilometers. China overtook the United States of America and became the first in the world (Pan Ruichun et al. 2012). There are many soft soil sites in the eastern coastal areas and the southern provinces of China. Most of these sites belong to the normal compaction soft clay with high compressibility, large settlement, and poor drainage consolidation stability. The new highway is often to pass through such sites. If not treated or handled properly, the soft foundation would result in lower quality roads, affecting the durability of the project (Kang Hong et al. 2009).

In recent years, many domestic and foreign experts and scholars of soft foundation treatment evaluation have put forward different soft foundation treatment plan evaluation methods successively, in which more research were done on qualitative evaluation method, statistical analysis method, fuzzy mathematics method, gray system theory, gray comprehensive evaluation, and so on. Each method has established a program evaluation model, and, combined with the actual project cases, verified the feasibility of the method (Pal et al. 2004). However, these decision models have their own defects, which need to continue to be optimized (Shi et al. 2012). Through the establishment of a hierarchy structure model, the weights of the indexes are assigned by AHP, and the gray theory was optimized. The optimized gray evaluation method was used in engineering practice and achieved a good result.

2 ESTABLISHING A PROGRAM EVALUATION MODEL

2.1 Establishing a hierarchy model

According to the system engineering and system-level principle, the various factors are classified that influence the soft ground treatment scheme by properties, and establish the hierarchical structure model of index system. It is generally divided into the target layer, the first index layer, and the second index layer.

2.2 Weight assignment

Weight assignment means assigning the factors' weight of primary and secondary index layers by the analytic hierarchy process (Peng 2016).

2.3 Determination of comparative data columns

It has been supposed that the alternative set consists of n program elements, and the evaluation indexes set of each program consists of m factors, such as construction period, difficulty of construction, and environmental impact. The second evaluation index data column of each program is called comparative data columns, expressed as

$$\{x_i(k)\} = \{x_i(1), x_i(2), \cdots, x_i(m)\} \tag{1}$$

In the formula, i = 1, 2,..., m. The value of each index, the highest and lowest values, can be obtained by the method of logical reasoning, test statistics, or expert evaluation, based on statistical data. In order to ensure the quality of the modeling and the correct result of the system analysis, we must transform and whiten to the collected original data, to eliminate the dimension and the data had comparability, such as the percentage transformation and interval value transformation (Liang et al. 2014, Chang et al. 1996, Tseng 2009).

2.4 Determination of the reference data column

The reference data column is an ideal optimal evaluation program, takes out the best index of the same evaluation indicator in all the programs, to the same evaluation indicators, and composes an ideal optimal evaluation scheme.

$$\{x_0\} = \{x_0(1), x_0(2), \cdots, x_0(m)\} \tag{2}$$

2.5 Calculation of correlation coefficient

Correlation coefficient is the difference of k-th element between comparative data columns and the optimal reference data column. The comparative measure of factors is calculated. The gray coefficient is the similarity of k-th element between the i-th options and reference data column, and it is represented by $\xi_{0i}(k)$.

$$\xi_{0i}(k) = \frac{\min_i \min_k |x_0(k) - x_i(k)| + \eta \max_i \max_k |x_0(k) - x_i(k)|}{|x_0(k) - x_i(k)| + \eta \max_i \max_k |x_0(k) - x_i(k)|} \tag{3}$$

where η is the resolution coefficient, generally taken as 0.5, the purpose is to reduce the maximum difference between the two grades, and avoid the maximum difference too large to affect the distortion of correlation coefficient.

2.6 Calculation of correlation degree

Owing to the large number of correlation coefficient and different relative importance of different factors in the same program, the comparison of various programs is very difficult, then you need to arrange the relative importance of each index factor, and propose the weight W_k of correlation coefficient to each factor, for example, a project schedule is tight, and hence, we need to give a higher weight to the factor of construction period. The correlation degree is calculated as follows:

$$R = \frac{1}{m} W \circ \xi \tag{4}$$

$$r_{0i} = \frac{1}{m} \sum_{k=1}^{m} W_k \circ \xi_{0i}(k) \tag{5}$$

3 CASE STUDY

3.1 Engineering situation

Take a highway project as an example, which is located in east coast of Tangshan City, Hebei Province. There are silt and muddy clay in some sections of the highway, and the section showing the flow of plastic and soft plastic state. In order to meet the design requirements, the paragraph foundation needs to be processed. Only if they meet the requirements, they can construct the next step. The project construction unit requires the soft foundation treatment program has the characteristics of short construction period, small influence on the environment, and obvious treatment effect. Depending on the construction experience, choose plastic drainage board, vibro gravel pile, powder mixing piles, and two ash pile four kinds of soft ground treatment program. In the program selection process, we must focus on the construction unit requirements, combined with

Table 1. The hierarchy model of evaluation index system for soft foundation treatment scheme.

Target layer	First-level index layer	Second-level index layer	Scheme layer
Total target A is the best solution	B_1 Economy	C_1 Cost, C_2 Maintenance	D_1 Vibro gravel pile D_2 Plastic drainage board D_3 Two ash pile D_4 Powder mixing piles
	B_2 Technical	C_3 Technical reliability, C_4 Treatment effect, C_5 Difficulty, C_6 Period	
	B_3 Environmental	C_7 Noise, C_8 Resource consumption, C_9 Environmental impact	

the actual situation of foundation, using the gray theory model to make evaluation and analyze the evaluation result.

3.2 *Determination of the hierarchical structure model*

According to the characteristics of the project, many factors affect the selection of the soft foundation treatment scheme, such as period, cost, and treatment effect. Hired ten experts, selected nine more important factors as the evaluation index system that the was hierarchical structure, the specific indicators in Table 1. Among them, the cost and time limit were obtained through calculation and other indicators were obtained by the expert scoring method.

3.3 *Assigning the weight of first-level and second-level index layer*

The AHP method is used to construct the judgment matrix, and the eigenvector W of each judgment matrix is the weight which is the relative importance of a certain layer factor relative to the factor of upper layer. The weight of each judgment matrix and the results of consistency test are given in Table 2.

3.4 *Determination of the index value of index layer*

3.4.1 *Quantitative indicator*

According to the preliminary design and construction design, the index of C_1 cost can be budgeted. The costs of Vibro gravel pile, plastic drainage board, two ash pile, and powder mixing piles were CNY 100 million and 85 million, 70 million, and 80 million. The index of period C_6 can be calculated by the design data, and the period of Vibro gravel pile was 145 d, plastic drainage board was 450 d, two ash pile was 150 d, and powder mixing piles was140 d.

3.4.2 *Qualitative indicator*

The qualitative indexes include C_2 maintenance, C_3 Technical reliability, C_4 treatment effect, C_5 difficulty, C_7 noise, C_8 resource consumption, and C_9 environmental impact, using the expert scoring method to

determine the qualitative indexes of each scheme. The index is divided into first, second, third, fourth, and fifth, a total of five grades; scoring criteria were 1~0.8, 0.8~0.6, 0.6~0.4, 0.4~0.2, and 0.2~0. The smaller the degrees of maintenance, noise, resource consumption, and environmental impact are, the more the front rank and the higher the score will be. The more reliable the technology is, the better the treatment will be, the easier the construction will be, the more the front rank will be and the higher the score will be. Conversely, the lower the score will be.

Employing 10 experts and technicians, according to the scoring standards of membership score, scored the four programs of qualitative indicators by 10 experts and technicians employed. Taking the average of 10 experts' score as the value of each index, the value of qualitative indicators is given in Table 3.

3.5 *Determination of comparative and reference data columns*

According to the theory given in Section 2.3, the qualitative indicators were processed by interval value transformation, and the quantitative indicators were processed by the percentage transformation; the

Table 3. Appraisal result of qualitative indexes.

Index	Average value				Maximum score	Minimum score
	D_1	D_2	D_3	D_4		
C_2	0.55	0.50	0.54	0.60	0.7	0.4
C_3	0.76	0.70	0.46	0.81	0.9	0.3
C_4	0.76	0.68	0.56	0.75	0.9	0.4
C_5	0.54	0.57	0.6	0.54	0.7	0.3
C_7	0.56	0.78	0.53	0.6	0.9	0.4
C_8	0.66	0.6	0.64	0.68	0.8	0.5
C_9	0.67	0.76	0.67	0.8	0.9	0.5

Table 4. The whitening value of technical and economic indexes that are processed by standard and reference data columns.

Index	Comparative data columns				Reference data column
	D_1	D_2	D_3	D_4	
C_1	1	0.85	0.7	0.8	0.7
C_2	0.5	0.67	0.53	0.33	0.33
C_3	0.77	0.67	0.27	0.85	0.85
C_4	0.72	0.56	0.32	0.7	0.72
C_5	0.4	0.33	0.25	0.4	0.25
C_6	0.32	1	0.33	0.31	0.31
C_7	0.68	0.24	0.74	0.6	0.24
C_8	0.47	0.67	0.53	0.4	0.4
C_9	0.58	0.35	0.58	0.25	0.25

Table 2. The weight of judgment matrix and the test result of consistency.

Judgment matrix	Lower factor weight W				consistency test
A	0.1358	0.2044	0.6598	—	Pass-test
B_1	0.8333	0.1667	—		Pass-test
B_2	0.0810	0.2969	0.0728	0.5493	Pass-test
B_3	0.2571	0.1275	0.6154	—	Pass-test

Table 5. The results of correlation coefficient and correlation degree.

k	1	2	3	4	5	6	7	8	9	r_{0i}
W_k	0.113	0.023	0.017	0.061	0.015	0.112	0.17	0.084	0.406	—
$i = 1$	0.535	0.67	0.812	1	0.697	0.972	0.439	0.831	0.511	0.069
$i = 2$	0.697	0.504	0.657	0.683	0.812	0.333	1	0.561	0.775	0.081
$i = 3$	1	0.633	0.373	0.463	1	0.945	0.408	0.726	0.511	0.069
$i = 4$	0.775	1	1	0.945	0.697	1	0.489	1	1	0.098

value of indicators were converted into the interval [0,1]. The whitening value of technical and economic indexes that are processed by standard and reference data column are listed in Table 4.

3.6 Calculation of the correlation degree

In Table 2, the index weight was relative to the upper layer index, the weights of second-level index need to multiply the weight that is first-level relative to the target layer and are given in Table 5. According to the theory given in sections 2.5 and 2.6, the correlation coefficient and correlation degree were calculated and the results are given in Table 5. Table 5 shows that D_4 powder mixing piles were the optimal solution. The second was D_2 plastic drainage board, and the final was the D_1 Vibro gravel pile and D_3 two ash pile.

4 CONCLUSION

The Analytic Hierarchy Process (AHP) has the characteristics of the systematic, practicality, and simplicity, and it can accurately obtain the weight coefficient of each evaluation index in the case of less quantitative data. The evaluation of the multi—hierarchy structure not only reflects the different levels of the evaluation factors, but also avoids the disadvantages of the weight distribution because of too many factors. Research shows that the optimal result of the gray comprehensive evaluation method conforms to the real project, and the gray comprehensive evaluation method is objective and feasible to select the scheme of soft ground treatment, and the principles of the model is easy to understand. The calculation is simple, which shows the value of popularization and application.

ACKNOWLEDGMENTS

The authors appreciate the financial support of the Langfang Science and Technology Program (Grant No. 2015013008) and the Science Foundation of North China Institute of Aerospace Engineering (Grant No. KY-2015-01).

REFERENCES

Chang Nibin, Wen C. G., Chen Y. L. et al. (1996). A grey fuzzy multiobjective programming approach for the optimal planning of a reservoir watershed. Part A: Theoretical development. *Water Research* 30(10): 2329–2334.

Kang Hong & Peng Zhenbin. 2009, Multi-level synthetic fuzzy evaluation of treatment method of soft foundation. *Journal of Natural Science of Hunan Normal University* 32(3): 120–124.

Liang Lidan, Zheng Da, Ju Nengpan, et al. (2014). Seismic safety evaluation of gravity retaining wall using grey correlation and fuzzy mathematics. *Journal of Engineering Geology.* 22(6): 1234–1240.

Pal S.K. & Shiu S.C.K. (2004). *Foundations of soft case-based reasoning.* Hoboken: John Wiley & Sons.

Pan Ruichun, Huang Ruizhang, Zhou Xinnian, et al. 2012. The research of the treatment scheme selection for the soft soil foundation of road engineering. *Highway traffic science and technology* 8(10): 23–26.

Peng Peng. (2016). Study of dam foundation curtain grouting scheme evaluation based on analytic hierarchy process. *Water Resources Planning and Design.* 29(3): 47–49.

Shi Huatang, Tang Ming & Liao Ronbin. (2012). The comparison and selection of construction organization scheme based on the grey AHP analysis. *Highway Engineering.* 37(6): 161–165.

Tseng M.L. (2009). A causal and effect decision making model of service quality expectation using grey-fuzzy dematel approach. *Expert systems with applications* 36(4): 7738–7748.

Advances in Materials Science, Energy Technology and Environmental Engineering – Patty & Zhou (Eds)
© 2017 Taylor & Francis Group, London, ISBN 978-1-138-19668-1

Research on biotrickling filters for wastewater and odor treatment

Jinghui Cui & Yangyang Zhang

School of Water Resources and Environment, China University of Geosciences, Beijing, China

ABSTRACT: The simultaneous treatment of sewage and odor was assessed in a Biotrickling Filter (BTF) packing biological ceramsite and volcanic rocks. Performance evaluation data of BTF were generated under different experimental conditions. The BTF was operated in a continuous mode at a liquid-to-gas ratio varying from 1:5 to 1:10 with a hydraulic loading of 2 $m^3/(m^2 \cdot d)$. The BTF had a better COD_{Cr} removal efficiency at a liquid-to-gas ratio of 1:5 in comparison with liquid-to-gas ratio 1:10. Steady-state H_2S removal efficiencies of 86.1% concomitant with wastewater COD_{Cr} removals of 76% were achieved at a hydraulic loading of 2 $m^3/(m^2 \cdot d)$ and at a liquid-to-gas ratio of 1:5. As can be concluded from the results, this study demonstrated good performance in BTF for the removal of wastewater and H_2S.

1 INTRODUCTION

Many studies have been carried out on the performances of the BTF in treating wastewater or odor for its advantages of high efficiency, low sludge production, environmentally friendly nature, and cost effectiveness (Kim S&Deshusses MA 2003, Baltzis B et al. 2001, Kim S& Deshusses MA 2008). Therefore, a suggestion of using BTF to simultaneously treat wastewater and odor was offered. In a BTF, a free aqueous phase was trickled over the packing; the chemicals in the gas were transported via convection along the vertical direction of the media bed inside the BTF. The pollutant undergoing treatment would be transferred first to the trickling liquid in a BTF and then wastewater and odor diffuse into the biofilm phase to be degraded by catabolism and anabolism (Lebrero R2014).

It could be understood that wastewater as a nutrient solution was supplied to the BTF to support growth of biofilms and metabolism of contaminants (Cox & Deshusses 2002, Mpanias & Baltzis 1998). Biofilms formed in BTF could be applied for treatment of many wastes simultaneously because multiple microbial species, with different functions, could coexist within biofilms (Wanner & Gujer 1984).

The aim of this work was to evaluate the feasibility of the simultaneous H_2S abatement and wastewater treatment in a BTF. The influence of different liquid-to-gas ratios on the removal of wastewater and H_2S treatment performance was investigated.

2 MATERIALS AND METHODS

2.1 Experimental apparatus and operating conditions

BTF was made of polypropylene and had an inner diameter of 200 mm and a height of 1200 mm. In the BTF packing biological ceramsite and volcanic rocks, the top layer was bioceramsite and the bottom layer was volcanic rocks; each layer had a height of 400 mm. H_2S from a company was controlled with an air rotameter and supplied with an air collector under concentration 2.95–6.14 mg/m^3. The inlet gas was fed to the bottom of BTF at different flow rates of 10 m^3/d and 20 m^3/d controlled via the rotameter, while the wastewater was trickled over the packing to provide optimum conditions to biofilms at a hydraulic loading of 2 $m^3/(m^2 \cdot d)$. The inlet and outlet COD_{Cr} and H_2S concentrations were determined periodically. The wastewater was continuously trickling through the media in the operation of BTF.

2.2 Packing material

The main characteristics of the packing materials used in this study are given in Table 1.

2.3 Wastewater characteristics

The wastewater was obtained from CUGB sewage outlet.

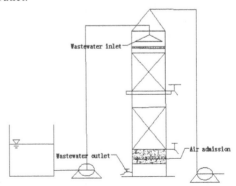

Figure 1. Schematic representation of the experimental set-up.

Table 1. Main characteristics of the packing materials used in this study.

Material	Volcanic rocks	Bioceramsite
Specific surface area (m²/m³)	22.34	15.62
Particle density (kg/m³)	0.63	0.74
Total porosity (%)	58	48
Water holding capacity (%)	0.9–1.2	1.1–1.4
Dimension (mm)	3–4	4–6

Table 2. Chemical characteristics of the wastewater.

Constituent	Mean	Maximum	Minimum
COD_{Cr} (mg/L)	302.5	450	255.4
pH	7.3	8.4	6.6

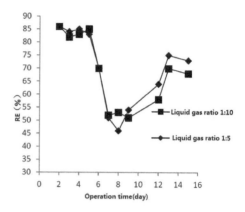

Figure 2. Removal rate of COD_{Cr} under different liquid-to-gas ratio conditions.

2.4 Analytical methods

The inlet and outlet H_2S concentrations were determined using a portable H_2S detector. Its lowest detection limit is 1 ppbv. Water samples from BTF were taken periodically for COD_{Cr} determination. The sulfate concentrations in the water samples were measured by ion chromatography (IC1010, China).

3 RESULTS AND DISCUSSION

3.1 Effect of different liquid-to-gas ratios on sewage treatment

The removal of COD_{Cr} under different liquid-to-gas ratio conditions was presented in Figure 2. In a BTF, COD_{Cr} removal relied mainly on aerobic heterotrophic microorganisms on the surface of the packing. When gas-to-liquid ratio increases, the amount of oxygen increases proportionally with the increase in volume, which will be more conducive to the growth of aerobic microorganisms, and thus can improve the removal rate of COD_{Cr}. However, during the initial period, the BTF achieved a Removal Efficiency (RE) of 86% and a decrease in the RE was observed. From day 5 onwards, the RE steadily decreased to 50%, and the removal efficiency of COD_{Cr} tended to increase and finally achieved a removal efficiency of 76% at a liquid-to-gas ratio 1:5, which had a slight advantage for the removal of H_2S at a liquid-to-gas ratio 1:10.

Specific analysis is as follows: The concentrations of SO4²⁻ increased steadily and came into the acidic environment as the gas-to-liquid ratio increased. A number of falling-off biofilms were observed at the bottom, which led to the decline of the removal rate of COD_{Cr}. As biofilms adapted to the changing acidic environment, the growth of biofilms promoted the removal efficiency of COD_{Cr}.

3.2 Effects of different liquid-to-gas ratios on the removal of H_2S

The process of H_2S could be explained by the double membrane theory, and the mechanism of H_2S transformation was visualized as follows:

Elemental sulfur was the main product:

$$2H_2S + O_2 \rightarrow 2H_2O + 2S + X \tag{1}$$

$$2H_2S + O_2 \rightarrow 2H_2O + 2S + X \tag{2}$$

$$S + 3O_3 + 2H_2O \rightarrow 12H_2SO_4 + X \tag{3}$$

Elemental sulfur was the main product:

$$7H_2S + O_2 + CO_2 \rightarrow SO_4^{2-} + H_2O \tag{4}$$

During the initial period, The BTF achieved a Removal Efficiency (RE) of 95% and a decrease in the RE was observed. From day 9 onwards, the RE steadily decreased to 85.3%, and removal rates of H_2S tended to be stable with the increase in operation time and maintained separately at 86.1% and 85%. BTF had a slight advantage for the removal of H_2S at a low liquid-to-gas ratio.

Specific analysis is as follows: the removal of H_2S mainly relied on adsorption before the 37th day, and subsequently, biodegradation played a major role. The concentrations of SO4²⁻ increased steadily and came into the acidic environment as the gas-to-liquid ratio increased. Due to these damaged microorganisms, the biofilm became detached easily. High

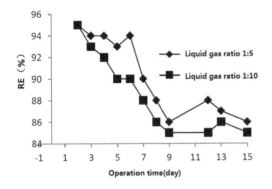

Figure 3. Removal rate of H₂S under different liquid-to-gas ratio conditions.

gas-to-liquid ratio was not favorable to the gas-liquid mass transfer and the liquid-solid mass transfer (Cho K-S, Ryu HW&Lee NY2000, Roshani B, Torkian A, Aslani H&Dehghanzadeh R 2012). In a similar study, Vivian Bermudez (2003) came to the conclusion that, under low flow rate, the mass transfer and the removal rate of gas pollutants would improve in the gas-liquid-solid three phase.

4 CONCLUSION

Low gas-to-liquid ratio would improve the mass transfer and the removal rate of COD_{Cr} and H_2S.

1. In the BTF, COD_{Cr} removal relied mainly on aerobic heterotrophic microorganisms on the surface of the packing. From day 5 onwards, the RE steadily decreased to 50%, and the removal efficiency of COD_{Cr} tended to increase and finally achieved a removal efficiency of 76% at a liquid-to-gas ratio 1:5, which had a slight advantage for the removal of H_2S at a liquid-to-gas ratio 1:10. While during the initial period, The BTF achieved Removal Efficiency (RE) of 95%(H_2S). From day 9 on, the RE steadily decreased to 85.3%, removal rates of H_2S tended to be stable with the increase of operation time and separately maintained at 86.1% and 85%. BTF had a slight advantage for the removal of H_2S at a low liquid-to-gas ratio.
2. The concentrations of SO_4^{2-} increased steadily and came into the acidic environment as the gas-to-liquid ratio increased. A number of falling-off biofilms, which lead to the decline of the removal rate of COD_{Cr} and H_2S, were observed. When the biofilms adapted to the changing acidic environment, their growth promoted the removal efficiency of COD_{Cr} and H_2S.
3. The BTF had a better COD_{Cr} removal efficiency at liquid-to-gas ratio 1:5 in comparison

with a liquid-to-gas ratio of 1:10. Steady-state H_2S removal efficiencies of 86.1% concomitant with wastewater COD_{Cr} removals of 76% were achieved at a hydraulic loading of 2 m³/(m²·d) and at a liquid-to-gas ratio of 1:5. This study demonstrates good performance in the BTF for simultaneous treatment of wastewater and odor, and the simultaneous treatment of sewage and odor is feasible in a biotrickling filter.

ACKNOWLEDGMENTS

This research was financially supported by the research and demonstration of ecological low-energy-consuming treatment technology and key equipment for processing rural domestic pollution from water resource region, and Hubei Hanjiang watershed water security key technology research and demonstration.

REFERENCES

Baltzis B, Mpanias C, Bhattacharya S.2001. Modeling the removal of VOC mixtures in biotrickling filters. Biotechnology and bioengineering. 72(4):389–401.

Bermudez V.2003. Biofiltration for control of H2S from Wastewater treatment plant gases.

Cho K, Ryu H, Lee N. 2000. Biological deodorization of hydrogen sulfide using porous lava as a carrier of Thiobacillus thiooxidans. Journal of Bioscience and Bioengineering. 90(1):25–31.

Cox H, Deshusses M. 2002. Co-treatment of H₂S and toluene in a biotrickling filter. Chemical Engineering Journal. 87(1):101–105.

Kim S, Deshusses M. 2003. Development and experimental validation of a conceptual model for biotrickling filtration of H2S. Environmental progress. 22(2):119–280.

Kim S, Deshusses M. 2008. Determination of mass transfer coefficients for packing materials used in biofilters and biotrickling filters for air pollution control. 1. Experimental results. Chemical Engineering Science. 63(4):841–850.

Lebrero R, Gondim A, Pérez R, et al. 2014. Comparative assessment of a biofilter, a biotrickling filter and a hollow fiber membrane bioreactor for odor treatment in wastewater treatment plants. Water research. 49:339–346.

Mpanias C, Baltzis B. 1998. An experimental and modeling study on the removal of mono-chlorobenzene vapor in biotricklingfilters. Biotechnology and bioengineering. 1998,59(3):328–334.

Roshani B, Torkian A, Aslani H. 2012. Dehghanzadeh R. Bed mixing and leachate recycling strategies to overcome pressure drop buildup in the biofiltration of hydrogen sulfide. Bioresource technology. 109:26–30.

Wanner O, Gujer W. 1984. Competition in biofilms. Wat Sci Tech. 17(2–3):27–44.

*Advances in Materials Science, Energy Technology
and Environmental Engineering – Patty & Zhou (Eds)*
© *2017 Taylor & Francis Group, London, ISBN 978-1-138-19668-1*

Diffraction imaging based on the plane-wave record

Xue Kong
College of Petroleum Engineering, in Shengli College China University of Petroleum, Dongying, China

Yue Teng, Xuan Zhang & Yunyan Shi
BRI, Bureau of Geophysical Prospecting INC., CNPC, Zhuozhou, China

ABSTRACT: The main objectives are structure elucidation, lithologic interpretation, and carbonate high precision imaging, such as faults, salts, karst caves, and fractures, which usually act as complicate diffractions on the seismic record. However, traditional seismic data processing and imaging always suppresses diffractions as interfering noise. Although migration can converge the diffractive waves, it might be hidden in the reflection energy, which is a disadvantage to the subsequent interpretation because of the weak amplitude and quick energy attenuation of the diffractions. According to the difference between reflection and diffraction on the plane-wave record, we use plane-wave destruction (PWD) filtering to obtain pseudo-hyperbolic diffraction, whose curvature is large by suppressing laterally continuous pseudo-linear reflection, which aims to enhance the imaging resolution of heterogeneous diffracting objectives. The 2D SEG/EAGE salt model test demonstrates that this method can effectively improve the imaging quality of heterogeneous structures, and it can enhance the interpretive precision together with conventional interpretation.

1 INTRODUCTION

Diffraction energy contains many heterogeneous structures, such as faults, salts, karst caves, and fractures, which are appropriate for hydro-carbon migration and accumulation. Therefore, diffracting heterogeneous structures become an important target in seismic interpretation. However, traditional data processing includes several reasons, which are unfavorable for diffraction imaging, as follows: a. during processing (such as NMO/DMO, velocity analysis, et al.), events not originating from a reflector are usually filtered out as interfering noise; b. conventional migration algorithms are designed to image specular reflection, even the classical diffraction stack tends to image reflections and ignore scatterings; and c. diffraction amplitudes are much weaker than reflections, and hence, although migration can converge diffractions, their energy may be usually hided by the energy of reflections.

According to the time lag and waveform difference between reflection and diffraction on prestack profile (such as common offset gather, common diffraction point section, common shot gather, and plane-wave sections), we make use of many ways including phase correction, correlation, dip filter, reflection focusing, weighted Radon transform, plane-wave destruction filters, and other methods to obtain image diffraction energy, in order to improve the imaging resolution of diffracting targets.

In this paper, we use plane-wave decomposition to generate plane-wave sections, and then realize diffraction extraction on plane-wave records based on PWD filtering technique, which can easily separate discontinuous events with heavy curvature from laterally continuous events. At last, we compare the imaging results before and after wave field separation to test this method.

2 DIFFRACTION GEOMETRY CHARACTERISTIC ANALYSIS

2.1 *Diffraction and reflection time-distance relation with a point source*

Assume that there is a dipping reflector with a hidden heterogeneous diffractor D in the subsurface, as depicted in Figure 1(a). Point O is the source location, and therefore, the reflection's time-distance (T-D) relation can be described as follows:

$$\frac{t^2}{(4h^2 - 4h^2 \sin^2 \varphi)\Big/v^2} - \frac{(x - 2h\sin\varphi)^2}{4h^2 - 4h^2 \sin^2 \varphi} = 1 \qquad (1)$$

and the T-D curve of the diffractor is:

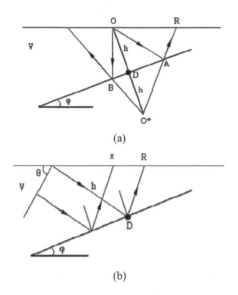

(a)

(b)

Figure 1. Ray path schemes of dipping interface and point diffractor (D). (a) Point source and (b) plane-wave source.

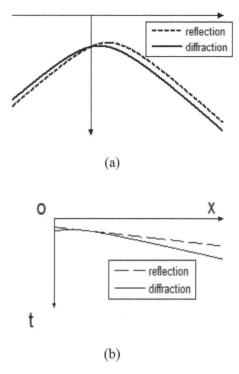

(a)

(b)

Figure 2. T-D relations of reflection and diffraction. (a) Point source and (b) plane-wave source.

$$\frac{(t-\frac{h}{v})^2}{h^2\cos^2\varphi/v^2}-\frac{(x-h\sin\varphi)^2}{h^2\cos^2\varphi}=1 \qquad (2)$$

Contrasting formulas (1) and (2), reflection and diffraction from the point source both appear on the seismic record in the form of hyperbolas (Figure 2(a)). They just have different apex locations but their curvature distinction is small; therefore, it is little difficult to pick up the diffracted energy on point source shot gathers.

2.2 Plane-wave destruction filters

Plane-wave destruction filters (Claerbout, 1992; Fomel, 2002) characterize seismic data by superposition of local plane waves. These filters can be thought of as a time-distance (T-X) analog of frequency-distance (F-X) prediction-error filters. They are constructed as finite-difference stencils for the local plane-wave differential equation.

With less adjustable parameters, the only quantity which PWD filters need to estimate has a clear physical meaning of the local plane-wave slope. These filters can estimate the local slope of smooth and continuous event perfectly and have insufficient estimation of the local slope of events whose continuity is bad and curvature is large. Therefore, using this character, diffraction waves whose curvature is large can be picked up from the synthetic plane-wave record.

Following the physical model of local plane waves, we can define the mathematical basis of the PWD filters as the local plane differential equation:

$$\frac{\partial P}{\partial x}+\sigma\frac{\partial P}{\partial t}=0 \qquad (3)$$

where P(t, x) is the wave field and σ is the local slope.

Considering time and space varying slopes, the solution of equation (3) is

$$\widetilde{P}_{x+1}(Z_t)=\widetilde{P}_x(Z_t)\frac{B(Z_t)}{B(1/Z_t)} \qquad (4)$$

where the ratio $B(Z_t)/B(1/Z_t)$ is an all-pass digital filter approximating the time-shift operator $e^{i\omega\sigma}$. The coefficients of filter $B(Z_t)$ can be determined by fitting the filter frequency response at low frequencies to the response of the phase-shift operator. Expression (5) is a three-point centered filter of $B(Z_t)$ yielded by the Taylor series technique.

$$B_3=\frac{(1-\sigma)(2-\sigma)}{12}Z_t^{-1}+\frac{(2+\sigma)(2-\sigma)}{6}$$
$$+\frac{(1-\sigma)(2+\sigma)}{12}Z_t \qquad (5)$$

Taking the variation in both t and x into consideration, the plane-wave information is predicted with the 2-D prediction filter (expression (6)).

$$A(Z_t, Z_x) = 1 - Z_x \frac{B(Z_t)}{B(1/Z_t)} \qquad (6)$$

To avoid the need for polynomial division, a modified version of the filter as shown in expression (6) is used by Fomel (2002):

$$C(Z_t, Z_x) = A(Z_t, Z_x) B(\frac{1}{Z_t}) = B(\frac{1}{Z_t}) - Z_x B(Z_t) \qquad (7)$$

By expression (5), we can notice that $C(Z_t, Z_x)$ is the function of local slope σ and can be denoted by $C(\sigma)$. Therefore, estimating the slope is a necessary step for applying the finite-difference plane-wave filters, and to determine the slope, we can define the least-squares goal:

$$C(\sigma)d \approx 0 \qquad (8)$$

3 NUMERICAL EXAMPLES

We choose the 2D SEG/EAGE salt model (Figure 3) to test the performance of this diffracting objecting imaging method. From the plane-wave record (Figure 4(a)), we can notice that the reflected energy produced by a gentle layer has good continuity on plane-wave record and the events are smooth. By contrast, the diffraction energy from heterogeneous structures, such as fault, small size breakpoint, and underpart of salt dome, is so weak that it is nearly covered by the reflected information.

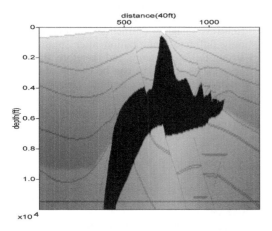

Figure 3. Velocity field of the 2D SEG/EAGE salt model.

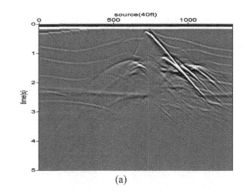

(a)

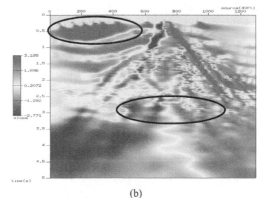

(b)

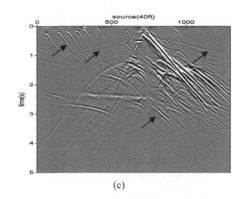

(c)

Figure 4. Procedure of wave field separation: (a) synthetic plane-wave record of the total wave field (p = 0); (b) local dip angle section; and (c) diffraction wave section after wave field separation.

We estimate the local slope of plane-wave record by PWD filters, as shown in Figure 4(b). We know that the PWD filters have better accuracy for local slope estimation of the smooth and continuous events, which correspond to the reflecting energy produced by gentle layer. Oversized local dip angle represents: a. the real steep events or b. insufficient dip angle estimation because of the large curvature and bad continuity of the event, as shown in t figure 4(b) by circle. Therefore, in the area where the diffracted wave is abundant and the event's continuity is bad, the error of local dip angle is big and residuals are large. This diffracted energy that contains the information of salt dome boundary, faults, and breakpoints is just what we need. Figure 4(c) shows the residuals after PWD filtering, and the weak diffraction energy produced by superficial small size breakpoint can be observed clearly in wave field after wave field separation in comparison with Figure 4(a). Large-dip-angle energy caused by high-steep salt body and discontinuous events generated by strong varying velocity in deep layers are kept because of insufficient dip angle estimation. Meanwhile, the direct wave is suppressed satisfactorily after wave field separation because its event is horizontal and linear in the plane-wave record.

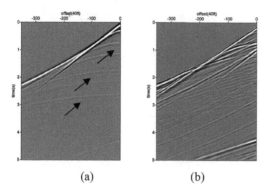

(a)   (b)

Figure 5. (a) Original point source shot record and (b) diffraction wave field after wave field separation.

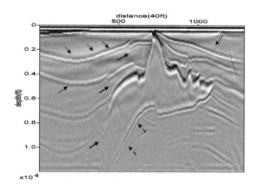

(a)

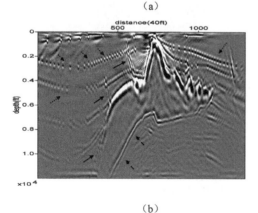

(b)

Figure 6. The results of prestack depth migration: (a) total wave field and (b) diffraction wave field after separation results.

4 CONCLUSION

Through wave field separation and imaging analysis of the 2D SEG/EAGE model, we recognize the following conclusions:

1. On point source and plane-wave records, diffracted energy is usually hidden by reflections because it is much weaker than reflected energy.
2. PWD filters can only naturally estimate smooth and continuous reflected event's local slop, but it has estimated error for diffracted events which contain heterogeneous structures, such as faults, small-scale breakpoints, and steep salt flanks. Residuals after filtering can be judged as diffracted wave field.
3. Full-wave field imaging is effective for reflected waves from large-scale structure imaging, but faults, breakpoints, and other diffracted information may be usually masked by large structures after migration because of the weak diffracted energy characteristics, hence these diffracted targets are difficult to identify on full-wave field imaging results. In contrast to full-wave imaging, diffraction imaging migrate only diffracted information after suppressing strong reflected energy, and therefore, the heterogeneous structures can be clearly observed on diffraction imaging sections. Therefore, diffraction imaging enhances imaging precision of diffracted targets, which could help interpreters to trace reservoir better.

REFERENCES

Bansal R. & M. G. Imhof. Diffraction enhancement in prestack seismic data. Geophysics, 2005, 70(3): V73–V79.
Berkovitch A., I. Belfer & Y. Hassin. Diffraction imaging by multifocusing. Geophysics, 2009, 74(6): WCA75-WCA81.

Fomel S.. Applications of plane-wave destruction filters: Geophysics, 2002, 67(6): 1946–1960.

Fomel S., E. Landa & M.T. Taner. Poststack velocity analysis by separation and imaging of seismic diffractions. Geophysics, 2007, 72(6): U89-U94.

Khaidukov V., E. Landa & T.J. Moser. Diffraction imaging by focusing-defocusing: An outlook on seismic super resolution. Geophysics, 2004, 69(6): 1478–1490.

Landa E., V. Shtivelman & B. Gelchinsky. A method for detection of diffracted waves on common-offset sections. Geophysical Prospecting, 1987, 35: 359–373

Moser T. & B.C. Howard. Diffraction imaging in depth: Geophysical Prospecting, 2008, 56: 627–641.

Nowak E.J. & M.G. Imhof. Diffractor localization via weighted Radon transforms. SEG Expanded Abstracts, 2004, 2108–2111.

Sa Li-ming, Yao Feng-chang & Di Bang-rang, et al. The theory and methods of fractured-vuggy reservoir seismic wave field identification. Beijing, Petroleum Industry Press, 2009.

Taner M.T., S. Fomel & E. Landa. Separation and imaging of seismic diffractions using plane-wave decomposition. SEG Expanded Abstracts, 2006, 2401–2405.

Advances in Materials Science, Energy Technology and Environmental Engineering – Patty & Zhou (Eds)
© 2017 Taylor & Francis Group, London, ISBN 978-1-138-19668-1

Screening of fosfomycin-degrading bacteria and its practical application in the CASS

Shiyue Liu, Lifang Zhao & Wencong Zhao
China University of Geosciences, Beijing, China

ABSTRACT: Pharmaceutical wastewater is a major threat to human health. Owing to the difficulty in antibiotics pharmaceutical wastewater treatment, the use of different methods of antibiotic wastewater has become extremely valuable. This study was directed to the difficult characteristics of antibiotic fosfomycin pharmaceutical wastewater treatment. The physical treatment, chemical treatment, and biological treatment are compared. Finally, the selection method came out with CASS-enhanced biological treatment of pharmaceutical wastewater from a pharmaceutical factory.

1 INTRODUCTION

Fosfomycin wastewater is an antibiotic-type chemical wastewater. During the time, split propyl phosphonate will be obtained with fosfomycin. Owing to enrichment of amin, phosphoric acid substances, fosfomycin pharmaceutical wastewater COD rises from tens of thousands to hundreds of thousands mg/L, Total Phosphorus (TP) up to ten thousand mg/L, and the high concentration of organic matter in wastewater is extremely difficult to be degradable with microbial toxicity. Owing to the big investment and high consumption of antibiotics pharmaceutical wastewater management, in most cases, antibiotics pharmaceutical wastewater is not fully the case of timely and effective treatment, which is directly discharged to the municipal pipe network or near the river. This water, mixed with many antibiotics pharmaceutical wastewater pollutants, does a lot of harm to water environment, water ecology, and human health. Therefore, the use of different methods of antibiotic wastewater has become extremely urgent with significance and practical value.

2 MATERIALS AND METHODS

2.1 Raw wastewater

This study was directed to the difficult characteristics of antibiotic fosfomycin pharmaceutical wastewater treatment, compared with the physical treatment, chemical treatment, and biological treatment. Finally, the selection method with CASS-enhanced biological treatment of pharmaceutical wastewater from a pharmaceutical factory comes out.

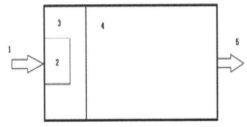

(1, influent; 2, biological reaction zone; 3, facultative anaerobic zone; 4, primary reaction zone; 5, effluent)

Figure 1. The process consisting figure of CASS reactor.

The CASS reactor mainly comprises biological selection area, oxygen reaction zone, and the main zone of three parts (Figure 1).

2.2 Experimental methods and equipment

Total DNA was extracted using Power Soil DNA Kit (Mo Bio Laboratories, Carlsbad, CA). Strains were identified and sequenced by Shanghai Biological Company.

Three bacterial strains, which could effectively degrade fosfomycin sodium, have been screened and isolated from activated sludge of a pharmaceutical factory SBR (sequencing batch reactor) in a pharmaceutical of China. The strains were identified by 16SrDNA. High-throughput data analysis is another method of analyzing the characteristics of two strains.

3 RESULTS AND DISCUSSION

The amplified PCR image of two strains has been obtained (Figure 2). The strains were identified by

257

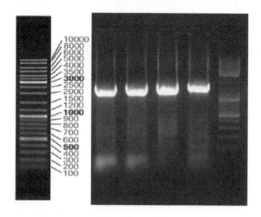

Figure 2.   PCR image of SY1 and SY2.

Figure 3.   Phylogenetic tree of SY1 and SY2.

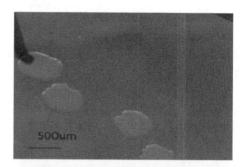

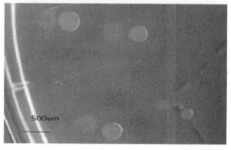

a: SY1 strains, b: SY2 strains

Figure 4.   Colony characteristics of two strains.

16SrDNA: SY1 as *Chryseobacterium* and SY2 as *Proteus* (Figure 3).

Besides, according to the observation of morphology of two strains by microscopy, the phylogenetic tree of two strains has been established (Figure 4).

Gradually increasing the concentration of fosfomycin sodium, the degradation by two strains have been varied intensely. When the concentration of fosfomycin sodium was set to 25ppm, the degradation by SY1 and SY2 were 40.45% and 32.05%. In addition, when the concentration was up to 50ppm, the degradation changed to 40.45% and 33.54%.

While the concentration had been increased to 75ppm, the degradation dropped greatly to 26.61% and 10.25%. Moreover, the growth of SY2 has been badly limited as the growth of SY1 reached its peak by degradation of 21.02%. Appropriate proportion of SY1 and SY2 in the activated sludge to achieve a high-efficiency degradation for the contaminants needs more research and this paper will provide a guidance for the next work.

Different proportions of fosfomycin are used in simulated pharmaceutical wastewater. In the

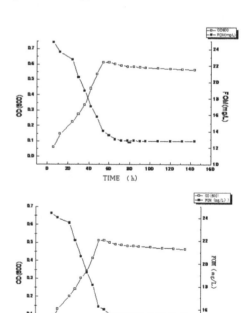

a: SY1 strains, b: SY2 strains

Figure 5.   The growth curves of two degrading bacteria in low concentrations of fosfomycin medium.

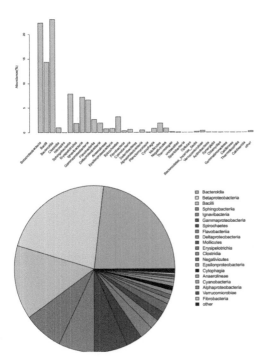

Figure 6.   Analysis of the reactor microflora.

case where COD was 500 ppm, the test substance is under the removal conditions of 0%:100%, 25%:75%, 50%: 50%, 75%:25%, and 100%:0%. It can be found that 50%: 50% of the optimal treatment effect was 53.67% (Figure 5).

Add two degrading bacterial suspension into a low-concentration fosfomycin medium, the growth characteristic curve is shown in figure 4 and figure 5. High-throughput sequencing technology is rapidly improving in quality, speed, and cost. It includes nucleic acid extraction from different environments, sample preparation, and high-throughput sequencing platforms. It also includes analyzing amplicon sequences, metagenome shotgun sequences, and metatranscriptome sequences. Therefore, the samples were sent to Shanghai Biological Company to perform the high-throughput data analysis and the results are shown in Figure 6.

4   CONCLUSIONS

In the conventional activated sludge circulation system (CASS) reactor, degrading bacteria is added for further processing an anaerobic baffled reactor (ABR) after a comprehensive wastewater pretreatment. The pharmaceutical wastewater of COD and ammonia is studied, as well as TP-related effect

characteristic mechanism, while the generic drug sludge CASS reactors were compared.

The results showed that adding degrading bacteria, COD is 232.86mg/L, ammonia was 18.19mg/L, and TP was 4.27mg/L, while for COD, ammonia, TP average removal rates were 90.66%, 91.15%, and 85.51% better than 83.23% of the original activated sludge, 84.16% and 70.31%.

Therefore, the degrading bacteria in a 1:1 ratio was added to the reactor in the actual treatment effect for COD, ammonia, TP, and other indicators having improved significantly. A high-throughput method for the analysis of bacteria in the reactor was found, and flora C1 reactor contents were 24% bacteroidia and 14% *Bacillus subtilis*.

Added degrading bacteria, C2 efficient mixing reactor contents were 38% Sphingobacteria and 12% is bacteroidia visible, having a significant impact on the populations of bacteria by adding efficient reactor after the bacteria, and thereby affecting the processing efficiency.

ACKNOWLEDGMENTS

The authors thanks "The research and demonstration of ecological low-energy-consuming treatment technology and key equipment for processing rural domestic pollution from water resource region" and "Hubei Hanjiang watershed water security key technology research and demonstration (no. 2012zx07205002-02-02)" projects for providing financial support to this study. They also thank Wang for patiently guiding them.

REFERENCES

Arjoon A, Olaniran A, Pillay B. 2013. Enhanced 1,2-dichloroethane degradation in heavy metal contaminated wastewater undergoing biostimulation and bioaugmentation. Chemosphere. 93(9): 1826–1834.

Inyang M, Gao B, Yao Y, et al. 2012. Removal of heavy metals from aqueous solution by biochars derived from anaerobically digested biomass. Bioresource Technology. 50–56.

Kangle Wang, Suiqiang Liu, Qiang Zhang, et al. 2009. Pharmaceutical wastewater treatment by internal micro-electrolysis-coagulation. biological treatment and activated carbon adsorption. Environmental Technology. 30(13): 1469–1474.

Kim S, Carlson K. 2007. Quantification of human and veterinary antibiotics in water and sediment using SPE/LC/MS/MS. Analytical and bioanalytical chemistry. 387(4): 1301–1315.

Lehmann J, Skjemstd J, Sohi S, et al. 2008. Australian climate-carbon cycle feedback reduced by soil black carbon. Nature Geoscience. 1(12): 832–835.

Li B, Zhang T. 2010. Biodegradation and adsorption of antibiotics in the activated sludge process. Environmental science & technology. 44(9): 3468–3473.

Mukherjee A, Zimmerman A, Harris W. 2011. Surface chemistry variations among a series of laboratory-produced biochars. Geoderma. 163 (3–4): 247–255.

Pimmata P, Reungsang A, Plangklang P. 2013. Comparative bioremediation of carbofuran contaminated soil by natural attenuation. bioaugmentation and biostimulation. International Biodeterioration & Biodegradation. 85: 196–204.

Popovic M, Steinort D, Pillai S, et al. 2010. Fosfomycin. European Journal of Clinical Microbiology& Infectious Diseases, 29(2): 127–142.

Shi A, Bian J, Han F, et al. 2010. Preparation and application of microorganism with highly effective water purification. China Brewing. 54–56.

Sho T, Takashi I, Toru H, et al. 2010. Molecular mechanisms of fosfomycin resistance in clinical isolates of Escherichia coli. International Journal of Antimicrobial Agents. 35(4): 333–337.

Wang J, He H, Wang M, et al. 2013. Bioaugmentation of activated sludge with Acinetobacter sp. TW enhances nicotine degradation in a synthetic tobacco wastewater treatment system. Bioresource Technology. 142(0): 445–453.

Advances in Materials Science, Energy Technology and Environmental Engineering – Patty & Zhou (Eds)
© 2017 Taylor & Francis Group, London, ISBN 978-1-138-19668-1

Regulations on low sulfur and revisions of marine fuel oil Standard GB/T 17411-2012

Sai Ye & Wenwen Feng
China Academy of Transportation Science, Beijing, China

ABSTRACT: Shipping industry is facing challenges to reduce emissions from ships. The International Maritime Organization has issued sulfur emission regulations. Low-sulfur fuel, clean fuel, shore power, and sulfur scrubbers are alternative solutions for shipping industry. Three new Domestic Emission Control Areas have been established in China. To lower the levels of ship-generated air pollution and reduce the sulfur content of fuels, Standardization Administration Committee of P.R.C. has launched the revisions of the marine fuel oils Standard GB/T 17411-2012 in 2014. The updated version GB 17411-2015 is a mandatory national standard. It classifies marine fuel oils according to the sulfur content of the fuels. With the use of low-sulfur marine fuel oils and the construction of emission control areas, the shipping sector can find a solution adequate to reduce air pollution from ships.

1 INTRODUCTION

Research on air pollution from marine vessels has intensified significantly since 2005, when revisions to the MARPOL (the International Convention for the Prevention of Pollution from Ships, 1973 as modified by the Protocol of 1978) Annex VI were implemented by the International Maritime Organization (IMO)'s Marine Environment Protection Committee at its 53rd session. Research has concentrated on emission inventories, the environmental impact of air pollution, engine performance, overall sector emissions, abatement, and relevant costs (Johan 2014).

1.1 Emission control areas

MARPOL Annex VI Regulation 14 relates to the sulfur oxides (SOx) and particulate matter. The sulfur content of any fuel oil used on board ships may not exceed 3.5% globally. It is also possible to designate a SECA (Sulphur Emission Control Areas) to protect a Sea Area specifically against sulfur emissions. IMO defines multiple times with different contents. For instance, in the North Sea area, the sulfur content may not exceed 0.1%. By 2020, the sulfur content of fuel must not exceed 0.5% globally. This global reduction to 0.5% will be reviewed in 2018 and probably postponed to 2025 (IMO 2013). Controls on the sulfur content are shown in Table 1.

As of March 2014, the SECAs established to limit SOx and particulate matter emissions are:

Table 1. MARPOL Annex VI—Fuel sulphur limits.

Outside an ECA established to limit SOx emission	Inside an ECA established to limit SOx emission
4.50% m/m prior to 1 January 2012	1.50% m/m prior to 1 July 2010
3.50% m/m on and after 1 January 2012	1.00% m/m on and after 1 July 2010
0.50% m/m on and after 1 January 2020*	0.10% m/m on and after 1 January 2015

1. Baltic Sea area—as defined in Annex I of MARPOL
2. North Sea area (including the English Channel) —as defined in Annex V of MARPOL
3. North American area (entered into force on 1st August 2012)—as defined in Annex VI of MARPOL; and
4. United States Caribbean Sea (entered into force on 1st January 2014)—as defined in Annex VI of MARPOL

1.2 Chinese new domestic emission control areas

Ministry of Transport of the People's Republic of China has published new regulations introducing sulfur emission control requirements within three Domestic Emission Control Areas (DECAs) (Ministry of Transport of the PRC. 2015): Pearl River Delta, Chang Jiang Delta, and Bohai Rim. These DECAs arise as a matter of Chinese domestic law and are not MARPOL Annex VI.

Table 2. Chinese DECAs—sulfur limits.

Date	Control areas	Sulfur content
1 January 2016	Where appropriate, the ports within the DECAs	≤0.50%
1 January 2017	At berth in the core ports of the DECAs (except during the first hour after berthing, and the last hour before departing)	≤0.50%
1 January 2018	At berth in all ports in the DECAs	≤0.50%
1 January 2019	Entering the DECAs	≤0.50%

The SOx emission controls can implement within three DECAs in phases, starting with the core ports, and then covering all waters.

Core ports have been designated within Shenzhen, Guangzhou, Zhuhai, Shanghai, Ningbo-Zhoushan, Suzhou, Nantong, Tianjin, Qinhuangdao, Tangshan, and Huanghua.

According to the regulations, the timeline is as follows:

From 1 January 2016, there is strict enforcement on sulfur oxides, particulate matter, and nitrogen oxides. Where appropriate, the ports within the DECAs may impose higher requirements including requiring vessels to use fuel of not more than 0.5% m/m sulfur content while berthing.

From 1 January 2017, vessels which are berthing at the core ports in the DECAs (excluding the first hour after arrival and the last hour before departure) shall use fuel containing 0.5% sulfur or less.

From 1 January 2018, vessels which are at berth in all ports in the DECAs shall use fuel containing 0.5% sulfur or less.

From 1 January 2019, vessels entering the DECAs shall use fuel containing 0.5% sulfur or less.

Before 31 December 2019, an assessment will be carried out by the authorities to decide the follow-up action.

These maximum sulfur contents are given in Table 2.

2 REGULATIONS ON LOW-SULFUR FUELS IN CHINA

2.1 Social background

Since the adoption of China's 12th Five-Year Plan for Environmental Protection (2011–2015), the Ministry of Environmental Protection of the PRC has issued a series of aggressive measures to tackle air pollution. Under the Dome, the documentary filmed by Chai Jing has caused the society to the misgiving of the ecosystem and environmental protection. Air pollution has become a social issue in China and poses a threat to public health and ecological environment. Reducing emissions in the shipping industry is on the cusp of a major revolution. High sulfur content in marine fuels can lead to serious metal corrosion and wear of storage systems, filter systems, and combustion system. The combustion products SOx have many negative impacts on human and the environment. Ships can meet the new requirements by using low-sulfur fuel oil such as Marine Gas Oil.

2.2 Demand analysis based on environmental protection

Ship types include domestic inland ships, domestic coastal ships, and international seagoing ships. Fuel categories and regional markets are different from others, as in Figure 1. Each category shall agree to apply within its own standard.

To decrease the sulfur content of the fuel used in inland vessels, general diesel fuels (GB 252-2011) shall be revised. To reduce the sulfur content of the fuel used in domestic coastal vessels, GB/T 17411-2012 shall be revised. However, if it is to reduce the level of sulfur content in the fuel oil used in the international seagoing ships, the international ocean shipping has full autonomy in the choice of the filling port. The reduction of sulfur content in the standard GB/T 17411-2012 cannot control international seagoing ships' SOx emission. We need to reduce the sulfur content of fuel oil in the international standard ISO 8217:2012. According to the problems above, sulfur content plays a key role in the prevention and control of air pollution on China's domestic coastal shipping.

2.3 Contribution in emissions inventory

Comparing the total emission inventory in Shanghai, the researchers of Shanghai Environmental Monitoring Center (Fu et al. 2012) have found

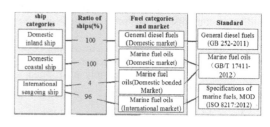

Figure 1. Marine fuel supply market and product standards.

that the share of the shipment contribution of the metropolis is equal to 12.0%, 9.0%, and 5.3% of the three key pollutants, including SO_2, NOx, and $PM_{2.5}$. Of all the kinds of shipment, the ocean-going vessels are responsible for the total emission inventory with the sulfur shares of 12.0%. Shanghai has remained the world's busiest container port for many years. Therefore, sulfur dioxide emissions from ships in Shanghai are larger than that in other port cities. Ocean going ship has been proven to be the primary source of air emissions from ships. The supply of marine fuel oil from the domestic market contributes more to SOx emission.

2.4 Approaching the international standard

Outside the emission control areas, the current limit for sulfur content of fuel oil is 3.50%. It may fall to 0.50% after 1 January 2020. The use of low—sulfur fuel oil is the main trend in the shipping industry in the world. Based on the results of market and industry research, the current shipping market can accept that the level of sulfur content is about 0.5%.

3 COMPARISON OF MARINE FUELS BETWEEN ISO AND CHINESE NATIONAL GB STANDARD

3.1 ISO standard for marine fuels

Most major suppliers that supply marine fuel oils in major ports conform to ISO 8217. The first version of 8217 standard for marine fuels was released in 1987, and it was not a mandatory standard. Thereafter, there have been four updated versions. The current ISO 8217-2012 standard specifies three different distillate grades (DM) and a number of residual grades (RM). New types of 0.1% sulfur fuels are entering the market in response to the 0.1% sulfur SECA limit (ISO 2012).

From ISO 8217:2010, the specifications include stronger clauses regarding any additional substances that may have been added to marine fuel, given that increasing consumer protection and safeguarding them from using fuels with any inappropriate additives. These clauses also address contaminants within the fuel that may affect the safety of the vessel, performance of machinery, human health, and the environment.

In fact, the 2010 and 2012 versions are extremely similar. There are only two small changes—one adjustment to include a more robust method for hydrogen sulfide (H_2S) testing and an amendment to confirm a correction to the DMX grade Distillate Pour Point Levels (Ram 2016).

3.2 Chinese national standard for marine fuels

3.2.1 Brief history of Chinese marine fuel oil standard

The first version of GB/T 17411 was released in 1998, and the second version of GB/T 17411 was released in 2012. They were both voluntary standards. The GB/T 17411-1998 specifications of marine fuels are equivalent to the ISO 8217-1996. GB/T 17411-2012 are equivalent to the ISO 8217-2010. GB 17411-2015 are equivalent to the ISO 8217-2012.

3.2.2 Main changes from GB/T 17411-2012
1. Mandatory standard
Hu (2014) pointed out that fuel oil should be a homogeneous mixture of hydrocarbons. While the competitive market of marine fuel oil is getting worse, low-quality raw materials such as coal tar, aromatic hydrocarbons, and other poor quality raw materials are blended in marine fuel oil. As coal tar has almost no sulfur content, the blending of marine fuel oil with coal tar can significantly reduce the former's sulfur content. However, these non-crude refinery marine fuels can bring risks and threats to fuel stability, oil compatibility, oil sedimentation, and economy performance.

The updated standard GB 17411-2015 has been published in December 31, 2015, and it will be put in practice in July 1, 2016. It is changed from voluntary standard into provision mandatory standard.
2. Sulfur content
According to the ISO 8217:2012 and the provisions on the SOx emission limits, Standard GB 17411-2015 adjusts the sulfur content of the marine fuel oil (General Administration of Quality Supervision, Inspection and Quarantine of the PRC. 2015).

Distillate marine fuels are divided into three grades according to the sulfur content. Sulfur content of DMA and DMZ change from 1.5% to 1.0% (See in Table 3).

RMA 10 and RMB 30 are divided into three grades according to the sulfur content. Other categories of residual marine fuels are divided into two grades according to the sulfur content, as given in Table 4.

Table 3. Sulfur content in distillate marine fuels.

Tier	Sulfur content (%)			
	DMX	DMA	DMZ	DMB
I	1.00	1.00	1.00	1.50
II	0.50	0.50	0.50	0.50
III	0.10	0.10	0.10	0.10

Table 4. Sulfur content in residual marine fuels.

| Tier | Sulfur content (%) | | |
	RMA 10	RMB 10	Others
I	3.50	3.50	3.50
II	0.50	0.50	0.50
III	0.10	0.10	

3. Net heating values

Heating value is the amount of heat release from a complete exothermic reaction during fuel burning. In most real combustion processes, the water vapor remains in the vapor state and hence the vaporization of the water uses some of the heat content of the fuel. Hence, net heating values of blended marine fuel are added in this standard.

4 CONCLUSIONS

To guarantee supply of ship fuel oil market through multiple modes and use by specifications, it is not enough to revise the marine fuel oil (GB/T 17411-2012), we also have to complete and perfect the application criteria of marine fuel. To ensure fuel quality, the fuel should comply with the new mandatory standard "GB 17411-2015".

The relatively higher fuel cost of low-sulfur fuel compared with other fuels is one of the major obstacles to its use in marine applications. Financial subsidies may be beneficial when they work together.

Ships may meet the SOx requirements by using approved equivalent methods, such as an apparatus or piece of equipment (for example, Exhaust Gas Cleaning Systems or "scrubbers", which "clean" the emissions before they are released into the atmosphere).

REFERENCES

Fu Q.Y. & Shen Y. & Zhang J. (2012). On the ship pollutant emission inventory in Shanghai port. Journal of Safety and Environment. 2012(05):57–64.

General Administration of Quality Supervision, Inspection and Quarantine of the People's Republic of China. (2015). Marine fuel oils GB 17411-2015.

Hu X.W. (2014). The use and control of the marine fuel oil of poor quality. China Water Transport. 14,8–11.

IMO (2013). Sulphur Oxides (SOx) – Regulation 14. URL: Available on line: http://www.imo.org/OurWork/Environment/PollutionPrevention/AirPollution/Pages/Sulphur-oxides-(SOx)-%E2%80%93-Regulation-14.aspx (accessed April 10, 2016)

ISO 8217(2012). Petroleum products—Fuels (class F) - Specifications of marine fuels. The fifth edition, 2012

Johan H, Zoi N, Linda Ramstedt. (2014). Modelling modal choice effects of regulation on low-sulphur marine fuels in Northern Europe. Transportation Research Part D.28, 62–73.

Ministry of Transport of the People's Republic of China. (2015). Implementation Plan on Domestic Emission Control Areas in Waters of the Pearl River Delta, the Yangtze River Delta and Bohai Rim. Available on line: http://zizhan.mot.gov.cn/zfxxgk/bzsdw/bhsj/201512/t20151204_1942434.html (accessed April 10, 2016)

Ram V. (2016). The Evolution of ISO 8217 standard from 2005 to 2016 Draft. Available on line: http://www.bunkerworld.com/community/blog/140741/Dr-Ram-Vis/The-Evolution-of-ISO-8217-standard-from-2005-to-2016-Draft. (accessed April 10, 2016)

*Advances in Materials Science, Energy Technology
and Environmental Engineering – Patty & Zhou (Eds)*
© *2017 Taylor & Francis Group, London, ISBN 978-1-138-19668-1*

Flexible nanoimprint mold based on a commercial SEBS thermoplastic elastomer

H.T. Chen, D. Trefilov, Y.S. Cui, C.S. Yuan & H.X. Ge
Nanjing University, Nanjing, Jiangsu, China

X. Hu
Changshu Institute of Technology, Changshu, Jiangsu, China

B. Cui
University of Waterloo, Waterloo, Ontario, Canada

ABSTRACT: A commercial thermoplastic block copolymer elastomer, Poly (Styrene-block-Ethylene-co-Butylene-block-Styrene) (SEBS), was investigated as flexible mold material for UV-curing nanoimprint lithography. The SEBS mold was fabricated by a thermal imprint technique on a SEBS sheet. After deposition of 10 nm-thick SiO_2 layer, the mold release agent, fluoroalkyltrichlorosilane, was able to covalently bind on the surface of the SEBS mold through the silanol groups on the SiO_2. The SEBS mold preserved its imprint fidelity over 10 repeated imprint cycles. Moreover, the SEBS was employed as an elastic support for a double-layer hybrid nanoimprint mold instead of PDMS. Gratings patterns with a pitch of 278 nm were successfully imprinted on the cylindrical surface of a microfiber, demonstrating that SEBS hybrid mold was capable of fabricating high-resolution nanostructures on non-planar substrates.

1 INTRODUCTION

Nanoimprint Lithography (NIL) has been demonstrated as a high-throughput and low cost lithography technology with sub-10 nm resolution (Chou et al. 1995). Various kinds of NIL techniques have been developed for different applications (Guo 2007). Among them, soft UV-NIL (Plachetka et al. 2004) which combines advantages of both UV-NIL (Bender et al. 2000, Colburn et al. 1999) to achieve a high resolution pattern transfer and soft lithography (Xia & Whitesides 1998) to enable patterning on non-planar surface, has been widely used as a simple and feasible method to pattern micro and nanostructures. The key component of this technology is a flexible or soft mold. The most commonly used mold (stamp) material in soft UV-NIL or soft lithography is the flexible polymer, poly (dimethylsiloxane) (PDMS), which allows an intimate physical contact with substrates without applying large external pressure. However, there are still shortcomings for PDMS as mold material. Its low modulus limits the achievable resolution. PDMS mold tends to absorb low viscous liquid resist during UV-imprinting process due to the intrinsic high permeability of PDMS for gases and organic molecules. Treating the PDMS surface by O_2 plasma for coating a mold release layer of

trichloro (1H, 1H, 2H, 2H-perfluorooctyl) silane can cause the spontaneous formation of disordered wavy patterns on it (Bodas & Khan-Malek 2007). The sticky property of the PDMS surface due to the residual uncured prepolymer could increase the risk of adhesion of particles and other contaminants and require great care in handling.

Several approaches have been developed to improve the flexible mold made of commercial soft PDMS. For instance, a higher modulus PDMS (hard PDMS) was developed by Schmid et al. (Schmid & Michel 2000) to achieve soft lithography with resolution into the sub-100-nm regime. Odom et al. (Odom et al. 2002) proposed a composite stamp consisting of two layers: a thin hard PDMS layer supported by a thick flexible PDMS layer, which combined the advantages of a rigid layer to achieve high resolution pattern transfer and an elastic support to enable conformal contact. In order to pattern high-resolution nanostructure on non-planar substrates, we developed a hybrid nanoimprint-soft mold composed of an ultra thin (100 ~ 200 nm) rigid patterning layer on a normal PDMS (thickness up to 2 mm) support (Li et al. 2009). The mold provided sub-15 nm spatial resolution free from cracks and fractures during conformal contact and mold release.

In contrast to thermoset elastomers based on cross-linking bonds such as PDMS, Thermoplastic Elastomers (TPE) are the other family of polymeric elastomers that combine the properties of rubber with the recyclability and thermal processing advantages of plastics (Holden et al. 1996). Poly(Styrene-block-Ethylene-co-Butylene-block-Styrene) (SEBS) is one of the most important TPE consisting of plastic blocks of polystyrene and rubber blocks of hydrogenated polybutadiene. Compared with PDMS.

SEBS can also offer transparency and flexibility while it is relatively economical and easy to use in manufacture by extrusion, molding, thermoforming and heat welding. SEBS has been attempted as stamp to replace PDMS in microcontact printing, a derived technique of soft lithography (Trimbach et al. 2003). Herein, we propose to use commercial SEBS as a rapid prototyping alternative for applications in flexible nanoimprint mold.

2 EXPERIMENTAL

Commercial SEBS pellets (1657, Kraton) were used as a starting material. Figure 1. shows the schematic fabrication procedure of the flexible nanoimprint mold. The polymer pellets were placed between two blank silicon wafers that both had been coated with self-assembled monolayers (SAMs) of trichloro(1H,1H,2H,2H-perfluorooctyl)silane. The sandwich was vacuumed in an imprint machine (ImprintNano, China), and then heated to 170°C under a pressure of 0.8 MPa for 5 min to form a smooth sheet. The desired thickness of the SEBS sheet was dependent on the amount of the pellets. After cooling down, a master nanoimprint mold was employed to replace one of the blank wafers and the abovementioned process was repeated. The patterned SEBS sheet was removed from the sandwich. A 10 nm thick layer of silicon dioxide was deposited on the patterned SEBS via the PECVD process with 710 sccm N_2O and 170 sccm SiH_4 at room temperature. Finally, a self-assembled monolayer of trichloro (1H,1H,2H,2H-perfluorooctyl) silane was coated on the SEBS by vapor-phase deposition and the flexible mold was ready to be used for imprint.

The soft UV-NIL process using the flexible mold was conducted manually with a broadband UV exposure. The homemade UV-curable resist was spin-coated onto a substrate. In order to minimize the amount of air trapped between the mold and the substrate, the mold was carefully placed on the substrate from one side to the other side to push air out. The imprinted sample was exposed to UV radiation in ambient atmosphere by using an Hg arc lamp with a dose of around 300 mJ/cm^2.

The blank SEBS sheet was also used to replace PDMS for elastic support of the hybrid nanoimprint mold consisting of a flexible support and cross-linked hard polymer atop. The details of fabrication of the hybrid molds using PDMS as support have been reported elsewhere (Li et al.

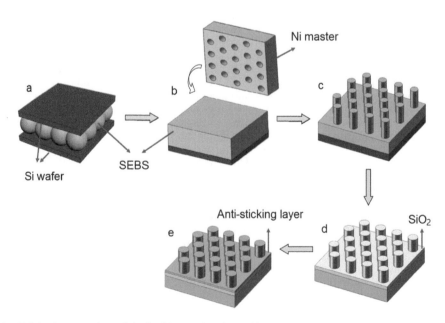

Figure 1. Fabrication procedure of the flexible nanoimprint mold.

2009). The prepared blank SEBS was placed onto a film of a silicon containing UV-curable resist spin-coated onto a silicon wafer. After 10 minutes, the SEBS, absorbed with liquid UV-curable resist, was separated from the Si wafer and then carefully placed onto a master imprint mold which had been spin-coated with the same resist film. Then, the sample was exposed to UV radiation and separated from the master mold. With a brief exposure to an O_2 plasma, the surface of the cured resist was oxidized to form inorganic silica and was then coated with a self-assembled monolayer of trichloro(1H,1H,2H,2H-perfluorooctyl) silane by vapor phase deposition. The resulting hybrid mold was attempted to pattern the cylindrical surface of a 7 μm-diameter optical fiber as a curved substrate through a double transfer imprint method (Shen et al. 2013). The UV-curable resist was spin coated on a silicon wafer and covered by the hybrid mold. Afterwards, the mold was removed from the wafer with some liquid resist adhered onto it, placed on the side of the optical fiber and UV cured by a mercury lamp under an air pressure of 0.03 MPa. When the mold was released, the resist was adhered to the fiber due to its higher surface energy, and as such, the nanopatterns were imprinted onto the cylindrical side of the fiber.

3 RESULTS AND DISCUSSION

The mechanical properties of the commercial PDMS, sylgard 184, and SEBS were estimated by Dynamical Mechanical Analysis (DMA) for the tensile modulus and elongationat break of their blank polymer sheets. The measured strain-stress curves are shown in Figure 2a. The tensile modulus can be calculated from the slope of the stress-strain curve in the linear region. The tensile modulus and elongation at break for the PDMS were 2.1 MPa and 180%, respectively. The modulus of SEBS was 4.9 MPa, while it was not ruptured even at the elongation of 300%. The modulus of SEBS was in the same order of magnitude as that of PDMS. SEBS possessed a higher strain at break, indicating its high toughness. Figure 2b shows the Ultraviolet-visible (UV-vis) transmission spectra of a 0.3 mm thick PDMS sheet and a 0.3 mm thick SEBS sheet. Although the transparency of SEBS was less than that of PDMS due to the styrene component in the backbone of SEBS, the SEBS transmittance for the 350–700 nm wavelength range was over 70%.

The flexible SEBS molds were fabricated via a thermal imprint process. To protect the master mold, the SEBS pellets were first thermally pressed into a smooth sheet by sandwiching them between a pair of fluoroalkyltrichlorosilane coated silicon wafers. The blank SEBS sheet was further ther-

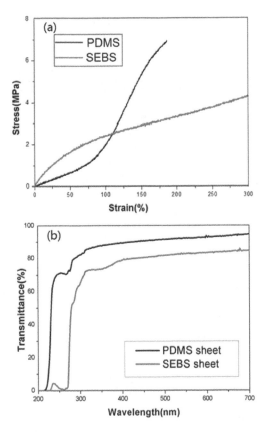

Figure 2. Strain-stress curves (a) and UV-vis transmission spectra (b) of SEBS and PDMS.

mally imprinted by the master mold. The mold release agent, fluoroalkyltrichlorosilane, could not directly react with SEBS due to its chemical inertia. A 10 nm thick SiO_2 layer was deposited on the patterned SEBS by a PECVD process. The fluoroalkyltrichlorosilane was able to covalently bind on its surface through the silanol groups on the SiO_2.

Figure 3a shows the SEM picture of the master mold with 600 nm pitch hexagonal array of holes. Figure 3b and 3c are the SEM image and the photographic image of SEBS replica, respectively. The pitch and diameter of the imprinted holes using the SEBS mold were identical to those of the master mold as seen in Figure 3d. To determine the durability of the SEBS mold, its imprint behavior was tested through repeated imprinting cycles. Figure 3e and 3f show the SEM images of the results of the 10th and 20th imprint with the same SEBS mold, respectively. The duplicated patterns of the 10th imprint were very close to that of the first imprint. For the 20th imprint, some imprinted holes were found deformed and faint irregular lines were observed on the surface of the patterns,

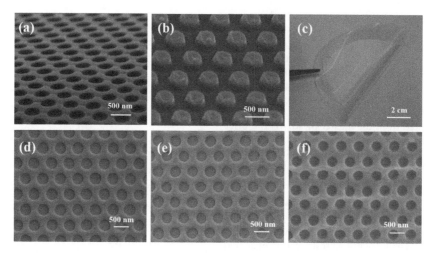

Figure 3. (a) SEM image of the master mold with 600 nm pitch hexagonal array of holes; (b) SEM image of the SEBS replica; (c) Photograph of 4" SEBS flexible mold; (d) SEM image of the first imprint result using the SEBS replica; (e) SEM image of the 10th imprint result; (f) SEM image of the 20th imprint result.

which were originated from the cracks formed in the hard SiO_2 film on the SEBS mold after multiple imprints.

The blank SEBS sheet was also tried as an elastic support for hybrid nanoimprint mold instead of PDMS. To demonstrate the utility and flexibility of the hybrid mold on SEBS, a microfiber with 7 μm diameter tapered from a standard optical fiber was employed as a highly curved substrate. Gratings with 278 nm pitch were patterned on the microfiber using the SEBS hybrid mold through a double transfer UV imprint method with air pressure assistance. Figure 4a illustrates the procedure for forming the nanogratings onto the cylindrical fiber. First, a thin, uniform liquid film of UV-curable resist was spin-coated onto a silicon wafer. The SEBS hybrid mold was then placed on the resist-covered silicon wafer. After a conformal contact between the hybrid mold and the silicon wafer, the mold was detached from the silicon wafer and a part of the liquid resist is adhered off by the mold. The mold was subsequently placed against the microfiber. A rather thin thickness of the elastic support was required for a conformal contact between the flexible mold and the fiber without external pressure, while the decrease of the elastic support thickness could result in decreasing the mechanical strength of the mold and increasing the difficulty in mold preparation. In this case, the thickness of the SEBS support was around 0.3 mm.

After the mold was placed on the fiber, only a small area on the top of the cylindrical surface was contacted with the hybrid mold. The capillary force of the liquid resist and own weight of the flexible mold could not bend and deform the

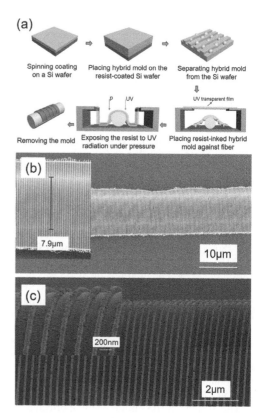

Figure 4. Forming the nanogratings onto the cylindrical fiber. (a) Schematic illustration of the fabrication process; (b) Top-view SEM image of a 278 nm pitch grating on the tapered fiber; (c) Side-view SEM image of a 278 nm pitch grating on the optical fiber.

mold enough to make an intimate contact with the curved substrate. To solve this problem, the sample was put into an imprint apparatus, top window of which was sealed with a transparent and flexible polymer sheet as shown in Figure 4a. A 0.03 MPa air pressure was applied on hybrid mold through the polymer sheet, and forced the mold to conform to the cylindrical contour of the fiber. After UV-curing and mold separation, the grating patterns were imprinted onto the surface of the fiber. Figure 4b and c show top-view and side-view SEM images of 278 nm pitch grating patterns on the tapered fiber, which was mounted on a silicon wafer. It was seen that the imprinted grating patterns cover the top half of the cylindrical surface.

4 CONCLUSION

A commercially available thermoplastic block copolymer elastomer, SEBS, could be applied as a promising mold material for UV-NIL. The SEBS mold could be fabricated by thermal imprint method. SEBS mold could be repeatedly used at least 10 times without damage. Moreover, SEBS could be used as an elastic support for a double-layer hybrid mold instead of PDMS. The imprinting result on the cylindrical microfiber indicated that the mold had capability of patterning highly curved substrate.

REFERENCES

Bender, M., Otto, M., Hadam, B., Vratzov, B., Spangenberg, B. & Kurz, H. (2000) Fabrication of Nanostructures using a UV-based imprint technique. *Microelectronic Engineering, 53*, 233–236.
Bodas, D. & Khan-Malek, C. (2007) Hydrophilization and hydrophobic recovery of PDMS by oxygen plasma and chemical treatment—An SEM investigation. *Sensors and Actuators B-Chemical, 123*, 368–373.
Chou, S. Y., Krauss, P. R. & Renstrom, P. J. (1995) Imprint of Sub-25 NM Vias and Trenches in Polymers. *Applied Physics Letters, 67*, 3114–3116.
Colburn, M., Johnson, S., Stewart, M., Damle, S., Bailey, T., Choi, B., Wedlake, M., Michaelson, T., Sreenivasan, S. V., Ekerdt, J. & Willson, C. G. (1999) Step and flash imprint lithography: A new approach to high-resolution patterning. in Vladimirsky, Y., (ed.) *Emerging Lithographic Technologies Iii, Pts 1 And 2.* 379–389.
Guo, L. J. (2007) Nanoimprint lithography: Methods and material requirements. *Advanced Materials, 19*(4), 495–513.
Holden, G., Legge, N. R. & Qurik, R. (1996). *TherMO-Plastic Elastomers.* Cincinnati: Rapra Technology Ltd.
Li, Z., Gu, Y., Wang, L., Ge, H., Wu, W., Xia, Q., Yuan, C., Chen, Y., Cui, B. & Williams, R. S. (2009) Hybrid Nanoimprint-Soft Lithography with Sub-15 nm Resolution. *Nano Letters, 9*(6), 2306–2310.
Odom, T. W., Love, J. C., Wolfe, D. B., Paul, K. E. & Whitesides, G. M. (2002) Improved pattern transfer in soft lithography using composite stamps. *Langmuir, 18*(13), 5314–5320.
Plachetka, U., Bender, M., Fuchs, A., Vratzov, B., Glinsner, T., Lindner, F. & Kurz, H. (2004) Wafer scale patterning by soft UV-nanoimprint lithography. *Microelectronic Engineering, 73–4*, 167–171.
Schmid, H. & Michel, B. (2000) Siloxane polymers for high-resolution, high-accuracy soft lithography. *Macromolecules, 33*(8), 3042–3049.
Shen, Y., Yao, L., Li, Z., Kou, J., Cui, Y., Bian, J., Yuan, C., Ge, H., Li, W.-D., Wu, W. and Chen, Y. (2013) Double transfer UV-curing nanoimprint lithography. *Nanotechnology, 24*(46).
Trimbach, D., Feldman, K., Spencer, N. D., Broer, D. J. and Bastiaansen, C. W. M. (2003) Block copolymer thermoplastic elastomers for microcontact printing. *Langmuir, 19*(26), 10957–10961.
Xia, Y. N. & Whitesides, G. M. (1998) Soft lithography. *Annual Review Of Materials Science, 28*, 153–184.

Advances in Materials Science, Energy Technology and Environmental Engineering – Patty & Zhou (Eds)
© 2017 Taylor & Francis Group, London, ISBN 978-1-138-19668-1

Construction of an urban water source management system and application of phenol paroxysmal pollution

X.L. Meng
School of Municipal and Environmental Engineering, Harbin Institute of Technology, Harbin, China
State Key Laboratory of Urban Water Resource and Environment, Harbin, China

X.Q. Ye, L. Zhang & J.H. Qu
School of Municipal and Environmental Engineering, Harbin Institute of Technology, Harbin, China

ABSTRACT: The emergency management system of urban water source is the important guarantee to drinking water safety. The framework of urban water source management system, including the early warning, emergency treatment technologies, and waste disposal, was built. An enterprise in the Heilongjiang province was chosen as the leaked source of phenol to study, and the mathematical model is used to predict phenol concentrations in water sources 1 and 2, respectively. According to the predicted concentration, the corresponding emergency measures are determined, i.e. the water source 1 has to close the water intake, as for water source 2, the activated carbon adsorption technology is suggested.

1 INTRODUCTION

At the situation that water pollution incidents are of common occurrence, the water source must establish a perfect urban water source emergency management system, otherwise, when water pollution incidents happen, the city will be helpless. For example, 2005 sudden nitrobenzene pollution in Songhua river caused that water supply of Harbin was cut off for four days. Therefore, the urban water source emergency management system is the necessary guarantee to drinking water safety.

In this paper, several factors are considered while constructing the emergency management system, such as urban water sources' position in the basin, available emergency countermeasures, and alternate source. Based on this, downstream water treatment plants can put forward an emergency countermeasure to ensure drinking water safety by analyzing the position relation of a certain phenol pollution source with the downstream urban water sources and the degree of impact while the phenol pollution source releases phenolic wastewater.

2 URBAN WATER SOURCE MANAGEMENT SYSTEM CONSTRUCTION

To reduce the influence of sudden water pollution incidents on urban water quality, the following aspects are necessary: First, through the control of pollution sources can decrease leakage, and then

according to water quality monitoring and model prediction, the pollutant concentrations in water sources must be obtained, which are the main basis of choosing emergency technologies such as dosing powder activated carbon, adding oxidant, improving filtering unit, even to stop water supply, and starting the alternate source (Zhang, 2007; Zhang, 2008). Although, by taking emergency technologies, the yielding water of water treatment plants can reach the standard, emergency waste will inevitably be produced. Improper disposal of emergency waste can also produce secondary pollution to the environment; therefore, it is necessary to dispose and manage emergency waste (Podkoscielny, 2007; Mckone, 2000; Gougar, 1996; Xing, 2007; Sun, 2015).

Based on water environmental pollution emergency research at home and abroad, considering the factors such as the transformation characteristics of pollutants in the water, the mature emergency techniques of phenol water pollution, and available disposal method of emergency waste containing phenol, constructed the urban water source emergency management system. Figure 1 shows its framework.

2.1 *The research object and background*

The research object of this research is a coal chemical company in Heilongjiang province. The company mainly uses coal as the raw material to produce gas for urban use by gasification. This company produces wastewater containing phenol

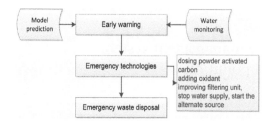

Figure 1. Framework of urban water source emergency management system.

in the process of gas washing. Normally, the phenol wastewater is preprocessed by phenol ammonia pretreatment system, and then, integrated into the wastewater treatment plant, finally, is discharged into the Songhua river after being treated. The initial concentration of phenolic wastewater is as high as 6500 mg/L, even by pretreatment; the concentration is still between 450 and 600 mg/L. Once the pretreatment system or integrated wastewater treatment plant does not work, this part of the high concentration phenolic wastewater will be discharged into the Songhua river causing negative influences to the downstream area. This paper explores the impacts of downstream urban water source 1 (110 km) and water source 2 (200 km) when the coal chemical company has an accidental emission.

2.2 The accidental situation

The set accidental situation: the concentration of wastewater containing phenol is 525 mg/L, flow rate is 140 m³/h, leak time is 1h, accident leakage M is 73.5 kg.

2.3 Concentration prediction of phenol in Songhua river

2.3.1 Prediction model

Owing to the side discharge of this company, although the Songhua river is a large river, it is necessary to consider the transverse dispersion and reflection from both sides of the river. This paper considers once reflection and ignore the horizontal velocity u_y (namely $u_y = 0$); the prediction model (Yu, 2007) is as follows:

$$C(x,y,t) =$$
$$\frac{M \exp(-Kt)\exp(-\frac{(x-u_xt)^2}{4D_xt})}{4\pi th\sqrt{D_xD_y}}$$
$$\left\{ \exp(-\frac{y^2}{4D_yt}) + \exp(-\frac{y^2}{4D_yt}) + \exp(-\frac{(2B+y)^2}{4D_yt}) \right\}$$

D_x is the longitudinal coefficient of diffusion, m²/s, $D_x = 2.51$ m²/s; D_y is the transverse diffusion coefficient, m²/s, $D_y = 0.38$ m²/s; u is the flow velocity, m/s, $u = 1.16$ m/s; B is the river width, m, $B = 490$ m; Q is the flow rate, m³/s, $Q = 1550$ m³/s; H is the depth, m, $H = 2.65$ m; C_h is the background value, mg/L, $C_h = 0$; and k is the degradation coefficient, d⁻¹, $k = 0.2$ d⁻¹.

2.3.2 Predicted results and discussions

Considering environment-sensitive plot Xo in the downstream position, in order to make pollutants concentration in Xo less than the concentration of the artificial setting threshold Cs (mg/L), the maximum emission (M*) can be calculated. It is easy to know that, when the pollution group reaches in plot Xo (t = Xo/u), the pollutant concentration is the highest. Therefore, M* can be calculated in advance according to the diffusion model, when a sudden water pollution incident happens, and through the comparison of M and M*, we can know whether the environment-sensitive plot Xo would be affected, and provide the basis for water treatment plants to choose what kinds of emergency countermeasures to take. In addition to this, through surveying and analyzing possible accidental emissions which are greater than M*, these companies should be paid more attention to avoid the occurrence of sudden water pollution incidents.

Generate t = x/u into the formula 1–1, the maximum emissions (M*) under different Cs can be calculated. Table 1 shows M* of urban water source 1 and urban water source 2 under different Cs. When Cs = 0.02 mg/L, M* of urban water source 1(38.386 kg) is less than M (73.5 kg), while M* of urban water source 1 (82.52 kg) is higher than M (73.5 kg). Thus, water treatment plant 1 needs to close its water intake, and start its alternate source or stop water supply, while water treatment plant 2 can dose powder activated carbon to achieve the purpose of removing phenol out of water without closing water intake.

Table 2 gives information that when water sources begin to be affected, how long it lasts, and how much the concentration is. Figure 2 describes

Table 1. Phenol wastewater quality.

	Concentration mg/L
Without any treatment	6500
After phenol ammonia pretreatment system	450–600

The distance between the company and water source Water source 1: 110 km; water source 2: 220 km

Table 2. M* of urban water source 1 and urban water source 2 under different Cs.

Cs	M* of urban water source 1	M* of urban water source 2
0.02	38.386	82.52
0.01	19.193	41.26
0.005	9.5965	20.63
0.002	3.8386	8.252

Table 3. M* of urban water source 1 and urban water source 2 under different Cs.

	Water source 1	Water source 2
location(km)	110	200
The time that begin to be affected (s)	93000	170000
The time that (s)	97000	175000
total harm time (ΔT)(s)	4000	5000
The time when Cmax appears (s)	94800	172800
Cmax (μg/L)	38.3	17.8
The time when pollutant concentration is higher than 10μg/L (s)	1945	1720

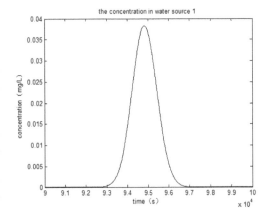

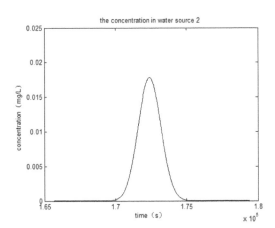

Figure 2. Concentration in the water sources 1 and 2.

the concentration change in the water sources 1 and 2, respectively.

2.4 Emergency countermeasures

According to section 2.3, as far as urban water source 1 is concerned, because the harm time (ΔT) is 4000 s (less than 2 h), there are two kinds of emergency countermeasures: ①close the water intake or stop water supply for two hours, and recover the water supply when it is lifted; ②close the water intake during the period that the concentration is more than 10 μg/L and adding activated carbon 30 mg/L (Ye, 2015) to water once the concentration decrease to less than 10 μg/L. Relatively, the urban water source 2 does not need to close the water intake. According to the time water flow from the intake to the water treatment plant, different additive quantities were chosen to guarantee the water quality, and the minimum dose of power activated carbon is 70 mg/L[10].

3 CONCLUSION

First, the framework of the urban water source management system including water quality early warning, emergency technology and countermeasures, and emergency waste disposal was built. Second, the case study results show that the total harm time to water source 1 is 4000 s, the maximum pollutant concentration is 38.3 μg/L, and the determined emergency measure is to close the water intake and avoid getting water during this period; as for urban water source 2, the maximum pollutant concentration is 17.8 μg/L, and total harm time is 5000 s. Yielding water quality can be guaranteed by dosing powder activated carbon to the water supply plant, and the minimum dose of powder activated carbon is 70 mg/L.

REFERENCES

Gougar MLD, Scheetz BE, Roy DM, 1996. Ettringite and c-s-h portlant cement phases for waste ion immobilization: A Review. J. Waste Management, Vol.4, No.16, 1996:295–303.
Mckone TE & Hannonds. KM, 2000. Managing the Health impacts of Waste Incineration. J. Environmental Science and Technology, No.280, 2000:387.
Podkoscielny P & Laszlo K, 2007. Heterogeneity of activated carbons in adsorption of aniline from aque-

ous solutions. J. Applied Surface Science, No.253, 2007:8762–8771.

Sun SF, Wenbo Jiang WB, Rui Guo. R. 2015. Cement Kiln Collaborative Disposal of Hazardous Waste Management and Technology Progress. J. Environmental Protection, Vol.1, No.43, 2015:41–44.

Wang DM. 2007. Study on The Whole Process Risk Assessment of The Sudden Fatal Cross-Border Water Pollution Events. D. Harbin Institute of Technology

Xing Y, Lv YL, Shi YJ, 2007. Assessment on PCBs Wastes Treatment Technologies Including Incineration, Cement Kiln and Secure Landfill. J. Environmental Science, Vol.3, No.28, 2007:673–678.

Ye XQ. 2015. Countermeasures of Water Treatment Plants to Deal With Phenol Paroxysmal pollution, J. Harbin Institute of Technology

Yu CJ, Xianlin Meng XL. 2007. Research on Environmental Warning System Model for Accident Happen in Water Environment. J. journal of Harbin University of Commerce (Natural Sciences Edition), Vol.1, 2007:75–79.

Zhang XJ, 2007. The Urban Water Supply Emergency Treatment Technology, J. Construction Science and Technology, 17–19.

Zhang XJ, 2008. Reflection on Water Pollution Emergency. J. Green Leaf, Vol.3,:86–90.

*Advances in Materials Science, Energy Technology
and Environmental Engineering – Patty & Zhou (Eds)*
© 2017 Taylor & Francis Group, London, ISBN 978-1-138-19668-1

Harvesting of microalgae *Chlorella pyrenoidosa* cultivated in municipal wastewater by electroflocculation

R.J. Tu, W.B. Jin, S.F. Han & W. Jiang
Shenzhen Graduate School, Harbin Institute of Technology, Shenzhen, China

ABSTRACT: Microalgae are a promising new source for biomass production. One of the major challenges with regard to cost effectiveness is the biomass harvest. High-energy input is required for the separation of the small algal cells from a large volume of surrounding media. Electroflocculation has been reported as a promising harvesting technique to improve the cost effectiveness within the downstream process. In this study, the electroflocculation method was developed for harvesting microalgae *C. pyrenoidosa* cells from municipal wastewater. The effect of several parameters such as the electrode gap, current density, time, and pH on harvesting efficiency was also determined. The maximum harvesting efficiency of this method was 98.07% at pH = 5.5 and current density of 0.0300 A/cm² during 10 minutes electroflocculation process in a 1200 ml tank. The maximum efficiency was achieved by aluminum electrodes with 1 cm distance between electrodes. It was concluded that electroflocculation is an efficient and cost-effective method for harvesting microalgae.

1 INTRODUCTION

Microalgae are considered as one of the most promising feedstocks for biodiesel production nowadays (Abomohra et al. 2014). However, the main drawback for economical biodiesel production from microalgae is the high cost of algal cultivation for the huge consumption of freshwater resources, nitrogen and phosphate, and CO_2 (Jiang et al. 2011). One possible solution to overcome this problem is to cultivate algae in municipal wastewater due to its abundance and enrichment of nutrients (Tu et al. 2015). It was found that the effluent after primary sedimentation is more conducive to the microalgal growth and accumulation of lipids compared with the effluent of secondary clarifier because of the rich nutrition. Although they are easy to cultivate, the bottleneck which often makes microalgal cultivation uneconomical is the downstream processing, which contributes to 20–30% of the biomass production costs (Uduman et al. 2010).

Separation of the cells (2–10 μm) from the surrounding growth media requires high energy inputs (Grima et al. 2003). Large volumes must be processed, as the concentration of cells is very low at approximately 0.1–1 g/L. Increasing the efficiency at low energy demands within the harvesting process is a major challenge. There are several methods for microalgae harvesting. The filtering method is a very effective method in laboratory-scale harvesting of microalgae, but several disadvantages of the filtering method are a challenge to its large-scale application. Centrifugation is a typical method that is widely used for microalgae separation from the culture medium, but it is time consuming, complicated, and expensive.

Electroflocculation is an almost new technology for wastewater treatment and removing of solid particles from water (Phalakornkule et al., 2009). The main advantages of electroflocculation over conventional methods are lack of secondary pollution and compact equipment. Electrocoagulation has advantages of simple equipment, high recovery efficiency, and the possibility of industrial application. Flocculation technology has been widely used in wastewater treatment and it also proved to be efficient in microalgae harvesting (Grima et al. 2003). Considering the mentioned problems in microalgae harvesting, the main aim of this research work was to suggest a method for microalgae harvesting with high efficiency and low cost. Therefore, the electroflocculation method was introduced for harvesting of microalgae and the effects of several factors, such as the electrode gap, current density, time, and pH, on harvesting efficiency were evaluated.

2 MATERIALS AND METHODS

2.1 *Algae strain and growth conditions*

C. pyrenoidosa obtained from Freshwater Algae Culture Collection at the Institute of Hydrobiology (strain number FACHB-9) was cultivated outdoors in a columnar photobioreactor made of Plexiglass with a working volume of 8 L (1200 mm in height, 140 mm in diameter) using municipal wastewater as the medium. The experiments were performed at Harbin Institute of Technology, Shenzhen, China (N 22° 35' 25", E113° 59' 01") during autumn. During the experiment, the weather was in a stable condition, and all were sunny days with maximum intensity of illumination from 70000 to 100000 lux and average temperature between 24°C and 27°C.

2.2 *Wastewater used in the experiment*

Wastewater was collected by pumping from sewage well at the campus of Harbin Institute of Technology, Shenzhen Graduate School, Shenzhen, China. The main source of the wastewater comes from student domes and teaching building complex. After settled in a plastic bucket, the supernatant was collected for the algal cultivation, and the water quality parameters were as follows: COD 130 ± 10 mg/L, TN 31 ± 5 mg/L, NH_3-N 28 ± 5 mg/L, TP 4.5 ± 0.5 mg/L, and pH 7.5 ± 0.8.

2.3 *Electroflocculation experiment*

The electroflocculation device mainly consists of constant pressure steady flow of power supply, an electrolytic tank (design by ourselves), an electrode, and a timer (Fig. 1). The highest voltage of the power supply is 60V and its maximum current is 5 A. The effective volume of electrolytic tank is $12 \times 10 \times 10$ cm. A pure aluminum electrode with a purity of 99.6% is used as the anode and cathode of the reaction, and the electrode area is 40 cm^2. The flocculation experiments were carried out at room temperature in a 1200 ml tank filled with 1000 ml of algae suspension. The initial cell concentration of the algae was 0.2 g/L. The electrode plates were cut from commercial-grade metal sheets. Before use, they were mechanically polished using abrasive paper and placed parallel and vertically into

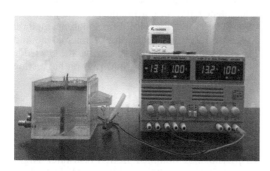

Figure 1. Electroflocculation device.

the algae suspension. The harvesting efficiency was calculated as follows:

$$R = \frac{c_f v_f}{c_0 v_0} \times 100\% \qquad (1)$$

where C_0 and C_f are the initial optical density before starting the electroflocculation and the optical density after electroflocculation (OD_{681}), respectively, and V_0 and V_f are the initial volume before starting the electroflocculation and the volume after electroflocculation, respectively.

3 RESULTS AND DISCUSSION

3.1 *Effect of electrode gap on harvesting efficiency*

The harvesting efficiency of microalgae *C. pyrenoidosa* suspension after electroflocculation was measured in six different electrode gaps (Fig. 2-a). When the electrode gap was increased to

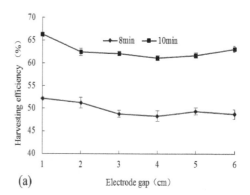

(a)

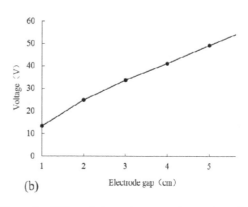

(b)

Figure 2. Effect of electrode gap on harvesting efficiency and voltage.

276

6 cm, the harvesting efficiency was less than 70% whether the electroflocculation time was 8 min or 10 min. The highest harvesting efficiency was achieved when the electrodes were closer to each other (Gap = 1 cm). This means that as the electrodes were adjusted closer to each other, and the flocculation process occurred at higher efficiency.

However, as the distance was increased, high potential energy was required for transferring the electrons between electrodes (Fig. 2-b). Therefore, shorter electrode gap was suitable for electroflocculation and it had the advantages of both low energy consumption and high harvesting efficiency.

3.2 Effect of current density on harvesting efficiency

Using different current densities showed significant increase in harvesting efficiency by increasing the current density up to 0.0275 A/cm², and then showed insignificant increase at 0.0275 and 0.0350 A/cm² (Fig. 3 a). When the current density was 0.0275 A/cm², the harvesting efficiency was above 95%. However, higher current density showed a significant effect on the voltage (Fig. 3 b). The conclusion shows that, although increasing the current density can improve the harvesting

efficiency, further increases in the current density will lead to increased power consumption; it is not recommended by only increasing the current density to improve the treatment effect. Therefore, the optimal current density of the microalgae harvesting technology was 0.0275 A/cm².

3.3 Effect of electroflocculation time on harvesting efficiency

The effect of electroflocculation time on harvesting efficiency was determined by measuring OD_{681} before and after electroflocculation at five different electroflocculation times including 5, 6, 7, 8 9, and 10 minutes (Fig. 4). The electrode gap and current density were 1 cm and 0.0275, 0.0300, 0.0325 A/cm², respectively. It was clear that, by increasing the time from 5 to 8 minutes, the harvesting efficiency was raised from 79.57% to 94.9%. After 8 minutes, the harvesting efficiency became stable and the electroflocculation process was stopped. By applying electricity to electrodes for 8 minutes, the maximum harvesting efficiency was achievable, and longer electroflocculation process was not necessary. Therefore, the optimal electroflocculation time of the microalgae harvesting technology was 8 minutes.

3.4 Effect of pH on harvesting efficiency

pH is an important factor which significantly affects the harvesting efficiency of microalgae C. pyrenoidosa (Fig. 5 a). A higher pH value showed significant effect on the voltage (Fig. 5 b), and further increases in the pH value will lead to increased power consumption. Consequently, pH = 5 was applied for further studies.

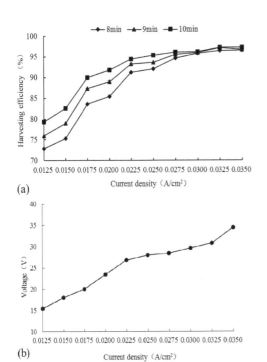

(a)

(b)

Figure 3. Effect of current density on harvesting efficiency and voltage.

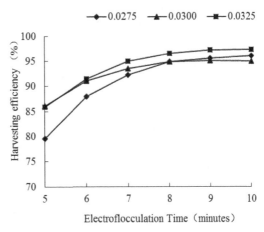

Figure 4. Effect of electroflocculation time on harvesting efficiency.

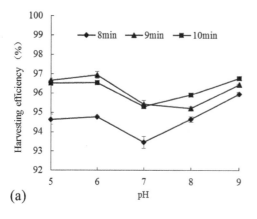

(a)

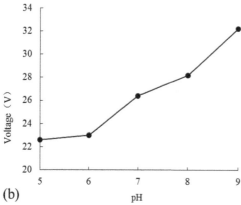

(b)

Figure 5.   Effect of pH on harvesting efficiency.

3.5   Analysis of orthogonal experiment results

According to the results of the single factor experiment, treatment of *C. pyrenoidosa* with 1 cm, 0.0275 A/cm², electroflocculation at pH of 5 for 8 min was nominated as optimum conditions of the single-factor experiment.

Through the analysis of the results of single-factor optimization, three levels of each factor were chosen as the orthogonal design experiment $L_9 (3^4)$ (Table 1). The results of the orthogonal test are given in Table 2. Considering the actual situation, we assessed the combination of four factors, the electrode gap, current density, time, and pH on the harvesting efficiency under the experimental conditions. We performed statistical analysis of the experimental results. By range and variance analysis, we determined that the optimal growth conditions were 1 cm, 0.0300 A/cm², electroflocculation at pH of 5.5 for 10 min. Under these conditions, the highest harvesting efficiency was 98% (Fig. 6).

Table 1.   Electroflocculation parameters setting.

	Factors			
Level	Current density (A/cm²)	Time (min)	pH	Electrode gap (cm)
1	0.0275	8	5.0	1
2	0.0300	9	5.5	2
3	0.0325	10	6.0	3

Table 2.   Orthogonal experiment results analysis.

	A	B	C	D	
Level	Current density (A/cm²)	Time (min)	pH	Electrode gap (cm)	Harvesting efficiency (%)
1	0.0275	8	5.0	1	95.22
2	0.0275	9	5.5	2	95.43
3	0.0275	10	6.0	3	95.22
4	0.0300	8	5.5	3	95.20
5	0.0300	9	6.0	1	96.01
6	0.0300	10	5.0	2	95.42
7	0.0325	8	6.0	2	95.12
8	0.0325	9	5.0	3	94.57
9	0.0325	10	5.5	1	96.04
Ran E	0.0030	0.0038	0.0049	0.0076	

Table 3.   Comparison and verification of experiment conditions.

Number	Current density (A/cm²)	Time (min)	pH	Electrode gap (cm)
1	0.0275	8	5.0	1
2	0.0300	10	5.5	1

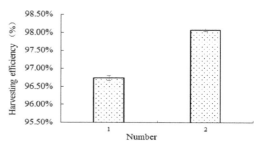

Figure 6.   Results of validation experiment.

Among the four factors assessed in the orthogonal test, the greatest influence on harvesting efficiency was the electrode gap, followed by pH, and current density; time was the least influential.

4 CONCLUSION

This paper discussed the efficiency of electroflocculation in harvesting *C. pyrenoidosa* as a feedstock for biodiesel. The present study suggests pretreatment of *C. pyrenoidosa* cells with 1 cm, 0.0275 A/cm^2, electroflocculation at pH of 5 for 8 min as for high harvesting efficiency. The harvesting efficiency of 98.07% is simply achievable by electroflocculation process. The distance between electrodes is a major factor in harvesting efficiency. Finally, it has been concluded that microalgae harvesting by electroflocculation process was an efficient method that could harvest microalgae from municipal wastewater during a short time, with high efficiency and without adding any extra chemical materials to municipal wastewater.

ACKNOWLEDGMENTS

This work was financially supported by Shenzhen Science and Technology Innovation projects (project number JCYJ20140417172417125).

REFERENCES

Abomohra, A., El-Sheekh, M., Hanelt, D. 2014. Pilot cultivation of the chlorophyte microalga *Scenedesmus obliquus* as a promising feedstock for biofuel. *Biomass Bioenergy* 64, 237–244.

Jiang, L., Luo, S., Fan, X., Yang, Z., Guo, R. 2011. Biomass and lipid production of marine microalgae using municipal wastewater and high concentration of CO$_2$. *Applied Energy*, 88, 3336–3341.

Tu, R. J., Jin, W. B., Xi, T. T., Yang, Q., Han, S. F., Abomohra, A. 2015. Effect of static magnetic field on the oxygen production of Scenedesmus obliquus cultivated in municipal wastewater. *Water Research*, 86, 132–138.

Uduman, N., Qi, Y., Danquah, M.K., Forde, G.M., Hoadley, A. 2010. Dewatering of microalgal cultures: a major bottleneck to algae-based fuels. *J Renew Sustain Energy* 2:15.

Grima, E.M., Belarbi, E.H., Fernandez, F.G.A., Medina, A.R., Chisti, Y. 2003. Recovery of microalgal biomass and metabolites: process options and economics. *Biotechnol Adv* 20:491–515.

Phalakornkule, C., Polgumhang, S., Tongdaung, W. 2009. Performance of electrocoagulation process in treating directy dye: batch and continuous up flow process. *World Academy Sci. Technol.*, 57: 277–282.

*Advances in Materials Science, Energy Technology
and Environmental Engineering – Patty & Zhou (Eds)*
© 2017 Taylor & Francis Group, London, ISBN 978-1-138-19668-1

Economic and environmental benefit analysis of energy performance contracting for a solar hot water system in Chinese universities

Hetao Yuan, Ke Yang, Xiaotong Yan, Qian Li, Hao Guo,
Teza Mwamulima & Changsheng Peng
*The Key Lab of Marine Environmental Science and Ecology, Ministry of Education,
Ocean University of China, Qingdao, China*

ABSTRACT: The aim of this paper is to analyze economic and environmental benefits of Energy Performance Contracting (EPC) for solar hot water system application in Chinese universities. In this paper, 11 universities using EPC for solar hot water system were investigated, and 5 types of boilers were demonstrated, to compare the energy-saving behavior. The results showed that the solar system could save energy as high as 49.5%. For 10, 000 students' bathing in a university campus using solar hot water system everyday, a plate solar collector of $7.7 \times 10^3 \text{m}^2$ was required, which could decrease the consumption of standard coal 2.5×10^3 kg/d, and thus reduce the emission of 5789 kg CO_2, 114 kg CO, 7 kg SO_2, 32 kg NO_x, and 675kg solid waste everyday. This paper provides an initial guide for a low-carbon campus building.

1 INTRODUCTION

China Statistical Bureau has recently published the "2014 National Economic and Social Development Statistical Bulletin". The bulletin showed that the number of undergraduate students reached 25.48 million and graduate students reached 1.85 million in Chinese universities. Therefore, at least 40 million campus students need hot water for bathing, including some high schools. At present, the main energy used in hot water boilers is coal, oil, gas, and electricity, while the coal is the most widely used in China.

Exhaust gas produced by coal-fired boiler contains large amount of harmful substances such as particulates, sulfur dioxide, and carbon monoxide. These particulates can enter the human through respiratory exposure; sulfur dioxide can lead to a series of health problems, such as severe lung failures, and can form acid rain to make a corrosive action on the ecosystem; carbon monoxide exposure can lead to extensive anemia and gastrointestinal disorders (Cai, 1999). In recent years, some novel technologies have been used to reduce the pollution of coal combustion in China, especially in Chinese universities; coal-fired boiler have been replaced by solar hot water system with the EPC model in some universities.

EPC was initially introduced as a market mechanism for energy conservation in America (Zhu, 2013), and then it was adopted in China in 1997 in joint efforts with the World Bank and Global Environment Fund to upgrade China's energy-intensive industries and renovate residential energy equipment. The project was jointly created by the Chinese government and the World Bank Global Environment Facility to promote China's energy conservation (Yan, 2012). Ever since China's Zhongshan University adopted EPC for hot water provision in 2004 for the first time, EPC has achieved leapfrog development; therefore, this study focuses on economic and environmental benefits analysis of EPC for solar hot water system in Chinese universities.

2 APPLICATION OF EPC OF SOLAR HOT WATER SYSTEM IN A UNIVERSITY

2.1 Structure of EPC project

There are many ways to structure an EPC model. As shown in Fig. 1, the general process of EPC mechanism typically includes the following steps: project identification, planning assessment, contractor selection, project design, financial arrangement and negotiation of EPC contract, construction and implementation, and measurement and verification of savings. The steps are divided into phases and tasks, based on solar hot water system construction. The most critical element to succeed in an EPC project is the development of a suitable design for the clients, which alone can affect the success of a project.

2.2 General situation of the university

In order to clearly describe the application of solar hot water system in Chinese universities, and analyze its benefits, a university, Qingdao campus of Shandong University, was selected as the example.

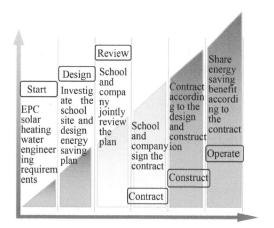

Figure 1. Flow chart of EPC solar heating water project in a university.

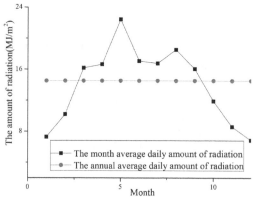

Figure 2. The amount of daily solar radiation in Qingdao.

The university has about 30, 000 students studying in campus, and about 10, 000 students need hot water for bathing everyday. The university is located in Qingdao (N36°03', E120°18'), a coastal city in the north of China, which is located in the north temperature zone over East Asian monsoon region with an annual average temperature of 12. 7°C.

According to the statistic date, the average solar radiation of Qingdao is about $1.5 \times 10^7 J/m^2 \cdot d$ (as shown in Fig. 2), which can meet the needs of solar energy development (Wang, 2001). Moreover, according to the local hydrological and meteorological conditions, the temperatures of inlet and outlet for the solar hot water system were designed to be 15°C and 55°C, Respectively.

2.3 Calculation of solar collector area

The area of a solar collector in a solar hot water system decides the amount of solar energy exploited. In other words, the solar collector area is the key component for the project; it accounts for about 40% of the total investments and thus directly affects project success and performance. Solar collector area is determined mainly by students' daily water consumption. Generally, the area can be calculated with the following formula based on GB50364-2005:

$$A_c = \frac{Q_w \rho_w C_w (t_{end} - t_i) f}{J_T \eta_{cd} (1 - \eta_L)} \qquad (1)$$

where A_C is the direct area of collectors, m²; Q_W is the average daily water consumption, here is 3.6×10^5 L; ρ_w is the water density, 1 kg/L; C_w is the specific heat at constant pressure, 4.18×10^3

J/kg·°C; t_{end} is the outlet temperature of reservoir, 55°C; t_i is the inlet temperature, 15°C; J_T is the average daily amount of solar radiation, 1.5×10^7 J/m²; J_T is the availability of the solar energy, 60%; η_{cd} is the efficiency of the solar collector, 68%; and η_L is the rate of heat loss of pipes and reservoir, taking experience value of 0. 2 (Li, 2013).

According to the calculation of the formula, the area of collectors should be 4.5×10^3 m². During winter season, the temperature is usually minus 10°C in Qingdao, and therefore, water can not be heated directly by a solar collector, instead, propylene glycol should be selected as a thermal conducting medium. Therefore, the indirect collecting area would be 7.7×10^3 m², 1.69 times greater than the direct collecting area due to convective heating and line loss during the indirect heating process.

2.4 Text and indenting the process of the solar hot water system

The system consists of a solar collector, a heating water tank, a heat storage water tank, an air source heat pump, an intelligent control system, a control cabinet, a heat exchanger, a circulating pump, main equipment, a fluid expansion tank box, and a piping system. The system has an average daily production of 3.6×10^5 L of hot water, and the process of the system is shown in Fig. 3. The solar collectors absorb energy and pass it to the heat exchanger, which in turn send the heat to the heating water tank through the heat-conducting medium. In rainy days, or when the solar radiation is not sufficient, the air heat pump comes in assistance and thus the water temperature is always maintained at 55°C. The solar collectors and the water tank are connected with a temperature sensor, which can lead to an automatic start of functioning of the circulating pump by the control

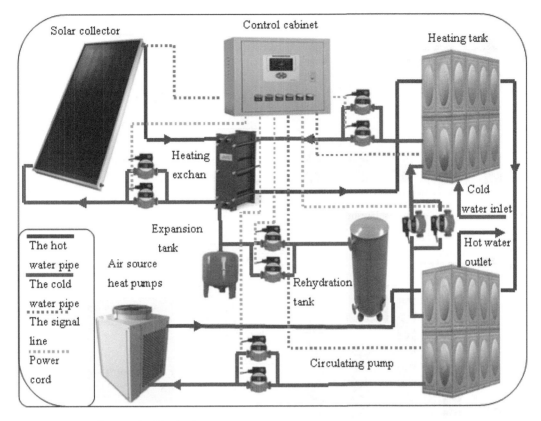

Figure 3. Flow diagram of EPC solar hot water system.

cabinet when the temperature differential is above 5°C, and it will automatically stop when the temperature differential is less than 5°C. The system also has the automatic replenishment and automatic heating function. If EPC solar hot water engineering system is to be used in South China with working environment temperature in the coldest months around 0 degrees Celsius, then the system would need to wait for the heat exchanger and the antifreeze; hence, the cost of operation would be more.

2.5 *Economic benefits of the solar hot water system*

According to the investigation and analysis, even in the cold north area, energy-saving benefits are very high. Here, we will compare the operating cost between the coal-fired boiler and the solar hot water system.

To maintain a supply of hot water for 10, 000 students, the total cost of coal-fired boiler would be 3.19 million USD for 15 years, considering the thermal efficiency of 60% in a coal-fired boiler and the

coal price 0. 08 USD/kg in 2015; while a solar hot water system would just cost 1. 61 million USD for 15 years, including the cost of electric heating when the solar radiation is not sufficient. Therefore, the operating cost of the solar system is almost the half that of a coal boiler in the long run.

As shown in Fig. 4 and 5, the cost of energy determines the total cost of hot water supply, and the long-term cost of equipment accounts for a very small part in the total cost. Based on the results obtained, the initial cost of solar hot water system is indeed lower than that of coal, gas, fuel, and electricity forms. Even under the condition of sufficient use of air source heat pump, power consumption is still minimum. In addition, countries using the EPC solar system are highly rewarded in terms of energy costs, making the long-term use of solar energy the most economic.

2.6 *Environmental benefits of EPC solar hot water project*

Take the solar system on Qingdao campus of Shandong University as an example again. The university

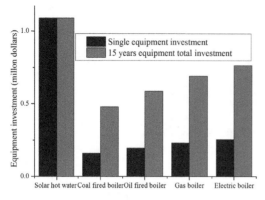

Figure 4.   Analysis of equipment investment.

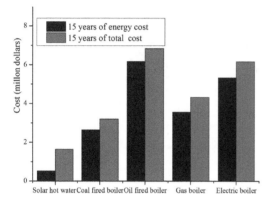

Figure 5.   Analysis of energy and operating cost of the project.

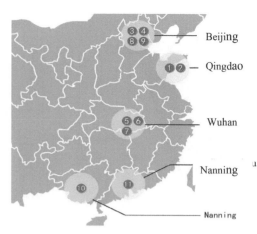

Figure 6. Geographical distribution of the 11 universities.

has an average number of ten thousand people taking baths daily and the solar system has a solar energy collecting area of 7.7×10^3 m^2. According to the data of the irradiance and the collection efficiency of the solar collector, the solar collectors of the system can change 7.4×10^{10} J/d solar energy into heat energy, which is equivalent to the combustion of standard coal 2.5×10^3 kg/d. As the components of Chinese power coal are usually 20% ash, 1% sulfur, 1.5% nitrogen, and 70% carbon (Wang, 2013), a large number of pollutants would be decreased everyday if coal-fired boilers were replaced by the solar system, such as CO_2 (5789kg), CO (114 kg), SO_2 (7 kg), NOx (32 kg), and solid waste (675 kg).

With the increasing EPC solar hot water system, the burning of coal will decrease and the emission of pollutants will decrease also, which is entirely consistent with the 2012 national "energy conservation and emission reduction of Twenty Five Year Plan of the new requirements", and thus solar energy brings in unparalleled environmental benefits.

3   THE PROSPECT OF EPC IN CHINESE UNIVERSITIES

To estimate the development prospect of EPC, 11 universities with EPC project were investigated here. As shown in Fig. 6, the 11 universities located in south of China (Guangzhou and Nanning), central China (Wuhan), and north of China (Beijing and Qingdao), which represent different types of meteorological conditions. Although the solar radiation in these places is different, they are all greater than 4.2×10^9 J/m^2·year, suitable for solar energy utilization. A large number of statistical and calculated data are given in Table 1. It can be seen from Table 1 that the EPC of solar hot water system have already been applied in these 11 universities, and about 1.8×10^4 kg standard coal could be saved everyday.

With the help of the successful experience of EPC in these 11 universities, we have made a reasonable prediction. If the technology of solar hot water system could be used in 10% of Chinese universities and high schools, at least 2.0×10^8 kg of standard coal could be saved, and then a large quantity pollutants could also be reduced as 4.6×10^8 kg CO_2, 9.1×10^6 kg CO, 5.9×10^5 kg SO_2, 2.6×10^6 kg NOx, and 5.4×10^7 kg solid waste every year, which would be great economic and environmental benefits.

Table 1. Statistical and calculated data of EPC solar hot water projects in the 11 Chinese Universities.

No	EPC project name*	Number of bath (people/day)	Solar collecting area (m²)	Alternative standard coal (kg/day)	Reducing emissions (kg/day)				
					CO_2	CO	SO_2	NO_X	Solid waste
1	SDU	10000	7722	2525	5789	114	7	32	675
2	OUC	1900	1532	447	1025	20	1	6	120
3	BJIPT	2400	1862	582	1335	26	2	7	156
4	BJIGC	1460	1125	352	807	16	1	4	94
5	CUG	1000	770	225	515	10	1	3	60
6	CSFU	8400	6482	1892	4337	85	6	24	506
7	WHU	4170	3212	937	2149	42	3	12	251
8	BJIT	9880	7612	2380	5457	107	7	30	636
9	BJNU	20200	15626	4886	11202	220	14	62	1306
10	GXU	8000	6236	1696	3888	76	5	21	453
11	ZSU	9090	7000	1887	4325	85	6	24	504
Total		76500	59179	17810	40830	802	52	225	4760

*SDU: Shandong University; OUC: Ocean University of China; BJIPT: Beijing Institute of Petrochemical Technology; BJIGC: Beijing Institute of Graphic Communication; CUG: China University of Geosciences; CSFU: Central South Financial University; WHU: Wuhan University; BJIT: Beijing Institute of Technology; BJNU: Beijing Normal University; GXU: Guangxi University; ZSU: Zhongshan University

4 CONCLUSIONS

EPC, as a market mechanism for energy conservation was introduced into China since 1997, and now developed quickly in Chinese universities for their solar hot water projects. An EPC project of solar system used in a university of Qingdao was selected as a specific example to analyze its economic and environmental benefits, and the results showed that the solar system could save energy as high as 49. 5%. Moreover, to estimate the development prospect of EPC, 11 universities with EPC project were also investigated, and about 1. 8×10^4 kg standard coal could be saved everyday in these universities. If EPC project of solar system could be used in 10% of Chinese universities and high schools, at least 2. 0×10^8 kg of standard coal could be saved, and about 4. 6×10^8 kg CO_2, 9. 1×10^6 kg CO, 5. 9×10^5 kg SO_2 2. 6×10^6 kg NOx, and 5. 4×10^7 kg solid waste could be reduced every year, according to our estimation.

REFERENCES

CaiYunzhong, Environmental hazards and control of industrial coal-fired boilers, Environment. 01(1999)1–2.

GB 50015-2003, Building water supply and drainage design specifications [S].
GB50364-2005, Civil construction technical code for solar water heating systems [S].
Li Guanyu, Li Qian, Solar thermal utilization project. Beijing: China Culture Publishing House. (2013)94–95.
Wang Wei, Mass balance estimates emissions from coal combustion, Shanxi Architecture, 36(2013)230–231.
Wang Yin, China meteorologica radiation data year book, Beijing: National Meteorological Center. 01(2001)19-20.
Yan Li, AHP-fuzzy evaluation on financing bottleneck in energy performance contracting in China, Energy Procedia. 14(2012)121–126.
Zhu Xiaokai, Energy-saving renovation of existing residential buildings with EPC model, Shenyang: Shenyang Jianzhu University. (2013)24
http://baike. so. com/doc/3541633-3725003. html(Last accessed on 1 February, 2016).
http://news. Science net. cn/htmlnews/2015/3/314504. shtm(Last accessed on 19 May, 2015).
http://www. zgmt. com. cn/. (Last accessed on 6 February, 2016).

Advances in Materials Science, Energy Technology and Environmental Engineering – Patty & Zhou (Eds)
© 2017 Taylor & Francis Group, London, ISBN 978-1-138-19668-1

Enhancement of the lipid production of *Scenedesmus obliquus* cultivated in municipal wastewater by light optimization outdoors

S.F. Han, W.B. Jin, R.J. Tu & X.Y. Wang
Shenzhen Graduate School, Harbin Institute of Technology, Shenzhen, China

ABSTRACT: Despite the significant breakthroughs in research on microalgae as a feedstock for biodiesel, its production cost is still high. One possible solution to overcome this problem is to cultivate algae using wastewater outdoors. During microalgae cultivation, the illumination condition is one of the most important factors that influence microalgae growth and lipid production. Therefore, in this study, light intensity and photoperiod were optimized in outdoor cultivation. In addition, a sunshine transmitter used in architectural lighting was introduced to enhance the light intensity at the regions far away from light incident surface, improving growth and lipid production by *Scenedesmus obliquus*. The results indicated that, on sunny days, proper shading was necessary. Biomass under 60% shading significantly increased by 45% with respect to control exposed to full sunlight. No additional light at night was needed. Furthermore, the sunshine transmitter provided a secondary light source for cells in light-deficient regions, and significantly increased lipid production of *Scenedesmus obliquus* by 15%.

1 INTRODUCTION

Microalgae are considered as one of the most promising feedstocks for biodiesel production nowadays (Abomohra et al. 2014, Han et al. 2015). However, the main drawback for economical biodiesel production from microalgae is the high cost of algal cultivation for the huge consumption of freshwater resources, nitrogen and phosphate, and CO_2 (Uri 2010, Jiang et al. 2011, Ketheesan & Nirmalakhandan 2011). One possible solution to overcome this problem is to cultivate algae in municipal wastewater outdoors. Therefore, wastewater can provide abundant nutrients for algae (Chen et al. 2015, Shi et al. 2007, Tu et al. 2015), and sunlight instead of artificial light can be used as the light source to reduce the cost for algae cultivation.

For the cultivation of microalgae outdoors, the growth and lipid content of microalgae are primarily affected by the light. Many studies demonstrated that the growth rate of microalgae increases with the increase in light intensity within a certain intensity range. If the light intensity exceeds the limits of the required, it produces inhibition and the photosynthetic rate decreased (Richmond et al. 2004). In addition, the lipid content of microalgae is influenced by the light intensity (Markou & Georgakakis 2011). Moreover, illumination time is equally important to algae growth and lipid production. A study shows that best growth of *Scenedesmus obliquus* is obtained with 18:6 h light and darkness cycle in laboratory cultivation (Khoeyi et al. 2012).

In practice, with the growth of algae, regions far from the light source receive insufficient light intensity to maintain growth conditions and prevent the accumulation of biomass. That is why typical cultivation depths are no more than 15 cm for most enclosed PBRs and 35 cm for open raceway ponds (Zhao & Su 2014), which limit the enlargement of algae cultivation. Various attempts have been made to alleviate the negative effect of light attenuation on microalgae growth. One approach is to expose microalgae cells to a sufficient number of photons by energy-intensive active stirring mechanisms or promote the mixing of microalgae suspension along light gradient by installing special static mixers in PBRs (Pruvost et al. 2008), which is energy consumption. Another way is to merely improve the incident light intensity, which may destroy the PS II complexes of the microalgae near the light source and subsequently decrease the biomass productivity of microalgae cultures (Carvalho et al. 2011). Thus, the optimization of transmitted light to microalgae suspension, especially to regions far away from light incident surface, is very important.

In this study, light intensity and photoperiod were optimized in outdoor cultivation. In addition, a sunshine transmitter used in architectural lighting was introduced to enhance the light intensity at the regions far away from light incident surface, improving growth and lipid production by *Scenedesmus obliquus*.

2 MATERIALS AND METHODS

2.1 Algae strain and growth conditions

Scenedesmus obliquus obtained from Freshwater Algae Culture Collection at the Institute of Hydrobiology (strain number FACHB-276) was cultivated outdoors using municipal wastewater as the medium. The experiments were performed at Harbin Institute of Technology, Shenzhen, China (N 22° 35′ 25″, E113° 59′ 01″) during autumn. During the experiment, light intensity and temperature were recorded using the agricultural environment monitoring instrument TNHY-5 made by Zhejiang top instrument Co., Ltd.

Figure 1. Sunshine transmitter device.

2.2 Wastewater used in the experiment

Wastewater was collected by pumping from sewage well at the campus of Harbin Institute of Technology, Shenzhen Graduate School, Shenzhen, China. The main source of the wastewater comes from student domes and teaching building complex. After settled in a plastic bucket, the supernatant was collected for the algal cultivation. The collected wastewater contained (mg L^{-1}) 50.00 ± 4.14 total nitrogen, 45.00 ± 8.28 ammonia-nitrogen, 3.00 ± 0.41 nitrate-nitrogen, and 5.00 ± 0.71 total phosphate with a COD of 240 ± 14 and pH of 7.00 ± 0.08.

2.3 Experiment set-up

When optimizing the light intensity, four kinds of sunshade net which can decrease light intensity by 40%, 60%, 80%, and 95% respectively were used. When optimizing the photoperiod, tubular fluorescent lamps (PHILIPS Master TL-D 85 W/840) with light intensity of 120 µmol/m²/s were used to increase the illumination time for 0 h, 2 h, 6 h, 10 h, and 14 h, respectively, at night. When enhancing the light intensity at the regions far away from the light incident surface, a sunshine transmitter showed in Figure 1 made by Nanjing Scientific Research Technology Co,. Ltd. was used. It can follow the sun to collect sunlight and transfer the sunlight by fiber into the algae culture to enhance the light intensity in the reactor. Total light transmission was 80%.

2.4 Biomass assay

Algal growth was monitored by measuring the culture Optical Density at 680 nm (OD_{680}) and by determination of algal Cellular Dry Weight (CDW). When measuring CDW, take a certain amount of algae fluid to centrifuge for 10 min under 4851 g, discard the supernatant, and transfer the algae with distilled water in a pre-weighed weighing bottle. Then, dry in oven at 105 °C until the weight was constant.

2.5 Lipid analysis

Extraction of lipids was done by the modified Folch method (1957). The lipid extracts were dried under a stream of argon. The pre-weighed glass vials containing the lipid extracts were dried at 80 °C for 30 min, cooled in a desiccator, and weighted.

2.6 Data analysis

All of the values were the means of three replicates. Data obtained were analyzed statistically to determine the degree of significance at P ≤ 0.05 using one-way analysis of variance (ANOVA) with SAS version 6.12

3 RESULTS AND DISCUSSION

3.1 Effect of light intensity on S. obliquus

To prevent light saturation during outdoor cultivation, sunshade nets were used during sunny days (light intensity of 200~2000 µmol/m²/s) to decrease the light intensity. Biomass under 60% shading was significantly increased by 45% when compared with that without shading (Fig. 2). After 5 days of cultivation, the three best-growing groups (60%, 80%, and 40% shading) were selected to measure the lipid production. Biomass and lipid production of *S. obliquus* under 60% shading had no significant difference when compared with that under 80% shading, but significantly increased by 34% and 29% when compared with that under 40% shading (Fig. 3). The results revealed that, on sunny days, proper shading was necessary, helping with avoiding the inhibition of light-saturated areas and promoting algae growth. This is in consistent with Ugwu et al. (2008) who found that shading of the tubular photobioreactor (PBR) surfaces that diminished solar irradiance to 70% led to higher biomass productivity compared with the values obtained when the PBR was completely exposed to full sunlight.

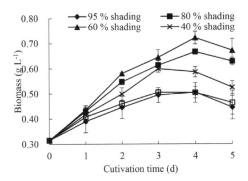

Figure 2. Growth curve of *S. obliquus* under different shading rates.

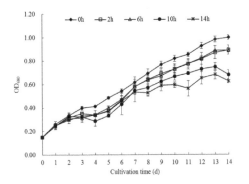

Figure 3. Biomass and lipid production of *S. obliquus* under different shading rates.

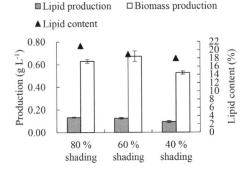

Figure 4. Growth curves of *S. obliquus* under different photoperiods.

3.2 *Effect of photoperiod on S. obliquus*

In cultivation of microalgae outdoors, the time of sunshine is limited. Therefore, artificial light was used at night to increase the illumination time by 0 h, 2 h, 6 h, 10 h, and 14 h, respectively to determine the best photoperiod of *S. obliquus*. Figure 4 shows that the additional light at night was not

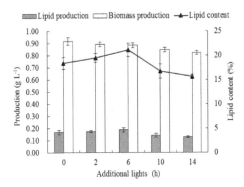

Figure 5. Biomass and lipid production of *S. obliquus* under different photoperiods.

conducive to the growth of *S. obliquus*. On the contrary, it decreased the growth of *S. obliquus* (Fig. 4). When the additional light increased to 2 h and 6 h at night, the biomass and lipid production had no significant difference compared with 0 h. When additional light increased to 10 h and 14 h at night, the biomass and lipid production decreased significantly compared with no light added at night (Fig. 5). Therefore, there was no need to increase the illumination time at night. The results showed that it was the shorter the light time, the better the growth. This is consistent with the results obtained by Khoeyi et al. (2012) who found that the maximum biomass was with 16:8 h light/dark photoperiod when cultivating *Chlorella vulgaris* in the laboratory.

3.3 *Effect of sunshine transmitter on S. obliquus*

To enhance the light intensity of the regions far away from light incident surface, a sunshine transmitter used in architectural lighting was introduced, located at 15 cm depth below the liquid level of the culture.

Using sunlight and the sunshine transmitter as the light source, and OD_{680} of the culture was 0.38. Light distribution characteristics are shown in Figure 6. The use of sunshine transmitter is similar to using a secondary light source and would provide better lighting conditions for microalgae cells in regions 10–40 cm from the light incident surface. Microalgae can grow in the light intensity range of 40~300 µmol/m²/s. When using only sunlight as the light source, within the range of 0 ~ 5 cm depth, the light intensity was about 1200~300 µmol/m²/s, which is light inhibition area; within the range of 5 ~ 15 cm depth, the light intensity was about 300~40 µmol/m²/s, which is the effective optical area; and below 15 cm, the light intensity was below 40 µmol/m²/s, which is the dark area. After the luminous end of the sunshine transmitter was placed at the depth of

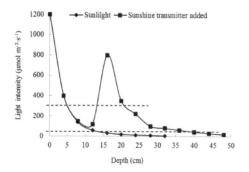

Figure 6. Distribution of light at different depths in the culture system.

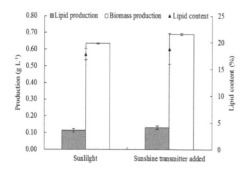

Figure 7. Biomass and lipid production of *S. obliquus* with and without sunshine transmitter.

15 cm, light distribution changed. The original light inhibition area basically had no change in the depth of 0 ~ 5 cm, but when added within the range of 13 ~ 21 cm depth, the light intensity was more than 300 μmol/m²/s. In addition, the effective illumination area extended to 40 cm. In addition, biomass and lipid production of *S. obliquus* were significantly increased by 9% and 15% separately (Fig. 7). In conclusion, importing the sunshine transmitter could significantly increase the light intensity in the culture system and increase lipid productivity.

4 CONCLUSION

In conclusion, this study demonstrated the feasibility of cultivation of *S. obliquus* using municipal wastewater outdoors. In addition, with proper light optimization such as shading on sunny days, biomass and lipid production of *S. obliquus* can be enhanced. In addition, this work provides an efficient approach to mitigating the adverse effects of heterogeneous light distribution on microalgae growth. By using a sunshine transmitter, biomass and lipid production of *S. obliquus* significantly increased.

ACKNOWLEDGMENTS

This work was financially supported by Shenzhen Science and Technology Innovation projects (project number JCYJ20140417172417125).

REFERENCES

Abomohra, A., El-Sheekh, M. & Hanelt, D. 2014. Pilot cultivation of the chlorophyte microalga Scenedesmus obliquus as a promising feedstock for biofuel. *Biomass Bioenerg.* 64: 237–244.

Carvalho, A.P., Silva, S.O., Baptista, J.M. & Malcata, F.X. 2011. Light requirements in microalgae photobioreactors: an overview of biophotonic aspects. *Appl. Microbiol. Biotechnol.* 89: 1275–1288.

Chen, G., Zhao, L. & Qi, Y. 2015. Enhancing the productivity of microalgae cultivated in wastewater toward biofuel production: A critical review. *Appl. Energ.* 137: 282–291.

Folch, J., Lees, M. & Stanley, G.H.S. 1957. A simple method for the isolation and purification of total lipids from animal tissues. *J. Biol. Chem.* 226: 497–509.

Han, S.F., Jin, W.B. Tu, R.J. & Wu, W.M. 2015. Biofuel production from microalgae as feedstock: current status and potential. *Crit. Rev. Biotechnol.* 35: 255–268.

Jiang, L., Luo, S., Fan, X., Yang, Z. & Guo, R. 2011. Biomass and lipid production of marine microalgae using municipal wastewater and high concentration of CO₂. *Appl. Energ.* 88: 3336–3341.

Ketheesan, B. & Nirmalakhandan, N. 2011. Development of a new airlift-driven raceway reactor for algal cultivation. *Appl. Energ.* 88(10): 3370–3376.

Khoeyi, Z.A., Seyfabadi, J. & Ramezanpour, Z. 2012. Effect of light intensity and photoperiod on biomass and fatty acid composition of the microalgae, Chlorella vulgaris. *Aquacult. Int.* 20: 41–49.

Markou, G. & Georgakakis, D. 2011. Cultivation of Filamentous cyanobacteria (blue-green algae) in agro-industrial wastes and wastewaters: a review. *Appl. Energ.* 88: 3389–3401.

Pruvost, J., Cornet, J.F. & Legrand, J. 2008. Hydrodynamics influence on light conversion in photobioreactors: an energetically consistent analysis. *Chem. Eng. Sci.* 63: 3679–3694.

Richmond, A. 2004. Principles for attaining maximal microalgal productivity in photobioreactors: an overview. *Hydrobiologia* 512: 3–37.

Shi, J., Podola, B. & Melkonian, M. 2007. Removal of nitrogen and phosphorus from wastewater using microalgae immobilized on twin layers: an experimental study. *J. Appl. Phys.* 19: 417–423.

Tu, R.J., Jin, W B., Xi, T.T., Yang, Q., Han, S.F. & Abomohra, A. 2015. Effect of static magnetic field on the oxygen production of Scenedesmus obliquus cultivated in municipal wastewater. *Water Res.* 86: 132–138.

Uri, P. 2010. Accumulation of triglycerides in green microalgae: a potential source for biodiesel. *FEBS J.* 277: 5–36.

Zhao, B. & Su, Y. 2014. Process effect of microalgal-carbon dioxide fixation and biomass production: a review. *Renew. Sust. Energy Rev.* 31: 121–132.

Advances in Materials Science, Energy Technology
and Environmental Engineering – Patty & Zhou (Eds)
© *2017 Taylor & Francis Group, London, ISBN 978-1-138-19668-1*

High Cu sedimentation locations in the Jiaozhou Bay

Dongfang Yang
Research Center for Karst Wetland Ecology, Guizhou Minzu University, Guizhou Guiyang, China
College of Chemistry and Environmental Science, Guizhou Minzu University, Shanghai, China
North China Sea Environmental Monitoring Center, SOA, Qingdao, China

Huazhong He, Fengyou Wang, Sixi Zhu & Ming Wang
Research Center for Karst Wetland Ecology, Guizhou Minzu University, Guizhou Guiyang, China
College of Chemistry and Environmental Science, Guizhou Minzu University, Shanghai, China

ABSTRACT: Based on investigation data on Cu in bottom waters in 1985, we analyzed the content, pollution, and horizontal distribution of Cu in bottom waters in the bay mouth of the Jiaozhou Bay, eastern China. Results indicated that Cu contents were 0.10–0.42 μg L^{-1}, and were confirmed with Grade I according to Chinese Sea Water Quality Standard (GB 3097–1997), indicating that this bay had not been polluted by Cu in 1985. High Cu sedimentation locations were occurring in different regions due to different Cu sources. In April and October, high Cu sedimentation location was in the open sea where marine current was the major source; while, in July, high Cu sedimentation location was in the bay mouth where stream flow was the major source. Generally, the spatial-temporal features of sedimentation of Cu were determined by pollution sources and the vertical water's effect.

1 INTRODUCTION

Cu is one of the necessary elements for organisms, yet the excessive intake is harmful. A large amount of anthropogenic Cu-containing waste gas and wastewater were discharged into the environment along with the rapid development of population and economy. The marine water was finally polluted by excess Cu (Yang & Miao & Song, et al., 2015; Yang & Miao & Cui, et al., 2015; Yang & Wang & Zhu, et al., 2015; Yang & Zhu & Wu, et al., 2015; Yang & Zhu & Wang, et al., 2015). Cu contents in sea waters were changing along with the transport process by means of vertical water's effect. Based on investigation data on Cu in bottom waters in 1985, we analyzed the content, pollution level, and transport feature of Cu in bottom waters in the Jiaozhou Bay, eastern China, and to provide scientific basis for researching on the existence and transfer of Cu in marine bay.

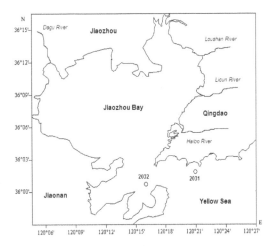

Figure 1. Geographic location and sampling sites in the Jiaozhou Bay.

2 MATERIALS AND METHODS

The Jiaozhou Bay (35°55′–36°18′ N, 120°04′–120°23′ E) is located in the south of Shandong Province, eastern China (Fig. 1). It is a semi-closed bay with the total area, average water depth, and bay mouth width of 446 km^2, 7 m, and 3 km, respectively, and is connected to the Yellow Sea in the south. There are more than ten inflow rivers (e.g. Haibo River, Licun River, Dagu River, and Loushan River), most of which have seasonal features (Yang & Chen & Gao, et al., 2005; Yang & Wang & Gao, et al., 2004). The data were provided by the North China Sea Environmental Monitoring Center. The survey was conducted in April, July, and October 1985. Bottom water samples in three

Table 1. Pb contents in the bottom water in the Jiaozhou Bay in April, July, and October 1985.

Time	April	July	October
Content/μg L⁻¹	0.10–0.12	0.19–0.42	0.19–0.30
Grade	I	I	I

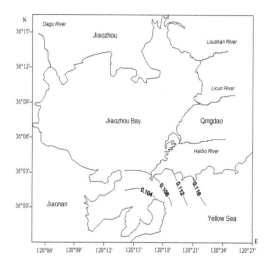

Figure 2. Horizontal distribution of Cu in bottom waters in the Jiaozhou Bay in April 1985/μg L⁻¹.

stations (i.e. 2031, 2032 and 2033) were collected and measured following the National Specification for Marine Monitoring (State Ocean Administration, 1991).

3 RESULTS AND DISCUSSION

3.1 *Contents of Cu in bottom waters*

The contents of Cu in April, July, and October 1985 in bottom waters in the Jiaozhou Bay and the open sea were 0.10–0.12 μg L⁻¹, 0.19–0.42 μg L⁻¹, and 0.19–0.30 μg L⁻¹, respectively. In comparison with Grade I (5.00 μg L⁻¹) of the National Sea Water Quality Standard (GB 3097–1997) for Cu, the Cu contents were very low. It could be concluded that the Jiaozhou Bay and the open sea had not been polluted by Cu in 1985.

3.2 *Distribution of Cu*

The three sampling Sites of 2031, 2032, and 2033 were located in the open sea, the bay mouth, and the inside of the bay mouth, respectively. In April, there was a high-value zone in the coastal waters in the outside of the bay mouth, and the highest value

(0.12 μg L⁻¹) was occurring in Site 2031 (Fig. 2). The contour lines were forming a series of parallel lines, which were decreasing from the open waters in the east to the bay mouth in the west (0.10 μg L⁻¹). In July, there was a high-value zone in the bay mouth, and the highest value (0.42 μg L⁻¹) was occurring in site 2032 (Fig. 3). The contour lines were forming a series of parallel lines, which were decreasing from the bay mouth to the open waters (0.19 μg L⁻¹). In October, there was a high-value zone in the coastal waters in the outside of the bay mouth, and the highest value (0.30 μg L⁻¹) was

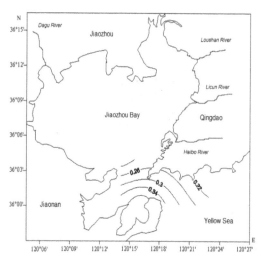

Figure 3. Horizontal distribution of Cu in bottom waters in the Jiaozhou Bay in July 1985/μg L⁻¹.

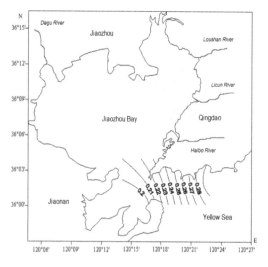

Figure 4. Horizontal distribution of Cu in bottom waters in the Jiaozhou Bay in October 1985/μg L⁻¹.

occurring in site 2031 (Fig. 4). The contour lines were forming a series of parallel lines, which were decreasing from the open waters in the east to the bay mouth in the west (0.19 μg L⁻¹). The horizontal distributions of Cu in bottom waters were showing seasonal variations, and were the products of pollution sources and the sedimentation of Cu.

3.3 *Sedimentation of Cu*

The contents of the substances were decreasing contentiously by means of water exchange (Yang & Miao & Xu, et al., 2013; Yang & Wu & Wang, et al., 2014). The Cu contents in April were 0.10–0.12 μg L⁻¹, and were increasing from the bay mouth to the open waters, because the marine current was the major Cu source in April, whose source strength was 0.39 μg L⁻¹, hence high sedimentation location was in the open waters, yet low sedimentation rate region was in the bay mouth. Cu contents in July were 0.19–0.42 μg L⁻¹, and were decreasing from the bay mouth to the open waters, as stream flow was the major Cu source in July, whose source strength was 0.37–0.42 μg L⁻¹, hence high sedimentation location was in the bay mouth, yet low sedimentation rate region was in the open sea. Cu contents in October were 0.19–0.30 μg L⁻¹, and were increasing from the bay mouth to the open waters, because marine current was the major Cu source in October, whose source strength was 0.39 μg L⁻¹, hence high sedimentation location was in the open waters, yet low sedimentation rate region was in the bay mouth. Generally, the spatial-temporal features of sedimentation of Cu were determined by pollution sources and the vertical water's effect.

4 CONCLUSIONS

The contents of Cu in April, July, and October 1985 in bottom waters in the Jiaozhou Bay and the open sea were 0.10–0.12 μg L⁻¹, 0.19–0.42 μg L⁻¹, and 0.19–0.30 μg L⁻¹, respectively, indicating that this bay had not been polluted by Cu in 1985. The horizontal distributions of Cu in bottom waters were showing seasonal variations, and were the products of pollution sources and the sedimentation of Cu. In April and October, high Cu sedimentation location was in the open sea where marine current was the major pollution source, while, in July, high Cu sedimentation location was in the

bay mouth where stream flow was the major pollution source. Generally, spatial-temporal features of sedimentation of Cu were determined by pollution sources and the vertical water's effect.

ACKNOWLEDGMENTS

This research was sponsored by Doctoral Degree Construction Library of Guizhou Nationalities University, Education Ministry's New Century Excellent Talents Supporting Plan (NCET-12–0659), the China National Natural Science Foundation (31560107), Major Project of Science and Technology of Guizhou Provincial ([2004]6007–01), Guizhou R & D Program for Social Development ([2014] 3036) and Research Projects of Guizhou Nationalities University ([2014]02), Research Projects of Guizhou Province Ministry of Education (KY [2014] 266), Research Projects of Guizhou Province Ministry of Science and Technology (LH [2014] 7376), and Comprehensive Reform Pilot Project of Environmental Science Specialty ([2013]446).

REFERENCES

State Ocean Administration. The specification for marine monitoring (HY003.4–91): Beijing, Ocean Precess, (1991). (in Chinese)
Yang DF, Chen Y, Gao ZH, et al.: Chinese Journal of Oceanology Limnoogy, Vil. 23(2005): 72–90.
Yang DF, Miao ZQ, Cui WL, et al.: Advances in intelligent systems research, (2015):17–20.
Yang DF, Miao ZQ, Song WP, et al.: Advanced Materials Research, Vol.1092–1093 (2015): 1013–1016.
Yang DF, Miao ZQ, Xu HZ, et al.: Marine Environmental Science, Vol. 32 (2013), p. 373–380. (in Chinese)
Yang DF, Wang F, Gao ZH, et al.: Maine Science, Vol. 28 (2004): 71–74. (in Chinese with English abstract)
Yang DF, Wang FY, He HZ, et al.: Proceedings of the 2015 international symposium on computers and informatics, Vol. (2015): 2655–2660.
Yang DF, Wang FY, Zhu SX, et al.: Advances in Engineering Research, 31(2015): 1284–1287.
Yang DF, Wu FY, Wang L, et al.:2014 IEEE workshop on electronics, computer and applications. Part C, (2014): 1008–1011.
Yang DF, Zhu SX, Wang FY, et al.: Advances in Computer Science Research, (2015): 1765–1769.
Yang DF, Zhu SX, Wu YJ, et al.: Advances in Engineering Research, 31(2015): 1288–1291.

Advances in Materials Science, Energy Technology
and Environmental Engineering – Patty & Zhou (Eds)
© 2017 Taylor & Francis Group, London, ISBN 978-1-138-19668-1

Seasonal variations of Pb sources in the Jiaozhou Bay during 1979–1983

Dongfang Yang
China College of Life Science, Shanghai Ocean University, Shanghai, China
North China Sea Environmental Monitoring Center, SOA, Qingdao, China

Danfeng Yang
College of Information Science and Engineering, Fudan University, Shanghai, China

Yinjiang Zhang & Xiancheng Qu
China College of Life Science, Shanghai Ocean University, Shanghai China

Yu Chen
College of Information, Shanghai Ocean University, Shanghai China

ABSTRACT: Based on investigation data on Pb in surface waters in the Jiaozhou Bay in September 1980, April 1981, June 1982, and October 1983, we analyzed the horizontal distributions of Pb contents in different seasons, and revealed the major Pb sources. The results indicated that the horizontal distributions of Pb were showing significant seasonal variations. Stream flow, overland runoff, atmosphere deposition, marine current, and the top of the island were the major Pb sources in spring; stream flow and marine current were the major Pb sources in summer; stream flow, marine current, and docks were the major Pb sources in autumn. Source treatment is one of the basic approaches to control Pb pollution, and relevant countermeasures should be taken for different pollution sources.

1 INTRODUCTION

Human activities have caused many environment pollution issues. Pb pollution in marine bay is one of the critical environmental problems in many countries and regions due to the lagging of waste treatment to the emission of Pb-containing waste (Yang & Su & Gao, et al., 2008; Yang & Guo & Zhang, et al., 2011; Yang & Zhu & Wang, et al., 2014; Yang & Geng & Chen, et al., 2014; Yang & Ge & Song, et al., 2014; Yang &Zhu & Wang, et al., 2014;). Since source treatment is one of the basic approaches for pollution control, understanding the seasonal variations of Pb sources is essential to environmental protection.

The Jiaozhou Bay is located in Shandong Province, China, and is surrounded by economic and agricultural developed regions of Qingdao, Jiaozhou, and Jiaonan. Previous studies showed that this bay had been polluted by various pollutants including Pb (Yang & Su & Gao, et al., 2008; Yang & Guo & Zhang, et al., 2011; Yang & Zhu & Wang, et al., 2014; Yang & Geng & Chen, et al., 2014; Yang & Ge & Song, et al., 2014; Yang & Zhu & Wang, et al., 2014;). This aim of this paper is to analyze the horizontal distributions of Pb contents and the seasonal variations of Pb sources in this bay, and to provide scientific basis for pollution control.

2 STUDY AREA

The Jiaozhou Bay (35°55′–36°18′N, 120°04′-120°23′ E) is located in the south of Shandong Province, eastern China (Fig. 1). It is a semi-closed bay with the total area and average water depth of 446 km^2 and 7 m, respectively. The bay mouth is located between Tuandao Island and Xuejiadao Island, and is connected to the Yellow Sea in the south. The width of the bay mouth is only 3 km. There are more than ten inflow rivers (e.g. Haibo River, Licun River, Dagu River, and Loushan River), most of which have seasonal features (Yang & Chen & Gao, et al., 2005; Yang & Wang & Gao, et al., 2004).

The data were provided by the North China Sea Environmental Monitoring Center. The investigations of Pb contents in surface waters were conducted in September 1980, April 1981, June 1982,

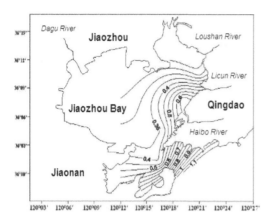

Figure 1. Horizontal distribution of Pb in surface waters in September 1980/μg L⁻¹.

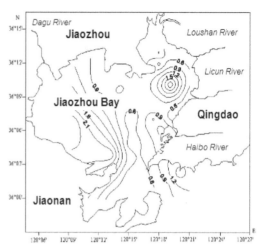

Figure 2. Horizontal distribution of Pb in surface waters in April 1981/μg L⁻¹.

and October 1983, respectively (Yang & Su & Gao, et al., 2008; Yang & Guo & Zhang, et al., 2011; Yang & Zhu & Wang, et al., 2014; Yang & Geng & Chen, et al., 2014; Yang & Ge & Song, et al., 2014; Yang & Zhu & Wang, et al., 2014;). Surface water samples were collected and measured following the National Specification for Marine Monitoring (State Ocean Administration, 1991). In study area, April, May, and June belong to spring; July, August, and September belong to summer; October, November and December belong to autumn.

3 RESULTS

The horizontal distributions of Pb in surface waters in the Jiaozhou Bay in September 1980, April 1981, June 1982, and October 1983 are shown in Fig. 1, Fig. 2, Fig. 3, and Fig. 4, respectively. In September 1980, there was a high-value region (0.76 μg L⁻¹) in the coastal waters between the estuaries of Licun River and Haibo River, and there were a series of parallel lines, which were decreasing along with the flow directions (Fig. 1). Meanwhile, another high-value region was occurring in the open waters (1.00 μg L⁻¹), and Pb contents were decreasing from the open waters to the bay mouth, as a series of parallel lines (Fig. 1). In April 1981, there was a high-value region (2.65 μg L⁻¹) in the coastal waters in the southwest of the bay, and Pb contents were decreasing from the high-value center to the center of the bay (0.60 μg L⁻¹) (Fig. 2). The was another high-value region (2.39 μg L⁻¹) in the northeast of the bay, and Pb contents were decreasing from the high-value center to the ambient-value center (0.60 μg L⁻¹)(Fig. 2). Moreover, there was the was also a high-value region (1.66 μg L⁻¹) in the open waters

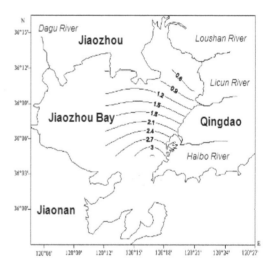

Figure 3. Horizontal distribution of Pb in surface waters in June 1982 /μg L⁻¹.

closed to the top of the island, and were decreasing from the high-value center to the ambient-value center (0.60 μg L⁻¹) (Fig. 2). In June 1982, Pb contents were decreasing from the bay mouth (3.35 μg L⁻¹) to the northeast of the bay (0.45 μg L⁻¹); the contour line was forming a series of parallel lines, which were decreasing along with the flow direction of the marine current (Fig. 3). In October 1983, there was a high-value region (2.34 μg L⁻¹) in the coastal waters between the estuaries of Licun River and Loushan River in the northeast of the bay, and there were a series of parallel lines, which

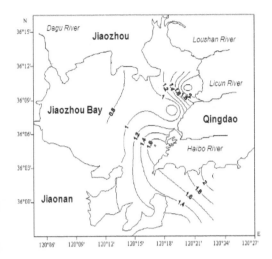

Figure 4. Horizontal distribution of Pb in surface waters in October 1983 /µg L⁻¹.

Table 1. Major Pb sources in different seasons.

Pb Source	Spring	Summer	Autumn
Stream flow	Yes	Yes	Yes
Overland runoff	Yes	No	No
atmosphere deposition	Yes	No	No
Marine current	Yes	Yes	Yes
Docks	No	No	Yes
the top of the island	Yes	No	No

were decreasing from the high-value center to the center of the bay (0.59 µg L⁻¹) (Fig. 4). There was another high-value region in the coastal waters in the east of the bay, and were decreasing from the high-value center (1.82 µg L⁻¹) to the northwest of the bay (0.67 µg L⁻¹) (Fig. 4). Meanwhile, another high-value were occurring in the open waters (2.22 µg L⁻¹), and Pb contents were decreasing from the open waters to the bay mouth (1.33 µg L⁻¹) (Fig. 4).

4 DISCUSSION

The horizontal distributions of Pb in surface waters in the Jiaozhou Bay were showing significant seasonal variations, indicating that the sources of Pb were also different in different seasons. In September 1980, high-Pb-content regions were occurring in the estuaries of Haibo River and Licun River, indicating that the stream flow was one of the major P sources. Meanwhile, there was another high-value region in the open waters, indicating that marine current was one of the major Pb sources. In April 1981, there was a high-value region in the coastal waters in the southwest of the bay, indicating that overland runoff was one of the major Pb sources. High-value region was occurring in the center of the northwest of the bay, form where the estuaries were far away, indicating that the atmospheric deposition was the major Pb source. Moreover, there was a high-value region in waters close to the top of the island, indicating that the top of the island was one of the major

Pb sources. In June 1982, there was a high-value region in the open waters, indicating that marine current was the major Pb source. In October 1983, a high-value region was occurring in the estuaries of Licun River and Loushan River, indicting that stream flow was the major P source. Another high-value region was occurring in the coastal water in the east of the bay, where there were docks of the bay, indicating that docks was the major Pb source. Moreover, there was also a high-value region in the open waters outside the bay, indicating that marine current was the major Pb source. Hence, it could be found that there were different major Pb sources in different seasons (Table 1). Source treatment is one of the basic approaches to control Pb pollution, and relevant countermeasures should be taken for different pollution sources. However, it should be noticed that marine current was one of the major Pb sources, implying that human activities had increased the pollution level in the ocean.

5 CONCLUSION

The horizontal distributions of Pb in surface waters in the Jiaozhou Bay were showing significant seasonal variations, indicating that the sources of Pb were also different in different seasons. Stream flow, overland runoff, atmosphere deposition, marine current, and the top of the island were the major Pb sources in spring; stream flow and marine current were the major Pb sources in summer; stream flow, marine current, and docks were the major Pb sources in autumn. Source treatment is one of the basic approaches to control Pb pollution, and relevant countermeasures should be taken for different pollution sources.

ACKNOWLEDGMENTS

This research was sponsored by Doctoral Degree Construction Library of Guizhou Nationalities University, Education Ministry's New Century Excellent Talents Supporting Plan (NCET-12-0659), the China National Natural Science Foundations (31560107 and 31500394), Research Projects of Guizhou Nationalities University ([2014]02), Research Projects of Guizhou Province Ministry of Education (KY [2014] 266), and Research Projects of Guizhou Province Ministry of Science and Technology (LH [2014] 7376).

REFERENCES

State Ocean Administration. The specification for marine monitoring: Beijing, Ocean Precess, (1991).

Yang D F, Su C, Gao Z H, et al.: Chin. J. Oceanol. Limnol., Vol. 26(2008): 296–299.

Yang DF, Chen Y, Gao ZH, Zhang J, et al.: Chinese Journal of Oceanology and Limnology, Vol. 23(2005): 72–90.

Yang DF, Ge HG, Song FM, et al.: Applied Mechanics and Materials, Vol. 651–653 (2014), p. 1492–1495.

Yang DF, Geng X, Chen ST, et al.: Applied Mechanics and Materials, Vol. 651–653 (2014), p. 1216–1219.

Yang DF, Guo JH, Zhang YJ, et al.: Journal of Water Resource and Protection, Vol. 3(2011): 41–49.

Yang DF, Wang F, Gao ZH, et al.: Marine Science, Vol. 28 (2004):71–74.

Yang DF, Zhu SX, Wang FY, et al.: Applied Mechanics and Materials, Vol. 651–653(2014), p. 1419–1422.

Yang DF, Zhu SX, Wang FY, et al.: Applied Mechanics and Materials, Vol.651–653 (2014), p. 1292–1294.

Advances in Materials Science, Energy Technology
and Environmental Engineering – Patty & Zhou (Eds)
© 2017 Taylor & Francis Group, London, ISBN 978-1-138-19668-1

Spatial-temporal variations of Pb sources in the Jiaozhou Bay during 1979–1983

Dongfang Yang
China College of Life Science, Shanghai Ocean University, Shanghai, China
North China Sea Environmental Monitoring Center, SOA, Qingdao, China

Yinjiang Zhang & Xiancheng Qu
China College of Life Science, Shanghai Ocean University, Shanghai, China

Danfeng Yang
College of Information Science and Engineering, Fudan University, Shanghai, China

Yu Chen
College of Information, Shanghai Ocean University, Shanghai, China

ABSTRACT: Based on the investigation of Pb contents in surface waters in different seasons in the Jiaozhou Bay during 1979–1983, this paper analyzed the Pb sources and the variations in the Jiaozhou Bay. The results indicated that Pb in the Jiaozhou Bay were mainly sourced from stream flow, overland runoff, atmosphere deposition, marine current, the top of the island, and docks, whose source strengths were 0–75–3.35 μg L-1, 2.65–3.30 μg L-1, 0.69–2.55 μg L-1, 1.47–1.82 μg L-1, 0.76–3.34 μg L-1, and 2.39–3.25 μg L-1, respectively. According to the National Sea Water Quality Standard (GB 3097–1997) for Pb, the source strengths of Pb in the Jiaozhou Bay during 1979–1983 were confirmed to be Grade II or III generally, indicating that the pollution level of Pb in this bay was slight or moderate in the early stage of reform and opening-up. However, the source strengths of Pb were showing increasing trends along with time, and the source treatment of Pb is necessary for Pb pollution control in this bay.

1 INTRODUCTION

Pb pollution is one of the critical environmental issues in many countries and regions, since the rapid development of industry, population, industrialization, and urbanization. Many marine bays have been polluted by Pb, since seawater is the sink of the pollutants (Yang & Su & Gao, et al., 2008; Yang & Guo & Zhang, et al., 2011; Yang & Zhu & Wang, et al., 2014; Yang & Geng & Chen, et al., 2014; Yang & Ge & Song, et al., 2014; Yang & Zhu & Wang, et al., 2014;). Understanding Pb sources and their spatial-temporal variations is essential for pollution control and environmental protection in marine bay.

The Jiaozhou Bay is located in Shandong Province, China, and is surrounded by economic and agricultural developed regions of Qingdao, Jiaozhou, and Jiaonan. This bay had been polluted by various pollutants including Pb, since the rapid development of economy of Chinese reform and opening-up (Yang & Su & Gao, et al., 2008; Yang & Guo & Zhang, et al., 2011; Yang & Zhu & Wang, et al., 2014; Yang & Geng & Chen, et al.,

2014; Yang & Ge & Song, et al., 2014; Yang & Zhu & Wang, et al., 2014;). This aim of this paper is to analyze the spatial-temporal variations in Pb sources in this bay, and to provide scientific basis for pollution control.

2 MATERIALS AND METHODS

The Jiaozhou Bay (35°55′-36°18′ N, 120°04′-120°23′ E) is located in the south of Shandong Province, eastern China. It is a semi-closed bay with the total area, average water depth, and bay mouth width of 446 km², 7 m, and 3 km, respectively, and is surrounded by Qingdao City, Jiaozhou City, and Jiaonan City in the east, north, and west, respectively. The bay mouth is located between Tuandao Island and Xuedao Island, and is connected to the Yellow Sea in the south (Fig. 1). There are more than ten inflow rivers (e.g. Haibo River, Licun River, Dagu River, and Loushan River), most of which have seasonal features (Yang & Chen & Gao, et al., 2005; Yang & Wang & Gao, et al., 2004).

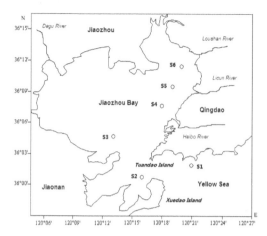

Figure 1. Geographic location and sampling sites in the Jiaozhou Bay.

The data were provided by the North China Sea Environmental Monitoring Center. The investigations of Pb contents in surface waters in six sampling sites (i.e. S1, S2, S3, S4, S5, and S6) were conducted in May, August, and October 1979; June, July, September, and October 1980; April, August, and November 1981; April, June, July, and October 1982; and May, September, and October 1983, respectively (Yang & Su & Gao, et al., 2008; Yang & Guo & Zhang, et al., 2011; Yang & Zhu & Wang, et al., 2014; Yang & Geng & Chen, et al., 2014; Yang & Ge & Song, et al., 2014; Yang & Zhu & Wang, et al., 2014;). Surface water samples were collected and measured following the National Specification for Marine Monitoring (State Ocean Administration, 1991). In the study area, April, May, and June belong to spring; July, August, and September belong to summer; October, November, and December belong to autumn.

3 RESULTS AND DISCUSSION

3.1 Spatial-temporal variations in high-Pb-content regions

Generally, the distributions of Pb contents in waters were mainly determined by Pb sources and source strengths, and the sources of Pb could be defined according to the spatial-temporal variations of high-Pb-content regions (Table 1). In May, August, and October 1979, high-Pb-content regions were occurring in estuaries of Licun and Haibo Rivers, and open waters and open waters, respectively. In June, July, September, and October 1980, high-Pb-content regions were occurring in

estuaries of Licun and Haibo Rivers; estuary of Licun River and Haibo River, and open waters; estuary of Haibo Rivers and open waters; and open waters, respectively. In April, August, and November 1981, high-Pb-content regions were occurring in coastal waters in the southwest of the bay, and center of the bay and open waters; the top of Xuedao Island, center of the bay, and open waters; and estuary of Haibo Rivers and coastal of the southwest, respectively. In April; June; July, and October 1982, high-Pb-content regions were occurring in center of the bay, open waters, and the top of Xuedao Island, respectively. In May; September, and October 1983, high-Pb-content regions were occurring in estuaries of Licun and Loushan Rivers, and coastal waters in the east of the bay, the top of Xuedao Island, estuaries of Licun and Loushan Rivers, coastal water in the east of the bay, and open waters, respectively. It could be defined that Pb in the Jiaozhou Bay were sourced from stream flow, overland runoff, atmosphere deposition, marine current, island top, and docks.

3.2 Spatial-temporal variations in Pb sources

Generally, the source strength of stream flow, overland runoff, atmosphere deposition, marine current, the top of the island, and docks were 0–75–3.35 µg L⁻¹, 2.65–3.30 µg L⁻¹, 0.69–2.55 µg L⁻¹, 1.47–1.82 µg L⁻¹, 0.76–3.34 µg L⁻¹, and 2.39–3.25 µg L⁻¹, respectively (Table 2). The variations in Pb sources in the Jiaozhou Bay were productions of anthropogenic activities, and were showing significant variations along with the rapid development of economic and population (Table 3). Marine current and stream flow were always the major source, and the source strengths were showing increasing trend during 1979–1983. Overland runoff was also one of the major Pb sources, yet were mainly dependent on rainfall-runoff, and was only responsible in 1981. Atmosphere deposition was responsible in 1981 and 1982, and the source strengths were relatively high. Island top was one of the processes and sources of Pb to the bay, and was responsible in 1981, 1982, and 1983, and there was no significant changing trend since the generation and emission of this source were also dependent on rainfall-runoff. Docks had been one of the major Pb sources since 1983, indicating that marine transport was responsible for Pb pollution in this bay due to the shipping oil pollution. According to the National Sea Water Quality Standard (GB 3097–1997) for Pb (Table 4), the source strengths of Pb in the Jiaozhou Bay during 1979–1983 were confirmed to be Grade II or III generally, indicating that the pollution level of Pb in this bay was slight or moderate in the early stage of reform and

Table 1. High-value positions of Pb content in the Jiaozhou Bay during 1979–1983.

Year	April	May	June	July	August	September	October	November
1979		Estuary			Open waters		Open waters	
1980			Estuary	Open waters		Open waters		
1981	Coastal waters, Bay center, Open waters				Island top, Bay center, Open waters			Estuary, coastal waters
1982	Bay center		Open waters	Open waters			Island Top	
1983		Estuary, coastal waters			Island Top		Estuary, coastal waters, Open waters	

Table 2. Source strengths of Pb in the Jiaozhou Bay during 1979–1983.

Pb source	Marine current	Overland runoff	Stream flow	Docks	Island top	Atmosphere deposition
Strength/μg L^{-1}	0.75–3.35	2.65–3.30	0.69–2.55	1.47–1.82	0.76–3.34	2.39–3.25
First appearance	1979	1981	1979	1983	1981	1981
Count	4	1	4	1	3	2

Table 3. Variations in Pb sources in the Jiaozhou Bay during 1979–1983/μg L^{-1}.

Year	Marine current	Overland runoff	Stream flow	Docks	The top of the island	Atmosphere deposition
1979	0.75–1.52		0.99			
1980	0.89–2.71		0.69–0.88			
1981		2.65–3.30	2.55		1.66–3.34	2.39
1982	2.67–3.35				0.76	3.25
1983	2.22		1.67–2.34	1.47–1.82	2.33	

Table 4. Guideline of Pb contents in the National Sea Water Quality Standard (GB 3097–1997).

Grade	I	II	III	IV
Guideline/μg L^{-1}	1.0	2.0	5.0	50.0

opening-up, and the source strengths were showing increasing trends.

4 CONCLUSION

Pb contents in the Jiaozhou Bay were mainly sourced from stream flow, overland runoff, atmosphere deposition, marine current, the top of the island, and docks. The pollution level of Pb in this bay was slight or moderate in the early stage of reform and opening-up, and the source strengths were showing increasing trends. Hence, the source treatment of Pb is necessary for Pb pollution control in this bay.

ACKNOWLEDGMENTS

This research was sponsored by Doctoral Degree Construction Library of Guizhou Nationalities University, Education Ministry's New Century Excellent Talents Supporting Plan (NCET-12–0659), the China National Natural Science Foundations (31560107 and 31500394), Research Projects of Guizhou Nationalities University ([2014]02), Research Projects of Guizhou Province Ministry of Education (KY [2014] 266), and Research Projects of Guizhou Province Ministry of Science and Technology (LH [2014] 7376).

REFERENCES

State Ocean Administration. The specification for marine monitoring: Beijing, Ocean Precess, (1991).

Yang D F, Su C, Gao Z H, et al.: Chin. J. Oceanol. Limnol., Vol. 26(2008): 296–299.

Yang DF, Chen Y, Gao ZH, Zhang J, et al.: Chinese Journal of Oceanology and Limnology, Vol. 23(2005): 72–90.

Yang DF, Ge HG, Song FM, et al.: Applied Mechanics and Materials, Vols. 651–653 (2014), pp. 1492–1495.

Yang DF, Geng X, Chen ST, et al.: Applied Mechanics and Materials, Vols. 651–653 (2014), pp. 1216–1219.

Yang DF, Guo JH, Zhang YJ, et al.: Journal of Water Resource and Protection, Vol. 3(2011): 41–49.

Yang DF, Wang F, Gao ZH, et al.: Marine Science, Vol. 28 (2004): 71–74.

Yang DF, Zhu SX, Wang FY, et al.: Applied Mechanics and Materials, Vols. 651–653 (2014),

Yang DF, Zhu SX, Wang FY, et al.: Applied Mechanics and Materials, Vols. 651–653(2014), pp. 1419–1422.

Materials science and materials processing

*Advances in Materials Science, Energy Technology
and Environmental Engineering – Patty & Zhou (Eds)*
© 2017 Taylor & Francis Group, London, ISBN 978-1-138-19668-1

Study on the surface roughness prediction model of TC4 titanium alloy

M.M. Hu & Y. Zhang
Kunming University of Science and Technology, Kunming, Yunnan, China

S.W. Yuan
Kunming Machine Tool, Kunming, Yunnan, China

ABSTRACT: Surface roughness is an important measure factor of workpiece surface quality. Establishing a TC4 titanium surface roughness prediction model has important significance in specific processing conditions. In this paper, the TC4 titanium surface roughness prediction model was established based on the RSM and orthogonal cutting tests by analyzing a large number of the milling test data. Through variance analysis, it can be concluded that the regression effect of the prediction model is highly significant. Then, the verification test was designed. The test results indicated that a predictive value is close to the test value. The results indicate that the prediction model is effective within a certain common range. The experiment proved that the prediction model has good practicability and it can provide some reference value for companies and researchers.

1 INTRODUCTION

Titanium alloy has a many advantages such as excellent corrosion resistance, small density, high specific strength, etc. Titanium alloy is found to have much more applications in aerospace, petrochemical industry, shipbuilding, automobile, medical departments, etc. Moreover, in the actual production of machining, efficiency and machining quality are low, as it is well known as a difficult-to-machining material. The surface roughness is an important aspect to measure the quality of TC4 titanium alloy used for processing. Forecasting the processing surface roughness and the optimizing the cutting parameters have important significance to improve the surface quality of parts (Long et al. 2005).

Most production enterprises have no mature experience formula and the data to select the cutting parameters for the actual processing. They mainly use the method of trial cut constantly to adjust the process parameters. Many international scholars have made many researches on the surface roughness prediction model. Zhang Hongzhou through the RSM made researches on turning TC11 titanium surface roughness prediction and optimization, and realized the high-efficient cutting (zhang et al. 2010). Yusuf Sahin used the quadratic regression method to establish the relationship between the surface roughness and the machining parameters (Yusuf et al. 2005). Lu Zesheng predicted the surface roughness produced by ultra-precision cutting process by genetic algorithm, and concluded the optimal cutting parameters (Lu et al. 2005). Conceieao Anotnio C A and Puaof Davim J established the surface roughness theory prediction model and analyzed the effects of cutting parameters on surface roughness (Conceicao & Paulo, 2002). Many scholars expect establishing a general prediction model. However, there are many factors affecting the surface roughness, and we cannot consider all the actual complex conditions in the process of cutting. Therefore, the error of the general prediction model is bigger. The prediction model established under the actual working condition is often more effective and accurate than the general prediction model. As research on the milling surface roughness prediction model for difficult-to-machine materials is less in number, the paper established the TC4 titanium surface roughness prediction model based on the orthogonal test, in order to lay a base for material process parameter optimization.

2 TEST CONDITIONS AND MATERIALS EQUIPMENT

2.1 *Workpiece material*

TC4 titanium alloy is the workpiece material, and its related parameters are shown in Table 1. Its physical properties at room temperature are shown in Table 2.

Table 1. Chemical composition of TC4 titanium alloy.

Workpiece	Plate: 460 mm × 460 mm × 80 mm		Brand: TC4		
Basic material:		Ti			
Other chemical compositions:	Fe ≤ 0.30 C ≤ 0.10 N ≤ 0.05	H ≤ 0.015	O ≤ 0.2	Al5.5~6.8	V3.5~4.5

Table 2. Physical properties of TC4 titanium alloy.

Hardness HRC	Tensile strength σb/Mpa	Yield strength σs/Mpa	Density (kg/m³)	Modulus of elasticity gpa	Poisson ratio	Elongation %
36	895	825	4.4	114	0.3	14

Table 3. STR milling tool parameters.

Cutter type	Xiamen Gesac STR—S4 – Φ 20 * 45
Cutter material: GU25UF Ultra-fine grain cemented carbide	Coating material: AlCrN
Tool diameter: 20 mm	Core diameter
Cutter teeth: 4	Radial rake Angle (a): 8 deg
Helix angle: 35–40 deg	Radial relief angle (b): 14 deg
Axial rake angle: 3 deg	Corner radius (Rc): 0.2 mm
Axial relief angle: 10 deg	Blade length: 45 mm

Figure 1. Perthom-M1CNOMO Roughmeter.

2.2 *Test tool*

According to the management process and tool maker's recommendations, selected four teeth milling cutter of Xiamen Gesac measures. The cutting tool parameters are shown in Table 3.

2.3 *Machine tool*

The machine tool is KHC-63/2 double location precision horizontal machining center manufactured by Kunming Machine Tool.

2.4 *Surface roughness tester*

In this experiment, the needle LSV principle of surface roughness measuring instrument is used. The model is German mahr Perthom-M1CNOMO Roughmeter (shown in Figure 1).

3 EXPERIMENTAL PROCEDURE

TC4 titanium alloy is a typical difficult-to-machine material, which is studied in this experiment. According to the orthogonal experiment method, the arrangement for the milling test was set up. Three sets of orthogonal test ranging from low speed to high speed were arranged, and each of them involves three factors (cutting speed, feed per tooth, and axial cutting depth) and four levels. Each set of orthogonal test has 16 groups of cutting parameters. The cutting was repeated three times for each group of cutting parameters. The average surface roughness was measured. The obtained experimental data of the surface roughness are shown in Table 4.

4 ESTABLISHING THE SURFACE ROUGHNESS PREDICTION MODEL

The response surface method uses the experience formula of the traditional model of the relationship between the output and the input to establish a two-order mathematical model. It could simulate multiple independent variables and a dependent variable of mathematical relationship under the confidence level of 99%. The cutting experiment in practice proved that the method can accurately describe more comprehensive influence of cutting parameters on the cutting force. Among them, the Dual Response Surface Method (DRSM) through a rigorous mathematical model in the cutting parameter optimization designs experimental parameters. At the same time, it considered the

Table 4.   Surface roughness of predicted value contrast and the experimental value.

	Number	Cutting parameters				Experimental predicted		
		n_0 (r/min)	vf (mm/min)	ae (mm)	ap (mm)	Ra (um)	Ra (um)	Error (%)
STRS4 (Low speed)	1	478	76	0.6	20	0.264	0.326	19.00%
	2	478	115	0.8	20	0.438	0.343	27.94%
	3	478	153	1.0	20	0.568	0.606	6.37%
	4	478	191	1.2	20	0.652	0.753	13.41%
	5	717	115	0.8	20	0.270	0.343	21.35%
	6	717	172	0.6	20	0.428	0.325	31.89%
	7	717	229	1.2	20	0.566	0.623	9.18%
	8	717	287	1.0	20	0.612	0.551	11.00%
	9	955	153	1.0	20	0.317	0.323	1.79%
	10	955	229	1.2	20	0.484	0.523	7.42%
	11	955	306	0.6	20	0.528	0.611	13.55%
	12	955	382	0.8	20	0.582	0.483	20.58%
	13	1195	191	1.2	20	0.406	0.316	28.62%
	14	1195	287	1.0	20	0.422	0.568	25.78%
	15	478	76	0.6	20	0.474	0.366	29.54%
	16	478	115	0.8	20	0.563	0.549	2.47%
STRS4 (Medium speed)	1	1433	229	0.1	20	0.264	0.280	5.72%
	2	1433	344	0.2	20	0.481	0.497	3.26%
	3	1433	459	0.3	20	0.611	0.669	8.64%
	4	1433	516	0.4	20	0.600	0.630	4.76%
	5	1672	268	0.2	20	0.281	0.260	7.88%
	6	1672	401	0.1	20	0.408	0.331	23.26%
	7	1672	535	0.4	20	0.479	0.323	48.24%
	8	1672	602	0.3	20	0.597	0.613	2.51%
	9	1911	306	0.3	20	0.285	0.240	18.71%
	10	1911	459	0.4	20	0.353	0.394	10.56%
	11	1911	612	0.1	20	0.505	0.580	12.92%
	12	1911	688	0.2	20	0.564	0.528	6.74%
	13	2150	344	0.4	20	0.276	0.320	13.66%
	14	2150	516	0.3	20	0.358	0.286	24.89%
	15	2150	688	0.2	20	0.450	0.405	11.11%
	16	2150	774	0.1	20	0.499	0.504	0.97%
STRS4 (High speed)	1	2070	331	0.1	20	0.202	0.228	11.27%
	2	2070	414	0.2	20	0.312	0.296	5.63%
	3	2070	497	0.3	20	0.366	0.360	1.69%
	4	2070	580	0.4	20	0.364	0.423	14.00%
	5	2229	357	0.2	20	0.254	0.329	22.78%
	6	2229	446	0.1	20	0.256	0.241	6.27%
	7	2229	535	0.4	20	0.323	0.362	10.86%
	8	2229	624	0.3	20	0.382	0.431	11.44%
	9	2388	382	0.3	20	0.285	0.282	1.13%
	10	2388	478	0.4	20	0.301	0.325	7.38%
	11	2388	573	0.1	20	0.296	0.277	6.84%
	12	2388	669	0.2	20	0.367	0.375	2.11%
	13	2548	408	0.4	20	0.297	0.235	26.39%
	14	2548	510	0.3	20	0.312	0.254	22.75%
	15	2548	612	0.2	20	0.320	0.400	19.88%
	16	2548	713	0.1	20	0.321	0.290	10.59%
			Mean error					13.64%

mutual effect between the controllable factors. The double response surface method of the quadratic regression model is described as follows:

$$u = \beta_0 + \beta_1 x_1 + \beta_2 x_2 + \beta_{12} x_1 x_2 + \beta_{11} x_1^2 + \beta_{22} x_2^2 + \varepsilon_u \tag{1}$$

$$\sigma^2 = \gamma_0 + \gamma_1 x_1 + \gamma_2 x_2 + \gamma_{12} x_1 x_2 + \gamma_{11} x_1^2 + \gamma_{22} x_2^2 + \varepsilon_\sigma^2 \tag{2}$$

where β and γ are the estimated coefficients, u is the mean, σ^2 is the variance, ε_u is the randomized trial error of expectation, and ε_σ^2 is the randomized trial error of variance. In order to improve the degree of fitting, people usually use $\ln \sigma^2$ replace σ^2. The first-order response between cutting parameters and the surface roughness, respectively, is shown in formula (3). The second-order response between cutting parameters and the surface roughness, respectively, is shown in formula (4).

$$y_1 = \beta_0 + \beta_1 x_1 + \beta_2 x_2 + \cdots + \beta_n x_n + \varepsilon \tag{3}$$

$$y_2 = \beta_0 + \sum_{i=1}^{n} \beta_i x_i + \sum_{i=1}^{n} \beta_{ii} x_i^2 + \sum_{i=1}^{n-1} \sum_{j=i+1}^{n} \beta_{ij} x_i y_j \tag{4}$$

where n_0 is the rotation speed, vf is the feeding speed, ae is the radial cutting depth, and ap is the axial radial cutting depth.

This paper chose the second-order response surface model. In this experiment, the cutting speed, each tooth feed, and radial cutting depth impact on the cutting force were mainly studied. In this experiment, three independent variables (n = 3), and then expansion formula (4) were considered.

$$y_2 = \beta_0 + \beta_1 x_1 + \beta_2 x_2 + \beta_3 x_3 + \beta_{11} x_1^2 + \beta_{22} x_2^2 + \beta_{33} x_3^2 + \beta_{12} x_1 x_2 + \beta_{13} x_1 x_3 + \beta_{23} x_2 x_3 \tag{5}$$

where $y_2 = R_a, x_1 = V_c, x_2 = f_z, x_3 = a_e \tag{6}$

$$\beta_0 = b_0, \beta_1 = b_1, \beta_2 = b_2, \beta_3 = b_3, \beta_{11} = b_4, \\ \beta_{22} = b_5, \beta_{33} = b_6, \beta_{12} = b_7, \beta_{13} = b_8, \beta_{23} = b_9 \tag{7}$$

where $V_c = (\pi d n_0)/1000$.

Importing formulas (6) and (7) into formula (5), we get

$$R_a = b_0 + b_1 V_c + b_2 f_z + b_3 a_e + b_4 V_c^2 + b_5 f_z^2 \\ + b_6 a_e^2 + b_7 V_c f_z + b_8 V_c a_e + b_9 f_z a_e \tag{8}$$

This experiment used the Taguchi method. Importing experimental data into formula (8), we obtain

$$R_{a1} = b_{1,0} + b_{1,1}V_{c1} + b_{1,2}f_{z1} + b_{1,3}a_{e1} + b_{1,4}V_{c1}^2 + b_{1,5}f_{z1}^2 \\ + b_{1,6}a_{e1}^2 + b_{1,7}V_{c1}f_{z1} + b_{1,8}V_{c1}a_{e1} + b_{1,9}f_{z1}a_{e1} \\ R_{a2} = b_{2,0} + b_{2,1}V_{c2} + b_{2,2}f_{z2} + b_{2,3}a_{e2} + b_{2,4}V_{c2}^2 \\ + b_{2,5}f_{z2}^2 + b_{2,6}a_{e2}^2 + b_{2,7}V_{c2}f_{z2} + b_{2,8}V_{c2}a_{e2} \\ + b_{2,9}f_{z2}a_{e2}$$

$\bullet$
$\bullet$
$\bullet$

$$R_{a48} = b_{48,0} + b_{48,1}V_{c48} + b_{48,2}f_{z48} + b_{48,3}a_{e48} + b_{48,4}V_{c48}^2 \\ + b_{48,5}f_{z48}^2 + b_{48,6}a_{e16}^2 + b_{48,7}V_{c48}f_{z48} + b_{48,8}V_{c48}a_{e48} \\ + b_{48,9}f_{z48}a_{e48} \tag{9}$$

$$R_a = \begin{bmatrix} R_{a1} \\ R_{a2} \\ \vdots \\ R_{a48} \end{bmatrix}, X = \begin{bmatrix} 1 & V_{c1} & f_{z1} & a_{e1} \\ 1 & V_{c2} & f_{z2} & a_{e2} \\ \cdots & \cdots & \cdots & \cdots \\ 1 & V_{c48} & f_{z48} & a_{e48} \end{bmatrix}, b = \begin{bmatrix} b_0 \\ b_1 \\ \vdots \\ b_9 \end{bmatrix} \tag{10}$$

The matrix form of formula (10) is as follows:

$$R_a = Xb + \varepsilon \tag{11}$$

$$b = (X^T X)^{-1} X^T Y \tag{12}$$

$$b = [-0.194994, 0.000212, 18.05455, \\ -0.302950, 0.000004 - 38.140700, \\ 0.326815, -0.057000, 0.002586, \\ -5.107510]^T \tag{13}$$

Importing the matrix form of b into formula (8), we get

$$R_a = -0.194994 + 0.000212 v_c + 18.05455 f_z \\ -0.30295 a_e + 0.000004 v_c^2 - 38.1407 f_z^2 \\ + 0.326815 a_e^2 - 0.057 v_c f_z + 0.002586 v_c a_e \\ -5.10751 f_z a_e \tag{14}$$

Importing cutting parameters into the prediction model, we get the predicted value. The error of the predicted values and experimental values are shown in Table 4.

Then, the residual error and the significance level of the prediction model were analyzed. Analysis of variance is as shown in Table 5.

The statistic of $F = S_R/S_e$ submits to the F-distribution: the first degree of freedom is f_R and the second degree of freedom is f_e. The effect of regression on the prediction model is highly significant under the level of significance $\alpha = 0.1$ (99% confidence level). It can be concluded that the surface roughness prediction model to predict the surface roughness is highly effective.

Table 5. Surface roughness prediction model (ANOVA table).

	Degree of freedom (f)	Sum of Squares	Mean square deviation (V)	Statistic (F)	F (3,44,0.01)	Significance
Regression (r)	3	0.675	0.225	45	26.4	High significance
Residual error (e)	44	0.204	0.005			
Sum total (T)	47	0.879				

5 CONCLUSIONS

This researcher mainly designed the orthogonal experiment of milling TC4 titanium alloy, and measured the data of surface roughness and established the prediction model based on the response surface method. The mean error between the experimental value and the predicted value is 13.64%. Through variance analysis, it can be concluded that the regression effect of the prediction model is highly significant. This prediction model is highly effective. This prediction model can be used in predicting the surface roughness and then optimizing the cutting parameters.

ACKNOWLEDGMENTS

This work was supported by the National Key Science and Technology Projects (No. 2012ZX04012-031).

Corresponding author: Zhang Yu, a professor researching on CIMS. The email is 498380267@qq.com.

REFERENCES

Conceicao Antonio C.A. & Paulo Davim J. (2002). Optimal cutting conditions in turning of particulate metal matrix composites based on experiment and a genetic search model. Composites. Part A, 213~219.

Long Zhenhai & Wang Xibin (2005). Analysis of variance about influence factor on surface roughness of difficult-to-cut material in high speed machining process. J. Tool Engineering. 39, 26~29.

Lu Zesheng & Wang Minghai (2005). Based on the genetic algorithm of ultra-precision cutting surface roughness prediction model parameter identification and optimization of cutting parameter. J. Journal of mechanical engineering. 41, 158–162.

Sahin, Y & Riza Motorcu, A (2005). Surface roughness model for machining mild steel. J. Materials and Design. 26, 321–326.

Zhang Hongzhou, Ming Weiwei & An Qinglong (2010). Response surface method in the application of the surface roughness prediction model and parameter optimization. J. Journal of Shanghai jiaotong university. 44, 447–451.

Advances in Materials Science, Energy Technology
and Environmental Engineering – Patty & Zhou (Eds)
© 2017 Taylor & Francis Group, London, ISBN 978-1-138-19668-1

Observation of a nickel aluminum bronze corrosion morphology

Z.H. Chen, X.F. Sun & Y.L. Huang
Academy of Armored Forces Engineering, Beijing, China

ABSTRACT: Nickel aluminum bronze is one of the commonly used materials to make ship propellers, which easily generate corrosion phenomenon under long-term wet and high-salinity working environment. The samples of this experiment are cast nickel aluminum bronze after salt fog corrosion for 75 days and nickel aluminum bronze corrosion sample under actual working condition. It utilized the optical microscope and the scanning electron microscope to observe their corrosion morphology. The results indicated that edge of the cast nickel aluminum bronze sample generated oxide skin and obvious erosion pit, flocculent corrosion products could be observed on its surface. The nickel aluminum bronze propeller sample generated regional corrosion, the corrosion layer thickness was 5.573 μm, and corrosion products on the propeller surface are shaped like islands.

1 INTRODUCTION

Nickel aluminum bronze is a kind of aluminum bronze, and its main elements are nickel, iron, and manganese. Since it has good mechanical properties and corrosion resistance, western ships manufacturing powerhouses began to use nickel aluminum bronze as the material for manufacturing propeller (Song, 2007). The propeller is the propulsion plant of ships, and it is easy to generate corrosion phenomenon under the long-time marine environment, and strict corrosion phenomenon is a serious threat to the safety of ships, and it reduces the working time of the propeller. Therefore, it is significant to observe the nickel aluminum bronze corrosion morphology to analyze the failure mechanism of the propeller and find effective solutions to repair the propeller.

2 EXPERIMENTAL METHODS

In this experiment, CuAl8Ni6 is used as the as-cast nickel aluminum bronze sample was processed into the size of 50 mm × 20 mm × 4 mm. The dirt on the surface of the sample was polished using abrasive paper and washed out with alcohol, and the sample was placed into salt fog workstation for 75 days.

The sample of nickel aluminum bronze propeller was processed into the size of 35 mm × 30 mm × 7 mm, and its surface was cleaned using an ultrasonic cleaner with acetone.

The corrosive samples were cut to expose the cross-section, and the cross-section was polished using abrasive paper and a polishing machine. Then, $FeCl_3$-HCl solution was used to etch the cross-section, and, after etching, it was cleaned with alcohol. Finally, the cross-section sample was dried using a blower.

The morphology of these samples was observed using OLYMPUS DSX100 and Nova Nano-SEM 650.

3 RESULTS AND DISCUSSION

3.1 Morphology of noncorrosive samples

The nickel aluminum bronze structure consists of α-Cu substrate and K-phase that is distributed on the substrate (Du, 2014). As shown in Fig. 1(a), α-phase can be observed by OM, and it appears to be yellow or turquoise in color with metal luster. In SEM picture, the α-phase appears as white patches. Black regions in these two pictures are K-phases that are irregularly distributed on the substrate.

3.2 Morphology of as-cast nickel aluminum bronze after salt fog corrosion

As shown in Fig. 2(a), after 75 days of salt fog corrosion, the oxide scale of certain thickness was generated on the edge of the as-cast nickel aluminum bronze sample. In Fig. 2(b), several elliptic erosion pits were generated on the surface of the sample. In Fig. 2(c), regions with different colors

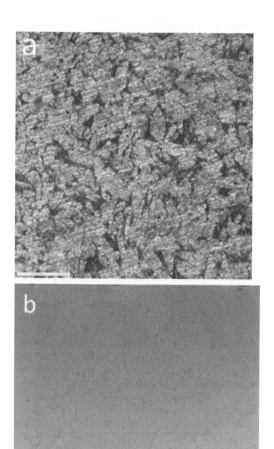

Figure 1. Microstructure of as-cast nickel aluminum bronze: a) OM and b) SEM.

are shown, because different elements produce different corrosion products. In Fig. 2(d), pitting corrosion can be obviously observed on the nickel aluminum bronze substrate.

As shown in Fig. 3(a), the erosion pit could be observed by SEM, and the corrosion products in the pit appear to be laminated. In Fig. 3(b), flocculent and porosity corrosion products were generated on the surface of the substrate.

3.3 Morphology of nickel aluminum bronze propeller after corrosion

As shown in Fig. 4(a), corrosion phenomena is generated in some regions, while it is not obvious

Figure 2. OM morphology of as-cast nickel aluminum after salt fog corrosion.

Figure 3. SEM microstructure of as-cast nickel aluminum after salt fog corrosion.

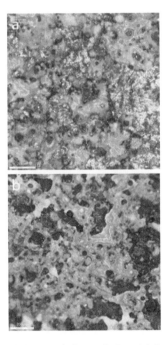

Figure 4. OM morphology of the nickel aluminum bronze propeller after corrosion under actual condition.

in a few regions. In Fig. 4(b), green corrosion products are distributed on milky white corrosion products, and it can also be observed that some yellow corrosion products appear sporadically.

As shown in Fig. 5(a), since the propeller rapidly rotates in marine environment, cavitation corrosion easily occurs on the surface of the propeller. In addition, the cavitation erosion pit can be observed in this picture. In Fig. 5(b), corrosion products appear on the propeller that is shaped like islands. In Fig. 5(c), the corrosion layer of the propeller can be observed, and it can be measured that the thickness of the layer was about 5.573 μm.

In order to reduce corrosion phenomenon, the as-cast nickel aluminum bronze material must

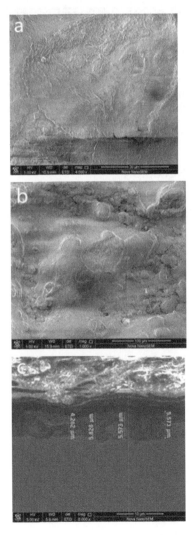

Figure 5. SEM microstructure of nickel aluminum bronze propeller after actual condition corrosion.

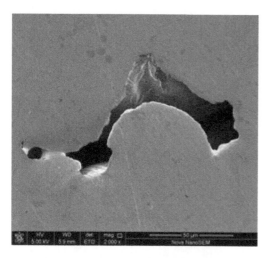

Figure 6. SEM microstructure of nickel aluminum bronze defect.

be compact to isolate the material from the corrosive medium. However, according to Fig. 6, as-cast nickel aluminum bronze contains defects. Sea water could infiltrate into the pores and generate corrosion phenomenon.

4 CONCLUSION

1. After 75 days of salt fog corrosion, as-cast nickel aluminum bronze generates denudation phenomenon, and the erosion pit that appeared to be laminated could be observed, and the corrosion products on the surface appear to be flocculent and have porosity.
2. Cavitation erosion pit can be observed. The corrosion phenomenon of the propeller is characterized to be regional. Corrosion products are distributed on the substrate like islands.
3. On the cross-section of the propeller, a corrosion layer with a relatively homogeneous thickness can be observed, and its thickness is approximately 5.573 μm.

REFERENCES

Du C.Y. 2014, Research on corrosion resistance of the nickel-aluminum bronze and the surface treatment on it.
Li X.Y., Yan Y.G. & Ma L., et al. 2004, Cavitation erosion and corrosion resistance of as-welded nickel aluminum bronze.
Shi X., Song D.J., Hu W.M. 2009. Research on microstructure of deforming nickel-aluminum bronze alloy.
Song D.J., Hu G.Y. & Lu H., et al. 2007, Survey of progress on the research and practice of nickel aluminum braze.
Zhang W.W., Tan W., Luo Z.Q. & Li Y.Y. 2012, Relationship between microstructure and brinell hardness of nickel-aluminum bronze by quantitative metallography analysis.

Advances in Materials Science, Energy Technology
and Environmental Engineering – Patty & Zhou (Eds)
© 2017 Taylor & Francis Group, London, ISBN 978-1-138-19668-1

Effect of deposition conditions on electrical properties of NiCr thin film

L.F. Lai, J.X. Wang & Y.N. Qiu

School of Materials Science and Engineering, Ningbo University of Technology, Ningbo, P.R. China

ABSTRACT: NiCr (80/20 at. %) alloy thin film was deposited on the copper foil substrate as Embedded Thin Film Resistor (ETFR) materials by DC magnetron sputtering. The structure and surface morphology of thin film were analyzed using XRD and SEM. The thermal stabilities of ETFR materials were detected by the detection probing station (HFSE-PB4). This study showed that the electrical properties of thin films were greatly affected by the sputtering powers and substrate temperatures. The results were useful in achieving the ETFR materials with expected properties.

1 INTRODUCTION

In recent years, with the miniaturization of electronic products, several million embedded passive components have been used in various types of electronic devices, which can save at least 40% space of Printed Circuit Board (PCB) compared with the traditional surface mounted technique (Wang et al. 2002). Embedded Thin Film Resistor (ETFR), one of the passive components, can be made by sandwiching the thin film resistor between the copper foil and dielectric layer, and pressed into PCB inside after etched, which possesses many merits, including low cost and electromagnetic interference, high performance and signal quality by reducing the wiring distance and the number of solder joints, etc. (Min 2005). The most common method to fabricate ETFR material was magnetron sputtering which can maintain good performance of thin film and compatible with the conventional metallization schemes used in integrated circuit technology (Lai et al. 2011). NiCr (80/20 at. %) alloy owing better reliability, high electrical resistivity and small Temperature Coefficient of Resistance (TCR) (Kazi et al. 2003) is a promising candidate for the ETFR material. Till now, the researches about the NiCr alloys deposited on glass, silicon, ceramic, stainless steel and so on (Yan et al. 2007, Vinayak et al. 2006) had been reported widely, however few studies have been reported about NiCr alloy deposited on copper foil as ETFR materials and investigating the thin film structure and properties according to different deposition conditions. Copper foil as substrate plays an important role in 3D electronics packaging technology since copper foil can be used as both multilayer core boards and electrode. Therefore, it is necessary to investigate the deposition conditions of the NiCr alloy deposited on copper foil.

In this paper, the electrical properties and structures of NiCr (80/20 at. %) ETFR materials were studied and two main sputtering parameters were discussed, i. e. the sputtering power and substrate temperature. The results are helpful for the ETFR materials to achieve expected properties.

2 EXPERIMENTAL DETAILS

NiCr (80/20 at. %) alloy target (99.99% purity, Φ74.90 mm × 6.15 mm) was used. High purity Ar (99.999% purity) was introduced as the sputtering gases. Very Low Profile (VLP) electrolytic copper foil (18 μm thickness) was selected as substrate. The Argon pressure of 0.85 Pa, the gas flow of 80 sccm, the base pressure of chamber of 8.5×10^{-4} Pa, the rotation speed of 20 rpm and the distance of 70 mm between the target and substrate, and the sputtered time of 4 min. Before sputtering, copper foil and float glass were ultrasonically cleaned in sequence of acetone, alcohol and deionized water, with 10 min for each step, respectively. The clean copper foil and flat glass was transferred into the sputtering chamber after completely flowed dry by nitrogen gun. To achieve good adhesion and large sheet resistor, the NiCr thin film was deposited on matte side of copper foil. The thickness of NiCr thin film on top of copper foil was determined by the alloy thickness on top of the float glass since the thin film on the copper foil is inconvenient to be measured for the soft and rough of copper foil (Lai et al. 2012).

The thin film thickness was determined by a calibrated surface profiler (Model XP-1). The crystal structure and the internal stress of ETFR were analyzed by XRD (X'pert PRO, NL). The surface morphology of ETFR material was examined by SEM (Hitachi S-4800). In order to analyze the thermal stability of ETFR, the ETFR was rapidly heated to 250°C at a rate of 40°C/min, then gradually cooled down at a rate of 20°C/min by the detection probing station (HFSE-PB4). The sheet resistance of ETFR was measured using a digital source meter (Keithley 2410) with the step of 10°C in the cooling process. The TCR values of the ETFR is defined as

$$TCR = \frac{(R_t - R_{t_0})}{R_{t_0}(t - t_0)} \times 10^6 \ (ppm/K) \qquad (1)$$

where R_{t_0} and R_t are the sheet resistance of ETFR measured at the temperature of t_0 (room temperature) and t, respectively. The stress vertical film surface is calculated by formula (Yang et al. 1991):

$$\sigma = \frac{E_f}{2\nu_f}\frac{d_0 - d}{d_0} \qquad (2)$$

where d_0 and d are the interplanar distance with stress or without stress, respectively. E_f is Young's modulus and ν_f is Poisson's ratio.

3 RESULTS AND DISCUSSION

3.1 Structural properties

Fig. 1 showed the XRD curves of NiCr thin film at different sputtering power and at substrate temperature 100°C. The sputtering power was selected at 80 W, 202 W, 405 W and 600 W, respectively. It was found that the diffraction peaks did not appear until the power was 202 W and the diffraction peaks of Ni (011), Ni (103), Cr (110) and Cr (200) became stronger with the power increasing, which illustrated the transition of thin film from amorphous to crystalline structure and the crystallization increasing. The stronger the sputtering power was, the greater the plasma energy of Ar was, and the higher kinetic energy of sputtered target atoms would be, thus the diffusion ability of the depos-

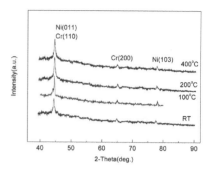

Figure 2. XRD spectra of NiCr thin films at different substrate temperature.

ited atoms along substrate surface was stronger. Atoms collided with each other to form atoms or groups of atoms, which can accelerate the nucleation and crystallization rate. Fig. 2 displayed the XRD patterns of the films prepared at different substrate temperatures and at sputtering power of 405 W. The substrate temperature was selected at room temperature (RT), 100°C, 200°C and 400°C, respectively. The effect of substrate temperature on the structure of the films was shown in Fig. 2. The diffraction peaks had been assigned to Ni (011), Ni (103), Cr (110) and Cr (110). Ni (011) peak and Cr (110) peak were almost overlapping and were dominant among the appeared diffraction peaks. It was further seen that the peak became more intense and sharper with increasing substrate temperature. This means that the crystallinity of the thin film was improved and the grain size of the crystallites became larger at a higher substrate temperature. That is, higher substrate temperature can increase the energy of deposited particles, improve the atomic diffusion ability, and provide nucleation work of critical nucleus (TIAN et al. 2006). The values of the crystal size at different sputtering power and at different substrate temperatures were also calculated by the Scherrer formula (Yang et al. 1991), respectively, and are listed in Table 1. The crystal size increased from 18.2 nm to 19.7 nm with sputtering power increasing from 202 W to 600 W, and increased from 19.3 nm to 20.1 nm with increasing substrate temperature from RT to 400°C.

3.2 Surface morphology

In order to confirm the effect of sputtering power and substrate temperature on the morphology of the films, the SEM surface images of the films deposited at different puttering power and substrate temperatures were investigated. Fig. 3 presented the SEM surface images of NiCr thin films at different puttering power and at substrate temperature 100°C. When the power was 80 W, the small particles were the dominant and the gaps between particles were big. The particles gradually

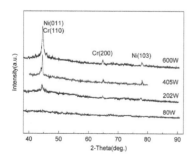

Figure 1. XRD spectra of NiCr thin films at different sputtering power.

Table 1. Crystal size and stress of thin film at different sputtering powers and substrate temperatures.

No.	d (× 10^{-1} nm)	Δd (nm)	Size (nm)	Stress (MPa)
202 W	2.0243	0.0091	18.2	1530
405 W	2.0309	0.0025	19.5	420
600 W	2.0231	0.0103	19.7	1730
RT	2.0229	0.0105	19.3	1770
100°C	2.0309	0.0025	19.5	420
200°C	2.0249	0.0085	19.8	1430
400°C	2.0242	0.0092	20.1	1550

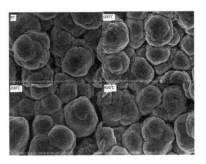

Figure 4. SEM surface images of NiCr thin films at different substrate temperature.

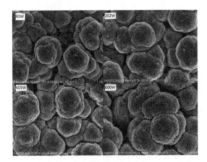

Figure 3. SEM surface images of NiCr thin films at different sputtering power.

became uniform and dense with the increasing of power up to 405 W. However, at 600 W of power, the particles became bigger, and the gaps were also bigger. It showed that the high sputtering power was advantageous to the diffusion of atom and thin film forming large particles and compact structure. However, the sputtering power cannot be too big, otherwise the thin film structure would be damaged and the thin film stress would be increased. Fig. 4 displayed the SEM surface images of NiCr thin films at different substrate temperature and at sputtering power 405 W. Up to 100°C, as the substrate temperature increases, the particles on film surface became larger. However, the particles on film surface seemed to decrease and became uniform at 200°C and 400°C substrate temperature. Especially at 400°C, the small particles on film surface seemed to connect blocks and become smooth, but the holes increased, which showed that the thin film surface had been oxidized partially. In addition, the atoms of substrate surface under a higher temperature were not easy to be adsorbed and agglomerated, which would increase the film defects and lead to the unstable performance of thin film.

3.3 Electrical properties

Fig. 5 presented the temperature dependence of the TCR under different sputtering power and at substrate temperature 100°C. The fluctuation of TCR was large when the power is 80 W and 600 W. The too low sputtering power would make the film structure loose and cause the lattice defects, and the too large sputtering power would damage the film structure and increase the thin film stress. When the sputtering power was 405 W, the internal stress of thin film was small, so the electrical property of thin film was stable. Fig. 6 presented the thermal test temperature dependence of the TCR under different substrate temperature and at sputtering power 405 W. It was found that the fluctuation of TCR curve was large, whether at the RT or at 400°C, but the TCR at 100°C was smallest among them. When the temperature increased from RT to 100°C, the diffusion of atoms was accelerated, the structure of thin film became closely and the defects in thin film were reduced, thus the stress was released and the electrical properties of thin film became stable. However, up to 400°C, the TCR became higher and unstable, which showed the stress increased. The data in Table 1 also proved the result. The stress decreased from 1770 MPa to 420 MPa with increasing substrate temperature from RT to 100°C, and increases from 420 MPa to 1550 MPa from 100°C to 400°C. High substrate temperature can improve the energy of deposited atomics and cause the deposited atomics to escape from the substrate surface. These high energy atomics would be deposited on the substrate surface after several rounds of impacts. The impacts between atomics and substrate can damage the forming film and caused new defects and additional stress.

Fig. 7 showed the sputtering power and substrate temperature dependence of the resistivity. The stronger the sputtering power was, the smaller the resistivity was. According to the general characteristics of metal and alloy, the grain boundary would be smaller as the grain was bigger (Ghosh et al. 2004). The small grain boundary could cause that the charge carriers scattering at grain boundary were weak, so the resistivity would

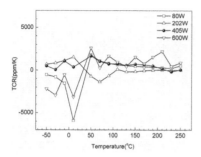

Figure 5. The temperature dependence of the TCR of NiCr thin films at different sputtering power.

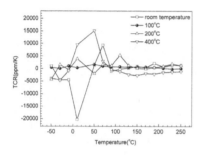

Figure 6. The temperature dependence of the TCR of NiCr thin films at different substrate temperature.

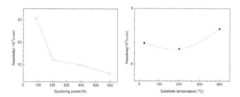

Figure 7. Effect of the sputtering power and substrate temperature on resistivity of NiCr thin film.

decrease. In addition, it was found that the resistivity decreased from RT to 200°C, and increased from 200 to 400°C. The resistivity of film deposited in the temperature range of RT-200°C should be affected by defect scattering during carrier movement, thus the resistivity decreased with the increasing temperature (Lu et al. 2002). However, the resistivity of thin film increased from 200°C to 400°C, which showed that the film surface might be oxidized (Khan et al. 2009).

4 CONCLUSIONS

The study investigated the influences of the sputtering power and substrate temperature on the structure, resistivity and TCR of NiCr ETFR materials. The structure and electrical properties of the films were strongly influenced by the sputtering power and substrate temperature. The resistivity decreased

with increasing sputtering power and substrate temperature duo to the shrink of grain boundary. The large sputtering power and substrate temperature could damage the thin film structure and led to unstable electrical properties. The TCR of NiCr ETFR materials were smaller under conditions of 405 W sputtering and 100°C substrate temperature.

ACKNOWLEDGMENTS

This work was financially supported by the National Natural Science Foundation of China (No. 51472126), Ningbo Natural Science Foundation (No. 2014 A610153), and Science and Technological Innovation Projects on College Students in Zhejiang Province (No. 2015R424016).

REFERENCES

Ghosh, J.R., D. Basak & S. Fujihara (2004). Effect of substrate-induced strain on the structural, electrical, and optical properties of polycrystalline ZnO thin films. J. Appl. Phys. 96, 2689–2692.

Kazi, I.H., P.M. Wild, T.N. Moore & M. Sayer (2003). The electromechanical behavior of nichrome (80/20 wt.%) film. Thin Solid Films 433, 337–343.

Khan, A.F., M. Mehmood, A.M. Rana & M.T. Bhatti (2009). Effect of annealing on electrical resistivity of rf-magnetron sputtered nanostructured SnO2 thin films. Appl. Surf. Sci. 255, 8562–8565.

Lai, L.F., R. Sun, T. Zhao, X.L. Zeng & Sh. Yu (2011). Processing Technology of Embedded Thin-Film Resistor Materials. 2011 International Symposium on APM, Xiamen, China, pp. 60–64.

Lai, L.F., W.J. Zeng, X.Z. Fu, R. Sun & R.X. Du (2012). Anealing effect on the electrical properties and microstructure of embedded Ni-Cr thin film resistor. J. Alloys Compd. 538, 125–130.

Lu, Y.M., W.S. Hwang & J.S. Yang (2002). Effects of substrate temperature on the resistivity of non-stoichiometric sputtered NiOx films. Surf. Coat. Technol. 155, 231–235.

Min, G. (2005). Embedded Passive Resistors Challenges and Opportunities for Conducting Polymers. Synthetic Metals 153, 49–52.

Tian, G.L., Y.M. Shen, J. Shen, J. d. Shao & Z.X. Fan (2006). Influence of Technological Conditions of Deposition Process on Microstructure of Thin Films. Chinese Journal of lasers 33, 673–678.

Vinayak, S., H.P. Vyas, K. Muraleedharan & V.D. Vankar (2006). Ni-Cr thin film resistor fabrication for GaAs monolithic microwave integrated circuits. Thin Solid Films 514, 52–57.

Wang, J.T., R. Hilburn, S. Clouser & B. Greenlee (2002). Manufacturing Embedded Resistors. Proceedings of the IPC Printed Circuits EXPO 2002, Louisiana, USA, pp. S03-4-1-S03-4-8.

Yan, J.W. & J.C. Zhou (2007). The oxidation and the electrical properties of Ni-Cr thin film after rapid thermal annealing. Int. J. Mod. Phys. B 21, 4561–4567.

Yang, L.Y., W.D. Guan & Z.M. Gu (1991). Material Surface Film Technology, Beijing: China Communications Press.

Advances in Materials Science, Energy Technology and Environmental Engineering – Patty & Zhou (Eds)
© *2017 Taylor & Francis Group, London, ISBN 978-1-138-19668-1*

Research on the cutting deformation of TC4 titanium alloy with simulation and experimental

Feng Kang, Chuankai Hu, Jun Lin, Xiangsheng Xia & Yanbin Wang
No. 59 Research Institute of China Ordnance Industry, Chongqing, China

ABSTRACT: Various features of titanium alloy material determine its low machinability and the cutting transformation simulation and experimental research for common titanium alloy material TC4 is conducted in the Thesis. The finite element method is used to reproduce the cutting deformation process and formation mechanism of cuttings. The simulation indicates that the material near nose of tool forms the softened area due to high temperature and high stress, thermoplastic instability occurs at the root of cuttings which are of tear type at the upper part and connected at the lower part in indented appearance. To further study the deformation mechanism of TC4 cuttings, the rapid roll setting test is conducted. The results indicate that what is near the nose of tool is deformed with relatively high temperature, the centralized shear slip occurs in cuttings with evident indented shape and the original criss-cross basket-weave structure turns into the ribbon pattern along cutting direction; there are small micro-cracks observed in scanning structure of cutting root and the cutting shape is fundamentally the same as simulation result. The research is of certain reference value to increase the machining quality and efficiency of titanium alloy.

1 INTRODUCTION

As most common titanium alloy material, the extremely low cutting rapid (generally below 50 m/min) is commonly adopted during cutting of TC4 as other common difficult-to-machine materials, so the production efficiency is very low (Yang, 2008). The reason is that cutting of titanium alloy is characterized with the following points:

1. Small deformation coefficient. The deformation coefficient of titanium alloy is less than or close to 1, and the distance of sliding friction for cuttings on front cutter face greatly increases which intensifies the wear of cutter (Komanduri, 2002; Liu, 2007; Mao, 2001).
2. High cutting temperature. Because the heat conductivity coefficient of titanium alloy is very small and the contact length of cuttings and front cutter face is extremely short, so the cutting heat cannot easily come out and is concentrated in the relatively small scope near cutting area and cutting edge resulting in very high cutting temperature (Geng, 2002; Qi, 2002).
3. The cutting force upon unit area is great. Because the length of cuttings and front cutter face is extremely short, the cutting force upon unit contact area greatly increases easily resulting in tipping. Meanwhile, because the elastic modulus of titanium alloy is small, bending and deformation can easily occur under the action

of radial force during processing which results in vibration, increases wear of cutter and affects the precision of workpiece.
4. Serious chill hardening. Because the chemical activity of titanium alloy is large, it is prone to absorb the oxygen and nitrogen in the air to form hard and crisp husk under high cutting temperature; meanwhile, the plastic deformation during cutting process can also result in hardening of the surface. The chilling hardening can not only reduce the fatigue strength of parts, but also intensify the wear of cutter (Wang, 2005; Lei, 2002).

To effectively cut the titanium alloy material, it is necessary to deeply study the mechanism of cutting deformation for TC4 titanium alloy according to the features of TC4 titanium alloy, correctly select the cutter material and optimize cutting parameters in order to realize effective cutting of titanium alloy under relative small influence on service life of cutter (Yang, 2006).

2 FINITE ELEMENT ANALYSIS OF CUTTING DEFORMATION OF TITANIUM ALLOY

The analog simulation of cutting plays a relatively great role in studying cutting behavior of material; the finite element simulation software Deform is adopted to conduct analog simulation for cutting process of TC4 titanium alloy material; for material model, the data import is obtained according to the performance test; the different mesh densities are adopted for mesh generation of the blank; the density of mesh generation near the cut-in end of cutter is large and the density far away is small.

For material model, besides defining the constitutive relation of plastic mechanics for material, it is necessary to define the failure criteria (fracture parameter) of material (Jawaida, 1999). The common failure criterion of Cockroft & Latham is adopted for crack formation and extension modeling (Jawaida, 1999), and it is considered that when the historically accumulated stress level and plastic strain reach the critical value of failure:

In the formula, $\overline{\varepsilon}_f$ is the effective fracture strain and σ1 is the maximum principal stress. Therefore, when the failure parameter value exceeds the critical failure value, i.e. D = Dcr, the indented cuttings begin to form. When the cutting speed is relatively high, the value of Dcr is 90; when the cutting speed is relatively low, the value of Dcr is 120.

The cutter parameters are shown in Table 1 and the simulation result is shown in Fig. 1.

It is shown in the simulation result that, for the cutting deformation simulation of titanium alloy material, the cutter cuts in the material to be cut, and firstly the material appears compressed plastic deformation and then the internal crystal framework slippage, followed by the plastic deformation; the deformation degree and temperature of material adjacent to the nose of tool are relatively high and the stressed area of the softening zone formed under the high temperature and high stress condition and the slippage area of cutting root is reduced and then the plastic instability appears; on the other hand, the flow velocity of cutting is not uniform, resulting in the tensile stress coming out of the external cutting surface to further make the instability; consequently, the plastic flow is concentrated and finally saw-tooth cuttings segment is formed. In the forming process of titanium alloy cuttings, the plastic deformation of material is relatively large and the working hardening increases the stress of cuttings on the cutting slippage surface, and the stress is concentrated on the blade, the shearing stress of metal adjacent to the blade exceeds the ultimate strength and then the cuttings are crushed upwards and still connected downwards, that is, the surface adjacent to the front cutter surface is relatively smooth and another surface is presented with saw-tooth cuttings.

Table 1. Tool parameters.

Front corner γ_o	Back corner α_o	Nose radius
14°	6°	0.3 mm

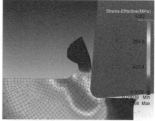

(a) Tool Going into 0.3mm

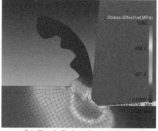

(b) Tool Going into 0.6mm

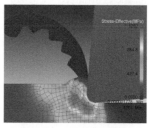

(c) Tool Going into 0.9mm

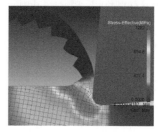

(d) Tool Going into 1.2mm

Figure 1. Chip Formation Process (v = 150 m/min, a_p = 0.5 mm, f = 0.05 mm/r).

3 RAPID ROLL SETTING TEST ON TITANIUM ALLOY MATERIAL

As the essential phenomenon of the cutting process, the cutting deformation is not an isolated phenomenon. The deformation occurred in cutting inevitably is accompanied with the function of cutting force and cutting heat, and the cutting force and cutting heat will not only act on the cuttings but also act on the workpiece surface, resulting in deformation. Therefore, the cutting deformation is the basis for the cutting force, cutting temperature, cutter wear, vibration, built-up edge, breaking cuttings and other cutting performance of materials in the cutting research. In order to further research the cutting deformation mechanism of TC4 titanium alloy, the rapid roll setting test has been conducted and the test condition is shown in Fig. 2 to Fig. 3. the central section of the cutter is located on the cutter rest through the round pin and the blade adjacent to the cutter bar is equipped with pressure spring and its another end is fixed on the blade rest and the another end of the cutter bar is contacted with the mobile locating pin; when the side-eat pin is rotated in testing, the vehicle bar can stably and rapidly separate itself from the material surface under the spring thrust.

Observe the micro-structures of the cuttings from the Titanium alloy material upon test as shown in Fig. 4 to Fig. 6 and those micro-structures are the cutting structure of TC4 titanium alloy and the surface structure of the cutting material. These figures show that, for the cuttings formation of titanium alloy, firstly when the cutting lay enters into the first deformation area, the cutter squeezes the layer to be cut and the deformation and temperature on the nose of tool are relatively high, with cuttings of local shear slipping and presenting obvious saw-tooth form; the original basket form with crossed distribution has become a striation along one direction and has certain inclination due to shearing slipping; the grain size adjacent to the nose of tool is obviously longer than that before processing and the fierce friction between the scanning structure of cuttings root and the front cutter face results in the cuttings bottom fiber lengthened along the direction of the cutter face. Fine micro-crack is also observed in the scanning structure

Figure 4.   Chip of TC4.

Figure 5.   Cutting surface of TC4.

Figure 6.   Microstructure in the root.

of cuttings root, for the material adjacent to the nose of tool is generated under the condition of concentrated stress and the nose of tool continues squeezing the layer to be cut and the crack continuously expands along the direction with lowest strain energy density to result in the material fracture and separate from the material surface, forming cuttings and this is relatively consistent with the that shown in the simulation result.

Figure 2.   Quick drop test device.

Figure 3.   Chip after drop test.

4   CONCLUSIONS

1. The finite element simulation of cutting process for TC4 titanium alloy material reproduces the cutting deformation process of titanium

alloy and the cuttings formation mechanism. The simulation shows, when the cutter cuts in the materials, the materials adjacent to the nose of tool forms a softening zone due to high temperature and high stress with thermoplastic instability of cutting root, and the cuttings are crushed upwards and still connected downwards, presented with saw-tooth cuttings.

2. The result of rapid roll setting test on TC4 titanium alloy shows that the cutting form is basically the same as that of the simulation result. When the cutter squeezes the layer to be cut, the cutter squeezes the layer to be cut and the deformation and temperature on the nose of tool are relatively high, with cuttings of local shear slipping and presenting obvious saw-tooth form; the original basket form with crossed distribution has become a striation along cutting direction. Fine micro-crack is also observed in the scanning structure of cuttings root.

3. The results of finite element simulation and test research on the cutting deformation process of TC4 titanium alloy are of certain reference value for further optimization of titanium allot cutting processing parameters, improvement of processing quality of titanium alloy materials, the processing efficiency of titanium alloy and the service life of cutter.

REFERENCES

Geng Guosheng, Xu Jiuhua. Milling For High Intensity Titanium alloy [J]. Machine Design and Manufacturing Engineering, 2002, (6):97–101.

Jawaida A., C.H. Che-Haron. Tool wear characteristics in turning of titanium alloy Ti-6246 [J]. Journal of Materials Processing Technology. 1999, 92(1): 329–334.

Komanduri R, Hou ZB. On thermoplastic Shear instability in the machining of a titanium alloy [J]. Metallurgical and Materials Ttansctions. A, 2002, (33A): 2995–3010.

Lei Shuting, Liu Weijie. High-speed machining of Titanium alloy using the driven rotary tool. Intemational Joumal of Machine Tools and Manufacture, 2002, 42(6):653–66.

Liu Sheng. FEM Simulation and Experiment Research of Cutting Temperature and Force in Orthogonal Cutting of Titanium Alloys [D]. Nanjing: Nanjing University of Aeronautics and Astronautics, 2007.

Mao Wenge. The Study on the cutting process of the Titanium Alloys [J]. Aeronautical Manufacturing Technology 2001, (1):64–66.

Qi Dexin. The Study of Cutting Features of BT20 Titanium Alloy [D]. Liaoning: Liaoning Technical University, 2002.

Wang, Z.G, Rahman, M.; Wong, Y.S.; Li, X.P., A hybrid cutting force model for high-speed milling of titanium alloys, CIRP Annals Manufacturing Technology, 2005, 54(1):71–74.

Yang Yong, Fang Qiang, KE Ying li. Mechanism of saw-tooth chip formation of titanium alloy based on finite element simulation [J]. Journal of Zhejiang University, 2008, 42(6):1010–1014.

Yang Yong, Ke Ying-lin, Dong Hui-yue. The finite element simulation of high-speed cutting [J]. Acta Aeronautica et Astronautica Sinica, 2006, 27(3):531–535.

Advances in Materials Science, Energy Technology and Environmental Engineering – Patty & Zhou (Eds)
© *2017 Taylor & Francis Group, London, ISBN 978-1-138-19668-1*

Assessment of zirconium species distributions in Zr-F-H aqueous solution based on thermodynamics

Dewei Mi
Huazhong University of Science and Technology, Wuhan, China
WWuhan Research Institute of Materials Protection, Wuhan, China

Changzhu Yang
Huazhong University of Science and Technology, Wuhan, China

ABSTRACT: The zirconium based conversion coatings can be formed by modified aqueous bath of hexafluorozirconic acid. Thus understanding the relationship of zirconium species distributions in aqueous solution at different conditions is important for controlling the properties of precursors in the aqueous chemistry of Zr-F-H. Many data can be found in the literature for the hydrolysis and complexation of zirconium. This paper presents a description of the distributions of zirconium inorganic species as a function of pH and free fluoride concentration involved by mathematical expression and graph. The objective of the assessment is to provide a simple quantitative mean of obtaining zirconium species distributions with the variations of pH and free F⁻, so that the zirconium species could be controlled.

1 INTRODUCTION

Due to these growing concerns over phosphates from phosphate conversion coating treatments, new environment-friendly surface conversion coatings need to be developed, without compromising on the corrosion protective performance (Stromberg, 2006; Unocic, 2011). Zirconium based conversion coating (ZrCC) formed on the metal surface by the aqueous bath based hexafluorozirconic acid was found to be mostly in the form of zirconia (ZrO_2). The results had indicated coating formation was found to proceed by precipitation of zirconium complexes out of solution initiated by an increase in interfacial pH resulting from cathodic water reduction reactions (Adhikari, 2011; Díaz, 2015; George; 2012). In the literature, many attentions have been paid to conversion coating that has particular characterizations and prosperities, such as coating morphology, chemical composition, corrosion resistance, electrochemical performances (Mohammadloo, 2012; Thomas, 2014).

Chemical conversion coating formation at metal surface are much complicated owing to the occurrences of spontaneous hydrolysis and precipitation reactions in the aqueous medium at the-the interface of solution and metal. These reactions depend on many parameters such as pH, concentration, or temperature. In the dilute hexafluorozirconic acid solution, several monomeric or complex solute zirconium species can exist simultaneously, so the real

chemical nature of the precursors is not known obviously. Moreover, the morphology, structure, and even chemical nature of the resulting conversion coating strongly depend on the presence of fluoride and pH (Verdier, 2006; Hosseini, 2014).

The speciation assessment provided a greatly improved, comprehensive view of inorganic complexation and hydrolysis in aqueous solution. Zirconium speciation exerted important controls on their chemical behaviors in aqueous solution. Assessments of the zirconium speciation in aqueous solution began with attempts to determine dominant chemical forms in aqueous solution based on available thermodynamic data. These data can be found in the literature for the hydrolysis and complexation of Zr^{4+} (Baes, 1976; Brown, 2005).

2 SPECIATION CALCULATIONS

The chemical behaviour of zirconium in aqueous systems is strongly influenced by the tendency of zirconium to form soluble complex ionic species with fluoride. Zirconium can form a series of stable complexes with fluoride in aqueous solution.

In the case of equilibrium involving the fluoride concentrations in aqueous solution, it can be conveniently written:

$$Zr^{4+} + iF \Leftrightarrow ZrF_i^{4-i} \tag{1}$$

The accumulated stability equilibrium constants of complexation are expressed in the form:

$$\beta_i^* = \frac{[ZrF_i^{(4-i)}]}{[Zr^{4+}][F^-]^i} \qquad (2)$$

Whereupon,

$$[ZrF_i^{(4-i)}] = [Zr^{4+}]\beta_i^*[F^-]^i \qquad (3)$$

Then, the total of the activity of all dissolved zirconium species is obtained:

$$\Sigma[Zr] = [Zr^{4+}]\sum_{i=0}^{6}\beta_i^*[F^-]^i \ (\beta_0^* = 1) \qquad (4)$$

And then, by substitution with Eq. (3) and Eq. (4). Thus:

$$\frac{[ZrF_i^{4-i}]}{\Sigma[Zr]} = \frac{\beta_i^*[F^-]^i}{\sum_{i=0}^{6}\beta_i^*[F^-]^i} \qquad (5)$$

In the case of a simple hydrolysis equilibria,

$$Zr^{4+} + jH_2O \Leftrightarrow Zr(OH)_j^{4-j} + jH^+ \qquad (6)$$

The accumulated stability equilibrium constants of hydrolysis are expressed in the form

$$\beta_j^* = \frac{[Zr(OH)_j^{(4-j)}][H^+]^j}{[Zr^{4+}]} \qquad (7)$$

Whereupon,

$$[Zr(OH)_j^{(4-j)}] = [Zr^{4+}]\beta_j^*[H^+]^{-j} \qquad (8)$$

The total concentration of zirconium $\Sigma[Zr]$ in the solution can be calculated by:

$$\Sigma[Zr] = [Zr^{4+}]\sum_{j=0}^{5}\beta_j^*[H^+]^{-j} \ (\beta_0^* = 1) \qquad (9)$$

And the distribution values of hydrolysis products are acquired:

$$\frac{[Zr(OH)_j^{(4-j)}]}{\Sigma[Zr]} = \frac{\beta_j^*[H^+]^{-j}}{\sum_{j=0}^{5}\beta_j^*[H^+]^{-j}} \qquad (10)$$

The accumulated stability equilibrium constants provide a simple means of assessing species concentration ratios as a function of free F^- or pH.

3 RESULT AND DISCUSSION

3.1 Theoretical distribution curves for zirconium-fluoride complexation system

Zirconium can form a series of stable complexes with fluoride in aqueous solution. The species in acidic aqueous solution that were taken into account for this calculation were Zr^{4+}; ZrF^{3+}; ZrF_2^{2+}; ZrF_3^+; ZrF_4; ZrF_5^-; ZrF_6^{2-}. By Eq (5), the distribution of zirconium fluoride species can be expressed in terms of free F^- and accumulated stability equilibrium constants β^*. The accumulated stability equilibrium constants β^* is a constant dependent on salinity, temperature, and pressure. If the free F^- is confirmed, the distribution will be obtained. Then using the equilibrium constants expressions of the fluoride-zirconium species[14], the distribution of the dissolved complex species formed in an aqueous system containing fluoride and zirconium can be represented graphically. At zero ionic strength, the concentration of the species in the solution is equivalent to the total analytical concentration of zirconium. For the purpose of this derivation, zero ionic strength is assumed. The results presented are considering pH values lower than 2 (no appreciable zirconium hydroxylfluoride species are formed).

The species distribution diagram in which the concentration of free F^- is plotted on the x-axis and the mole fraction of the fluoride-zirconium species are plotted on the y-axis. The fluoride-zirconium species in aqueous is determined by the concentration of free F^-. The concentration of the fluoride in aqueous solution varied, the proportion of the mole of fluoride and zirconium is varied, and the dominant forms of zirconium species are also changed.

3.2 Theoretical distribution curves for the species in the Zr-OH system

In the diluted aqueous solution of zirconium mononuclear hydrolysis could occur, and there are

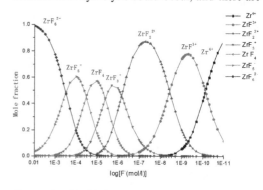

Figure 1. Equilibrium distribution of zirconium-fluoride complexes in strong acid solution.

different hydroxyl hydrolyzed products under different pH conditions. The hydrolysis of Zr ions in solution plays an important role in the chemical properties of the solution. Baes and Mesmer reviewed the early results about Zr hydrolysis, which indicated that the mononuclear species of Zr $(Zr(OH)_j^{4-j}, j = 1–5)$ were dominant only at low Zr concentration $(< 10^{-4}$ M) [12]. According to the accumulated stability constant of zirconium hydrolysis[13], if only mononuclear species are present in solution, the distribution of different hydrolyzed products can be calculated as shown in Figure 2.

The pH values were plotted on the x-axis and mole fraction of the zirconium species were plotted on the y-axis in this species distributions diagram. The species distributions of zirconium hydrolysis products were determined by the pH values. When the pH values changed, and the dominant forms of zirconium species were also varied. After this critical evaluation, it is now known that zirconium speciation is strongly influenced by pH.

3.3 Theoretical distribution curves for the species in the Zr-F-H system

By extension of the procedures already given, it is possible to evaluate all the possible dissolved species of zirconium by considering simultaneously the complexation reactions involving the concentration of free F^- and free H^+, and to express the results in one diagram, provided that the concentrations of only two components are allowed to vary at once. To cover any significant range of conditions a very large number of diagrams would be required, and the ones of particular interest could easily be prepared in a computer programming. On a three-dimensional diagram or model, three variables could be handled at once.

By making the simplifying assumption that F^- and H^+ are present in considerable excess, the activity of free F^- becomes nearly equal to the total

concentration at zero ionic strength. It is now possible to evaluate the relative relations of complexation and hydrolysis over a considerable range.

Consider Eq. (4) and Eq. (9), then

$$\sum[Zr] = [Zr^{4+}]\left(1 + \sum_{i=1}^{6}\beta_i[F^-]^i + \sum_{j=1}^{5}\beta_j[H^+]^{-j}\right) \quad (11)$$

Then the distribution values of zirconium fluoride species could be calculated by:

$$\frac{[ZrF_i^{4-i}]}{\sum[Zr]} = \beta_j[F^-]^i / \left(1 + \sum_{i=1}^{6}\beta_i[F^-]^i + \sum_{j=1}^{5}\beta_j[H^+]^{-j}\right) \quad (12)$$

And the distribution values of hydrolyzed product species are acquired:

$$\frac{[Zr(OH^-)_j^{4-j}]}{\sum[Zr]} = \beta_j[H^+]^{-j} / \left(1 + \sum_{i=1}^{6}\beta_i[F^-]^i + \sum_{j=1}^{5}\beta_j[H^+]^{-j}\right) \quad (13)$$

The species distribution values of zirconium fluoride and hydrolysis products could be calculated by Eq. (12) and Eq. (13).

The distribution of ZrF_6^{2-} species are calculated as followed:

$$\frac{[ZrF_6^{2-}]}{\sum[Zr]} = \beta_j[F^-]^6 / \left(1 + \sum_{i=1}^{6}\beta_i[F^-]^i + \sum_{j=1}^{5}\beta_j[H^+]^{-j}\right) \quad (14)$$

The MATLAB is used to draw the figure in Figure 3. The distribution diagram showed that the ZrF_6^{2-} species distribution depended on the concentration of free F^- rather than pH.

Then the distribution of $Zr(OH)_4$ are calculated as followed:

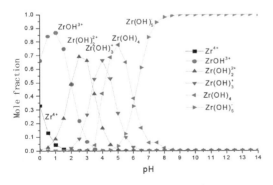

Figure 2. Species distribution of mononuclear zirconium hydrolysis products as a function of pH in an aqueous solution.

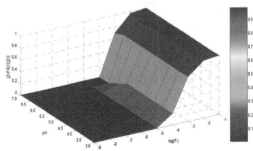

Figure 3. Species distribution of ZrF_6^{2-} as a function of pH and free fluoride in an aqueous solution.

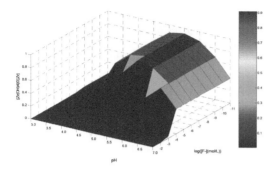

Figure 4. Species distribution of $Zr(OH)_4$ as a function of pH and free fluoride in an aqueous solution.

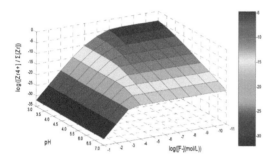

Figure 5. Species distribution of free Zr^{4+} as a function of pH and free fluoride in an aqueous solution.

$$\frac{[Zr(OH)_4]}{\sum[Zr]} = \beta_j[H^+]^{-4}/(1+\sum_{i=1}^{6}\beta_i[F^-]^i + \sum_{j=1}^{5}\beta_j[H^+]^{-j}) \quad (15)$$

The $Zr(OH)_4$ distribution can be drawn in Figure 4. Then both the concentration of free F^- and pH can influence the ratio of zirconium hydroxide to the total dissolved zirconium in aqueous solution.when the free fluoride is fixed, the ratio of zirconium hydroxide to total dissolved zirconium is highest at pH 5.5–6.0.

From the Eq. (11), the free Zr^{4+} distribution can be obtained as followed:

$$\frac{[Zr^{4+}]}{\sum[Zr]} = 1/\left(1+\sum_{i=1}^{6}\beta_i[F^-]^i + \sum_{j=1}^{5}\beta_j[H^+]^{-j}\right) \quad (16)$$

Then the species distribution of free Zr^{4+} can be shown in Figure 5.

4 CONCLUSIONS

Assuming chemical equilibrium, mathematical treatment can be used to find the predominant dissolved species in systems, where any or all of the principal complexation ligands occurs; it can also be used to give the proportion of complexed zirconium in solution. The results of the calculations could be presented in the form of graphs which can be used to evaluate directly the degree of complexation or hydrolysis. The purposes of the calculations and the graphs presented here is to ascertain the equilibrium activity of hydrolyzed, uncomplexed or complexed zirconium species for a wide span of solution conditions, and the relationship of the concentration of difference zirconium species and free F^- at specified pH.

REFERENCES

Baes, C.F., R. E. Mesmer. The Hydrolysis of Cations[M] John Wiley & Sons, New York, 1976.

Brown, P. L., E. Curti, B. Grambow, et al. Chemical thermodynamics Series, vol.8:chemical thermodynamics of zirconium[M] Elsevier Science Publishers, North-Holland, Amsterdam, 2005.

Díaz, B., L. Freire, M. Mojío, X.R. Nóvoa. Optimization of conversion coatings based on zinc phosphate on high strength steels, with enhanced barrier properties [J] Journal of Electroanalytical Chemistry. 2015, 737, 174–183.

George, F.O., P. Skeldon, G.E. Thompson. Formation of zirconium-based conversion coatings on aluminium and Al–Cu alloys [J]. Corrosion Science, 2012, 65:231–237.

Hossein Eivaz Mohammadloo, Ali Asghar Sarabi, Ali Asghar Sabbagh Alvani, et al. Nano-ceramic hexafluorozirconic acid based conversion thin film: Surface characterization and electrochemical study[J]. Surface & Coatings Technology, 2012, 206: 4132–4139.

Lostak Thomas, Maljusch Artjom, Klink Björn, Krebs Stefan, Kimpel Matthias, Flock Jörg, Schulz Stephan, Schuhmann Wolfgang. Zr-based conversion layer on Zn-Al-Mg alloy coated steel sheets: insights into the formation mechanism[J]. Electrochimica Acta, 2014, 137, 65–74.

Mohammad Hosseini, R., A.A. Sarabi and H. Eivaz Mohammadloo. The performance improvement of Zr conversion coating through Mn incorporation: With and without organic coating[J]. Surface & Coatings Technology. 2014, 258, 437–446.

Saikat Adhikari, K.A. Unocic, Y. Zhai, et al. Hexafluorozirconic acid based surface pretreatments: Characterization and performance assessment[J], Electrochimica Acta, 2011, 56: 1912–1924.

Stromberg, C., P. Thissen, I. Klueppel, et al. Synthesis and characterisation of surface gradient thin conversion films on zinc coated steel[J]. Electrochim. Acta, 2006, 52 (78):804–815.

Unocic, K.A., Y. Zhai, et al. Hexafluorozirconic acid based surface pretreatments: Characterization and performance assessment[J]. Electrochimica Acta, 2011, 56: 1912–1924.

Verdier, S., N. van der Laak, and F. Dalard. An electrochemical and SEM study of the mechanism of formation, morphology, and composition of titanium or zirconium fluoride-based coatings[J]. Surf. Coat. Technol. 2006, 200(9), 2955–2962.

*Advances in Materials Science, Energy Technology
and Environmental Engineering – Patty & Zhou (Eds)*
© 2017 Taylor & Francis Group, London, ISBN 978-1-138-19668-1

Performance of piperazine-urea-N-(2-hydroxyethyl) piperazine ternary system on simultaneous desulfurization and denitration

M. Jin, G.X. Yu & P. Lu

*Hubei Key Laboratory of Industrial Fume and Dust Pollution Control, Jianghan University, Wuhan,
Hubei, P.R. China*

ABSTRACT: To overcome the pollution of SO_2 and NO_x in the industrial waste gas, we proposed an absorbent composed of piperazine-urea-N-(2-hydroxyethyl) piperazine ternary system to promote the absorption efficacy. To assess its efficacy, we have measured the performances of simultaneous desulfurization and denitration from N_2-NO-SO_2 simulated flue gas with the ternary system absorbent in a static absorption experiment. Furthermore, the influences of the concentration of N-(2-hydroxyethyl) piperazine, pH value, the concentrations of NO and SO_2 in the initial simulated flue gas and the flow rate of simulated gas on the absorption behavior were discussed systematically. A better simultaneous desulfurization and denitration performance could be obtained under pH value of 5, the concentration of N-(2-hydroxyethyl) piperazine of 0.3 mol/L, the flow rate of 500 mL/min, and the concentrations of NO and SO_2 in the initial simulated flue gas of 1200 mg/m^3 and 2500 mg/m^3, respectively. Under the above-mentioned operating condition, simultaneous NO removal efficiency of 78.92% and SO_2 removal efficiency of 98.20% could be obtained.

1 INTRODUCTION

As is well-known, air pollution is becoming worse with the development of chemical industrialization in the world, especially the pollution of SO_2 and NO_x in the industrial waste gas (Skalska et al. 2010). To overcome the problem, more efforts have been put forward to develop new cleaner technologies to reduce the emissions of SO_2 and NO_x. Among the technologies, simultaneous desulfurization and denitration by chemical method is regarded as a typical process technology, which can be classified into dry process and wet scrubbing process (Fang et al. 2011). Between these two process technologies, the wet scrubbing technology appears to offer a practical alternative one. In order to obtain a better efficacy of simultaneous desulfurization and denitration, some researchers proposed ammonia, piperazine (PZ), urea, and so on, as absorbents or additives (Lee et al. 2005, Wei et al. 2009, Wang et al. 2014).

In this study, the performances of simultaneous desulfurization and denitration from N_2-NO-SO_2 simulated flue gas by using PZ-urea-N-(2-hydroxyethyl) piperazine ternary system as absorbent were investigated experimentally in a static absorption experiment. In addition, the absorption conditions, such as the concentration of N-(2-hydroxyethyl) piperazine, pH value, the flow rate of simulated gas, the concentrations of NO and SO_2 in initial simulated flue gas, were discussed systematically.

2 EXPERIMENT

2.1 Materials

Analytical reagent urea was obtained from Shanghai Hengli Fine Chemicals Co., Ltd., PZ was supplied by Wuhan Organic Industrial Co., Ltd., and N-(2-hydroxyethyl) piperazine was provided by Beijing Iark Technology Co., Ltd..

SO_2 (99.90%) and NO (99.90%) were supplied by Shanghai Flextronics standard gas Co., Ltd., and N_2 (99.99%) was obtained from Wuhan Tianci standard gas Co., Ltd..

2.2 Static absorption experiment

Figure 1 shows a schematic diagram of the absorption experimental setup, and the absorption operating conditions are listed in Table 1. In each run, 50 mL absorbent composed of PZ-urea-N-(2-hydroxyethyl) piperazine with a fixed proportion was taken into an absorption tube filled with glass beads, and then N_2-NO-SO_2 simulated flue gas was introduced via rotameter from the bottom of the absorption tube and reacted with absorbent. After that, the tail gas composed of N_2 and unabsorbed NO and SO_2 was led into the bottles filled with $KMnO_4$ solution and NaOH solution in turn. The inlet and outlet gas was collected and analyzed. HJ479–2009 standard naphthylethy lenediamine hydrochloride colorimetric method and iodometry method was adopted for the determination of NO and SO_2, respectively. According to the analysis results, the removal efficiency of NO and SO_2 could be calculated, respectively.

3 RESULTS AND DISCUSSION

3.1 Mechanisms of desulfurization and denitration

It is well known, the mechanism of simultaneous desulfurization and denitration by using urea is that urea reacts with NO and SO_2 simultaneously, which can be expressed as equations (1) and (2). While the mechanism of desulfurization and denitration for PZ can be denoted as equations (3)–(6). Different from the mechanism for urea, the formation of SO_3^{2-} in equation (5) can lead to a promotion of the removal efficiency of NO. Therefore, it can be conducted that there is a synergistic effect between desulfurization and denitration by using PZ as absorbent. Similar to the role of PZ in absorbent, N-(2-hydroxyethyl) piperazine also plays a positive effect on the simultaneous removal of NO and SO_2, and the mechanism of desulfurization and denitration using N-(2-hydroxyethyl) piperazine as absorbent is represented as equations (7)–(11).

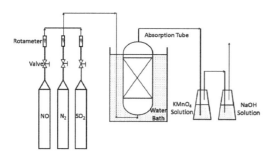

Figure 1. Schematic diagram of static absorption experiment.

$$SO_2+CO(NH_2)_2+1/2O_2+2H_2O \rightarrow (NH_4)_2SO_4+CO_2 \tag{1}$$

$$6NO+2CO(NH_2)_2 \rightarrow 5N_2+2CO_2+4H_2O \tag{2}$$

$$SO_2+H_2O \rightarrow H_2SO_3 \tag{3}$$

$$H_2SO_3+R_2R_3NH \rightarrow R_2R_3NH_2^{+}+HSO_3^{-} \tag{4}$$

$$HSO_3^{-} \rightarrow H^{+}+SO_3^{2-} \tag{5}$$

$$2HN\!\!\!\diagdown\!\!\!N\!\!-\!\!OH + H_2O+SO_2 \longrightarrow 2H_2N^{+}\!\!\!\diagdown\!\!\!N\!\!-\!\!OH + SO_3^{2-} \tag{7}$$

$$H_2O+SO_2+SO_3^{2-} \rightarrow 2HSO_3^{-} \tag{8}$$

$$H_2N^{+}\!\!\!\diagdown\!\!\!N\!\!-\!\!OH + H_2O+SO_2 \longrightarrow H_2N^{+}\!\!\!\diagdown\!\!\!N^{+}H\!\!-\!\!OH + HSO_3^{-} \tag{9}$$

$$HSO_3^{-} \rightarrow H^{+}+SO_3^{2-} \tag{10}$$

$$2SO_3^{2-}+2NO \rightarrow 2SO_4^{2-}+N_2 \tag{11}$$

3.2 Absorption performance of ternary system

The experimental results of the simultaneous desulfurization and denitration by using PZ—urea—N-(2-hydroxyethyl) piperazine ternary system were shown in Figure 2 under the condition of Case 1 listed in Table 1.

Compared with the experimental result by using PZ-urea binary system, the simultaneous removal efficiencies of SO_2 and NO are increased by using ternary system. Different from the molecular structures of PZ and urea, a steric-hinerance effect derived from hydroxyethyl group in N-(2-hydroxyethyl) piperazine can generate two amino groups with different properties. One is an amino group with weaken alkaline and the other is an amino group with enhanced alkaline. It happens that there is a strong binding ability between the amino group with enhanced alkaline and H^+ in absorbent solution. As a result, based on the mechanism of each absorbent, it can be deduced the ternary system absorbent can provide a better performance of

Table 1. Experiment conditions.

Case	1	2	3	4	5
Temperature			40 °C		
Pressure			Atmosphere		
PZ /mol/L			0.01		
Urea /mol/L			0.3		
N-(2-hydroxyethyl) PZ /mol/L	0–1.0	0.3	0.3	0.3	0.3
pH	5	4–8	5	5	5
NO /mg/m³	1200	1200	480–2400	1200	1200
SO_2 /mg/m³	2500	2500	2500	750–5000	2500
Gas Flow rate / mL/min	500	500	500	500	400–1000

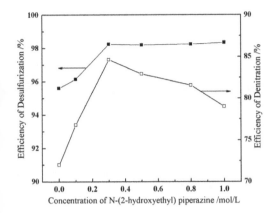

Figure 2. Influence of the concentration of N-(2-hydroxyethyl) piperazine on the simultaneous desulfurization and denitration.

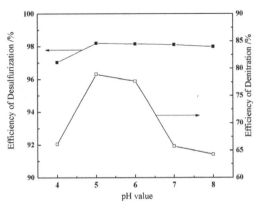

Figure 3. Influence of pH value on the simultaneous removal of SO₂ and NO.

simultaneous desulfurization and denitration than that of PZ-urea binary system.

It can also be seen from Figure 2 that the variation in the concentration of N-(2-hydroxyethyl) piperazine plays a different effect on the absorption performance. When the concentration of N-(2-hydroxyethyl) piperazine is less than 0.3 mol/L, the addition of N-(2-hydroxyethyl) piperazine in absorbent is beneficial to simultaneous absorption due to the steric-hinerance effect of hydroxyethyl group in N-(2-hydroxyethyl) piperazine. As a result, the removal efficiencies of SO_2 and NO are all increased. When the concentration of N-(2-hydroxyethyl) piperazine is more than 0.3 mol/L, due to a high solubility of SO_2, the removal efficiency of SO_2 keeps at a high value about 98.20%. As for the absorption of NO, on the one hand, there is a competition between the formation of HSO_3^- and the reaction illustrated in equation (6); on the other hand, the solubility of NO becomes lower in a viscous absorbent consisted of a higher concentration of N-(2-hydroxyethyl) piperazine. Accordingly, the removal efficiency of NO decreases along with the increase concentration of N-(2-hydroxyethyl) piperazine.

3.3 Influence of pH on the absorption behavior

Figure 3 shows the simultaneous absorption experiment results under the operating condition of Case 2 listed in Table 1.

In Figure 3, along with increasing of pH value, it can be seen the removal efficiency of SO_2 increases firstly and then maintains a constant value, and the removal efficiency of NO increases firstly and then decreases sharply. Since SO_2 has a high solubility in absorbent and it is a kind of acidic gas which is easily absorbed in an alkaline absorbent, then the

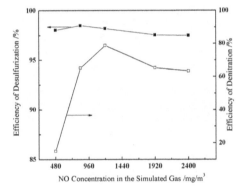

Figure 4. Effect of the initial NO concentration in the simulated flue gas on the absorption performance.

efficiency of desulfurization is always in a higher level with the increase pH value. However, urea is easily occurred hydrolysis in a higher pH solution, especially pH value between 7 and 8, which have an passive impact on denitration. Therefore, pH value of 5 can give a better simultaneous desulfurization and denitration efficacy.

3.4 Effect of initial NO concentration in the simulated flue gas on the absorption behavior

The simultaneous desulfurization and denitration were conducted under Case 3 operating condition illustrated in Table 1, and the results are shown in Figure 4. In Figure 4, although the results show the variation of the initial NO concentration has a slightly effect on the desulfurization behavior, there has a great effect on the denitration performance. Because of the same reason of high solubility of SO_2, the removal efficiency of SO_2 always keeps in a high value. Different from the absorption behav-

329

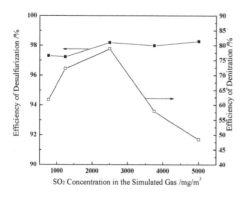

Figure 5. Effect of the initial SO_2 concentration in the simulated flue gas on the absorption performance.

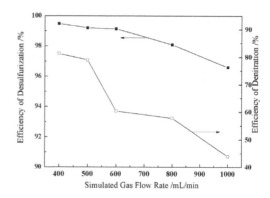

Figure 6. Influence of the flowrate of the siimulated gas on the simultaneous of desulfurization and denitration.

ior of SO_2, the removal efficiency of NO increases firstly and then decreases with an increase in the initial NO concentration. When the initial NO concentration is less than 1200 mg/m³, the amount of absorbent is sufficient for the absorption with NO dissolved in the solution, as a result, the removal efficiency of NO shows a trend of rising up to 78.92%. While after that, the amount of NO dissolved in absorbent reaches a saturation condition, which causes the excessive amounts of NO can not be absorbed. Therefore, the removal efficiency of NO appears a decreasing trend. In terms of comprehensive, the initial NO concentration of 1200 mg/m³ is an appropriate value.

3.5 *Effect of initial SO_2 concentration in the simulated flue gas on the absorption behavior*

The absorption experimental results were shown in Figure 5 under the absorption condition of Case 4 illustrated in Table 1. It can be seen in Figure 5, along with increasing of the initial SO_2 concentration, the removal efficiency of SO_2 shows an increase firstly and then keeps in a constant line because of a high solubility of SO_2 in absorbent, while the removal efficiency of NO increases firstly and then decreases. A better simultaneous absorption behavior of SO_2 and NO can be found at the initial SO_2 concentration of 2500 mg/m³. According to the mechanism of denitration expressed as equation (6), there is a lower concentration of SO_3^{2-} in the solution under a lower initial concentration of SO_2, then the promotion of the synergistic effects of simultaneous desulfurization and denitration resulted from PZ and N-(2-hydroxyethyl) piperazine is weaker, which causes a lower removal efficiency of NO. When the initial concentration of SO_2 is more than 2500 mg/m³, due to a lower solubility of NO in absorbent, a lower removal efficiency of NO can be obtained.

3.6 *Influence of the simulated flue gas flow rate on the absorption performance*

Figure 6 shows the variation of the removal efficiency of SO_2 and NO with the simulated gas flow rate under the condition of Case 5 listed in Table 1.

As we all know, a higher flow rate of gas means a shorter contact time between the gas and the liquid, which suggests a poor mass transfer condition between the gas and the liquid. Hence, the removal efficiency of SO_2 and NO exhibits a downward trend with the increase of the simulated flue gas flow rate, respectively. Compared with the lower solubility of NO in the solution, the solubility of SO_2 is higher, accordingly, the removal efficiency of SO_2 is better than that of NO under the same absorption condition. Therefore, in theory, an appropriate simulated gas flow rate can be in a lower value. Nevertheless, during the practical production process, the cost and the capacity of industrial production should be considered. Hence, the simulated gas flow rate of 500 mL/min can give a better simultaneous removal behavior.

4 CONCLUSIONS

The influences of the absorption conditions, such as the concentration of N-(2-hydroxyethyl) piperazine, pH value, the initial concentration of NO and SO_2 in the simulated gas and the simulated gas flow rate, on the simultaneous desulfurization and denitration were investigated in detail by using Urea-PZ-N-(2-hydroxyethyl) piperazine ternary system as absorbent. According to a relatively higher solubility of SO_2 in absorbent, it was found that the above-mentioned influence factors had a slightly effect on the removal efficiency of SO_2. As for the absorption of NO, there was a different absorption performance due to the synergistic effect in the ternary system and the lower solubility of NO in the solution. SO_2 removal efficiency of 98.20% and NO removal efficiency of 78.92% could be obtained under pH value of 5, concentration of N-(2-hydroxyethyl) piperazine of 0.3 mol/L, simulated flue gas flow rate of 500 mL/min, initial concentration of NO and SO_2 in the simulated flue gas of 1200 mg/m^3 and 2500 mg/m^3, respectively.

ACKNOWLEDGMENTS

The work was supported by Wuhan Science and Technology Project (20150617010111597).

REFERENCES

Fang, P., C.P. Cen, Z.X. Tang, P.Y. Zhong, D.S. Chen & Z.H. Chen (2011). Simultaneous removal of SO_2 and NO_x by wet scrbbing using urea solution. *Chem. Eng. J.* 168, 52–59.

Lee, S., K. Park, J.W. Park & B.H. Kim (2005). Characteristics of reducing NO using urea and alkaline additives.*Combustion and Flame.* 141, 200–203.

Skalska, K., J.S. Miller & S. Ledakowicz (2010). Trends in NOx abatement: a review. *Science of the Total Environment.* 408, 3976–3989.

Wang, F., T. Chen, M. Jin & P. Lu (2014). Simultaneous Desulfurization and Denitrification from Flue Gas Using Urea/Piperazine Solution. *Adv. Mater. Research,* 881–883, 641–644.

Wei, J.C., Y.B. Luo, P. Yu & H.Z. Tan (2009). Removal of NO from flue gas by wet scrubbing with $NaClO_2$/$(NH_2)_2CO$ solutions. *J. Ind. Eng. Chem.* 15, 16–22.

Advances in Materials Science, Energy Technology
and Environmental Engineering – Patty & Zhou (Eds)
© 2017 Taylor & Francis Group, London, ISBN 978-1-138-19668-1

The antioxidant activitives and application of miracle fruit leaves

Yu-Ge Liu, Xiu-Mei Zhang, Fei-Yue Ma & Qiong Fu
Key Laboratory of Tropical Fruit Biology, Ministry of Agriculture, South Subtropical Crop Research Institute,
Chinese Academy of Tropical Agricultural Science (CATAS), Zhanjiang, Guangdong, China

ABSTRACT: In the present work, the antioxidant activities of miracle fruit leaves were evaluated by investigating their extract with different concenrations. The extract was also used for the cultivation of *Caenorhabditis elegans*. Results showed that miracle fruit leaves possessed a high total phenolic content of 87 mg GAE/g DW (Dried Weight) and ferric reducing power, and they were excellent free radical inhibitor. The total phenol content in the extract was highly correlated with DPPH antioxidant activity. The extract can improve the tolerance of *Caenorhabditis elegans* to paraquat. The research here demonstrated that the leaves of miracle fruit can not only be a resource of antioxidants but also possess big potential in medical application.

1 INTRODUCTION

Synsepalum dulcificum Daniell (Sapotaceae), also known as miracle fruit or mysterious fruit, is an evergreen tropical plant natively grown in West Africa. This plant is quite famous for the property of remarkably altering the sour taste into sweet taste (Kurihara and Beidler, 1968).

Polyphenols are secondary metabolites found in higher plants like fruits and vegetables (Manach et al, 2004). These compounds can scavenge and prevent the formation of reactive oxygen and nitrogen species which may cause many diseases to human beings (Robards et al, 1999; Eberhardt et al, 2000; Kim et al, 2002). Therefore, much attention has been paid to the antioxidant activities of polyphenols of plants in recent years. The polyphenolic compounds exist in all parts of plants, including fruit, leaf, root, stem and so on. The phenolic content and antioxidant properties of miracle fruits have been investigated in detail (Du et al, 2014; Inglett et al, 2011). However, there have been few reports on the antioxidant activities of miracle fruit leaves.

In the present work, the antioxidant activities of leaves miracle fruit were evaluated by investigating their extract. As an application, the extract was also used for the cultivation of *Caenorhabditis elegans*. The research here demonstrated that the leaves of miracle fruit can not only be a resource of antioxidants but also possess big potential in medical application.

2 MATERIALS AND METHODS

2.1 Materials and reagents

The leaves used in this paper were collected just from the trees of miracle fruit planted in South Subtropical Crop Research Institute. Mature leaves were first cleared and dried at 50°C, and then ground using a stainless-steel grinder. They were stored in vacuum-packaged polyethylene pouches at –20°C until required for analysis.

Folin–Ciocalteu's (FC) phenol reagent and Gallic Acid (GA) were purchased from Fluka. The 2,2'-diphenyl-2-picrylhydrazyl (DPPH) radical was received from Sigma-Aldrich. *Caenorhabditis elegans* (CL4176) was received from Link C, University of Colorado, Boulder, CO.

2.2 Sample preparation

Dried powder sample (1.00 g) was weighed and refluxed with 30 ml of 70% methanol at 60°C for 2 h under magnetic stirring. The filtrate was separated by centrifugation, and the extraction was repeated for 3 times. All the filtrate was collected and concentrated under reduced pressure at 40°C. The dried residue was resolved in 30 ml of methanol and used for antioxidant tests.

2.3 Determination of total phenolic content in the extract

The Total Phenol Content (TPC) was determined using the FC assay described before with some modifications (Singleton and Rossi, 1965). Typically, 0.025 ml of the extract (diluted 10 times with methanol in advance) were introduced into test tubes and followed by the addition of 2.0 ml of Folin-Ciocalteu's reagent (diluted 10 times with water in advance) and 5.975 ml of water. The solutions were allowed to stand 5 min at room temperature before the addition 2 ml of sodium carbonate solution (7.5% w/v). After reacting in dark for 30 min at room temperature, the absorbance of the solutions were measured at 760 nm on a UV–vis spectrophotometer (Shimadzu UV-2700, Japan). The calibration curve was prepared using a standard solution of gallic acid. The results were expressed as milligram Gallic Acid Equivalents (GAE)/g Dry Weight (DW).

2.4 DPPH radical scavenging activity assay

The free radical scavenging activity of the extract was performed by measuring the decrease in absorbance of DPPH solution at 517 nm in the presence of the extracts by the method proposed by Liyana-Pathirana et al (2010) with minor changes. The solution of 0.5 mM was prepared by dissolving DPPH in methanol. For the evaluation of free radical scavenging activity, 3 ml of DPPH was added into 0.5 ml of the extracts with different concentrations. The mixture was then allowed to stand at room temperature for 30 min in dark before the absorbance at 517 nm was read. The control was prepared as above without extract. The antioxidant activity could be expressed as the following equation:

$$\text{Scavenging activity} = \frac{A_0 - A_s}{A_0} \times 100\%$$

where A_0 and A_s were the absorbance at 517 nm of the control and sample solution, respectively.

2.5 Ferric reducing power

The antioxidant capacity of the extract was also tested by the ferric reducing power. This assay was performed according to a modified method described by Juntachote and Berghofer (2005). Sample extracts of 0.5 ml with different concentrations were added into 0.5 ml of 0.2 M phosphate buffer (pH 6.6) and 1.5 ml of potassium ferricyanide (0.3%). The mixtures were incubated at 50°C for 30 min, and then 1 ml of trichloroacetic acid (10%) was added. After centrifuging at 5000 rpm for 10 min, 1 ml of ferric trichloride (0.3%) was added into the mixture. The absorbance of the solution at 700 nm was measured after standing for 30 min. The assay was run in triplicate and the increase in absorbance of the reaction indicated the reducing power of the samples.

2.6 Tolerance of Caenorhabditis elegans to paraquat

The solution of extract used in this part was dissolved by water.

Synchronized *Caenorhabditis elegans* (worms) were processed by the extract at three different

concentrations as the procedure described by Ye et al (2014) and rinsed to EP tubes by M9 buffer. The tubes were then flushed with M9 buffer to remove escherichia coli OP50. The worms were then treated by paraquat of 50 mM and their death was tested by prodding with the worm picker. Vitamin C of 10 ug/ml was used as positive control.

2.7 Data analysis

The significant differences among paralysis assay treatments were conducted by Log-rank (Mantel-Cox) Test, where * means $p < 0.05$, ** means $p < 0.01$ and *** means $p < 0.001$, respectively.

3 RESULT AND DISCUSSION

3.1 Content of total phenols

The yield of the extract was about 0.3 g from 1.0 g of miracle fruit leaves. The content of total phenolics determined by Folin-Ciocalteu's method was 87 mg GAE/g DW. The value of TPC in the leaves was much higher than that of strawberry (264, 368 and 83.9 mg GAE/100 g Fresh Weight(FW)), blackberry (660 mg GAE/100 g FW), blueberry (348, and 531–795 mg GAE/100 g FW), cranberry (709 mg GAE/100 g FW), chokeberries (8563–10804.1 GAE/1 kg FW) which were reported antioxidant-rich substances (Brat et al, 2006; Wu et al, 2004; Heinonen, et al, 1998; León-González et al, 2013). All this indicates that the leaf of miracle fruits is rich in antioxidant phytochemicals and may possess big potential in the field of antioxidants.

3.2 The antioxidant activities of miracle fruit leaf extract and correlation between antioxidant capacity and total phenolic content

Due to its operating simplicity, DPPH is one of the most popular methods employed for the evaluation of antioxidant ability, especially in plant extract. DPPH is a kind of stable organic radical. In the radical form, the molecule of DPPH has an absorbance at 517 nm, which will disappear after the acceptance of an electron or hydrogen radical from an antioxidant in the solution to become a stable diamagnetic molecule (Matthäus, 2002).

Solutions with different concentration of the extract were used in this paper. The relationship between the scavenging activity and the concentration is shown in Figure 1. It can be clearly seen that that the scavenging activity of the extract was concentration-dependent. With the increase of the amount extracted into solution, the scavenging activity increased accordingly. IC50 was defined as the concentration of the methanol extracts to

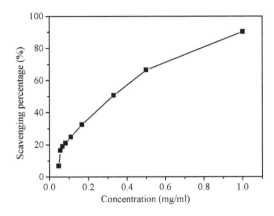

Figure 1. DPPH scavenging ability of the extract of miracle fruit leaves at different concentrations.

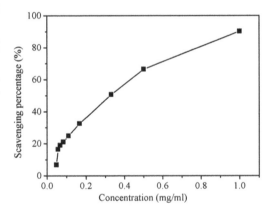

Figure 2. Ferric reducing power of the extract of miracle fruit leaves at different concentrations.

quench 50% of DPPH in the solution under the chosen experimental conditions. The IC50 for the extract derived from the figure of scavenging activity and concentrations was 0.34 mg/ml.

Correction between the antioxidant activity and the concentration of total phenolics was studied to determine the function of those polyphenolic compounds. Results showed that total phenol content was highly correlated with DPPH antioxidant activity and the relative correction coefficient was 0.995.

The Fe^{3+}–Fe^{2+} reducing power is widely used to determine the electron donating capacity of antioxidants. It can also serve as a significant indicator of potential antioxidant activity. During this assay, the presence of reductants (antioxidants) in the sloution would reduce Fe^{3+}/ferricyanide complex to the ferrous form (Fe^{2+}). The resulted Fe^{2+} can then be monitored by measuring the formation of Perl's Prussian blue at 700 nm. The reducing power

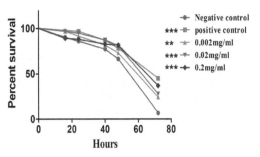

Figure 3. Impact of the extract of miracle fruit leaves to tolerance of *Caenorhabditis elegans* to paraquat.

of the extract was shown in Figure 2. Similar to that of the DPPH assay, the reducing power of the extract was also concentration-dependent.

It had been reported that the antioxidant activity may be concomitant with the development of the reducing power. It can be seen from Figures 1 and 2 that both the scavenging activity and the reducing power increased with the increase of extract concentration. The correlation coefficients between antioxidant activity and reducing power for the extract was quite high ($R^2 = 0.995$), indicating that antioxidant properties were concomitant with the development of reducing power. This means that the reducing power of the extract of miracle fruit leaves was also highly correlated with the total phenolics. This result is in agreement with the extracts of Holy basil, Galangal, longan leaves and barks (Juntachote and Berghofer, 2005; Liu et al, 2012)

3.3 Improvement of tolerance of Caenorhabditis elegans to paraquat

Paraquat is a herbicide that popularly used worldwide. It is believed that the antioxidant system of organisms can be damaged by uptake of this herbicide, and *Caenorhabditis elegans* is often used as a model in investigating this.

The impact of the extract of miracle fruit leaves on the tolerance of *Caenorhabditis elegans* to paraquat was shown in Figure 3. It can be seen that after treated by the extract of different concentraton, the survival percent of *Caenorhabditis elegans* was greatly improved as compared with negative control. This implied that miracle fruit leaves possessed giant potential in synthesizing medicines to improve human body's antioxidant system.

4 CONCLUSIONS

The antioxidant activities of miracle fruit leaves were evaluated by investigating their extract with

DPPH and ferric reducing power assay. Results showed that the leaves possessed high total phenolic content and ferric reducing power, and they were excellent free radical inhibitor. Correlation between the antioxidant capacities and total phenol content indicated that phenolic compounds may be the main contributors to the antioxidant activities. The correlation between radical scavenging activities and reducing power for the extract means that the reducing power of the extract of miracle fruit leaves was also highly correlated with the total phenolics. The extract of miracle fruit leaves can also improve the tolerance of *Caenorhabditis elegans* to paraquat. The research here indicated the giant potential of miracle fruit leaves as natural antioxidants resource and in medicinal application.

ACKNOWLEDGEMENTS

The research was supported by the National Natural Science Foundation of Hainan province (No. 20153122) and the Fund on Basic Scientific Research Project of Nonprofit Central Research Institutions (No. 1630062014016).

REFERENCES

Brat, P., S. George, A. Bellamy, L. D. Chaffaut, A. Scalbert, L. Mennen, N. Arnault, & M. J. Amiot (2006). Daily polyphenol intake in France from fruit and vegetables. *J. Nutr. 136(9)*, 2368–2373.

Du, L., Y. Shen, X. Zhang, W. Prinyawiwatkul, & Z. Xu (2014). Antioxidant-rich phytochemicals in miracle berry (Synsepalum dulcificum) and antioxidant activity of its extracts. *Food Chem. 153*, 279–284.

Eberhardt, M. V., C. Y. Lee, & R. H. Liu (2000). Antioxidant activity of fresh apples. *Nature 405*, 903–904.

Heinonen, I. M., P. J. Lehtonen, & A. I. Hopia (1998). Antioxidant activity of berry and fruit wines and liquors. *J. Agric. Food Chem. 46(1)*, 25–31.

Inglett, G. E. & D. Chen (2011). Contents of Phenolics and Flavonoids and Antioxidant Activities in Skin, Pulp, and Seeds of Miracle Fruit. *J. Food Sci. 76(3)*, C479-C482.

Juntachote, T. & E. Berghofer (2005). Antioxidative properties and stability of ethanolic extracts of Holybasil and Galangal. *Food Chem. 92*, 193–202.

Kim, D., K. W. Lee, H. J. Lee, & C. Y. Lee (2002). Vitamin C equivalent antioxidant capacity (VCEAC) of phenolic phytochemicals. *J. Agric. Food Chem. 50*, 3713–3717.

Kurihara, K. & L. M. Beidler (1968). Taste-modifying proteinfrom miracle fruit. *Science 161*, 1241–1243.

León-González, A. J., P. Truchado, F. A. Tomás-Barberán, M. López-Lázaro, M. C. D. Barradas, & C. Martín-Cordero (2013). Phenolic acids, flavonols and anthocyanins in Corema album (L.) D. Don berries. *J. Food Comp. Anal. 29(1)*, 58–63.

Liu, Y., L. Liu, Y. Mo, C. Wei, L. Lv, & P. Luo (2012). Antioxidant activity of longan (Dimocarpus longan) barks and leaves. *African J. Biotechnol. 11(27)*, 7038–7045.

Liyana-Pathirana, C. M. & F. Shahidi (2005). Antioxidant activity of commerical soft and hard wheat (Triticum aestivum L.) as affected by gastric pH conditions. *J. Agric. Food Chem. 53(7)*, 2433–2440.

Matthäus, B (2002). Antioxidant activity of extracts obtained from residues of different oilseeds. *J. Agric. Food Chem. 50*, 3444–3452.

Robards, K., P. D. Prenzler, G. Tucker, P. Swatsitang, & W. Glover (1999). Phenolic compounds and their role in oxidative processes in fruits. *Food Chem. 66(4)*, 401–436.

Singleton, V. L. & J. A. Rossi (1965). Colorimetry of total phenolics with phosphomolybdic-phosphotungstic acid reagents. Am. *J. Enol. Vitic. 16*, 144–158.

Wu, X., G. R. Beecher, J. M. Holden, D. B. Haytowitz, S. E. Gebhardt, & R. L. Prior (2004). Lipophilic and hydrophilic antioxidant capacities of common foods in the United States. *J. Agric. Food Chem. 52*, 4026–4037.

Ye, X., J. M. Linton, N. J. Schork, L. B. Buck, & M. Petrascheck (2014). A pharmacological network for lifespan extension in Caenorhabditis elegans. *Aging Cell 13(2)*, 206–215.

Advances in Materials Science, Energy Technology and Environmental Engineering – Patty & Zhou (Eds)
© *2017 Taylor & Francis Group, London, ISBN 978-1-138-19668-1*

Comparison of the determination methods for total flavonoids in *Manilkara zapota* leaves

Haifang Chen, Qiong Fu, Xiumei Zhang, Yuge Liu & Feiyue Ma
Key Laboratory of Tropical Fruit Biology, Ministry of Agriculture, South Subtropical Crop Research Institute, Chinese Academy of Tropical Agricultural Science (CATAS), Zhanjiang, Guangdong, China

ABSTRACT: $Al(NO_3)_3$-$NaNO_2$-$NaOH$, boric acid-citric acid, and $AlCl_3$ colorimetric assays were compared to determine the total flavonoids in *Manilkara zapota* leaves. The results indicated that $Al(NO_3)_3$-$NaNO_2$-$NaOH$ colorimetric assay was the most appropriate for the determination of total flavonoids in *Manilkara zapota* leaves. The stability, precision, and repeatability were very good, with the RSD being 1.94%, 1.22%, and 1.36%, respectively. The content of total flavonoids in *Manilkara zapota* leaves was 12.38%. The method is appropriate for the quantitative determination of total flavonoids in *Manilkara zapota* leaves and their preparations.

1 INTRODUCTION

Manilkara zapota is a nutritious, fleshy berry, with a scurfy brown peel and light brown, brownish yellow to reddish brown pulp, with a texture varying from gritty to smooth. The pulp has a very sweet pleasant flavor. The fruit and its peel contain high amounts of saponin, which has astringent properties similar to tannin (Devatkal, 2014). *Manilkara zapota* had the highest oxygen radical absorbance capacity and total phenols content among many fruits (Isabelle, 2010). The aqueous extract of *Manilkara zapota* has a significantly higher reducing power (Jamuna, 2013). Among them, the flavonoids, which are the secondary metabolites of *Manilkara zapota*, are one of the active compounds. However, there is scant report on extracting active compounds from *Manilkara zapota* leaves.

Flavonoids, a large category of plant polyphenol secondary metabolites, are widely distributed in medicinal herbs, fruits, teas, etc. (Uckoo, 2011). It possessed particular interest with regard to human health effects. The flavonoids are classified into two types, glycosides and aglycones, and the predominant form of the naturally occurring flavonoids is flavonoid glycoside (Liu, 2009). Modern phytochemical and pharmacological investigations reveal that flavonoids possess various biological activities, such as anti-oxidant, anti-tumor, anti-viral, and anti-bacterial effects (Wu, 2009; Luo, 2010; Wu, 2011; Kong, 2010). Our preliminary investigation indicated that flavonoids were abundant in *Manilkara zapota* leaves. Both flavonoids have great potentials to be used as clinical therapeutic agents, food additives, and nutraceutical products (Balerdi, 2000).

The present work aims to develop a fast and efficient determination method for total flavonoids in *Manilkara zapota* leaves. The results indicated that the $Al(NO_3)_3$-$NaNO_2$-$NaOH$ colorimetric assay was the most appropriate for the determination of total flavonoids in *Manilkara zapota* leaves. The stability, precision, and repeatability were very good. It can be referenced for the quality control of *Manilkara zapota* leaves. The present research would be helpful for the further exploration and full use of the renewable resource leaves of *Manilkara zapota*.

2 MATERIALS AND METHODS

2.1 *Materials and reagents*

The leaves of *Manilkara zapota* were collected from South Subtropical Crop Research Institute. The material was dried in the shade and powdered using a disintegrator (HX-200 A, Yongkang Hardware and Medical Instrument Plant, China). Ethanol obtained from Tianjin Chemical Reagents Co. (Tianjin, China) was of analytical grade. Rutin and vitexin were purchased from Sigma–Aldrich (Steinheim, Germany).

2.2 *Sample preparation and the determination methods for total flavonoids*

A sample of pulverized *Manilkara zapota* leaves (2 g) was extracted with 60 mL of 50% ethanol solution in an ultrasonic bath (Kunshan Ultrasonic Instrument, China) for 40 min. The extraction solution was filtered by membrane filtration. The filtrate was concentrated to remove the excess ethanol using a rotary evaporator. The total flavonoids yield of the samples was determined by three different methods using rutin and vitexin as standard.

2.2.1 $Al(NO_3)_3$-$NaNO_2$-$NaOH$ colorimetric assays

Extracts' sample solution (1 mL) or standard solutions were mixed with 0.3 mL of 10% $NaNO_2$ solution in a 10 mL volumetric flask. After 6 min, 0.3 mL 10% $AlCl_3$ solution was added, and the mixture was allowed to stand for another 6 min. Then, 4 mL 4% NaOH was added. The volume was made up to the mark by adding 50% ethanol solution. The reaction solution was mixed well, kept for 15 min, and the absorbance was determined at 510 nm (Al-Matani, 2015; Wang, 2011).

2.2.2 Boric acid-citric acid (BCC) colorimetric assays

Extracts' sample solution (2 mL) or standard solutions were mixed with 2.5 mL of 10% acetone solution of citric acid and 2.5 mL of 0.8% boric acid solutions in a 25 mL volumetric flask. The volume was made up to the mark by adding acetone. The reaction solution was mixed well, and the absorbance was determined at 400 nm.

2.2.3 $AlCl_3$ colorimetric assays

Extracts' sample solution (1 mL) or standard solutions were mixed with 2 mL of 0.1 mol·L^{-1} $AlCl_3$ solution and 3 mL of 1 mol·L^{-1} potassium acetate solutions in a 10 mL volumetric flask. The volume was made up to the mark by adding 50% ethanol solution. The reaction solution was mixed well, kept for 30 min, and the absorbance was determined at 420 nm.

Total flavonoids yield was calculated using a standard curve.

2.3 Standard curve and method validation

Accurately weighed rutin and vitexin were dissolved in methanol as standard stock solutions (1 mg/mL). A series of proper concentrations in order to make the calibration curves were obtained by accurately drawn standard stock solutions (0, 0.1, 0.2, 0.3, 0.4, 0.5 mL). All the standard solutions were stored at 4 °C in the dark before analysis.

3 RESULT AND DISCUSSION

3.1 Method validation: Standard curve

After the yield of total flavonoids was determined, validation tests were performed for linearity, accuracy, precision, and recovery.

The linearity of the calibration curve was tested by diluting the stock solution into a series of concentrations.

The curve was plotted for various standard concentrations and absorbances. The different working calibration curves were obtained based on the reference compounds of rutin and vitexin, respectively. The results of total flavonoids using three methods are shown in Figure 1. The flavonoids are classified into two sorts, glycosides and aglycones, and the predominant form of naturally occurring flavonoids is flavonoid glycoside. Even though vitexin was a C-glycosides compound, from figure 1 it was observed that rutin was appropriate as a reference compound. The absorbances were the average values of three replicate injections. The regression lines for $Al(NO_3)_3$-$NaNO_2$-$NaOH$, boric acid-citric acid and $AlCl_3$ colorimetric assays were $Y = 1.2311X + 0.00005$ ($R^2 = 0.9991$), $Y = 2.3657X + 0.0029$ ($R^2 = 0.9994$), and $Y = 3.2877X + 0.003$ ($R^2 = 0.9990$), where Y is the absorbance and X is the concentration of rutin. All calibration curves exhibited excellent a linear behavior ($R^2 > 0.999$) in a relatively wide range of concentration.

3.2 Method validation: Stability

In order to select the appropriate method, stability, precision, and reproducibility were investigated. RSDs were calculated to assess the standard.

For the stability test (Table 1), the sample solution was analyzed from 0 min to 180 min, and the sample solution was found to be rather stable within 180 min (R.S.D.<6.0%). However, the RSD values of boric acid-citric acid colorimetric assays were relatively obvious. The RSD values of three methods were 1.94% ($Al(NO_3)_3$), 5.71% (boric acid-citric acid), and 2.97% ($AlCl_3$), respectively.

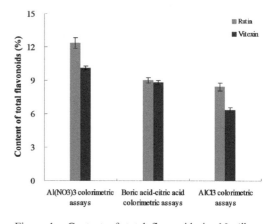

Figure 1. Content of total flavonoids in *Manilkara zapota* leaves by three methods.

Table 1. Stability experiments of three methods.

Method	Time (min) 0	10	20	30	60	180	RSD (%)
Al(NO₃)₃	0.122	0.121	0.120	0.119	0.117	0.116	1.94
BCC	0.174	0.183	0.178	0.169	0.161	0.158	5.71
AlCl₃	0.225	0.227	0.221	0.216	0.212	0.212	2.97

Table 2. Precision experiments of three methods.

No.	Al(NO₃)₃ Absorbance	RSD (%)	BBC Absorbance	RSD (%)	AlCl₃ Absorbance	RSD (%)
1	0.121		0.174		0.229	
2	0.119		0.171		0.223	
3	0.121	1.22	0.176	1.70	0.224	1.27
4	0.123		0.169		0.229	
5	0.122		0.170		0.228	

Table 3. Repeatability experiments of three methods.

No.	Al(NO₃)₃ Absorbance	RSD (%)	BBC Absorbance	RSD (%)	AlCl₃ Absorbance	RSD (%)
1	0.122		0.172		0.223	
2	0.122		0.178		0.231	
3	0.119	1.36	0.175	1.89	0.224	2.20
4	0.123		0.171		0.221	
5	0.120		0.170		0.232	

3.3 Method validation: Precision

The precision test was carried out by determining the same sample solution 5 times for each sample. In Table 2, it was indicated that the results of three methods were in the error range, and the data could be credible. The RSD values of three method were 1.22% (Al(NO₃)₃), 1.70% (boric acid-citric acid), and 1.27% (AlCl₃), respectively.

3.4 Method validation—reproducibility

For repeatability test, determination was performed for five times on the same day under the same conditions. A good repeatability was obtained for the determination of total flavonoids, and the RSD values of three methods were less than 3%. It can be observed that the results of that three methods were in the error range. The RSD values of the three methods were 1.36% (Al(NO₃)₃), 1.89% (boric acid-citric acid), and 2.20% (AlCl₃), respectively. The results are shown in Table 3.

Therefore, the Al(NO₃)₃-NaNO₂-NaOH colorimetric assays method was precise, accurate, and sensitive enough for quantitative evaluation of total flavonoids in *Manilkara zapota* leaves.

3.5 Application of the Al(NO₃)₃-NaNO₂-NaOH colorimetric assays method

The developed Al(NO₃)₃-NaNO₂-NaOH colorimetric assays method was applied to determine total flavonoids and analyze their contents in *Manilkara zapota* leaves. Under these conditions, the total flavonoids yield reached 12.38%. The developed method exhibited a sufficiently high sensitivity and accuracy for the analysis of flavonoids in *Manilkara zapota* leaves. From these results, the total flavonoids yields were high in *Manilkara zapota* leaves. It has a significantly higher reducing power and antioxidant activity. Consequently, *Manilkara zapota* leaves total flavonoids extraction represents a promising alternative for the further investigation of biologically active compounds.

4 CONCLUSIONS

In the present study, compared with the three determination methods of total flavonoids, the Al(NO₃)₃-NaNO₂-NaOH colorimetric assay was precise, accurate, and sensitive enough for the quantitative evaluation of total flavonoids in *Manilkara zapota* leaves. The developed Al(NO₃)₃-NaNO₂-NaOH colorimetric assay revealed good stability (RSD≤1.94%), precision (RSD≤1.22%), and reproducibility (RSD≤1.36%). The results above clearly indicated that the Al(NO₃)₃-NaNO₂-NaOH colorimetric assays method was reliable for the determination of total flavonoids in *Manilkara zapota* leaves. It can be referenced for the quality control of *Manilkara zapota* leaves. This work provides a promising alternative for food and medical studies and applications of flavonoids from *Manilkara zapota* leaves.

ACKNOWLEDGMENTS

This work was financially supported by the Fund on Basic Scientific Research Project of Nonprofit Central Research Institutions (1630062015011, 1630062014016) and Waste Utilization of Tropical crops Research and Demonstration.

REFERENCES

Al-Matani S.K., Al-Wahaibi R. N. S., Hossain M. A.: Karbala Int.: J. Mod. Sci. (2015) 1:166–171.

Balerdi C. F., Crane J. H., Institute of Food and Agricultural Sciences, University of Florida: Homestead, 2000.

Devatkal S. K., Kamboj R., Paul D.: J. Food Sci Technol, (February 2014) 51(2):387–391.

Isabelle M, Bee L.L, Meng T.L, Woon-Puay K, Dejian H, Choon N.O: Food Chem. (2010) 123:77–84.

Jamuna K.S., Ramesh C.K, Srinivasa T.R, Raghu K.L: J Pharm Res (2010) 3:2378–2380

Kong Y., Fu Y.J., Zu Y.G., Chang F.R., Chen Y.H., Liu X.L., Steltend J., Schiebele H.M.: Food Chem. 121 (2010) 1150–1155.

Liu W., Fu Y., Zu Y.G., Kong Y., Zhang L.B., Zu Y.G., Efferth T., J. Chromatogr. A 2009, 1216, 3841–3850.

Luo M., Liu X., Zu Y.G., Fu Y.J., Zhang S., Yao L.P., Efferth T.: Chem. Biol. Interact. 188 (2010) 151–160.

Uckoo R.M., Jayaprakasha G.K., Patil B.S.: Sep. Purif. Technol. 81 (2011) 151–158.

Wang Y., Zu Y.G., Long J.J., Fu Y.J, Li S.M, Zhang D.Y, Li J., Wink M., Efferth T.: Food Chem. 126 (2011) 1178–1185.

Wu N., Fu K., Fu Y.J., Zu Y.G., Chang F.R., Chen Y.H., Liu X.L., Kong Y., Liu W., Gu C.B.: Molecules 14 (2009) 1032–1043.

Wu N., Kong Y., Zu Y.G., Fu Y.J., Liu Z.G., Meng R.H., Liu X., Efferth T.: Phytomedicine 18 (2011) 110–118.

Advances in Materials Science, Energy Technology
and Environmental Engineering – Patty & Zhou (Eds)
© 2017 Taylor & Francis Group, London, ISBN 978-1-138-19668-1

Classification and progress of emergency cutting technology based on energetic materials

Wen-Tong Xin, Yu-Hua Liu, Li-Ming Sun & Xiu-Mei Cui
Department of Vehicles and Electrical Engineering, Mechanical Engineering College, Shijiazhuang, China

ABSTRACT: Emergency cutting technology based on energetic materials includes pyrotechnic cutting, manual SHS cutting, and manual thermite cutting. Characteristics, research, and problem of this technology are summarized. Besides, the paper discusses the future prospect of manual thermite cutting.

1 INTRODUCTION

Emergency cutting in the wild is an important technology for emergency repair in special environments such as disaster relief, field repairs, etc. Not only the thermal cutting method including plasma cutting or gas cutting, but also cold cutting method including hand sawing is currently utilized. With low efficiency, high work intensity, and high technical requirements, hand sawing is only applied to low-intensity small and thin structures without surface hardening. Plasma cutting which can be used to cut a variety of metals is a highly efficient thermal cutting method using a high-temperature plasma arc. The cutting efficiency of plasma cutting is several hundred times higher than the one of hand sawing. Nevertheless, plasma cutting requires expensive power supply and heavy equipment, which seriously restricts its flexibility in the wild. Gas cutting is a common thermal cutting method by using combustion reaction. Although in recent years, technical staff have made a lot of painstaking work on new combustible gas research and development as well as gas equipment miniaturization, regardless of flammable gas or liquid, hyperbaric oxygen devices are required. The damage or breakage of a hyperbaric oxygen tube is very dangerous in disaster relief environment (for example, fire). Therefore, in order to meet the requirements of the emergency cutting, the research and development on new thermite cutting technology based on energetic materials without power, gas supply, and equipment is necessary. Currently, there exists pyrotechnic cutting technology, manual SHS (Self-propagating High-temperature Synthesis) cutting technology, and manual thermite cutting technology.

2 PYROTECHNIC CUTTING TECHNOLOGY

Pyrotechnic cutting technology achieves the purpose of cutting material by a high-temperature molten metal or corrosive gases from the redox reaction of pyrotechnic compositions packed in a pyrotechnic cutting torch.

2.1 Pyrotechnic cutting torches

Pyrotechnic cutting torches are used for filling pyrotechnic composition, the structure of which will directly affect the cutting effect. There are many kinds of pyrotechnic cutting torches, such as amphibious pyrotechnic cutting torch invented by Phillips et al (1973), cutting torch with a nozzle invented by Proctor et al (1986), and cutting torch with a stand invented by Sery et al (1991). Torches are shown in Fig. 1

Observing these structures, we find some common characteristics given as follows: the torch mainly consists of three parts: metal tube, nozzle, and ignition devices. The metal tube is filled with pyrotechnic composition. The nozzle made of a refractory material, for example, graphite, is mounted at one end of the metal tube and ignition devices are mounted at the other end of the tube.

In China, a few establishments have systematically studied the development of cutting torch used for loading pyrotechnic. According to the relevant literature in recent years, we find that Bo-Qin Ruan et al (2003) invented a pipe-cutter using thermite and Qiang Gao et al (2009) invented an annular cutting device for a non-explosive metal tube. Molten metals are produced by the combustion of pyrotechnic jet to the surrounding in the

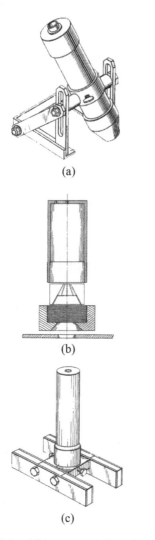

(a)

(b)

(c)

Figure 1. (a) Amphibious pyrotechnic cutting torch, (b) cutting torch with a nozzle, and (c) cutting torch with a stand.

direction perpendicular to the wall instead of from one end of the tube, which is suitable for cutting metal pipe.

2.2 Pyrotechnic composition

The key of pyrotechnic cutting technology is the pyrotechnic composition. Early studies on this were conducted in the USA. Helms and coworkers (1972) first developed a pyrotechnic cutting agent, which was made of nickel, metal oxides selected from the group consisting of Fe_2O_3, Fe_3O_4, and Cr_2O_3, combustible metal powder selected from the group consisting of aluminum or magnesium, beryllium, boron, cadmium, and bismuth, and a

small amount of the gas generating agent composition. After three years, they (1975) improved the previous formula and achieved a wider mix of ingredients. Mueller et al (1982) selected sulfate (calcium sulfate or magnesium sulfate or strontium sulfate) instead of metal oxides, aluminum, magnesium, or magnesium alloy as the reducing agent, halogenated polymer as the binder, and sulfur as the accelerant. Christopher and copartners (1984) invented pyrotechnic formulations used to cut the tube, which comprised metal powder including aluminum magnesium, niobium, titanium, metal oxides from a group of Fe_2O_3, FeO, CuO, Cr_2O_3, and PTFE as the gas-producing additive. In the same year, pyrotechnic composition invented by Kennedy et al (1984) was made up of thermite, copolymer of vinylidene fluoride and hexafluoropropylene, and a small amount of graphite. Woytek et al (1984) chose salts containing NF_4BF_4 as the main ingredient of the pyrotechnic, a mixed gas of which consisting of F2 and NF3 produced by combustion reaction acted to corrode metal and rock. Halcomb et al (1990) invented a stable thermite composition, which could be used in a broad range of environmental conditions, such as at very low to very high temperatures, under water, and in space. The thermite composition includes an oxidizable metal, an oxidizing reagent, and a high-temperature-stable gas-producing additive selected from metal carbides and metal nitrides. Carter et al (2003) invented a solid thermite fuel for pyrotechnics with high heat transfer efficiency, which consists of the strongly reducible metal oxide, for example, CuO, the strong reducing agent, for example, Mg, the diluent, and a certain amount of gas generating compounds taken from the group consisting of metal carbides, metal nitrides, and nitrates. Hlavacek et al (2004) chose polyvinyl chloride or Teflon as main pyrotechnic formulations, by which chlorine or fluorine gas would react with the incised work-pieces.

In China, few researches on pyrotechnic composition are conducted and only Chinese People's Armed Police Forces Academy technical personnel made comparatively deep research. Peng Wang and co-workers (2011) analyzed the theoretical basis of pyrotechnic cutting technology from heat transfer and mechanical. Pyrotechnic composition was designed based on the homogenous experimental design and developed for the melting metal material. Seven iron sheets of 0.5 mm thickness were melted by the molten alloy jet, which was generated by the combustion of 25 g of pyrotechnic.

2.3 Drawbacks of pyrotechnic composition

The pyrotechnic used for pyrotechnic cutting technology burns severely and molten slags are spattered, which warns a person not to operate manually close. Therefore, pyrotechnic is usually ignited by spark and the pyrotechnic cutting torch is fixed, resulting in its poor flexibility. Besides, while burning pyrotechnic, the combustion wave moves from front to back along the axis. As a result, the inner wall of the torch is unevenly heated and the tip is ablated for a long time. Thus, the main body of torch, for example, the metal tube is made of a thick high-temperature-resistant high-strength steel. This makes the torch complex in structure, and leads to high cost of manufacturing and use.

3 MANUAL SHS CUTTING TECHNOLOGY

Manual SHS cutting technology is essentially a thermal fuse cutting technology aiming at cutting the metal. It heats and melts the part of the work-piece by heat released by a self-propagating cutter pen while burning, and blows slag and molten metal by the blowing force.

3.1 Researches

Manual SHS cutting technology is developed on the basis of manual SHS welding technology by Wen-Tong Xin and co-workers from Mechanical Engineering College. The core of Manual SHS cutting technology is to develop an energy-efficient cutting pen.

Mechanical Engineering College (Yong-Sheng WU et al 2009) established a rapid assessment system of the manual SHS cutting technology from three respects of the cutting pen operability, cut quality, and slag; researched on the ratio of primary reaction system and formulations optimization; designed and produced the cutting pen the grain of which comprised a thermite mixture consisting of CuO, Fe_2O_3, and Al as main combustion reaction system, $NaNO_3$ as the gasification agent, and Cr_2O_3 as slagging agent, and the structure of cutting pen is shown in Fig. 2. The information of

the kerf is revealed in Fig. 3 and the upper, middle, and lower parts affected by different factors were found. The cutting process parameters including cutting pen diameter of 12 mm, cutting speed of 7 mm/s~8 mm/s, and cutting inclination of 80°~90° when cutting industrial Q235 steel with 5 millimeter thickness were confirmed. These studies provide a theoretical basis and technical support for the wide use of cutting pen. This method implements emergency cutting for industrial Q235 steel with 5 millimeter thickness without external power supply, gas supply, and equipment.

3.2 Discussions

As the cutting pen combustion is open burning, heat loss is large, and the energy density is small. Hand-cutting technology has a low self-propagating cutting efficiency, and it is difficult to cut thick steel and other issues using this technology. In addition, the production cost of a cutting pen is also higher.

4 MANUAL THERMITE CUTTING TECHNOLOGY

Manual thermite cutting technology achieves the purpose of cutting metal by using cutting ammunition filled cutting agent. When cutting, the operator loads cutting ammunition into a special torch and a hand-held torch with a cutting member. After the bomb is lighted, high temperature melts and gases produced by the combustion of cutting agent spray through a nozzle-shaped compression. Then, the part of a work-piece is melted down and blown. The cutting ammunition structure shown in Fig. 4 is made of five parts including the nozzle, cutting agents, cartridge case, primer, and base (Yi-Ying WU et al 2013).

At present, combustion-cutting bomb of 25 mm diameter developed by Wen-Tong XIN and co-

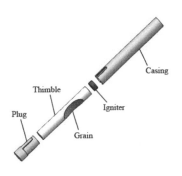

Figure 2. Structure of a cutting pen.

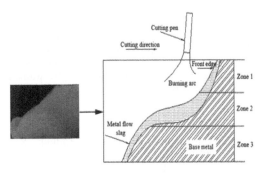

Figure 3. Formation of the kerf by manual SHS cutting technology.

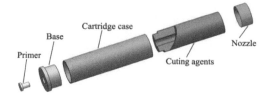

Figure 4. Structure of a cutting bomb.

workers from Mechanical Engineering College can be used in cutting a 20 mm-thick steel structure component without external power supply, gas supply, and equipment.

4.1 *Cutting ammunition preparation*

Mechanical Engineering College graduate Sen Wang (2012) studied how thermite has an effect on the cutting performance of cutting ammunitions. After conducting many experiments, he chose $CuO+Al$ and Fe_2O_3+Al as thermites. He found lower $CuO+Al$, lower system temperature, and longer working hours and determined the optimal mass fraction of $CuO+Al$ in thermites to be 13.6%~22.8%. In addition, he found that contents of KNO_3 would affect the burning time of cutting ammunition and the macroscopic appearance of cut holes. When the mass fraction of KNO_3 is in the range of 4.8%~7.8%, burning time of cutting ammunition increases to 4 seconds and the taper angle of the hole is 14°~17°. The burning rate of the cutting agent is adjusted by Al_2O_3, for which when the mass ratio is 1.5%, the burning speed is moderate.

Mechanical Engineering College graduate Yi-Ying Wu (2013) conducted researches on the influence of the particle size of Al powder on the cutting speed and cutting performance, and found that when the average particle size of Al powder is 53.5 um, the burning rate is relatively stable. He also studied how structural parameters of cutting ammunition affected the cutting performance of cutting ammunition from three respects of charge inner diameter d, charge outer diameter D, and charge length L. While d = 6 mm, D≤32 mm, and L = 80 mm, the cutting ammunition can cut a steel plate with 20 mm thickness. Besides, he studied many characteristics of Al powder.

4.2 *Cutting mechanics*

Mechanical Engineering College graduate Sen Wang (2013) made researches on the manual thermite cutting heat and calculated the maximum power density of cutting ammunition as 2.7×105 W/cm^2, which is equivalent to that of

plasma arc heat. In addition, they analyzed low combustion of the cutting agent and found that the cutting agent burns in a radial direction, spreading from the solid heating zone I, through the condensed phase reaction zone II, to the mixed phase zone III.

4.3 *Process parameters*

Mechanical Engineering College technical personnel studied (Xiu-Mei CUI et al 2014) the variation in burning rate of cutting ammunition under different charge densities, finding that the greater the charge density, the faster the burning. Mechanical Engineering College student Xiumei Cui and partners, who studied the influence of the distance between nozzle and work-piece l, cutting speed v, and cutting angle θ determined l, v, θ, respectively, to be in the optimum range 5 mm-10 mm, 14 mm/s~16 mm /s, and 80°~90°.

5 CONCLUSIONS

Comprehensively, because of its characteristics of simple operation, portability, and high efficiency, manual thermite cutting technology without external power supply, gas supply, and equipment would be widely applied to special circumstances such as disaster relief and emergency repair in the field. Further researches are required for better population and application of manual thermite cutting. The combustion process of cutting agent and the influence of burning rate as well as cutting effectiveness on manual operation should be studied further. The noise would be reduced in a new method in which the fuse is inserted from the front of the cutting bomb. The cutting bomb would be loaded from the front instead of the rear end of the torch. Hence, not only will manufacturing costs be declined because of torch without any complicated locking mechanism, but also ablation pollution of torch will be reduced resulting in improvement in the service life of the torch. The cutting bomb structure would be a solid double-grain structure, the center of which would be filled with a fast burning agent, in case that cutting bomb is damaged in the storage, transportation, and use.

REFERENCES

Bo-Qin Ruan, Xi-Hua Lin & Jin-Ren Liu, et al (2003). Apparatus for severing conduits: China, CN 1544795 A.
Carter & Greg JR (2003). Portable metal cutting pyrotechnic torch: US20030145752.
Christopher, Glenn. B (1984). Pyrotechnic compositions for severing conduits: US4424086.

Halcomb, Danny L Mohler & Jonathan H (1990). High and low-temperature-stable termite composition for producing high-pressure, high velocity gases: US4963203.

Helms, Jr, Horace H Rosner & Alexander G (1972). Pyrotechnic composition: US3695951.

Helms, Jr, Horace H Rosner & Alexander G (1975). Pyrotechnic composition: US3890174.

Hlavacek Vladimir & Pranda Pavol (2004). High speed chemical drill: US20040206451.

Kennedy, Katherine L Proctor & Paul W Dow (1984). Pyrotechnic composition for cutting torch: US4432816.

Kurt F. Mueller, Marguerite S & Farncomb (1982). Metal-cutting pyrotechnic composition: US4349396.

Peng Wang & Jing Zhang (2011). Thermo dynamic analysis, composition design and experimental study on metal cutting pyrotechnic composition. *Energetic Materials*. 19, 459–463 In Chinese.

Phillips & Ralph, O (1973). Project cutter for under water or land use: US3724372.

Pual W Proctor & Robert L Dow (1986). Nozzle for self-contained cutting torches: US4601761.

Qiang Gao, Sen-Yuan Li & Chao-Min Zheng, et al (2009). Charge unit and preparation methods of annular cutting device for non-explosive metal pipe: China, CN101619007.

Sen Wang & Wen-Tong Xin, et al (2013).Study on Cutting Process of Manual Thermite Cutting Technology. *Advanced Materials Research*. 621, 211–215 In Chinese.

Sen Wang, Wen-Tong Xin, Yong-Sheng Wu & Li-Feng Qu (2012). Effect of thermit on cutting capability of combustion cutting ammunition. *Hot Working Technology*. 4, 202–204 In Chinese.

Sen Wang, Wen-Tong Xin & Yong-Sheng Wu (2012). Study on influence of KNO_3 on cutting capability of combustion cutting ammunition. *Initiators & Pyrotechnics*. 2, 14–17 In Chinese.

Sery, Robert S Rozner, Alexander G Waldron, et al (1991). Apparatus for attaching ordnance to barrier targets: USH865.

Woytek, Andrew J Lileck & John T Steigerwalt (1984). Method and apparatus for perforating or cutting with a solid fueled gas mixture: US4446920.

Xiu-Mei Cui, Wen-Tong Xin & Li-Ming Sun (2014). Effect of operating technics of portable special breaking technology on breaking capability. *Manufacturing Automation*. 36, 56–58 In Chinese.

Xiu-Mei Cui, Wen-Tong Xin, Li-Ming Sun & Sen WANG (2014).Effect of Al_2O_3 on breaking capability of portable special breaking technology. *Initiators & Pyrotechnics*. 1, 5–7 In Chinese.

Yi-Ying Wu, Rui-Lin Wang, Wen-Tong Xin & Sen WANG (2014). Research on grain design of cutting ammunition. *Hot Working Technology*. 42, 166–169 In Chinese.

Yi-Ying Wu, Rui-Lin Wang, Wen-Tong Xin & Sen Wang (2013). Research on summarization of cutting ammunition based on firework cutting technology. *Hot Working Technology*. 3, 215–218 In Chinese.

Yi-Ying Wu, Rui-Lin Wang, Wen-Tong Xin, Sen Wang & Xiu-Mei CUI (2014). Combustion and cutting efficiency of cutting ammunition influencing by charge diameter. *Electric Welding Machine*. 44, 26–30 In Chinese.

Yi-Ying Wu, Rui-Lin Wang, Wen-Tong Xin, Sen Wang (2013). Influence of charge length on cutting property of cutting ammunition. *Initiators &Pyrotechnics*. 1, 13–16 In Chinese.

Yong-Sheng Wu, Wen-Tong Xin, Zhi-Zun Li & Bao-Feng LI (2009). Study on thermit of combustion welding rod for cutting. *Hot Working Technology*. 38, 144–146 In Chinese.

Advances in Materials Science, Energy Technology
and Environmental Engineering – Patty & Zhou (Eds)
© 2017 Taylor & Francis Group, London, ISBN 978-1-138-19668-1

Studies on the HPLC fingerprint of a methanolic extract of *Schisandra chinensis*

Jing Bai

Research Center of Life Scineces and Environmental Sciences, Harbin University of Commerce, Harbin, China

ABSTRACT: HPLC fingerprints were developed for the quality control of *Schisandra chinensis* (Turcz.) Baill. Chromatographic separation was performed on a Kromasil C_{18} column (250 mm × 4.6 mm, 5 µm) using acetonitrile and water as the mobile phase in a gradient program. The wavelength of the detector was monitored at 218 nm and the column temperature was maintained at 30°C. In fingerprint analysis, 10 characteristic peaks in the chromatograms were selected as the common peaks. The new method was validated and was successfully applied for the chromatographic fingerprint analysis. All the results indicated that the HPLC fingerprint assay in combination with multi-component determination is suitable for the comprehensive quality control of crude herbs.

1 INTRODUCTION

Fructus Schisandrae chinensis, originated from the dried ripe fruits of *Schisandra chinensis* (Turcz.) Baill. And was ranked as a high-grade herbal drug in the ancient medical book 'Shen Nong Bencao Classics'. Lignans were the most abundant and active components isolated from the fruit of *Schisandra chinensis*, which has demonstrated a wide range of pharmacological activities, including hepatoprotective and neuroprotective, anti-oxidant, anti-inflammatory, immunomodulating effects, etc. (Xu, 2012; Tae, 2012; Pan, 2008; Li, 2013; Rahul, 2012; Chang, 2013; Guo, 2015; Wu, 2014; Gu, 2010; Ba, 2015). In the past few years, chromatographic techniques have been successfully applied to determine the contents of bioactive components in herbal medicines (Wu, 2015; Yi, 2016).

In our study, *Schisandra chinensis* were extracted with water, 75% ethanol, and methanol, respectively, in order to perform screening of the active fraction. The result of the sedative and hypnotic experiments indicated that the methanol extract showed the highest pharmacological activities than that of the other extracts. The results showed that the methanol extract of *Schisandra chinensis* can shorten sleep latency significantly, and increase the sleeping time. Herein, we developed methods for a systematic identification of active fractions of *Schisandra chinensis*. In this study, a HPLC-DAD method has been developed and validated for the fingerprints of Fructus Schisandrae chinensis. To some extent, the effective substances of prescription compatibility were clarified.

2 EXPERIMENTAL SECTION

2.1 *Instrumentation*

HPLC analyses were performed using a Shimadzu RP-HPLC system (Shimadzu Corporation, Japan) consisting of an SCL-10 AVP system controller, LC-10 ATvp infusion pumps, SPD-10 Avp UV detector, ANASTAR chromatography workstation, and Mettler Analytical Balance AE240 (Mettler-Toledo Instruments Co., Ltd).

2.2 *Drugs*

Schizandrol A (lot number: 0857-200304), deoxyschisandrin (lot number: 0764-200107), and schisandrin B (lot number: 0765-200205) are used. A standard reference was also obtained from the National Institute for the Control of Pharmaceutical and Biological Products, China. HPLC-grade acetonitrile was obtained from Fisher Scientific (Pittsburgh, PA), and *Schisandra chinensis* was collected from Heilongjiang, China. Ultra pure water was used.

2.3 *Preparation of sample solution*

About 1 g of the sample was accurately weighed and ultrasonically extracted with 50 ml menthol for 30 min. The weight loss was made up, the contents were shaken, the supernatant was filtered (0.45 µm), and the test solution was prepared.

2.4 *Chromatographic conditions*

The column was Kromasil C18 (5 µm, 4.6 mm × 250 mm). A binary gradient elution system consisting of acetonitrile (A) and water (B) was used with the following gradient programs: 0–55 min, 70% B–55% B; detection wavelength was set at 218 nm, and the flow rate was 1.0 mL/min. The injection volume was 20 μl and the column temperature was maintained at 30°C.

2.5 HPLC fingerprint analysis

The HPLC chromatogram from ten batches of *Schisandra chinensis* was compared with a professional software "Similarity Evaluation System for Chromatographic Fingerprint of Traditional Chinese Medicine" recommended by SFDA. The correlative coefficients between different chromatograms were calculated against the reference fingerprint via the same software. Peak 1 (schizandrol A) was assigned as a reference peak from which the Relative Retention Time (RRT) and Relative Peak Area (RPA) of the other common peaks were calculated.

3 RESULTS AND DISCUSSION

3.1 Method validation of HPLC fingerprint

For the method validation study, the precision, stability, and repeatability experiments were performed on the methanol extract sample. The precision was determined by assaying the same samples for five times. The RSD of the relative retention time of each common peak was found to range between 0.1% and 0.8%, and the RSD of the relative peak area of each common peak were between 0.2% and 2.5% (Tables 1 and 2). The stability test was determined by replicate analysis of the same

Table 1. Precision of the determination of relative retention time of *Schisandra chinensis*.

Peak No.	1	2	3	4	5	RSD/%
1	1.000	1.000	1.000	1.000	1.000	0
2	1.072	1.072	1.073	1.072	1.074	0.1
3	1.139	1.141	1.138	1.140	1.143	0.2
4	1.225	1.226	1.228	1.224	1.230	0.2
5	1.416	1.417	1.419	1.415	1.426	0.3
6	1.618	1.619	1.621	1.616	1.632	0.4
7	2.340	2.342	2.349	2.338	2.359	0.4
8	2.946	2.949	2.952	2.943	2.981	0.5
9	3.102	3.109	3.116	3.101	3.139	0.5
10	3.308	3.321	3.323	3.304	3.352	0.6
11	3.904	3.918	3.926	3.901	3.953	0.5
12	4.452	4.461	4.476	4.447	4.483	0.3
13	4.625	4.638	4.649	4.621	4.697	0.6
14	4.745	4.766	4.773	4.769	4.798	0.4
15	4.880	4.898	4.912	4.871	4.927	0.5
16	5.488	5.497	5.523	5.479	5.594	0.8

Table 2. Precision of the determination of relative peak area of *Schisandra chinensis*.

Peak No.	1	2	3	4	5	RSD /%
1	1.000	1.000	1.000	1.000	1.000	0
2	0.136	0.139	0.138	0.132	0.141	2.5
3	0.175	0.176	0.182	0.173	0.177	1.9
4	0.966	0.969	0.971	0.962	0.973	0.5
5	0.070	0.071	0.072	0.069	0.073	2.4
6	0.282	0.285	0.283	0.281	0.289	1.1
7	0.196	0.197	0.199	0.193	0.201	1.5
8	0.230	0.232	0.236	0.231	0.239	1.6
9	0.104	0.105	0.106	0.104	0.109	2.0
10	0.069	0.068	0.07	0.067	0.071	2.3
11	0.118	0.119	0.12	0.116	0.121	1.6
12	0.816	0.818	0.819	0.812	0.822	0.5
13	0.213	0.218	0.219	0.211	0.221	1.9
14	0.225	0.229	0.228	0.223	0.227	1.0
15	1.109	1.108	1.112	1.107	1.113	0.2
16	0.457	0.458	0.459	0.453	0.462	0.7

Table 3. Relative retention time of stability test of *Schisandra chinensis*.

Peak No.	0 h	6 h	12 h	24 h	48 h	RSD/%
1	1.000	1.000	1.000	1.000	1.000	0
2	1.074	1.075	1.076	1.078	1.081	0.2
3	1.143	1.146	1.149	1.151	1.153	0.3
4	1.231	1.233	1.232	1.235	1.239	0.3
5	1.425	1.429	1.428	1.413	1.427	0.5
6	1.623	1.625	1.629	1.638	1.641	0.5
7	2.351	2.358	2.357	2.364	2.361	0.2
8	2.945	2.957	2.966	2.969	2.976	0.4
9	3.115	3.122	3.127	3.129	3.138	0.3
10	3.336	3.328	3.321	3.311	3.347	0.4
11	3.932	3.945	3.937	3.961	3.978	0.5
12	4.462	4.474	4.483	4.499	4.435	0.5
13	4.693	4.626	4.631	4.694	4.702	0.8
14	4.772	4.713	4.783	4.792	4.725	0.7
15	4.925	4.865	4.881	4.906	4.918	0.5
16	5.565	5.508	5.598	5.587	5.621	0.8

sample solution during 48 hours (0, 6, 12, 24, 48h). The RSD of relative retention time and relative peak area were no more than 0.8 and 2.6%, respectively (Tables 3 and 4). The repeatability test was analyzed by injecting five independently prepared samples, and the RSD of relative retention time and relative peak area were no more than 1.2 and 2.4%, respectively (Tables 5 and 6). All the RSD of precision, stability, and repeatability of the sample were less than 3%, according to the regulation.

3.2 HPLC fingerprints of Schisandra chinensis

To standardize the fingerprints, we analyzed ten samples using the optimized HPLC-UV method.

Table 4. Relative peak area of stability test of *Schisandra chinensis*.

Peak No.	1	2	3	4	5	RSD/%
1	1.000	1.000	1.000	1.000	1.000	0
2	0.129	0.126	0.131	0.133	0.128	2.1
3	0.171	0.169	0.179	0.177	0.175	2.3
4	0.961	0.968	0.973	0.959	0.982	1.0
5	0.067	0.069	0.071	0.07	0.068	2.3
6	0.281	0.276	0.277	0.286	0.273	1.8
7	0.189	0.186	0.197	0.193	0.194	2.3
8	0.226	0.238	0.229	0.232	0.236	2.1
9	0.113	0.111	0.108	0.107	0.112	2.3
10	0.059	0.06	0.058	0.062	0.061	2.6
11	0.116	0.118	0.122	0.119	0.117	1.9
12	0.809	0.811	0.813	0.819	0.823	0.7
13	0.209	0.211	0.216	0.214	0.217	1.6
14	0.219	0.223	0.227	0.224	0.229	1.7
15	1.113	1.109	1.112	1.117	1.126	0.6
16	0.449	0.456	0.451	0.458	0.453	0.8

Table 5. Relative Retention Time of repeatability test of *Schisandra chinensis* (n = 5).

Peak No.	1	2	3	4	5	RSD/%
1	1.000	1.000	1.000	1.000	1.000	0
2	1.086	1.083	1.091	1.076	1.085	0.5
3	1.156	1.152	1.163	1.168	1.171	0.7
4	1.237	1.241	1.232	1.249	1.247	0.6
5	1.426	1.432	1.438	1.441	1.434	0.4
6	1.636	1.639	1.642	1.652	1.663	0.7
7	2.353	2.369	2.371	2.378	2.382	0.5
8	2.952	2.967	2.973	2.981	2.992	0.5
9	3.126	3.133	3.147	3.145	3.154	0.4
10	3.338	3.362	3.357	3.369	3.381	0.5
11	3.965	3.973	3.981	3.992	3.997	0.3
12	4.476	4.481	4.498	4.491	4.463	0.3
13	4.703	4.681	4.628	4.711	4.726	0.8
14	4.801	4.765	4.723	4.812	4.768	0.7
15	4.936	4.912	4.864	4.835	4.867	0.8
16	5.634	5.727	5.536	5.608	5.597	1.2

Peaks were found for all ten samples. There were 16 common peaks in all ten batches and the area sum of them accounted for above 90% of the overall peak area. Peaks 1, 11, and 16 were identified as schizandrol A, deoxyschisandrin, and schisandrin B by comparing with the corresponding chemical references under the same condition. The software of Similarity Evaluation System for Chromatographic Fingerprint of Traditional Chinese Medicine was used to evaluate these chromatograms. The correlation coefficients of similarity were greater than 0.993. Therefore, the detection of these common peaks in HPLC fingerprints is useful in assessing the quality of *Schisandra chinensis*. The results are shown in Fig. 1 and Fig. 2 and presented in Table 7.

Table 6. Relative peak area of repeatability test of *Schisandra chinensis* (n = 5).

Peak No.	1	2	3	4	5	RSD/%
1	1.000	1.000	1.000	1.000	1.000	0
2	0.137	0.135	0.142	0.14	0.138	1.9
3	0.181	0.178	0.179	0.176	0.183	1.5
4	0.976	0.971	0.983	0.989	0.987	0.8
5	0.072	0.076	0.076	0.074	0.073	2.4
6	0.286	0.291	0.293	0.285	0.28	1.8
7	0.195	0.201	0.193	0.189	0.194	2.2
8	0.237	0.235	0.243	0.241	0.242	1.4
9	0.119	0.117	0.121	0.118	0.12	1.3
10	0.071	0.072	0.069	0.07	0.071	1.6
11	0.122	0.121	0.117	0.118	0.123	2.2
12	0.824	0.826	0.814	0.825	0.827	0.6
13	0.227	0.219	0.223	0.224	0.221	1.4
14	0.228	0.231	0.232	0.239	0.238	2.0
15	1.119	1.125	1.128	1.131	1.139	0.7
16	0.461	0.469	0.468	0.463	0.452	1.5

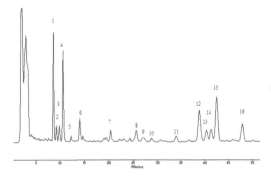

Figure 1. HPLC fingerprints of *Schisandra chinensis*.

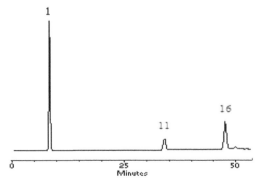

Figure 2. HPLC chromatograms of reference substances. 1, schizandrol A; 11, deoxyschisandrin; 16, schisandrin B.

349

Table 7. The correlation coefficients of similarity in fingerprint (n = 3).

No	1	2	3	4	5
similarity	0.926	0.925	0.978	0.914	0.951
No	6	7	8	9	10
similarity	0.929	0.954	0.982	0.940	0.983

4 CONCLUSION

Under above optimized chromatographic conditions, 16 chromatographic peaks were well separated. The results indicated that the samples from different batches shared a satisfactory similarity, and the similarities of 10 batches samples were higher than 0.90. The convenient and high specific method could be used to identify and evaluate the quality of *Schisandra chinensis*. The developed method also lays a solid foundation for a further comprehensive study on the pharmaco-dynamic-pharmacokinetic effects of *Schisandra chinensis*.

REFERENCES

Ba, Q. 2015. Schisandrin B shows neuroprotective effect in 6-OHDA-induced Parkinson's disease via inhibiting the negative modulation of miR-34a on Nrf2 pathway. *Biomed Pharmacother* 75: 165–172.

Chang, R.M. 2013. Protective role of deoxyschizandrin and schisantherin A against myocardial ischemia-reperfusion injury in rats. *PLoS One* 8 (4): e61590.

Gu, B.H. 2010. Deoxyschisandrin inhibits H_2O_2-induced apoptotic cell death in intestinal epithelial cells through nuclear factor-kappaB. *International Journal of Molecular Medicine* 26(3): 401–406.

Guo C.X. 2015. Schisandrin A and B induce organic anion transporting polypeptide 1B1 transporter activity. *Pharmazie.* 70(1): 29–32.

Li, L.B. 2013. Schisandrin B attenuates acetaminophen-induced hepatic injury through overexpression of heat shock protein 27 and 70 in mice. *Journal of Gastroenterology and Hepatology* 29 (3): 640–647.

Pan, S.Y. 2008. Schisandrin B from Schisandra chinensis reduces hepatic lipid contents in hypercholesterolaemic mice. *Journal of Pharmacy and Pharmacology* 60 (3): 399–403.

Rahul, C. 2012. Schisandrin B exhibits anti-inflammatory activity through modulation of the redox-sensitive transcription factors Nrf2 and NF-κB. *Free Radical Biology and Medicine* 53 (7): 1421–1430.

Tae, H.L. 2012. Neuroprotective effects of Schisandrin B against transient focal cerebral ischemia in Sprague-Dawley rats. *Food and Chemical Toxicology* 50 (12): 4239–4245.

Wu, J. 2014. Deoxyschizandrin, a naturally occurring lignan, is a specific probe substrate of human cytochrome P450 3A. *Drug Metab Dispos* 42(1): 94–104.

Wu, X. D. 2015. Studies on Chromatographic Fingerprint and Fingerprinting Profile-Efficacy Relationship of Saxifraga stolonifera Meerb. *Molecules* 20 (12): 22781–22798.

Xu, X. 2012. Schizandrin prevents dexamethasone-induced cognitive deficits. *Neuroscience Bulletin* 28 (5): 532–540.

Yi, J. 2016. Quality evaluation of the leaves of Magnolia officinalis var. biloba using high-performance liquid chromatography fingerprint analysis of phenolic compounds. *Journal of Separation Science* 39 (4): 784–792.

Advances in Materials Science, Energy Technology and Environmental Engineering – Patty & Zhou (Eds)
© 2017 Taylor & Francis Group, London, ISBN 978-1-138-19668-1

Effect of rosemary extracts on the antimicrobial activity and quality of minced pork during refrigerated storage

Yifei Wang, Hong Wang & Erbing Hua
School of Biological Engineering, Tianjin University of Science and Technology, Tianjin, China

Zhenou Sun, Huali Wang & Hao Wang
Key Laboratory of Food Nutrition and Safety, Ministry of Education, Tianjin University of Science and Technology, Tianjin, China

ABSTRACT: Rosemary is well known as a very promising dietary antioxidant. The antimicrobial activities of 80% ethanol crude extract and four subfractions (petroleum ether fraction, ethyl acetate fraction, butanol fraction, and aqueous fraction) from rosemary to test strains were investigated by the Oxford-cup assay. The results revealed that petroleum ether fraction, ethyl acetate fraction, and crude extract exhibited a good antimicrobial activity to test strains. Petroleum ether fraction had the best antimicrobial activity, while butanol fraction and aqueous fraction only had an antimicrobial action on *Proteus vulgaris* Hauser. Besides, the effect of petroleum ether fraction, ethyl acetate fraction, and crude extract on minced pork stored for 15 days at 4°C were compared, respectively. In addition, sensory evaluation, Total Bacterial Count (TBC), pH, the values of Malondialdehyde (MDA), and Conjugated Dienes (CD) were used as a criterion. The results evidenced that petroleum ether fraction treatment was obviously better than other subfractions in minced pork. Five major compounds were identified by High-Performance Liquid Chromatography-Mass Spectroscopy (HPLC-MS). It has been concluded that petroleum ether fraction had better antimicrobial, preservative, and antioxidant capacities in minced pork, which might be at least partially associated with its interactions with high contents of carnosol and carnosic acid.

1 INTRODUCTION

Rosemary (*Rosmarinus officinalis L.*) is a woody, evergreen, perennial herb with fragrance and it grows in Europe, Asia, and Africa and is native to the Mediterranean basin (Mulinacci et al. 2011, Stansbury 2014). It is well known as a culinary spice, as well as for the increasing interest of its drug activity (Edwards et al. 2015), and hence, it has been used as a kind of medicinal ingredient (Ojeda-Sana et al. 2013, RaMilanovi & Mikov 2014). It contains a series of potentially and biologically active compounds attributed mainly to phenolic diterpenes, carnosol, and carnosic acid (Lo et al. 2002, Bauer et al. 2012).

Rosemary extracts had been proven as an antioxidant of lipids in foods (Ojeda-Sana et al. 2013). In addition to their antioxidant activity, they were also used to inhibit the growth of pathogens, foodborne microorganisms, and moulds (Elgayyar et al. 2001, Campo & Amiot 2000, Hosni et al. 2013). Although the antimicrobial activity of rosemary extract in meat storage has long been acknowledged (Jayasena & Jo 2013, Ojeda-Sana et al. 2013), limited information is available about which component was associated with its antimicrobial activity.

Therefore, the objective of this work was to measure and compare the antimicrobial activity of different components of rosemary extracts from different solvents, and the preservative activity in minced pork. Besides, the major compounds from rosemary extracts were identified using HPLC-MS.

2 MATERIALS AND METHODS

2.1 Materials

Ground rosemary leaves were purchased from Jian Feng Natural Products Company, Tianjin, China. All chemicals were of analytical grade. Formic acid, acetonitrile, and methyl alcohol for HPLC were chromatographically pure.

Proteus vulgaris Hauser (CMCC 49027), *Bacillus subtilis* (CMCC 63501), *Bacillus cereus* (CMCC 63301), *Escherichia coli* (ATCC 25922), *Staphylococcus aureus* (ATCC 43300), *Salmonella* (CMCC 50071), and *Micrococcus luteus* (CMCC 28001) were obtained from the College of Food Engineering and Biotechnology in Tianjin University of Science and Technology, China.

2.2 Preparation of extracts

Dried ground rosemary leaves were refluxed with 80% ethanol (1:10, w/v) for 2 hours at 80°C. The supernatant and the sediment were separated by vacuum filtration (Greatwall, Zhengzhou, China). The residues were extracted another two times following the same procedure. The supernatants were mixed and half of the supernatants were then sequentially extracted with petroleum ether, ethyl acetate, butanol, and ultra-pure water (1:10, w/v). The solvent was then evaporated in a vacuum freeze drier (Thermo Fisher Scientific, USA) and the powder was stored in amber flasks in the dark at 4°C until utilization.

2.3 HPLC-MS

The extracts were diluted with a methanol-water (2:1, v/v) solvent and filtered using a 0.5 μm filter before HPLC analysis. The analysis was performed on a Finnigan Surveyor HPLC-MS system (Thermo, USA). The stationary phase was a Venusil XBP C18 column (3 μm i.d., 2.1 × 150 mm, Bonna-Agela Technologies Inc., Tianjin, China) at a flow rate of 0.2 mL/min anda constant temperature of 30°C. Eluent A was 0.1% aqueous methanol solution while eluent B was acetonitrile. Starting at 70% B. A gradient was followed to 90% B at 45 min and then 20% B at 55.1 min until the end of 70 min analysis. Peaks were detected at 230 nm. The injection volume was 10 μL. MS analysis was run in negative ionization mode using an Electrospray Ionization (ESI) source. The ESI spray voltage, capillary voltage, capillary temperature, tube lens offset voltage, sheath gas (N_2) flow, and auxiliary gas were 4.5 kV, −10 V, 275°C, −50.0 V, 30 arb, and 5 arb, respectively. Data were collected using the full scan mode over a mass range of m/z 100–500.

2.4 Antimicrobial activity analysis

The Oxford-cup assay was used to determine the antimicrobial activity of rosemary extract (Lan & Zesheng 2005). The procedure is as follows: A solid broth medium was sterilized, cooled to 50°C, aseptically poured into a sterile Petri dish (10 mL per plate), and then permitted to solidify. Sterilized Oxford cups were placed on the surface of the solid broth medium and then the cooled semi-solid medium with test bacteria was poured on the solid broth medium (15 mL of broth medium for every plate). A sample solution of 60 μL, which was diluted with dimethyl sulfoxide to 100 mg/mL, was poured into an agar well after the Oxford cups

were removed. Plates were incubated for 24 to 48 h at the appropriate temperature (37°C) and then the diameter of the resulting zone of inhibition was measured. The control well was added with dimethyl sulfoxide to exclude its effect. All experiments were carried out in triplicate.

2.5 Assay for minced pork during refrigerated storage

2.5.1 Preparation of minced pork samples
Fresh minced pork (1000 g) was obtained from the local market, added salt (1.5% w/w) and soybean oil (1.5% w/w), and divided into four different treatments in order to prepare the experimental treatments. Except for one part, which served as the control, the other parts were added petroleum ether fraction, ethyl acetate fraction, and 80% ethanol crude extract, respectively (at a concentration of 300 mg/kg) (Georgantelis et al. 2007, Vareltzis et al. 1997). Each mixture (approximately 135 g) was divided into three parts of 6 cm diameter and 1 cm height for sensory evaluation and the remaining part was analyzed for TBC, pH, the value of MDA and CD. Throughout the procedure, the temperature of the meat blend was kept constant at 4–8°C.

2.5.2 Sensory evaluation
At days 0, 3, 6, 9, 12, and 15 during storage, the minced pork was served to a sensory panel which consisted of six food science-majored students. The color, odor, elasticity, and viscosity were determined using 1–20 point hedonic scale and CIE L* (lightness), a* (redness), and b* (yellowness) values were calculated using a colorimeter (WSC-S, Shanghai Precision & Scientific Instrument Co. Ltd, China). L*, a*, b* were measured on three spots on each sample. Visual assessment was imprinted in high-resolution pictures taken on each sampling day.

2.5.3 Total bacterial count
At days 0, 3, 6, 9, 12, and 15 during storage, a 1.0 g portion from each treatment was homogenized in 9 mL sterilized peptone water. Serial decimal dilutions were prepared using peptone water. Duplicate 1 ml volume of three suitable dilutions spread onto the dried surface of solid broth medium plates, incubated at 37°C for 24 h (Georgantelis et al. 2007, Soultos et al. 2009). All results were reported as log 10 CFU/g. Each experiment was carried out in triplicate.

2.5.4 Determination of pH
The samples were homogenized 3 times with distilled water at 1:4 ratio in a high-speed blender (5000 r/min, 20 s per time), centrifuged at 5000 r/min for 10 min (Georgantelis et al. 2007), and then the pH of the samples was measured using a pH meter (PHS-3C, INESA, Shanghai, China).

Means of three measurements were recorded for each different sample.

2.5.5 *Measurement of MDA*

In brief, 1.0 g portion from each sample was homogenized in the presence of 10 mL of 5 g/100 mL aqueous trichloroacetic acid and 5 mL of 0.8 g/100 mL butylated hydroxytoluene in hexane, and the mixture was centrifuged at 3000 r/min for 10 min. The top layer was discarded, and a 2.5 mL aliquot from the bottom layer was mixed with 1.5 mL of 0.8 g/100 mL aqueous 2-thiobarbituric acid to be further incubated at 70°C for 30 min. Following incubation, the mixture was cooled and submitted to conventional spectrophotometry (Shimadzu, UV-1800, Tokyo, Japan) at 521.5 nm and MDA concentration (mg/100 g) in the samples was calculated (Georgantelis et al. 2007).

2.5.6 *CD values determination*

A portion of 5.0 g from each sample was homogenized with 3.0 mL water, 5.0 mL methyl alcohol, and 10.0 mL trichloromethane, and then centrifuged at 5000 r/min for 10 min. Trichloromethane layer was condensed to obtain lipid in minced pork, diluted with 2,2,4-trimethylpentane, and analyzed at 232 nm (Georgantelis et al. 2007).

2.6 *Statistical analysis*

The results were expressed as means ± Standard Deviation (SD). One-way Analysis of Variance (ANOVA) was used for statistical evaluation of differences among the groups (SPASS version 11.0, statistical Package for the Social Sciences software, SPSS, Chicago, USA). P-values less than 0.05 were considered statistically significant.

3 RESULTS

3.1 *HPLC-MS analysis of rosemary extract*

Chromatograms of 80% ethanol crude extract, petroleum ether fraction, ethyl acetate fraction, butanol fraction, and aqueous fraction from rosemary are shown in Figure 1. The analysis identified five major elements: rosmarinic acid, rosmanol, carnosol, carnosic, and methyl carnosate. Mass spectral data of five major compounds are summarized in Table 1. The relative molecular mass was 360, 346, 330, 332, and 346, respectively. The retention time was 9.78, 18.14, 25.26, 28.77, and 31.15 min, respectively. Their first-order mass spectrums are shown in Figure 2. Major compounds in different fractions are shown in Table 2 expressed as percentages. Five compounds were identified in 80% ethanol crude extract and ethyl acetate fraction. Except rosmarinic acid, the other four compounds were identified in petroleum ether fraction, which had higher content of carnosol and carnosic acid. Interestingly, related substances were not identified in butanol fraction and aqueous fraction.

3.2 *Antimicrobial activity*

The antimicrobial activity of rosemary extracts was expressed and compared by the diameter of the

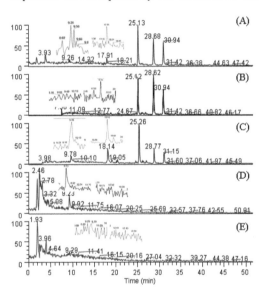

Figure 1. HPLC chromatograms of 80% crude ethanol extract (A), petroleum ether fraction (B), ethyl acetate fraction (C), butanol fraction (D), and aqueous fraction (E) from rosemary.

Table 1. MS characteristics of the five major compounds identified from rosemary extracts.

Peak number	The name of compound	Retention time (min)	[M-H]⁻ (m/z)	Main fragment ions	Molecular formula	Relative molecular mass
1	Rosmarinic acid	9.78	359	161	$C_{18}H_{16}O_8$	360
2	Rosmanol	18.14	345	301, 283	$C_{20}H_{26}O_5$	346
3	Carnosol	25.26	329	285	$C_{20}H_{26}O_4$	330
4	Carnosic acid	28.77	331	287	$C_{20}H_{28}O_4$	332
5	Methyl carnosate	31.15	345	301, 286	$C_{21}H_{30}O_4$	346

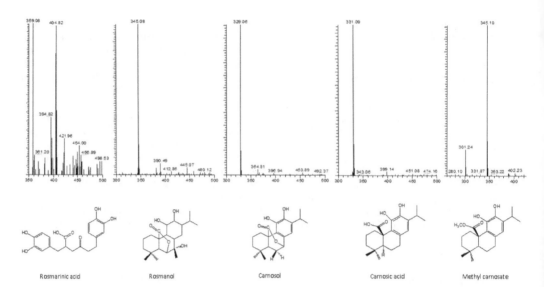

Figure 2. The chemical structures and first-order mass spectrum of the five major compounds.

Table 2. The weight percentage of major compounds in different fractions.

Fraction	Compound	Weight percentage (%)
80% ethanol crude extract	Rosmarinic acid	1.5
	Rosmanol	7.2
	Carnosol	33.2
	Carnosic acid	30.3
	Methyl carnosate	27.8
Petroleum ether fraction	Rosmarinic acid	ND
	Rosmanol	2.0
	Carnosol	28.8
	Carnosic acid	41.4
	Methyl carnosate	27.7
Ethyl acetate fraction	Rosmarinic acid	10.8
	Rosmanol	17.4
	Carnosol	44.3
	Carnosic acid	14.1
	Methyl carnosate	13.5

ND: not determined.

Table 3. Antibacterial activity of rosemary extract fraction (100 mg/mL) against bacteria (units: mm).

Microorganism	80% ethanol crude extract	Petroleum ether fraction	Ethyl acetate fraction	Butanol fraction	Aqueous fraction
E. coli	15	16	11	–	–
Staphylococcus aureus	20	26	18	–	–
Salmonella	13	14	–	–	9
Bacillus subtilis	18	20	9	–	–
Bacillus cereus	13	19	12	–	–
Proteus vulgaris Hauser	18	20	12	22	23
Micrococcus luteus	29	32	22	–	–

The number in the table is the diameter of the resulting zone of inhibition. –, no inhibition.

resulting zone of inhibition (Table 3). The results showed that petroleum ether fraction, ethyl acetate fraction, and 80% ethanol crude extract had extensive inhibition effects on not only gram-positive strains including *Bacillus subtilis*, *Bacillus cereus*, *Staphylococcus aureus*, and *Micrococcus luteus*, but also Gram-negative strains including *Proteus vulgaris* Hauser, *Salmonella*, and *E. coli*. In addition, petroleum ether fraction had higher bacteriostatic efficiency than the other fractions. However, butanol fraction and aqueous fraction did not have inhibition effects against most of the test strains.

Table 4. Evolution of sensory evaluation values of minced pork containing different fractions.

Attribute	Storage time (days)	Control group	80% ethanol crude extract	Petroleum ether fraction	Ethyl acetate fraction
Color	0	20.00 ± 0.00	20.00 ± 0.00	20.00 ± 0.00	20.00 ± 0.00
	3	17.00 ± 0.00	17.22 ± 0.44**	19.22 ± 0.44**	18.22 ± 0.44**
	6	14.44 ± 0.53	15.22 ± 0.44**	18.44 ± 0.53**	17.22 ± 0.44**
	9	10.22 ± 0.44	11.78 ± 0.45**	16.11 ± 0.33**	15.22 ± 0.44**
	12	6.78 ± 0.44	7.78 ± 0.44**	14.56 ± 0.53**	13.56 ± 0.53**
	15	2.89 ± 0.33	5.33 ± 0.71**	12.00 ± 0.50**	11.11 ± 0.93**
Odor	0	20.00 ± 0.00	20.00 ± 0.00	20.00 ± 0.00	20.00 ± 0.00
	3	16.78 ± 0.44	17.33 ± 0.50*	19.11 ± 0.33**	18.33 ± 0.50**
	6	12.11 ± 0.60	14.56 ± 0.73**	17.56 ± 0.53**	16.44 ± 0.53**
	9	8.33 ± 0.50	11.33 ± 0.87**	15.33 ± 0.50**	14.00 ± 0.71**
	12	5.11 ± 0.33	7.67 ± 0.50**	13.00 ± 0.50**	12.00 ± 0.71**
	15	2.11 ± 0.33	5.00 ± 0.50**	8.78 ± 0.67**	8.11 ± 0.33**
Elasticity	0	20.00 ± 0.00	20.00 ± 0.00	20.00 ± 0.00	20.00 ± 0.00
	3	18.22 ± 0.44	18.78 ± 0.44**	19.56 ± 0.53**	19.33 ± 0.50**
	6	15.44 ± 0.53	16.22 ± 0.44**	18.56 ± 0.53**	17.22 ± 0.44**
	9	9.44 ± 0.53	11.89 ± 0.60**	16.67 ± 0.50**	14.67 ± 0.50**
	12	7.33 ± 0.50	9.78 ± 0.44**	11.56 ± 0.53**	10.33 ± 0.71**
	15	4.33 ± 0.50	5.67 ± 0.50**	9.67 ± 0.71**	8.33 ± 0.50**
Viscosity	0	20.00 ± 0.00	20.00 ± 0.00	20.00 ± 0.00	20.00 ± 0.00
	3	17.22 ± 0.44	18.22 ± 0.44**	19.56 ± 0.53**	18.78 ± 0.44**
	6	13.22 ± 0.44	14.67 ± 0.50**	18.78 ± 0.44**	17.44 ± 0.53**
	9	9.67 ± 0.50	10.11 ± 0.33*	16.22 ± 0.44**	14.78 ± 0.44**
	12	6.67 ± 0.50	7.22 ± 0.44*	15.11 ± 0.60**	14.44 ± 0.53**
	15	4.22 ± 0.44	5.44 ± 0.53**	13.22 ± 0.44**	12.22 ± 0.44**
L*	0	39.86 ± 0.56	40.05 ± 0.90	40.13 ± 0.44	39.86 ± 0.72
	3	39.75 ± 0.53	39.93 ± 0.45	40.06 ± 0.49	40.11 ± 0.49
	6	39.54 ± 0.47	39.66 ± 0.47	39.81 ± 0.36	39.95 ± 0.64
	9	39.50 ± 0.34	39.72 ± 0.24	39.84 ± 0.24*	39.76 ± 0.39
	12	39.16 ± 0.35	39.54 ± 0.20**	39.48 ± 0.34*	39.39 ± 0.33
	15	39.02 ± 0.21	39.28 ± 0.20**	39.27 ± 0.35*	39.15 ± 0.38
a*	0	13.40 ± 0.34	13.37 ± 0.24	13.33 ± 0.27	13.17 ± 0.19*
	3	12.65 ± 0.39	12.23 ± 0.23**	12.49 ± 0.43	12.39 ± 0.54
	6	11.81 ± 0.44	11.66 ± 0.44	11.43 ± 0.38*	11.30 ± 0.31**
	9	11.39 ± 0.79	11.25 ± 0.31	11.59 ± 0.45	11.36 ± 0.59
	12	9.63 ± 0.35	10.05 ± 0.22**	9.86 ± 0.06*	10.05 ± 0.37*
	15	8.79 ± 0.16	9.06 ± 0.27*	8.83 ± 0.97	9.11 ± 0.35*
b*	0	14.43 ± 0.45	14.72 ± 0.33	15.17 ± 0.48**	14.43 ± 0.45
	3	15.04 ± 0.39	14.77 ± 0.34	15.33 ± 0.34	15.40 ± 0.10**
	6	12.96 ± 0.44	13.68 ± 0.51**	13.71 ± 0.59**	13.82 ± 0.37*
	9	11.66 ± 0.26	13.11 ± 0.49**	13.51 ± 0.35**	12.36 ± 0.18*
	12	10.66 ± 0.10	12.38 ± 0.14**	12.71 ± 0.34**	11.32 ± 0.33*
	15	10.01 ± 0.52	11.06 ± 0.58**	11.98 ± 0.61**	10.40 ± 0.31*

L* values: 0 = black and 100 = white; a* values: −60 = green and 60 = red; b* values: −60 = blue and 60 = yellow. Data are expressed as mean ± SD. * $P < 0.05$; ** $P < 0.01$ compared with the value of control group.

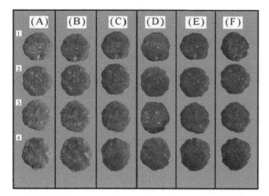

Figure 3. Appearance of the experimental minced pork burgers during storage at: (A) 0 days, (B) 3 days, (C) 6 days, (D) 9 days, (E) 12 days, and (F) 15 days. (1) Control; (2) 80% ethanol crude extract; (3) petroleum ether fraction; and (4) ethyl acetate fraction.

Therefore, we did more research on petroleum ether fraction, ethyl acetate fraction, and 80% ethanol crude extract in the follow-up tests.

3.3 Sensory evaluation

The results of the sensory evaluation of minced pork during the whole storage period are presented in Table 4. A decreasing trend was observed with regard to the score of sensory evaluation. The L^* value of petroleum ether fraction treatment after 9 days of storage was significantly ($P < 0.05$) higher compared with the control. Between day 0 and day 6, a^* values of all treatments were more green than the control group, while a^* values of petroleum ether fraction treatment after 9 days of storage were higher than the control. B^* values of all treatments after 6 days of storage were significantly ($P < 0.05$) higher than control.

3.4 Total bacterial count

All treatments had comparatively lower microbial counts during the storage period (Figure 4), while those with 80% ethanol crude extract added had no significant difference compared with the control group ($P > 0.05$). It was clear that ethyl acetate fraction and petroleum ether fraction could effectively inhibit bacterial reproduction.

3.5 pH values

Changes in pH values during storage period are presented in Figure 5. The pH value slightly decreased until day 3 for petroleum ether fraction and ethyl acetate fraction, whereas after day 3, there was a gradual increase. All treatments increased slowly and were lower compared with the control group.

3.6 MDA values

The extent of lipid oxidation was measured by MDA formation. In this study, the MDA values of the samples with extracts added were lower than that of the control. At day 3, the MDA value of the control increased from 0.13 to 0.42 mg/100 g, while those of treated samples were 0.36, 0.2, and 0.18 mg/100 g for ethyl acetate fraction, 80% ethanol crude extract, and petroleum ether fraction, respectively, and were significantly lower than the control ($P < 0.05$). After day 6 of storage, the MDA values decreased for the control group, the petroleum ether fraction-treated group, and the ethyl acetate fraction-treated group. The MDA values of control decreased to 0.17 mg/100 g at day 15, while the MDA values of the samples ranged from 0.12 to 0.19 mg/100 g for the petroleum ether fraction-treated group, the 80% ethanol crude extract-treated group, and the ethyl acetate fraction-treated group (Figure 6).

3.7 CD values

The conjugated dienes value was used as an index of oil oxidant (Elisia et al. 2013). Figure 7 illustrated

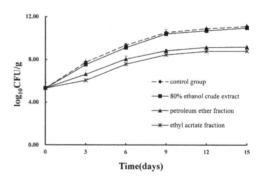

Figure 4. Evolution of total bacterial count of minced pork containing different fractions.

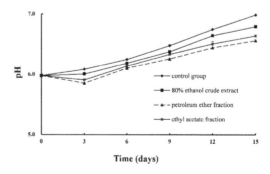

Figure 5. pH values of minced pork during the 15 days storage period.

356

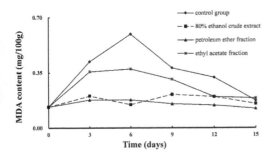

Figure 6.   Curves of MDA changes in minced pork containing different fractions.

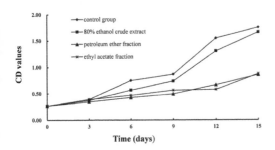

Figure 7.   Curves of CD value changes in minced pork containing different fractions.

that CD value of all the samples increased during the storage period. In addition, all treatments had significantly ($P < 0.01$) lower CD values compared with the control group. At day 15, the control had significantly higher CD values of 1.752, compared with that of the samples with rosemary extracts added (1.662, 0.875, and 0.859 for 80% ethanol crude extract treatment, ethyl acetate fraction treatment, and petroleum ether fraction treatment).

4 DISCUSSION

Rosemary has been effectively applied in many fields (Crego et al. 2004). Using in food has increased notably and the ability of preventing fats and oils from oxidation is important in the food industry (Wanasundara et al. 2005).

In our study, addition of petroleum ether fraction, ethyl acetate fraction, and 80% ethanol crude extract from rosemary resulted in lower microbial counts, pH, MDA, and CD values of the samples, but higher sensory evaluation. Our results indicated that rosemary extracts have certain activity of antimicrobial, preservative, and inhibiting oxidation in minced pork.

With analysis of HPLC-MS, we identified five major compounds: Rosmarinic acid, rosmanol,

carnosol, carnosic, and methyl carnosate (Table 1). It was in agreement with the study conducted by Vaquero, et al, who found that a total of 15 phenolic compounds were present in the rosemary extract (Vaquero et al. 2012). Table 2 illustrated the content of five compounds in different fractions. Rosmarinic acid is hydrosoluble (Jordán et al. 2012) and a strong polar compound while rosmanol and carnosol are liposoluble and low-polar (Leonardi 2005). In addition, carnosic acid and methyl carnosate are less polar and liposoluble components (Trojakova et al. 2001). Ethanol and butanol are polar solvents, petroleum ether is a non-polar solvent, ethyl acetate is a medium polar solvent, and water is a strong polar solvent (Zhang et al. 2007, Wanasundara et al. 2005). Owing to those characteristics, there were more contents of carnosol and carnosic acid in petroleum ether fraction.

Table 3 illustrated that all treatments had broad-spectrum antimicrobial activities. Petroleum ether fraction had higher bacteriostatic efficiency than 80% ethanol crude extract fraction, and ethyl acetate fraction treatment (Figure 4). This result was probably because of high contents of carnosic acid and carnosol. It was in agreement with Bernardes et al, who reported that the antimicrobial activity of the extract from rosemary leaves against oral pathogens may be ascribed mainly to the action of carnosic acid and carnosol (Bernardes et al. 2010).

The results of sensory evaluation showed that the effect of petroleum ether fraction is better (Table 4) generally. Odor releases from the process of pork corruption due to the increasing concentration of putrescine, tyramine, and cadaverine during storage (Edwards et al. 1987). The special flavor of rosemary could be causing interference in the detection of rancid aroma and partly improving flavor during storage (Liu et al. 2009).

During the storage time, the pH values increased gradually illustrating that minced pork went into autolysis and corruption period (Edwards et al. 1987). The change rate of minced pork with treatment increased slowly than that of control group especially the petroleum ether fraction-treated group. It was obvious that these three rosemary extracts could effectively inhibit meat corruption.

These decreases in MDA values (Figure 6) were probably due to the ability of antioxidation and retarding lipid oxidation of rosemary (Liu et al. 2009). The positive effects of rosemary to prevent lipid oxidation have been well documented (Doolaegea et al. 2012). Yu et al, reported that rosemary extracts added were effective in delaying lipid oxidation (Yu et al. 2002). According to other studies (Nassu et al. 2003, Liu et al. 2009), the same results were found in the fermented goat meat sausage processing and fresh chicken sausage during refrigerated storage.

The oxidation process of oils was also investigated by measuring the formation of conjugated dienes (Abdalla & Roozen 1999, Elisia et al. 2013). The results indicated that the petroleum ether fraction had higher anti-lipid peroxidation ability than 80% ethanol crude extract fraction and ethyl acetate fraction. Morán, et al and María J. Jordán, et al, reported that carnosic acid added in lamb feeding seemed to be useful to delay lipid peroxidation and increase the shelf life of the meat (Jordán et al. 2014, Morán et al. 2012). Carnosol and carnosic acid were the most potent antioxidants in inhibiting oil oxidation (Frankel et al. 1996, Aruoma et al. 1992). Therefore, the anti-lipid peroxidation activity may be ascribed mainly to the high contents of carnosic acid and carnosol.

In conclusion, rosemary extracts have certain activity of antimicrobial, preservative, and antioxidant in minced pork, which might be at least partially associated with its interactions with high contents of carnosol and carnosic acid.

REFERENCES

Abdalla, A.E. and Roozen, J.P. (1999). Effect of plant extracts on the oxidative stability of sunflower oil and emulsion. *Food Chemistry*, 64(3), 323–329.

Aruoma, O.I., Halliwell, B., Aeschbach, R. and Löligers, J. (1992). Antioxidant and pro-oxidant properties of active rosemary constituents: carnosol and carnosic acid. *Xenobiotica*, 22(2), 257–268.

Bauer, J., Kuehnl, S., Rollinger, J.M., Scherer, O., Northoff, H., Stuppner, H., Werz, O. and Koeberle, A. (2012). Carnosol and carnosic acids from Salvia officinalis inhibit microsomal prostaglandin E2 synthase-1. *Journal of Pharmacology and Experimental Therapeutics*, 342(1), 169–176.

Bernardes, W.A., Lucarini, R., Tozatti, M.G., Souza, M. G., Andrade Silva, M.L., Da Silva Filho, A.A., Martins, C.H.G., Miller Crotti, A.E., Pauletti, P.M. and Groppo, M. (2010). Antimicrobial activity of Rosmarinus officinalis against oral pathogens: relevance of carnosic acid and carnosol. *Chemistry & biodiversity*, 7(7), 1835–1840.

Campo, J.D. and Amiot, M.J. (2000). Antimicrobial effect of rosemary extracts. *Journal of Food Protection*, 63(10), 1359–1368.

Crego, A.L., Ibµæez, E., García, E., de Pablos, R.R., Seæørµns, F.J., Reglero, G. and Cifuentes, A. (2004). Capillary electrophoresis separation of rosemary antioxidants from subcritical water extracts. *European Food Research and Technology*, 219(5), 549–556.

Doolaegea, E.H.A., Vossena, E., Raesb, K., De Meulenaerc, B., Verhéd, R., Paelincke, H. and De Smet, S. (2012). Effect of rosemary extract dose on lipid oxidation, colour stability and antioxidant concentrations, in reduced nitrite liver pâtés. *Meat science*, 90(4), 925–931.

Edwards, R.A., Dainty, R.H., Hibbard, C.M. and Ramantanis, S.V. (1987). Amines in fresh beef of normal pH and the role of bacteria in changes in concentration observed during storage in vacuum packs at chill temperatures. *Journal of Applied Bacteriology*, 63(5), 427–434.

Edwards, S.E., Rocha, I.D.C., Williamson, E.M. and Heinrich, M. (2015). Rosemary Rosmarinus officinalis L. *Phytopharmacy: An Evidence-Based Guide to Herbal Medicinal Products*, 328.

Elgayyar, M., Draughon, F., Golden, D. and Mount, J. (2001). Antimicrobial activity of essential oils from plants against selected pathogenic and saprophytic microorganisms. *Journal of Food Protection*, 64(7), 1019–24.

Elisia, I., Young, J.W., Yuan, Y.V. and Kitts, D.D. (2013). Association between tocopherol isoform composition and lipid oxidation in selected multiple edible oils. *Food Research International*, 52(2), 508–514.

Frankel, E.N., Huang, S., Aeschbach, R. and Prior, E. (1996). Antioxidant activity of a rosemary extract and its constituents, carnosic acid, carnosol, and rosmarinic acid, in bulk oil and oil-in-water emulsion. *Journal of Agricultural and Food Chemistry*, 44(1), 131–135.

Georgantelis, D., Ambrosiadis, I., Katikou, P., Blekas, G. and Georgakis, S. A. (2007). Effect of rosemary extract, chitosan and α-tocopherol on microbiological parameters and lipid oxidation of fresh pork sausages stored at 4°C. *Meat Science*, 76(1), 172–181.

Georgantelis, D., Blekas, G., Katikou, P., Ambrosiadis, I. and Fletouris, D.J. (2007). Effect of rosemary extract, chitosan and α-tocopherol on lipid oxidation and colour stability during frozen storage of beef burgers. *Meat Science*, 75(2), 256–264.

Hosni, K., Hassen, I., Chaâbane, H., Jemli, M., Dallali, S., Sebei, H. and Casabianca, H. (2013). Enzyme-assisted extraction of essential oils from thyme (Thymus capitatus L.) and rosemary (Rosmarinus officinalis L.): Impact on yield, chemical composition and antimicrobial activity. *Industrial Crops and Products*, 47, 291–299.

Jayasena, D.D. and Jo, C. (2013). Essential oils as potential antimicrobial agents in meat and meat products: A review. *Trends in Food Science & Technology*, 34(2), 96–108.

Jordán, M.J., Castillo, J., Bañón, S., Martínez-Conesa, C. and Sotomayor, J.A. (2014). Relevance of the carnosic acid/carnosol ratio for the level of rosemary diterpene transfer and for improving lamb meat antioxidant status. *Food chemistry*, 151, 212–218.

Jordán, M.J., Lax, V., Rota, M.C., Lorán, S. and Sotomayor, J.A. (2012). Relevance of carnosic acid, carnosol, and rosmarinic acid concentrations in the in vitro antioxidant and antimicrobial activities of Rosmarinus officinalis (L.) methanolic extracts. *Journal of agricultural and food chemistry*, 60(38), 9603–9608.

Lan, W. and Zesheng, Z. (2005). Extraction and Examination of Chlorogenic Acid from Flos Lonicerae. *Food Science*, 6, 037.

Leonardi, M. (2005). Studio di una nuova miscela di oli vegetali per frittura. *Rivista Italiana delle Sostanze Grasse*, 82(2), 71–81.

Liu, D., Tsau, R., Lin, Y., Jan, S. and Tan, F. (2009). Effect of various levels of rosemary or Chinese mahogany on the quality of fresh chicken sausage during refrigerated storage. *Food Chemistry*, 117(1), 106–113.

Lo, A., Liang, Y., Lin-Shiau, S., Ho, C. and Lin, J. (2002). Carnosol, an antioxidant in rosemary, suppresses inducible nitric oxide synthase through down-regulating nuclear factor-κB in mouse macrophages. *Carcinogenesis*, 23(6), 983–991.

Morán, L., Andrés, S., Bodas, R., Prieto, N. and Giráldez, F. J. (2012). Meat texture and antioxidant status are improved when carnosic acid is included in the diet of fattening lambs. *Meat science*, 91(4), 430–434.

Mulinacci, N., Innocenti, M., Bellumori, M., Giaccherini, C., Martini, V. and Michelozzi, M. (2011). Storage method, drying processes and extraction procedures strongly affect the phenolic fraction of rosemary leaves: An HPLC/DAD/MS study. *Talanta*, 85(1), 167–176.

Nassu, R.T., Gonçalves, L.A.G., Da Silva, M.A.A.P. and Beserra, F.J. (2003). Oxidative stability of fermented goat meat sausage with different levels of natural antioxidant. *Meat Science*, 63(1), 43–49.

Ojeda-Sana, A.M., van Baren, C.M., Elechosa, M.A., Juárez, M.A. and Moreno, S. (2013). New insights into antibacterial and antioxidant activities of rosemary essential oils and their main components. *Food Control*, 31(1), 189–195.

Ra, A., Milanovi, I. and Mikov, M. (2014). Antioxidant activity of rosemary (Rosmarinus officinalis L.) essential oil and its hepatoprotective potential. *BMC complementary and alternative medicine*, 14(1), 225.

Soultos, N., Tzikas, Z., Christaki, E., Papageorgiou, K. and Sterisa, V. (2009). The effect of dietary oregano essential oil on microbial growth of rabbit carcasses during refrigerated storage. *Meat science*, 81(3), 474–478.

Stansbury, J. (2014). Rosmarinic Acid as a Novel Agent in the Treatment of Allergies and Asthma. *Journal of Restorative Medicine*, 3(1), 121–126.

Trojakova, L., Reblova, Z., Nguyen, H. and Okorny, J. (2001). Antioxidant activity of rosemary and sage extracts in rapeseed oil. *Journal of Food Lipids*, 8(1), 1–13.

Vaquero, M. R., Yáñez-Gascón, M., Villalba, R.G., Larrosa, M., Fromentin, E., Ibarra, A., Roller, M., Tomás-Barberán, F., de Gea, J.C.E. and García-Conesa, M. (2012). Inhibition of gastric lipase as a mechanism for body weight and plasma lipids reduction in Zucker rats fed a rosemary extract rich in carnosic acid. *PloS one*, 7(6), e39773.

Vareltzis, K., Koufidis, D., Gavriilidou, E., Papavergou, E. and Vasiliadou, S. (1997). Effectiveness of a natural rosemary (Rosmarinus officinalis) extract on the stability of filleted and minced fish during frozen storage. *Zeitschrift für Lebensmitteluntersuchung und-Forschung A*, 205(2), 93–96.

Wanasundara, U., Wanasundara, P.K.J.P. and Shahidi, F. (2005). Novel Separation Techniques for Isolation and Purification of Fatty Acids and Oil By-Products. *Wiley Online Library*.

Yu, L., Scanlin, L., Wilson, J. and Schmidt, G. (2002). Rosemary extracts as inhibitors of lipid oxidation and color change in cooked turkey products during refrigerated storage. *Journal of Food Science*, 67(2), 582–585.

Zhang, Z., Li, D., Wang, L., Ozkanc, N., Chen, X., Mao, Z. and Yang, H. (2007). Optimization of ethanol–water extraction of lignans from flaxseed. *Separation and Purification Technology*, 57(1), 17–24.

Advances in Materials Science, Energy Technology
and Environmental Engineering – Patty & Zhou (Eds)
© 2017 Taylor & Francis Group, London, ISBN 978-1-138-19668-1

Research advances in sorghum awn traits

Chunming Bai, Chengguang Tao & Xiaochun Lu
Liaoning Academy of Agricultural Sciences, Shenyang, China

Yifei Liu & Lijun Zhang
Shenyang Agricultural University, Shenyang, China

ABSTRACT: Wild sorghum seeds are characterized by their long awns densely covered with barbs, as easy for dissemination, propagation, and bird trouble prevention. However, cultivated varieties in breeding are generally sorghum strains with short awns or even without awns, for it is conducive to harvesting, storage and processing. The transition from awned plants to awnless ones is a crucial event in sorghum domestication. Nevertheless, molecular mechanisms, which regulate such a crucial transition, still remain unclear. Discussion of biological characteristics for sorghum awns, as well as gene cloning progress in other graminaceous crop awns are presented hereby, at the significant prospect of controlling genes cloning for awn presence to molecular mark assisted breeding as associated with sorghum awnedness, to sorghum evolution molecular mechanism, and to awn development molecular mechanism likewise.

1 BIOLOGICAL CHARACTERISTICS OF AWNS

Seeds serve as the key carrier for plant dissemination and heritage across generations. Normally, seeds of wild plants display manifold adaptions in their each own morphology. Awns are the protrusion on the floret palae/lemma for these gramineous plants, and appear as the extension from midrib at floret palae/lemma top. Awns insert themselves at seed glume top and are densely covered with sharp barbs. Long barbs play dominant roles in seed field survival and dissemination. Firstly, the barbs enable seeds to cling to animal fur for long distance propagation, and the propagation distance and location are likely related to awn length, awn presence, even adhesion ratio to animal fur. Secondly, long awns will enable seed to fall into the soil in equilibrium, when one end of the seed shall touch down the earth, and bury itself into the soil. Since the seed germ is at the tip, when the germ buried into the soil, it will be easy for further germination, during which awns shall perform counter-force for roots to be anchored in the soil, and for seedlings to set up furthermore. Thirdly, long barbs prove significantly deterrent to birds, preventing seeds from birds or animal predation. Therefore, long barbs are considered as a sort of vital adaptive evolution for seeds against natural environment, and it is beneficial to their own survival and dissemination.

There are vast variations among different species of gramineous plants in terms of awn presence, awn length and distribution. Moreover, both awn presence and length are restricted by such environment factors as photoperiod, temperature and abiotic stress (drought, for example). Normally, awns are classified as two types: hygroscopic awns and rigid awns based on their own functions (Peart, 1979). There is at least one geniculate body usually for hygroscopic awns in dry conditions. Section from the geniculate body to the seed emerges as a spiral twist, while section from geniculate body to awn tip emerges as a straight tail densely covered with barbs (Qing et al., 2007), as seen in wild sorghum and the similar. Compared with hygroscopic awns, rigid awns (wheat, for example) are firm, straight without any geniculate body, and they shall not be under any deformation in spite of humidity changes (Stinson et al., 1979).

As studies on awn cellular structure have revealed, "hygroscopic movements" are due to lignin content or cellulose orientation disparity at awn base for different cell population. Such hygroscopic movements are of adaptive importance to seed self-burial and germination, or awn absorbency results from plant adaption to arid, semi-arid conditions during a long-term range.

Awn barbs are regarded as products from single epidermal cell specialization. Studies on epidermal cells are greatly profound, with focus mostly on the genetic regulation mechanism of leaf epidermal hair and root hair. Barbs are not only to provide horizontal friction, but provide vertical adhesion as well, so that seeds may be anchored into the soil during self-burial and germination to prevent them being forced out of soil. The unique arrangement of awn barbs has ratchet effects, which are ready to converse disproportional environmental fluctuation into directional movement in favor of seed self-burial. Thus it is obvious that awn barbs are indispensable structures for awns to exert their biological functions, but few reports are available on awn barbs. Awns for barley and wheat are a kind of organs with high photosynthetic efficiency and low transpiration. Both of them share a triangular cross-section with two zones of chlorenchema cells and three vascular bundles. There are stomata on the surface (Yuo et al., 2012). It is supposed that long awns of barley and wheat are their own adaption to arid environment, for long awns are able to facilitate water and photoassimilate transportation towards grains to achieve yield increase. Contribution from awns to yield amounts to 10%. In contrast, rice awns are typical of one vascular bundle without any zones of chlorenchema cells inside, so these awns are not able to perform photosynthesis (Toriba et al., 2010), and this is why both barley and wheat retain a certain length of awns during domestication.

2 AWN REGULATORY GENE CLONING AND FUNCTION STUDIES

Copy the template file B2ProcA4.dot (if you print on A4 size paper) or B2ProcLe.dot (for Letter size paper) to the template directory. Awns are one of the key traits for gramineous plants. What changes in genome have led to awn presence or length? Studies on cereal crop awns were conducted early in 1898, and it was discovered that awns were able to fix CO_2 with their stomata. Subsequently, more studies on barley and rice awn gene among cereal crops were conducted (Xiong et al., 1999; Hu et al., 2011; Thomson et al., 2003).

The first cloned controlling gene for awn length in grass family is Lks2 for barley. There are mainly two genes to control barley awn presence and length, namely Lks1 at 2HL and Lks2 at 7HL, of which Lks1 is a dominant awnless gene, and Lks2 is a recessive short-awn gene. Lks2, which encodes a transcription factor from SHI family, is specifically expressed in the awn and pistil, and such mutation makes awn length get shorter. As histological research has indicated, short awns are due to less awn cell proliferation. Natural variations for Lks2 gene generally include three types: Lks22.b1 and Lks22.b2 are restricted to Easter China, Korea and Japan, while varieties of Lks22.b3 as allele carrier mostly share their distribution in India, Nepal and Tibet. Lks22.b3 is greatly different from the former two types in terms of sequence, a fact suggesting its independent origins. Lks2 mutation is

only observed in some barley varieties, so this variation is due to domestication. Lks2 is widely distributed in barely of Eastern Asian (China, Japan, Korea and India, for example), and this is might be barley's adaption to humid climate in Eastern Asia, for short-awned varieties are under artificial selection are likely to reduce lodging (Yuo et al., 2012). Rice awn is a critical domestication trait, whereas it is not as consistent as other domestication traits in cultivated rice, for awn presence and length show continuous variations in cultivated rice. Most local varieties and cultivated rice is awnless, and only about 40% of cultivated varieties retain awns, while vast majority of awned cultivated varieties retain shorter awns, or they are even barbless.

QTL/gene as controlling awn traits in rice is distributed in 12 rice chromosomes, of which Awn-1 at chromosome 4 was of the highest detection frequency among different mapping population. It is reported that there are more than 30 gene sites controlling rice awns (http://www.gramene.org/), but there are few under fine mapping and cloning as a matter of fact.

Han Bin and his research group cloned An-1 gene which controlled awn development of wild rice through map-based cloning, and such a gene encoded *bHLH* transcription regulatory factors. As genetic transformation and near-isogenic line construction have confirmed, such a gene enabled a recovery to long-awn phenotype for awnless cultivated rice, which was responsible for longer grains, but for grain number dwindle per panicle at the same time. Further experiments indicated that the up-regulated expression of An-1 gene reduced yield per plant, nevertheless its down-regulated expression increased yield per plant remarkably ((Luo et al., 2013).

Long awn barbs are critical to wild rice in terms of its field survival and dissemination. Sun Chuanqing and his research group spotted *LABA1* (LONG AND BARBED AWN 1), which controlled long awns of wild rice. It was located at the long arm of rice chromosome 4, and encoded some activating enzyme for cytokinine. *LABA1* specific expression at primordium improved active cell mitogen content at awn primordium and enhanced cell division activity, both of which led to awn elongation and barb formation. Nucleotide diversity of cultivated rice at the site of *LABA1* declined severely as compared with that of wild rice, a fact indicating that *LABA1* was an important target under artificial selection during rice domestication (Lei et al., 2015). Additionally, rice mutant studies observed some genes related to awn development, *DL* and *SHL2* (Toriba et al, 2014) for instance. Those mutants were unable to maintain meristem activity, or to establish adaxial-abaxial polarity in a sense, and they may throw light on the molecular mechanisms for awn development.

Wheat is classified into three types in light of awns, namely awnless varieties, tip awned ones and full-awned ones. Watkin and Ellerton believed that five gene decided wheat awn development: B1, B2, B3, A and Hd. Gene B1, B2 and B3 were awn suppressor genes, among which B1 showed the strongest effects, awnless as expressed even in spite of heterozygous status, and it was followed by B2, B3 and Hd showed the weakest effects. Gene A served to accelerate awn growth, while gene Hd is the one for barb formation. Genes as effective on awn length were located on chromosome 2A, 3A, 4A, 1B, 2B, 3B, 5B, 2D, 3D, 6D and 7D, and the awn barb controlling gene was on chromosome 4B, gene suppressing awns at chromosome 5A, 4B, 5B, 6B and 7D (Du et al., 2010). However, there is no report on gene cloning as concerned. Sorghum ranks the fifth among grain crop in the world, only second to wheat, rice, corn and barley. Awns on the panicle glumes are a favorable trait for wild sorghum, since it is convenient for propagation and dissemination. Though awns are not to affect sorghum yield and they may prevent bird predation in a way (Lu Q S, 1999), modern cultivated varieties are usually awnless to facilitate sorghum harvesting, storage and processing.

As indicated from the performance of hybrids between awned varieties and awnless varieties, awnlessness is dominant against awnedness for sorghum. Sorghum awns are classified into 4 grades on the whole: AA for awnless, aa for strong awn, aat for short awn, atat for tip awn (Vinall et al., 1921). Awnless is almost dominant completely against strong awn, or tip awn, while strong awn is not totally dominant against tip awn. Strong awn means long awn as we commonly refer to. In 2014, Zhai Guowei reported of fine mapping of sorghum awn genes, which were located within the range of 115 Kb on chromosome 3 (Zhai et al., 2014), but no gene was cloned at all.

3 SIGNIFICANCE OF RESEARCH ON SORGHUM AWN GENE CLONING

Cultivated sorghum (*Sorghum bicolor* (L.) Moench) evolves from wild sorghum upon natural selection and artificial selection. Snowden believed that sorghum cultivars experienced multiple origins in Africa, mainly from three wild species, namely *S. arundinaceum*, *S. verticilliflorun* and *S. aethiopicum*. Many distinct traits are sensed among wild and cultivated population, for instance, natural ability to spread seeds and dormancy among morphological and physiological traits. These typical traits help to distinguish cultivars

from their wild ancestors. Though trait changes will reduce plant adaption to wild environment remarkably, they are more likely to satisfy human being, and finally enable crops to become steady source of food supply, so these traits are deemed as domestication traits. Related gene cloning and domestication are of priority in research on crop origin and evolution, including heredity of domestication traits, molecular basis, origin and spread of functional variation sites, artificial selection effects and other major issues. Then understanding of heredity basis of crop domestication and differentiation shall be instructive to crop breeding process undoubtedly.

Wild sorghum, as the cultivar origin, is characterized with long awns densely covered with barbs. Since these barbs not only hamper manual harvesting of sorghum seeds with awns, but affect sorghum seed storage and processing as well, awn length of cultivars is notably reduced or they even become awnless. Hence the transition from awned plants to awnless ones is a crucial event in sorghum domestication. Nevertheless, molecular mechanisms, which regulate such a crucial transition, still remain unclear. Sorghum awn gene cloning is of great importance to both molecular mechanisms in sorghum evolution and molecular mechanisms for awn development. During our previous study, 418 sorghum accessions were adopted to simplify genome sequencing to clone genes controlling sorghum awn presence. Awn presence genes were located in genome-wide association study, and based on candidate gene sequencing analysis, a candidate gene associated with awn presence was eventually identified within the mapping interval (which is under further studies, so no statistics were reported concerning its mapping range, though), and it was named as *SbAn-1*, not included in the range of 115 Kb as reported by Zhai and others (2014). Such a gene was not the transcription factor of the same kind as regulatory genes for rice awn presence as reported. *An-1*, as controlling rice awn development, encoded the transcriptional regulatory factor of *bHLH* (Luo et al., 2013).

To verify if cloned awns are functional in respect of *SbAn-1*, based on our research results, we constructed near-isogenic line, conducted gene expression and transformation study. We also made analysis for SbAn-1 gene classification, proportion in different sorghum cultivars, wild accessions, along with the genetic diversity analysis for the genome around such a locus under artificial selection. Therefore, our studies shall elucidate such gene functions and domestication. The research hereby is not only helpful to reveal the important role by this gene during awn development, but also significant for further understanding of sorghum domestication process and molecular mechanisms.

ACKNOWLEDGEMENTS

This work was financially supported by Natural Science Foundation of Liaoning Province (CN) (2014027018), National Natural Science Foundation of China (CN) (31301842) and Cultivation Plan for Youth Agricultural Science and Technology Innovative Talents of Liaoning Province (2014043).

REFERENCES

Du B, Cui F, Wang H G, Li X F. Characterization and Genetic Analysis of Near-isogenic Lines of Common Wheat for Awn-inhibitor Gene B1. Molecular Plant Breeding, 2010, 8(2): 259–264.

Hu G L, Zhang D L, Pan H Q, Li B, Wu J T, Zhou X Y, Zhang Q Y, Zhou L, Yao G X, Li J Z, Li J J, Zhang H L, Li Z C. Fine mapping of the awn gene on chromosome 4 in rice by association and linkage analyses. Chinese Science Bulletin, 2011, 56 (9): 835–839.

Lei H, Wang D R., Tan L, Fu Y C, Liu F X, Xiao L T, Zhu Z F, Fu Q, Sun X Y, Gu P, Cai H W, McCouch S R, and Sun C Q. LABA1, a domestication gene associated with long, barbed awns in wild rece. The Plant Cell, 2015, 27(7):1875–1888.

Lu Q S. Sorghum Science. Beijing: Agriculture Press, 1999: 466.

Luo J H, Liu H, Zhou T Y, Gu B G, Huang X H, Shangguan Y Y, Zhu J J, Li Y, Zhao Y, Wang Y C, Zhao Q, Wang A, Wang Z Q, Sang T, Wang Z X, Han B. An-1 encodes a basic helix-loop-helix protein that regulates awn development, grain size, and grain number in rice. Plant Cell, 2013, 25: 3360–3376.

Paterson A H, Bowers J E & Bruggmann R et al. The Sorghum bicolor genome and the diversification of grasses. Nature, 2009, 457: 551–556.

Peart M H. Experiments on the biological significance of the seed-dispersal units in grasses. Journal of Ecology, 1979, 67(3): 843–846.

Qing X L, Bai Y F. Review on morphology and adaptative significance of trypanophorous diaspore. Acta Ecologica Sinica, 2007, 27(6): 2547–2553.

Stinson R H and Peterson R L. On sowing wild oats. Canadian Journal of Botany, 1979, 57(11): 1292–1295.

Thomson M J, Tai T H, McClung A M, Lai X-H, Hinga M E, Lobos K B, Xu Y, Martinez C P, McCouch S R. Mapping quantitative trait loci for yield, yield components and morphological traits in an advanced backcross population between Oryza rufipogon and the Oryza sativa cultivar Jefferson. Theoretical and Applied Genetics, 2003, 107(3): 479–493.

Toriba T, Hirano H Y. The DROOPING LEAF and OsETTIN2 genes promote awn development in rice. The Plant Journal, 2014, 77(4): 616–626.

Toriba T, Suzaki T, Yamaguchi T, Ohmori Y, Tsukaya H, Hirano H Y. Distinct regulation of adaxial-abaxial polarity in anther patterning in rice. Plant Cell, 2010, 22(5): 1452–1462.

Vinall H N, Cron A B. Improvement of sorghums by hybridization. Journal of Heredity, 1921, 12(10): 435–443.

Xiong L Z, Liu K D, Dai X K, Xu C G, Zhang Q. Identification of genetic factors controlling domestication-related traits of rice using an F2 population of a cross between Oryza sativa and O.rufipogon. Theoretical and Applied Genetics, 1999, 98(2): 243–251.

Yuo T, Yamashita Y, Kanamori H, Matsumoto T, Lundqvist U, Sato K, Ichii M, Jobling S A, Taketa S. A SHORT INTERNODES (SHI) family transcription factor gene regulates awn elongation and pistil morphology in barley. Journal of Experimental Botany, 2012, 63(14): 5223–5232.

Zhai G W, Wang H, Zou G H, Shao J F, Tao Y Z. Fine mapping of awn gene in Sorghum bicolor (L.) Moench. Jiangsu J. of Agr.Sci., 2014, 30(3): 486–490.

Advances in Materials Science, Energy Technology and Environmental Engineering – Patty & Zhou (Eds)
© 2017 Taylor & Francis Group, London, ISBN 978-1-138-19668-1

Experimental study on the mechanical behavior of strain hardening cementitious composites using local ingredients

Zhihua Li, Weikang Chen, Xu Zhou & Fengquan Chen
Building Science and Engineering College, Yangzhou University, Yangzhou, P.R. China

ABSTRACT: In order to reduce the cost of conventional Strain Hardening Cementitious Composite (SHCC) and then increase its field application, a ductile SHCC material is developed using domestic PVA fibers as well as other local ingredients, including cement, fly ash, silica fume and sand. To better understand its mechanical behavior, four-point bending test, uniaxial compression and tension test were carried out to characterize the bending, compressive and tensile behavior of the newly developed composites. The test results reveal that the newly developed SHCC using local ingredients exhibits deflection hardening accompanied with multiple cracks in flexure. In addition, complete stress-strain curves in uniaxial compression and tension were obtained. Based on the test results, the compressive parameters, including the elastic modulus, the strain at the peak stress and the Poisson's ratio, were evaluated.

1 INTRODUCTION

Strain Hardening Cementitious Composite (SHCC) has been developed in the past two decades as an alternative infrastructure material to brittle concrete. SHCC belongs to the family of High Performance Fiber Reinforced Cementitious Composite (HPFRCC) designed for high ductility based on micromechanics concept and develops macroscopic pseudo strain hardening behavior under uniaxial tension, as opposed to single crack and tension softening of ordinary concrete. Multiple cracking behavior of SHCC results in large strain capacity of 2%–6% and tight crack width, which in turn can enhance the formation capacity and energy dissipation, damage tolerance and collapse prevention of concrete structures. Therefore, SHCC is beneficial to solve the durability and structural safety problem of reinforced concrete structures and has been applied to full-scale building, transportation, water and energy infrastructures in Asia, Europe and the US.

Despite its superior tensile ductility and tight micro crack width, the application of SHCC in China is somewhat hindered by the high cost, which is mainly driven by the high cost of the imported PVA fiber. In order to reduce the cost of SHCC and promote its application in China, it is necessary to employ local materials, especially domestic PVA fibers to develop SHCC materials with lower cost. As the physical/mechanical properties of the domestic PVA fibers are considerably different from those of commonly used PVA fibers from Japan, several research studies should be conducted to reveal the mechanical properties of SHCC with domestic PVA fibers before its application.

The aim of this study is to present some preliminary experimental results of recently developed SHCC material using local materials, such as cement, fly ash, silica fume, sand and PVA fibers. Appropriate mix proportions were determined through experiment, utilizing the knowledge obtained from development of SHCC with Japanese PVA fiber. Mechanical tests, including four-point bending test, uniaxial compression and tension test, were then carried out to characterize the flexural, compressive and tensile behavior of the newly developed SHCC.

2 EXPERIMENTAL PROGRAM

2.1 Materials and mix proportions

The cement used in this study was ordinary Portland cement P.O 42.5. Its 3-day bending and compressive strength is 6.06 MPa and 30.94 MPa respectively and 28-day strength is 9.37 MPa and 48.43 MPa respectively. Class-I fly ash, silica fume, fine silica sand, PVA fibers, and a polycarboxylate-based High Rang Water Reducer (HRWR) were also used. Two types of PVA fibers were also used: one is local PVA fibers (PVA1) and the other one is imported PVA fibers (PVA2) from Japan. The properties of different PVA fibers used are listed in Table 1.

The mix proportions are given in Table 2. A fiber volume of 2% was adopted for all the mixes, and 20% of silica fume by binder weight was used for fiber dispersion. Various water/binder ratios,

Table 1. Properties of different PVA fibers.

Fiber	Length/mm	Diameter/μm	Modulus/GPa	Elongation/%	Tensile strength/MPa	Density/(g/cm³)
PVA1	12	12–18	35	6–8	1200	1.3
PVA2	12	40	41	6.5	1560	1.3

Table 2. Mix proportion of SHCC mixture.

Mix number	Cement	Fly ash	Silica fume	Sand/binder	Water/binder	HRWR	PVA(vol.%)	PVA type
M1	0.2	0.6	0.2	0.36	0.4	0.004	2	PVA1
M1'	0.2	0.6	0.2	0.36	0.4	0.004	2	PVA2
M2	0.2	0.6	0.2	0.36	0.35	0.004	2	PVA1
M3	0.2	0.6	0.2	0.36	0.3	0.004	2	PVA1
M4	0.4	0.4	0.2	0.36	0.4	0.004	2	PVA1
M5	0.6	0.2	0.2	0.36	0.4	0.004	2	PVA1

sand/binder ratios and fly ash content were considered in SHCC mixes with domestic PVA fibers, and a constant water/binder ratio of 0.4 and fly ash content of 0.6 were considered in SHCC mix with PVA fibers imported from Japan.

2.2 Mixing and testing

The cement, fly ash and silica fume were first added to the mixer and mixed at low speed for 1 min. Then, the water and HRWR were added and followed by wet mixing at low speed for 1 min and at high speed for another 2 min. Finally, fibers were added and mixing continued at high speed for 6–10 min. The fresh SHCC was cast in 100 mm × 100 mm × 400 mm prismatic molds for four-point bending test, in 40 mm × 40 mm × 160 mm prismatic molds for uniaxial compression test and in 15 mm × 40 mm × 160 mm prismatic molds for uniaxial tension test separately. All specimens were demolded after 24 h and then cured for 28 days in rigorous conditions.

Three specimens were tested for each mixture in the four-point test. This test was carried out on a WDW-100 testing machine at a loading rate of 0.2 mm/min according to Chinese standard GB/T 50081–2002. The span of four-point bending test was 300 mm with a middle span of 100 mm. The center deflection was measured using a displacement transducer.

Six specimens were prepared for the mixture in the uniaxial compression test, of which three were tested to obtain the peak load and the other three were examined to obtain elastic modulus, Poisson's ration and the stress-strain relationship. A WDW-100 testing machine was used to conduct the tests. Two gauges were centered approximately midheight of the specimen to measure the longitudinal and

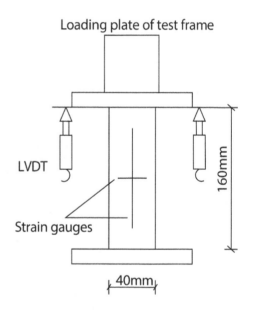

Figure 1. The schematic diagram of uniaxial compression test setup.

transverse strain respectively as shown in Figure 1. Two LVDTS were attached symmetrically to the top end of the specimen to measure the whole longitudinal deformation. Before specimens were formally loaded, three cycles of preloading were performed. When the preloading value reached 40% of the peak load, the specimen were unloaded to 2 kN (1.25 MPa). After three cycles, the specimens were then subjected to formal loading. During the whole loading process, the loading rate was set to be 0.2 mm/min to obtain a stable softening stage.

Figure 2. The specimens used in uniaxial tension test.

Three specimens were tested for the mixture in the uniaxial tension test. This test was conducted using a DNS-100 testing machine under displacement control with a loading rate of 0.2 mm/min. Before testing, four pieces of 50×40 carbon fiber cloth were first attached to both specimen ends using epoxy resin and then 50 mm $\times$ 40 mm $\times$ 0.9 mm aluminum sheet was attached to carbon fiber cloth as shown in Figure 2 to ensure that the tensile load is evenly applied to the specimen, which is beneficial to avoid the fracture at the ends of specimen due to local compression. It has to be noticed that tensile loads should be always introduced to the tensile specimens along the central axis during testing. Two extensometers were used for deformation measurement as well as for test control. The displacement of the center 50 mm of the specimen was measured by means of two extensometers and the tensile strain was calculated by dividing this measured displacement by the reference length of 50 mm.

3 RESULTS AND DISCUSSION

3.1 Flexural performance

Figure 3 presents the typical crack pattern after failure of PVA1-SHCC. It can be seen that multiple-cracking behavior occurred for the mixtures with PVA1, which is different from the continuous widening of localized cracks in ordinary concrete. The load versus midspan deflection curves of the SHCC beams is illustrated in Figure 4. It can be found that all specimens showed deflection hardening behavior. The first cracking strength (f_c), flexural strength (f_u) and their corresponding deflection, namely first cracking deflection (δ_c)

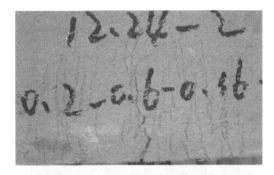

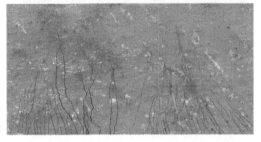

Figure 3. Typical crack patterns of PVA1-SHCC after failure.

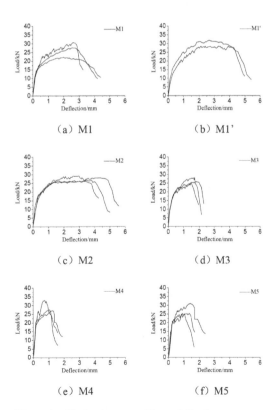

(a) M1

(b) M1'

(c) M2

(d) M3

(e) M4

(f) M5

Figure 4. The load versus midspan deflection curves.

367

and flexural deflection capacity (δ_u), are listed in Table 3. The mixture M1' presented the highest flexural strength and M2 showed the best deflection capacity at peak load. Mixture M2 also exhibited basically similar bending ductility to M1'. However, three mixtures M3, M4 and M5, exhibit relatively low deflection capacity.

Figure 5. Failure of prismatic specimens under compression.

3.2 Uniaxial Compressive performance

Figure 5 shows the failure modes of the mixture M2. (Other mixtures with PVA1 were not tested in uniaxial compression given the relatively low bending ductility). As load increased, micro cracks continued to appear and propagate in the middle of the specimen. After the peak load, as the cracks extended, a major inclined shear crack formed along the specimens with the cracking plane approximately 15°–30° from vertical plane instead of vertical splitting due to the bridging effect of fibers. Finally, the SHCC specimens failed in ductile shear failure rather than brittle splitting failure.

Figure 6 illustrates the stress-strain curves for SHCC specimens. The compressive properties of SHCC, including the peak stress (f'_c), the peak strain (ε_0, corresponding to the peak load), the elastic modulus (E_0) and Poisson's ratio at the elastic stage (v_0), were exhibited in Table 3. Similar to the stress-strain curve of normal concrete, the curves of SHCC specimen could be divided into four stages: linear elastic ascending stage, nonlinear ascending stage, descent stage and residual softening stage. The peak strain of SHCC could reach 0.006 which was considerably larger than that of normal concrete. Beyond the peak stress, the curves dropped to 20% of the ultimate strength quickly and then the stress decreased stably with the deformation until the failure occurred.

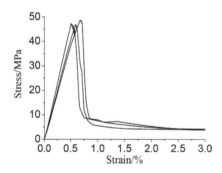

Figure 6. Stress-strain curves of SHCC specimens.

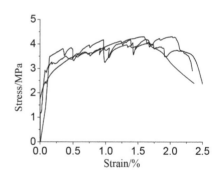

Figure 7. Tensile stress—strain curves of M2.

Table 3. Mechanical properties of PVA-SHCC.

Mix number	f_c /Mpa	δ_c /mm	f_u /Mpa	δ_u /mm	f_c /Mpm	ε_0 /$\mu\varepsilon$	E_0 /GPa	v_0
M1	5.31	0.45	8.08	2.40	/	/	/	/
M1'	6.57	0.84	9.03	3.32	/	/	/	/
M2	5.90	0.49	8.22	3.40	47.63	6000	19.3	0.26
M3	456.	0.58	7.91	1.67	/	/	/	/
M4	6.23	0.21	8.72	0.99	/	/	/	/
M5	5.75	0.25	8.17	1.24	/	/	/	/

3.3 Uniaxial tensile performance

Figure 7 shows the 28 days tensile stress-strain curves of the mixture M2. (Other mixtures with PVA1 were not tested in uniaxial tension given the relatively low bending ductility). The mixture exhibited strain hardening behavior accompanied with multiple cracking. The average values of tensile strength and tensile strain at peak load of the mixture M2 are 4.21 MPa and 1.76%.

4 CONCLUSIONS

This paper presents the results of experimental investigation on the mechanical behavior of newly developed SHCC material using local materials under four-point bending, uniaxial compression and tension. From the test results, it was demonstrated that deflection hardening behaviour accompanied by multiple cracking can be achieved in bending test for all the mixtures with PVA1. Specifically, the mixture M2 with 2% PVA1, and a water/binder ratio of 0.35, sand/binder ratio of 0.36 and 60% replacement of binder by fly ash, exhibited the best deflection capacity at peak load, and basically similar bending ductility to M1' with PVA2. The failure mode of the mixture M2 changes from brittle splitting failure to ductile shear failure under uniaxial compression, and the compressive peak strain can be up to 0.006. M2 also showed tensile strain hardening behavior, a tensile strength of 4.21 MPa and a tensile strain at peak load of 1.76%. While the data presented in this paper are encouraging, much more research work should be conducted to better understand the fundamental properties of SHCC produced with domestic PVA fibers.

REFERENCES

Billington, SL. 2004. Damage-tolerant cement-based materials for performance based earthquake engineering design: research needs. *Proceedings of International Conference on Fracture Mechanics of Concrete and Concrete Structures (Framcos-5)*, Ia-FraMCos: 53–60.

Chinese Standard. 2002. GB/T50081-2002 Standard for test method of mechanical properties on ordinary concrete. Beijing: China Building Industry Press.

Li VC. 2003. On engineered cementitious composites (ECC)—a review of the material and its application. *Journal of Advanced Concrete Technology* 1(3): 215–30.

Maalej M, Li VC. 1994. Flexural/tensile strength ratio in engineered cementitious composites. *Journal of Materials in Civil Engineering* 6(4): 513–528.

Qian SZ, 2012. Zhang ZG. Development of engineered cementious composites with local ingredients. *Journal of Southeast University (English Edition)* 28(3): 327–330.

Advances in Materials Science, Energy Technology
and Environmental Engineering – Patty & Zhou (Eds)
© 2017 Taylor & Francis Group, London, ISBN 978-1-138-19668-1

Characteristics of red pitaya wine fermented by different yeast strains

Junwei Yin
College of Food Science and Technology, HuaZhong Agricultural University, Wuhan, China
Chinese Academy of Tropical Agricultural Sciences, Agricultural Products Processing
Research Institute, Zhanjiang, China

Yajun Li
Guangxia (Yinchuan) Industrial Co. Ltd., Yinchuan, China

Shujian Li
School of Chemistry and Chemical Engineering, Lingnan Normal University, Zhanjiang, China

Xiaofang Wang, Xiao Gong, Yangyang Liu, Lijing Lin, Jihua Li
Chinese Academy of Tropical Agricultural Sciences, Agricultural Products Processing Research Institute,
Zhanjiang, China

ABSTRACT: Red pitaya wine fermentation by different yeast strains including *Saccharomyces cerevisiae* CC18, INVGC1, SC5, SC8, Y5128, SUN1, EC1118, and UV-434 were done in order to add value to the fruit. This study evaluated the fermentation characteristics of red pitaya wines. The fermentation process was monitored daily by analyzing the concentration of soluble solids and the pH. At the end of fermentation, the wines were subjected to sensory analysis. The wines showed variations in the quality. The sensory analysis indicated that the wines fermented by the strain of EC1118, SC8, UV-434, and CC18 had rates of acceptance above 80%. Based on these analyses, EC1118, SC8, UV-434, and CC18 were picked out to study the relationship between the quantity of yeast inoculation and the quality of red pitaya wine. The study showed that the wine produced by inoculation with UV-434 proved to be good. UV-434 is most suitable for the production of red pitaya wine.

1 INTRODUCTION

Wine is one of the most popular beverages among people all over the world. Although wine is traditionally made from the fermented juice of grapes, many countries, especially in Europe, produce fruit wines. Fruit wines have considerable economic potential due to the trend of increasing acceptance in consumer surveys and the contribution they make to reduce postharvest losses of perishable fruits. Thus, many research groups, mainly in tropical countries, have employed the same processes used in winemaking to prepare fruit wines from banana (B. Cheirsilp and K. Umsakul 2008), mango (Reddy and Reddy 2005), lychee (Dai Chen and Shao-Quan Liu 2014), cherry (Sun, S.Y. 2010), kiwifruit (Towantakavanit et al. 2010), and orange (Kelebek H. et al. 2009).

Another tropical fruit that can be used for the production of fermented alcoholic beverage is the red pitaya (*Hylocereus polyrhizus*).

Red pitaya (*Hylocereus polyrhizus*) native to Thailand, Vietnam, Taiwan, South America, and some other parts of the world has recently drawn much attention of growers worldwide, not only because of its unique appearance, red-purple color, and economic value as food products, but also for their anti-oxidative and antimicrobial activity from the beta-cyanin contents (Gian C.T. et al. 2012, Junsei Taira et al. 2015).

Red pitaya (*Hylocereus polyrhizus*) is usually consumed fresh and is limited by its fruiting season and short shelf life at room temperature. Therefore, the production of a fruit wine from red pitaya pulp offers an alternative way of preservation as well as adding value. Red pitaya wine can be a potential fruit wine for niche markets.

Wine fermentation is a complex microbiological process where yeast convert sugars to ethanol and carbon dioxide, with numerous by-products such as alcohols, ketones, acids, and esters accounting for the characteristics such as aroma and flavor of the wine. Most of the compounds are affected by the type and the concentration of yeast strains that were used for wine fermentation (Dai Chen & Shao-Quan Liu 2014).

The aim of this study is to evaluate the effects of inoculation with different strains of *S. cerevisiae* on the concentration of soluble solids, pH, and sensory characteristics of the wines and then to find out the suitable yeast for red pitaya winemaking.

2 MATERIALS AND METHODS

2.1 Yeast strains and culture media

Eight *S. cerevisiae* wine yeast strains, namely CC18, INVGCl, SC5, SC8, Y5128, SUN1, EC1118, and UV-434, were used in the red pitaya fermentation. The yeast cultures were propagated in sterile nutrient broth (2% w/v Peptone, 2% w/v glucose, 1% yeast extract) for up to 24 h at 35 °C. and stored at −80 °C before use.

2.2 Preparation of red pitaya juice

Red pitaya imported from the USA was purchased at a supermarket in China. These fruits were selected for uniformity of shape and color, and unhealthy fruits were discarded. After being cleaned by water, the pulp was extracted manually. The pulps were then packed into plastic bags and stored at −25 °C. To prepare the juice (24 L), pulps were thawed at room temperature, crushed in a blender, and filtered in an industrial sieve to conduct laboratory-scale fermentation. Red pitaya juice (a nature °Brix of 11.4% and pH 4.78) was adjusted to °Brix of 20.5 by adding 5% food-grade sucrose. The juice was filter-sterilized through 0.65 μm and 0.45 μm filter membranes (Hangzhou Cobetter Filtration Equipment, Hangzhou, China). Potassium metabisulfite ($K_2S_2O_5$) of 200 mg/L was added to the juice to obtain 100 mg/L of residual sulfur dioxide (SO_2) for reducing the microbial load without affecting the activity of fermentation yeasts and preventing oxidation reactions.

2.3 Fermentation conditions

A preculture medium was prepared from sterilized red pitaya juice (100 mL), which was inoculated with 10% (v/v) of thawed yeast broth culture (stored at −80 °C). The inoculated juice was cultured at 35 °C for 24 h under static condition.

The pretreated red pitaya juice (24 L) was taken in 24 separate sterile Erlenmeyer flasks (1 L for each flask). Each three flasks were inoculated with 2% (v/v) of a juice preculture of the respective yeast strains for triplicate fermentations. The fermentation occurred at 20 °C under static conditions for 6 days.

2.4 Quantity of yeast inoculation analysis

EC1118, SC8, UV-434, and CC18 strains were taken to study the relationship between the quantity of yeast inoculation and the quality of red pitaya wine.

The fermentation condition was the same as described in section 2.3.

2.5 Physicochemical analysis

The samples were analyzed for the concentration of soluble solids using a refractometer (ATAGO, Tokyo, Japan) and pH by a pH meter (METTLER TOLEDO, Switzerland).

2.6 Sensory evaluation

To test the acceptability of the wines, a group of five experienced testers (two males and three females) and consumers of some kind of wine were selected. Each of the testers sampled 20 mL of each wine served separately in transparent disposable cups at room temperature. For sensory evaluations of the acceptability of the wines in relation to the attributes of appearance, aroma, and flavor and to the overall acceptance of each wine, the panelists completed an evaluation from comprising a 9-point hedonic scale ranging from 1 ("I extremely dislike it") to 9 ("I extremely like it") (Dias et al. 2007).

2.7 Statistical analysis

Mean values and standard deviation were calculated from the date obtained from triplicate fermentations.

3 RESULTS

3.1 Analysis of juices during fermentation

The time taken for fermentation is 6 days for the samples fermented with all the yeasts.

The concentration of total soluble solids was reduced from about 20.5 to around 18 for all red pitaya wine samples after fermentation (Figure 1, Table 1). All resultant red pitaya wines had significantly higher pH values after fermentation and the pH value of red pitaya wine inoculated with strain INVGCL reached approximately 3.92 (Figure 2, Table 1).

3.2 Sensory evaluation of red pitaya wines

When soluble solids no longer change, pomace precipitates at the bottom of the bottle, which can be considered as the end of fermentation. After the fermentations were completed, the juice was filter-sterilized, and then sensory evaluation was carried out.

Table 2 shows a statistically significant difference in the degree of acceptability among the red pitaya wines across all parameters.

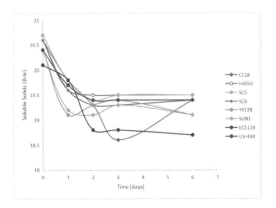

Figure 1. Consumption of soluble solids (°Brix) during the fermentation of red pitaya wines inoculated with different yeast strains (CC18, INVGCl, SC5, SC8, Y5128, SUN1, EC1118, and UV-434).

Table 1. Characteristics of red pitaya juice and red pitaya wines fermented with different yeasts.

Samples	Soluble Solid	pH
Juice	20.5 ± 0	4.65
CC18	19.4 ± 0.06	3.97
INVGCL	19.5 ± 0.2	3.92
SC5	19.5 ± 0.06	3.94
SC8	19.4 ± 0.06	3.92
Y5128	19.1 ± 0.15	4.02
SUN1	19.1 ± 0.06	3.95
EC1118	19.4 ± 0.12	4.00
UV-434	18.7 ± 0.1	3.99

The wines fermented by EC1118, SC8, UV-434, and CC18 showed the highest scores, thereby indicating their greater acceptance among the testers. However, there was no significant difference between the estimation of the attributes of acceptability for the other four yeast strains.

3.3 Effect of quantity of yeast inoculation on red pitaya wines

EC1118, SC8, UV-434, and CC18 strains were taken to study the relationship between the quantity of yeast inoculation and the quality of red pitaya wines.

Fermentation occurred in 3–7 days. The time taken for fermentation has significant correlation with the quantity of yeast inoculation. The concentration of total soluble solids decreases quickly, and almost stays unchanged from the fourth day when the quantity of inoculation is 8% (Figure 3).

The concentration of total soluble solids was reduced approximately from 20.5 to 5 for all red

Table 2. Results of the degree of acceptability of red pitaya wines by five experienced testers expressed as mean scores for each sensory attribute evaluated.

Wines	Attributes			
	Appearance	Aroma	Taste	Overall
CC18	5.31	5.51	7.22	6.37
INVGCL	5.15	5.45	5.85	5.57
SC5	5.15	5.34	6.45	5.87
SC8	5.36	5.56	6.03	5.84
Y5128	5.23	5.54	6.13	5.32
SUN1	5.07	5.28	5.72	5.36
EC1118	7.55	5.26	6.55	7.03
UV-434	5.08	5.29	7.28	6.25

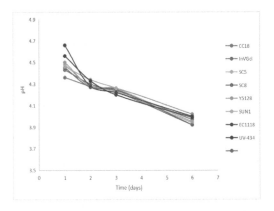

Figure 2. Variation in pH values during the fermentation of red pitaya wines with different yeast strains (CC18, INVGCl, SC5, SC8, Y5128, SUN1, EC1118, and UV-434).

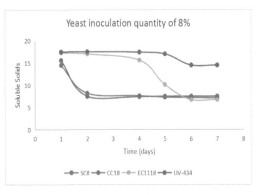

Figure 3. Consumption of soluble solids (°Brix) during the fermentation of red pitaya wines inoculated with 8% CC18, SC8, EC1118, and UV-434.

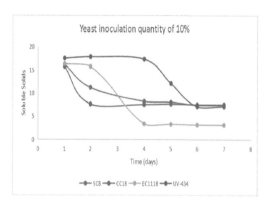

Figure 4. Consumption of soluble solids (°Brix) during the fermentation of red pitaya wines inoculated with 10% CC18, SC8, EC1118, and UV-434.

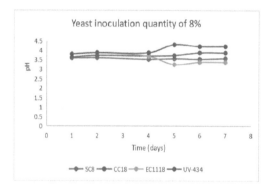

Figure 5. Variation of pH values during the fermentation of red pitaya wines inoculated with 8% CC18, SC8, EC1118, and UV-434.

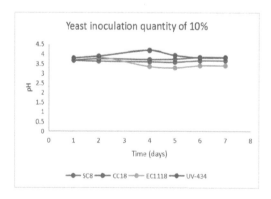

Figure 6. Variation of pH values during the fermentation of red pitaya wines inoculated with 10% CC18, SC8, EC1118, and UV-434.

pitaya wine samples after fermentation (Figure 3, Figure 4). All red pitaya wines had significantly higher pH values after fermentation and

the pH value of red pitaya wine inoculated with strain UV-434 (10%) increased at the late period of fermentation, which is different from others (Figure 5). The rise may result from esterification of some acids with alcohols.

On the seventh day, all wines were filter-sterilized for sensory evaluation.

4 CONCLUSION

The chemical and sensory characteristics of red pitaya wines fermented with eight *S. cerevisiae* strains were evaluated. The chemical composition of the red pitaya wine fermented with UV-434 was different from other red pitaya wines in terms of pH, concentration of soluble solids, and sensory characteristic.

The results could provide a basis to select yeast strains for different characters of red pitaya wines.

REFERENCES

Cheirsilp B. and K. Umsakul. 2008. Processing of banana-based wine product using pectinase and α-amylase. *Journal of Food Process Engineering* 31: 78–90.

Dai Chen & Shao-Quan Liu. 2014. Chemical and volatile composition of lychee wines fermented with four commercial *Saccharomyces cerevisiae* yeast strains. *International Journal of Food Science and Technology* 49: 521–530.

Dias D.R., Schwan R.F., Freire E.S., Serodio R.S. 2007. Elaboration of a fruit wine from cocoa (*Thebroma cacao L.*) pulp. *Int J Food Sci Technol* 42(3): 19–29.

Gian Carlo Tenore, Ettore Novellino, Adriana Basile. 2012. Nutraceutical potential and antioxidant benefits of red pitaya (*Hylocereus polyrhizus*) extracts. *Journal of functional foods* 4: 129–136.

Junsei Taira, etal. 2015. Antioxidant capacity of beta-cyanins as radical scavengers for peroxyl radical and nitric oxide. *Food Chemistry* 166: 531–536.

Kelebek, H., Selli, S., Canbas, A., & Cabaroglu, T. 2009. HPLC determination of organic acids, sugars, phenolic compositions and antioxidant capacity of orange juice and orange wine made from a Turkish cv. Kozan. *Microchemical Journal* 91:187–192.

Li-chen Wu, etal. 2006. Antioxidant and antiproliferative activities of red pitaya. *Food Chemistry* 95: 319–327.

Reddy L.V.A., Reddy O.V.S. 2005. Production and characterization of wine from mango fruit (*Mangifera indica L*). *World J Microbiol Biotechnol* 21(13): 45–50.

Sun, S.Y., Jiang, W.G., & Zhao, Y.P. 2011. Evaluation of different *Saccharomyces cerevisiae* strains on the profile of volatile compounds and polyphenols in cherry wines. *Food Chemistry* 127: 547–555.

Towantakavanit, K., Park, Y.S., & Gorinstein, S. 2011. Bioactivity of wine prepared from ripened and over-ripened kiwifruit. *Central European Journal of Biology* 6, 205–315.

Advances in Materials Science, Energy Technology
and Environmental Engineering – Patty & Zhou (Eds)
© 2017 Taylor & Francis Group, London, ISBN 978-1-138-19668-1

Effect of translation energy on the reaction $N(^2D) + D_2(v = 0, j = 0)$

Xianfang Yue

Department of Physics and Information Engineering, Jining University, Qufu, Shandong Province, China
Beijing Computational Science Research Center, Beijing, China

ABSTRACT: Effect of translational energy on the reaction $N(^2D) + D_2(v = 0, j = 0) \rightarrow ND + D$ is investigated by employing Quasiclassical Trajectory (QCT) method. The QCT calculated integral cross sections are in good agreement with previous wave packet results. It is also found that differential cross sections display both the forward and backward scatterings for all the collision energies, which means that the reaction mechanism of title reaction is the direct dynamical process. The product rotational angular momentum j' is slightly aligned along the direction perpendicular to the reagent relative velocity k, and the orientation is along both the negative and positive directions of the y-axis. As the translations energy increasing, the product alignment and orientation is weakly and non-monotonically changed. The generalized polarization-dependent differential cross-sections distribution demonstrates that the product angular distributions are anisotropic.

1   INTRODUCTION

Molecular dynamics have acquired exciting progress with the advances in molecular beam and laser spectroscopic technique, as well as in theoretical methodologies and computer capabilities in the past decades. The Quasi-Classical Trajectory (QCT) method has been verified to be a powerful tool to study the chemical reaction dynamics. Especially for the benchmark three-atom reactions $H + H_2$, $F + H_2$, and $Cl + H_2$, the QCT calculated results on accurate potential energy surfaces (PESs) are in excellent agreement with quantum ones. These systems are direct abstraction reactions which have no well on the minimum energy path. In recent years, much attention has been paid to the complex reactions of three atoms which occur on PESs with a deep potential well between reactants and products. $O(^1D) + H_2$, $C(^1D) + H_2$, $S(^1D) + H_2$, and $N(^2D) + H_2$ reactions belong to this category. These reaction intermediates are the bound species H_2O, H_2C, H_2S, and H_2N, respectively. In particular, considerable theoretical and experimental studies are carrying out currently on the reaction $N(^2D) + H_2$ and its isotopic variants. This may be due to the important role that the $N(^2D) + H_2$ reaction plays in the combustion of nitrogen containing fuels and atmospheric chemistry.

It is well known that the accuracy of the theoretical results depends significantly on the precision of the PES. Studies focusing on the refinement and improvement of the $1^2 A''$ PES of NH_2 have been implemented. Ho et al. reported a new Reproduc-ing Kernel Hilbert Space (RKHS) PES for the $1^2A''$ state of NH_2 based on 2715 Multireference Configuration Interaction (MRCI) points. Varandas and Poveda and Qu et al. calculated PES for the same system from internally contracted MRCI calculations using an aug-cc-pVQZ basis set. Based on these accurate PESs, both QM and QCT calculations have been performed for the $N(^2D) + H_2/D_2/HD$ reactions. Chu et al. investigated the reaction probabilities and rate constants of the $N(^2D) + H_2(v = 0, j = 0-5) \rightarrow NH + H$ reaction using the time-dependent quantum wave packet method. Castillo et al. investigated the $N(^2D) + H_2(v = 0, j = 0) \rightarrow NH + H$ reaction and its D_2 and HD isotopic variants by means of QM real wave packet and wave packet with split operator and QCT methodologies. Rao and Mahapatra calculated the initial state-selected total reaction probabilities, integral cross sections, and thermal rate constants of the $N + H_2$ reaction. Most of these studies focused on the calculations of the reaction probabilities, the integral cross sections and the thermal rate constants. Very few investigations paid attention to the vector properties calculations for the $N(^2D) + H_2/D_2/HD$ reactions. Vector properties, such as velocities and angular momenta can provide the valuable information about chemical reaction stereodynamics. Therefore, it is necessary to study the vector properties for fully understanding the dynamics of the title reactions.

In the present work, the QCT calculations of the $N(^2D) + D_2$ reaction in the translational energy ranging from 1.7 to 40 kcal/mol are investigated based on an accurate $1^2A''$ state PES by Ho et al. The vector correlation between reagents and products velocities and angular moment are presented.

2 THEORY

In the calculation, batches of 10 000 trajectories are run for each reaction and the integration step size is chosen to be 0.1 femtosecond (fs). The trajectories start at an initial distance of 15 Å between the N atom and the Center of Mass (CM) of the D_2 molecules. The collision energy is chosen to be from 1.4 to 40 kcal/mol.

The CM frame was used as the reference frame in the present study. The reagent relative velocity vector k is parallel to the z-axis. The x-z plane is the scattering plane which contains the initial and final relative velocity vectors, k and k'. θ_t is the angle between the reagent relative velocity and product relative velocity (so-called scattering angle). θ_r and ϕ_r are the polar and azimuthal angles of the final rotational angular momentum j'.

The distribution function $P(\theta_r)$ describing the k-j' correlation can be expanded in a series of Legendre polynomials as

$$P(\theta_r) = \frac{1}{2}\sum_k [k] a_0^k P_k(\cos\theta_r) \tag{1}$$

where $[k] = 2k + 1$. The a_0^k coefficients are given by

$$a_0^k = \langle P_k(\cos\theta_r)\rangle \tag{2}$$

The expanding coefficients a_0^k are called orientation (k is odd) and alignment (k' is even) parameters.

The dihedral angle distribution function $P(\phi_r)$ describing k-k'-j' correlation can be expanded in Fourier series as

$$P(\phi_r) = \frac{1}{2\pi}(1 + \sum_{even,n\geq2} a_n \cos n\phi_r + \sum_{odd,n\geq1} b_n \sin n\phi_r) \tag{3}$$

where $a_n = 2\langle\cos n\phi_r\rangle$, and $b_n = 2\langle\sin n\phi_r\rangle$.

The joint probability density function of angles θ_r and ϕ_r, which determine the direction of j', can be written as

$$P(\theta_r,\phi_r) = \frac{1}{4\pi}\sum_{kq}[k] a_q^k C_{kq}(\theta_r,\phi_r)^*$$

$$= \frac{1}{4\pi}\sum_k \sum_{q\geq0} [a_{q\pm}^k \cos q\phi_r - a_{q\mp}^k i\sin q\phi_r] C_{kq}(\theta_r,0) \tag{4}$$

$C_{kq}(\theta_r,\phi_r)$ are modified spherical harmonics. The polarization parameter is evaluated as

$$a_{q\pm}^k = 2\langle C_{k|q|}(\theta_r,0)\cos q\phi_r\rangle, \text{ k is even,} \tag{5}$$

$$a_{q\pm}^k = 2i\langle C_{k|q|}(\theta_r,0)\sin q\phi_r\rangle, \text{ k is odd.} \tag{6}$$

The full three-dimensional angular distribution associated with k-k'-j' correlation can be represented by a set of generalized Polarization-Dependent Differential Cross-Sections (PDDCSs) in the CM frame. The fully correlated CM angular distribution is written as

$$P(\omega_t,\omega_r) = \sum_{kq} \frac{[k]}{4\pi} \frac{1}{\sigma} \frac{d\sigma_{kq}}{d\omega_t} C_{kq}(\theta_r,\phi_r)^* \tag{7}$$

The angles $\omega_t = \theta_t,\phi_t$ and $\omega_r = \theta_r,\phi_r$. σ is the integral $\frac{1}{\sigma}\frac{d\sigma_{kq}}{d\omega_t}$ is a generalized PDDCS. In the present work, PDDCS$_{00}$ ($\frac{2\pi}{\sigma}\frac{d\sigma_{00}}{d\omega_t}$), PDDCS$_{20}$ ($\frac{2\pi}{\sigma}\frac{d\sigma_{20}}{d\omega_t}$), PDDCS$_{22+}$ ($\frac{2\pi}{\sigma}\frac{d\sigma_{22+}}{d\omega_t}$), and PDDCS$_{21-}$ ($\frac{2\pi}{\sigma}\frac{d\sigma_{21-}}{d\omega_t}$) are calculated. In the calculations, $P(\theta_r)$, $P(\phi_r)$, $P(\theta_r,\phi_r)$, and PDDCSs are expanded up to $k = 18$, n = 24, $k = 7$ and $k = 7$, respectively, which shows a good convergence.

3 RESULTS AND DISCUSSION

Figure 1 displays the integral cross sections of the title reaction. The solid squares and bare circle represent the present QCT result and Castillo's wave packet ones. Obviously, the present results are in good agreement with previous reports.

The product $P(\theta_r)$ distribution describes the k-j' vector correlation with $k \cdot j' = cos\theta_r$. Figure 2 shows the calculated product $P(\theta_r)$ distribution of the reaction $N(^2D) + D_2(v = 0, j = 0) \rightarrow ND + D$. Obviously, each $P(\theta_r)$ distribution is symmetric about $\theta_r = 90°$, and illustrates a distinct peak at $\theta_r = 90°$. This indicates that the product rotational angular momentum vector j' is aligned perpendicular to the reagent relative velocity direction k. The product rotational alignment of the reaction $N(^2D) + D_2(v = 0, j = 0)$ at collision energy of 1.7 kcal/mol is strongest. With the increasing collision energy, the product rotational alignment becomes weak. However, this variation is not monotonic for the title reaction in all of collision energies.

The $P(\phi_r)$ distribution describes the k-k'-j' vector correlation and can provide both product alignment and orientation information. Figure 3 illustrates the $P(\phi_r)$ distributions for the $N(^2D) + D_2(v = 0, j = 0) \rightarrow ND + D$ reaction. The $P(\phi_r)$ distributions appear double small and broad peaks at about $\phi_r = 270°$ and $\phi_r = 90°$, respectively.

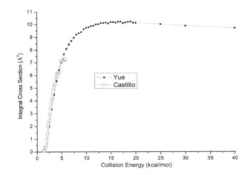

Figure 1. A comparison of the integral cross sections between the present QCT-computed results and previous wave packet ones for the title reaciton.

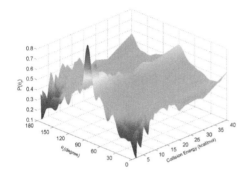

Figure 2. Distributions of the $P(\theta_r)$ as a function of the polar angle θ_r and collision energies ranging from 1.7 to 40.0 kcal/mol for the reaction $N(^2D) + D_2(v = 0, j = 0) \rightarrow ND + D$.

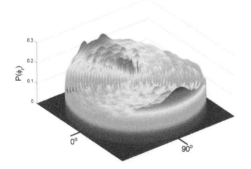

Figure 3. Distributions of the $P(\phi_r)$ as a function of the dihedral angle ϕ_r for reactions $N(^2D) + D_2(v = 0, j = 0) \rightarrow ND + D$ at collision energies of 1.7, 2.0, 2.5, 3.0, 4.0, 5.0, 6.0,7.0,8.0,9.0,10.0, 11.0, 12.0, 13.0, 14.0, 15.0, 17.0, 18.0, 19.0, 20.0, 25.0, 30.0 and 40.0 kcal/mol (from inner to outer).

This means that the orientation of the product rotational angular momentum tends to point to the both the negative and positive directions of y-axis.

That is to say, the product molecules prefer both clockwise and counterclockwise rotations (see from the positive direction of y axis, hereafter, which will not be mentioned as the same visual angle of the product molecule rotation.) in the plane parallel to the scattering plane. As the collision energy increasing, the orientation becomes stronger.

The generalized PDDCSs describe the k-k'-j' correlation and the scattering direction of the product molecule. Figure 4 shows the calculated results of the PDDCSs for the title reaction. The PDDCS$_{00}$ is simply proportional to the Differential Cross-Section (DCS), and only describes the k-k' correlation or the product angular distributions. Figure 4a plots the PDDCS$_{00}$ as a function of scattering angle θ_t for the above reactions. As clearly seen in Fig. 4a, the PDDCS$_{00}$ distribution shows both forward and backward scatterings for the $N(^2D) + D_2(v = 0, j = 0) \rightarrow ND + D$ reaction at different collision energies. This discloses a direct dynamical process. The PDDCS$_{00}$ exhibit a strong backward scattering in the lower and higher collision energies. This features an indirect reaction mechanism.

For title reaction, some PDDCS$_{20}$ values are negative at larger and smaller scattering angle, however, others are positive at medium scattering angles. The results suggest that the product rotational angular momentum j' polarizes along the both directions to perpendicular and parallel to k. Obviously, the PDDCS$_{20}$ values are more negative for the $N(^2D) + D_2(v = 0, j = 0) \rightarrow ND + D$ in the lower collision energies.

Figure 4c and 4d depict the PDDCSs distributions with $q \neq 0$. All of the PDDCSs with $q \neq 0$ are equal to zero at the extremities of forward and backward scatterings. At these limiting scattering angles, the k-k' scattering plane is not determined and the

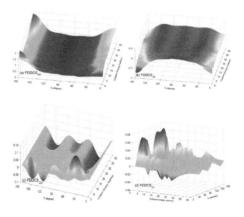

Figure 4. Panel (a) and (b) show the PDDCSs with $(k, q) = (0, 0)$ and $(2, 0)$, respectively. Panel (c) and (d) show the PDDCSs with $(k, q\pm) = (2, 2+)$ and $(2, 1-)$, respectively.

value of these PDDCSs with $q \neq 0$ must be zero. The behavior of PDDCSs with $q \neq 0$ at scattering angles away from extreme forward and backward direction is more interesting, which can provide detailed information about the product rotational alignment and orientation. The PDDCS$_{22+}$ is related to $\langle \sin^2 \theta_r \cos 2\phi_r \rangle$. The negative values of the PDDCS$_{22+}$ correspond with the product rotational alignment along the y-axis, while the positive values with the rotational alignment along the x-axis. The larger the absolute value, the stronger the product rotational alignment is along the corresponding axis. As shown in Fig. 4c, the PDDCS$_{22+}$ distribution show the negative values for the reaction under the higher collision energies. For the lowest collision energy of 1.4 kcal/mol, some values are positive. This demonstrates that the product j' alignment is preferable along the y axis, which is consistent with the $P(\phi_r)$ distributions described in Fig. 3. The PDDCS$_{21-}$ is related to $\langle -\sin 2\theta_r \cos \phi_r \rangle$, and its behavior is similar with that of the PDDCS$_{22+}$. The PDDCS$_{21-}$ is positive or negative, corresponding to the product rotational angular momentum j' along the directions of vector x-z or x+z. As shown in Fig. 4d, the PDDCS$_{21-}$ values vary with the different scattering angles, which imply that the product angular momentum distributions are anisotropic.

Figure 4c and 4d depict the PDDCSs distributions with $q \neq 0$. All of the PDDCSs with $q \neq 0$ are equal to zero at the extremities of forward and backward scatterings. At these limiting scattering angles, the k-k' scattering plane is not determined and the value of these PDDCSs with $q \neq 0$ must be zero. The behavior of PDDCSs with $q \neq 0$ at scattering angles away from extreme forward and backward direction is more interesting, which can provide detailed information about the product rotational alignment and orientation. The PDDCS$_{22+}$ is related to $\langle \sin^2 \theta_r \cos 2\phi_r \rangle$. The negative values of the PDDCS$_{22+}$ correspond with the product rotational alignment along the y-axis, while the positive values with the rotational alignment along the x-axis. The larger the absolute value, the stronger the product rotational alignment is along the corresponding axis. As shown in Fig. 4c, the PDDCS$_{22+}$ distribution show the negative values for the reaction under the higher collision energies. For the lowest collision energy of 1.4 kcal/mol, some values are positive. This demonstrates that the product j' alignment is preferable along the y axis, which is consistent with the $P(\phi_r)$ distributions described in Fig. 3. The PDDCS$_{21-}$ is related to $\langle -\sin 2\theta_r \cos \phi_r \rangle$, and its behavior is similar with that of the PDDCS$_{22+}$. The PDDCS$_{21-}$ is positive or negative, corresponding to the product rotational angular momentum j' along the directions of vector x-z or x+z. As shown in Fig. 4d, the PDDCS$_{21-}$ values vary with the different scattering angles, which imply that the product angular momentum distributions are anisotropic.

4 CONCLUSIONS

In this paper, a quasiclassical trajectory calculation is carried out for the N(^{2}D) + D$_2$($v = 0, j = 0$) $\rightarrow$ ND + D reaction under large range of collision energy from 1.4 to 40 kcal/mol. The integral cross sections, $P(\theta_r), P(\phi_r)$ and polarized-dependent differential cross sections are explored. The $P(\theta_r)$ distribution shows a weak peak at $\theta_r = 90°$, and the $P(\phi_r)$ distribution displays broad and small peaks at $\phi_r = 270°$ and $\phi_r = 90°$ for the title reaction. This indicates that the product rotational angular momentum j' is aligned along the direction perpendicular to the reagent relative velocity k. The product orientation is along both the negative and positive directions of the y-axis. Differential cross sections illustrate both forward and backward scatterings for the title reaction in different collision energies. The PDDCS$_{21-}$ distribution demonstrates that the product angular distributions are anisotropic.

ACKNOWLEDGEMENTS

The author gratefully acknowledges the financial support provided by the National Natural Science Foundation of China (Grant No. 11447014), the Project of Shandong Province Higher Educational Science and Technology Program (Grant No. J14 LJ09) and the China Postdoctoral Science Foundation (Grant No. 2014M550595).

REFERENCES

Castillo, J.F., Bulut, N., Banares, L. & Gogtas, F. 2007. Wave packet and quasiclassical trajectory calculations for the N(^{2}D) + H$_2$ reaction and its isotopic variants. *Chem. Phys.* 332: 119–131.

Chen, M.D., Han, K.L. & Lou N.Q. 2003. Theoretical study of stereodynamcis for the reactions Cl+H$_2$/HD/D$_2$. *J. Chem. Phys.* 118(10): 4463–4470.

Chu, T.S., Han, K.L. & Varandas A.J.C. 2006. A quantum wave packet dynamics study of the N(^{2}D) + D$_2$ reaction. *J. Chem. Phys.* 110: 1666–1671.

Ge, M.H., Zheng, Y.J. 2011. Quasi-classical trajectory study of the stereodynamics of a Ne+H$_2^+$---$\rightarrow$NeH$^+$+H reaction. *Chin. Phys. B* 20(8): 083401.

Kang, L.H. 2011. A comparision of the stereodynamics between C(^{1}D) + H$_2$ and C(^{1}D) + HD reactions. *Inter. J. Quantum. Chem.* 111: 117–122.

Yue, X.F. & Feng, P., 2012. A quasi-classical trajectory study on stereodynamics of the F+HCl ($v = 0, j = 0$)$\rightarrow$HF+Cl reactions. *J. Theor. Comput. Chem.* 11(3): 663–674.

Yue, X.F. & Miao, X., 2011. A quasi-classical trajectory analysis of stereodynamics of the H+FCl ($v = 0$–3, $j = 0$–3) $\rightarrow$HCl+F reaction. *J. Chem. Sci.* 123(1): 21–27.

Yue, X.F. 2013. Product polarization and mechanism of Li+HF ($v = 0, j = 0$)$\rightarrow$LiF(v', j')+H collision reaction. *Chin. Phys. B* 22(11): 113401.

Advances in Materials Science, Energy Technology
and Environmental Engineering – Patty & Zhou (Eds)
© 2017 Taylor & Francis Group, London, ISBN 978-1-138-19668-1

Adsorption properties of modified powdered activated carbon on salicylic acid wastewater

X. Tan, B. Liu, W. Hong, G.F. Chang & X.F. Zou
Key Laboratory of Shandong Academy of Environmental Science, Jinan, China

ABSTRACT: Powdered Activated Carbons (PAC) were treated with nitric acid as a modifier by immersion and immersion-thermal method, respectively. The results of Boehm titration and acid/base titrations showed that the acid-modified PAC has strong surface acidity, and, after being thermally modified, the surface acidic groups are removed due to their decomposition. The results of nitrogen adsorption and iodine adsorption indicated that surface areas and micropore volumes of PACs decreased after immersion, but increased after the subsequent sudden heat, and the resulting decomposition of acid groups affects the structure of PAC and creates new micropores and channels for adsorption. The adsorption kinetics studies were carried out with salicylic acid methyl ester wastewater. The law of kinetics for modified PACs can be described by second-order kinetic equation better and the correlation coefficients were all above 0.99. The PAC modified by the immersion-thermal method behaved best, and the maximum adsorption capacity of COD_{Cr} is 30.67 mg/g.

1 INTRODUCTION

Activated carbon, which has a large surface area and a microporous structure, has been widely applied in wastewater treatment (Liang & Wang 2011). Generally, the modification of surface chemical characterization is done by redox to modify the relative content of surface hydrophilicity, and accordingly modify the adsorption capacity of polar molecules. Thermal modification can improve the adsorption capacity of organics (Mohammad, Wan & Wan 2011). The adjustment of the porous structure and surface properties of AC through modification methods will be able to obtain the specific products for performance needs.

The accepted commercial process for salicylic acid methyl ester is based on salicylic acid and methanol, in which the concentrated sulfuric acid acts as a catalyst. In China, a batch reactor is used by most of the manufacturers, and the reaction conditions need high purity of acid and high dosage of sulfuric acid, thus producing large amounts of wastewater with high organic concentration, strong acidity, high salinity, and toxicity. The adsorption of salicylic acid methyl ester wastewater with AC will reduce the salinity and toxicity, and remove some amount of COD_{Cr}, which will improve the biochemical capability of wastewater.

In this study, Powdered Activated Carbons (PAC) were treated with nitric acid as a modifier by immersion and immersion-thermal method,

respectively. Low-temperature nitrogen adsorption, iodine adsorption, and acid/base titrations were performed to describe the surface area, microporous structure, and surface acidity and basicity. The modified PAC products were used in the adsorption of salicylic acid methyl ester wastewater, and the adsorption capacity was measured, respectively.

2 MATERIALS AND METHODS

2.1 Materials

The PAC used in this study was commercially used coconut shell activated carbon, the size of which was 100~200 mesh, and salicylic acid methyl ester wastewater was obtained from a chemical plant in Shandong province. The pH of the wastewater was 1.1~1.3, TDS was 84000~86000 mg/L, and COD_{Cr} was 4640~4990 mg/L.

2.2 Preparation of modified PAC

The PAC was first washed with deionized water to remove impurities, vacuum filtered, and dried at 105°C for 24h. This kind of PAC was marked as PAC-0 and 20 g of PAC-0 was impregnated with 100 mL 10 wt% HNO_3 solution for 2h, then vacuum filtered and washed with water until the pH becomes neutral, and dried at 60°C for 24h. This kind of PAC was marked as PAC-1 and 10 g of dried PAC-1 was placed into a muffle furnace in

600°C for sudden heat and in the absence of air for 5 min. This kind of PAC was marked as PAC-2.

2.3 *Characterization of modified PAC*

Several techniques have been used for the characterization of PACs. The pore structure including surface area, pore volume, and size was determined by nitrogen adsorption/desorption isotherms at −196°C using a surface area analyzer (JW-BK122 W, Beijing JWGB Sci. & Tech. Co., Ltd., China). The capacity of iodine adsorption was detected by the method described in GB 12496.8.

The concentration of the acidic functional group on PACs was determined by Boehm titration (Boehm, 1994). The surface acidity and basicity was determined as follows: a series of 0.2 g PAC-0, PAC-1, and PAC-2 were added into conical flasks containing 50 mL deionized water and stirred with a magnetic stirrer. Added 50μL of 0.5M NaOH or HCl solution each time, measured, and recorded the steady pH in sequence (Song, Yue & Nie 2009).

2.4 *Batch adsorption experiment*

The adsorption properties of the three kinds of modified PAC were performed in a series of flasks under ambient conditions. The adsorption capacity of each PAC was determined using 0.5 g modified PAC/200 mL salicylic acid methyl ester wastewater in flasks, while the contact time was controlled. The flasks were conditioned at 90 rpm at 20°C and the residual concentration of COD_{Cr} was determined subsequently. The samples adsorption capacity at time t (q_t) was calculated by:

$$q_t = (C_0 - C_t)V/W \qquad (1)$$

where q_t is the amount of COD_{Cr} adsorbed per gram of PAC (mg/g), C_0 is the initial concentration (mg/L) of wastewater COD_{Cr}, and C_t is the equilibrium concentration (mg/L) of wastewater at time t after adsorption, V is the volume of solution (L) in the flask, and W is the weight (g) of the PAC.

3 RESULTS AND DISCUSSION

3.1 *Boehm titration analysis*

A direct acid-base titration technique was performed to determine the concentration of acidic surface functional groups. The concentrations of the total acidic functional groups on PACs followed the order of PAC-1>PAC-2>PAC-0 (Table 1). It can be seen that PAC-1 and PAC-2 contained less acidic functional groups than PAC-0, indicating that HNO_3 benefit the creation of acidic functional groups on modified PACs and the thermal method decreases the amount of acidic functional groups.

3.2 *Surface acidity and basicity of modified PAC*

The results of acid/base titrations for modified PAC are shown in Fig. 1. When HCl was added, the pH of the solution with PAC-1 and PAC-2 decreased faster than PAC-0, and the pH of the solution with PAC-1 was lower than that with of the solution PAC-2. This result shows that when impregnated with HNO_3 solution, the surface acidity of PAC-1 and PAC-2 were increased and after oxidation, the maximum amount of total surface

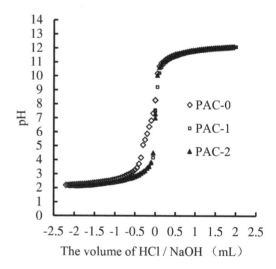

Figure 1.   Acid/base titrations for PACs.

Table 1.   Boehm titration of PACs.

PAC	Carboxyl (mmol/g)	Lactone (mmol/g)	Phenolic hydroxyl (mmol/g)	Total groups (mmol/g)
PAC-0	0.075	0.225	0.025	0.325
PAC-1	0.845	0.195	0.180	1.220
PAC-2	0.385	0.325	0.070	0.780

acidic functional groups increased. While after sudden heat, the surface acidic groups of PAC-2 were partly removed due to their decomposition.

3.3 Iodine adsorption of modified PAC

The results of iodine adsorption for three kinds of modified PAC was shown in Table 2. It is shown that the iodine adsorption of acid-impregnated PAC-1 is lower than that of washed out PAC-0, and the acid-impregnated-thermal method PAC-2 is the best one. This result is in agreement with the previous studies (Wu, Sun & Zhang 2011.), indicating that after impregnation with HNO_3 solution, some micropores and channels were broken and the adsorption capacity of PAC decreased. While after thermal process, the surface structure of PACs had been changed and adsorption capacity of iodine increased.

3.4 Properties of surface structure of PACs

The results of nitrogen adsorption are shown in Fig. 2 and the structural parameters of PACs are listed in Table 3. As can be seen, PAC-2 has the highest BET surface area of 660 m²/g and a total pore volume of 0.577 cm³/g, suggesting that the pore structure of PAC was better developed, especially for micropore. PAC-1 has the lowest BET surface area of 596 m²/g and a total pore volume of 0.505 cm³/g, indicating that the addition of HNO_3 solution decreased the pore development of PAC compared with PAC-0 and PAC-2. Although the addition of HNO_3 solution resulted in a decrease in the total volume, the surface of micropore volume had been increased. This result indicated that after impregnation with HNO_3, the sudden heat created new micropores and channels for adsorption.

3.5 Kinetic studies of modified PAC

Fig. 3 shows the change in the uptake of COD_{Cr} by the given modified PAC as a function of time. Results of experiments revealed the adsorption stage with high initial rates followed by lower rates near equilibrium. It could be seen that more than 50% of the total uptake of COD_{Cr} could be achieved within 120 min and it would take about 600 min to reach the adsorption equilibrium. The data in Fig. 3 also shows that, at the initial stage

Table 2. Iodine adsorption capacity of different kinds of PACs.

Adsorbent	PAC-0	PAC-1	PAC-2
Iodine adsorption capacity(mg/g)	736.6	546.1	775.6

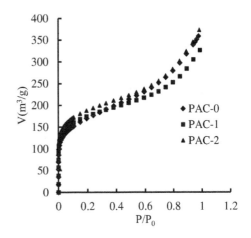

Figure 2. N_2 adsorption isotherms of PACs.

Table 3. Porous structure parameters of three PACs.

Type	S_{BET} (m²/g)	S_{mic}	%	V_{tot} (cm³/g)	V_{mic} (cm³/g)	D_p (nm)
PAC-0	614	306	49.8	0.555	0.2501	0.6544
PAC-1	596	402	67.4	0.505	0.2602	0.6473
PAC-2	660	449	68.0	0.577	0.2796	0.669

S_{BET}, BET surface area; S_{mic}, micropore surface area; V_{tot}, total pore volume; V_{mic}, micropore volume; Dp, mean pore diameter.

of adsorption, the COD_{Cr} adsorption on PAC-2 is quicker than that on PAC-0 and PAC-1. When it comes to the equilibrium stage, the adsorption capacity of PAC-2 is higher than that of PAC-0, and PAC-0 is higher than PAC-1. This result is in agreement with the previous iodine adsorption studies (PAC-2<PAC-0<PAC-1).

The pseudo first-order equation and second-order equation were considered for interpreting the experimental data. In the theory of pseudo first-order equation, the rate of adsorption is directly proportional to the concentration of the adsorbent, and mass transfer resistance is the limiting factor in the adsorption process. Pseudo first-order equation:

$$dq_t/dt = k_1(q_e - q_t) \qquad (2)$$

where q_t (mg/g) is the concentration of COD_{Cr} at time t, q_e (mg/g) is the adsorption capacity of COD_{Cr} at equilibrium, and k_1 is the equilibrium rate constant of pseudo first-order equation (g (mg min)⁻¹).

The pseudo second-order model assumes that the limit factor in the adsorption is adsorption mechanism, not the mass transfer resistance. The pseudo second-order equation:

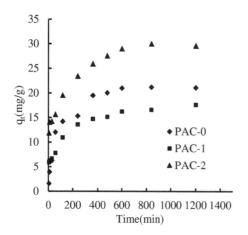

Figure 3. Adsorption of COD_{Cr} on PACs.

Table 4. The kinetic parameters of the pseudo first-order equation and second-order equation for three kinds of PACs.

Adsorbent	Pseudo first-order equation			Pseudo second-order equation		
	q_e	k_1	R^2	q_e	k_2	R^2
PAC-0	14.32	2.0×10^{-3}	0.9488	22.52	6.77×10^{-4}	0.9974
PAC-1	10.67	3.1×10^{-3}	0.9667	17.92	9.74×10^{-4}	0.9949
PAC-2	16.37	3.6×10^{-3}	0.9827	30.67	7.59×10^{-4}	0.9965

$$dq_t/dt = k_2(q_e - q_t)^2 \qquad (3)$$

where q_t (mg/g) is the concentration of COD_{Cr} at time t, q_e (mg/g) is the adsorption capacity of COD_{Cr} at equilibrium, and k_2 is the equilibrium rate constant of the pseudo second-order (g (mg min)$^{-1}$) equation.

The kinetic parameters of the pseudo first-order equation and second-order equation for three kinds of PAC were calculated and given in Table 4.

The coefficient of determinations (R^2) for the pseudo second-order kinetic model for three kinds of PAC are all above 0.99 and higher than that for the pseudo first-order kinetic model, and the calculated q_e values for the pseudo second-order kinetic model are in good agreement with the experimental q_e values. This result indicates that the adsorption for three kinds of PAC fits better to the pseudo second-order kinetic model, and chemical adsorption is in dominating station (Sun, Yue, Mao, Gao, Gao & Huang 2014). The equilibrium adsorption capacity of PAC-1 is 17.92 mg/g, lower than that of PAC-0 (22.52 mg/g), and the PAC-2 is the highest (30.67 mg/g), twice the PAC-1. This result indicates that after impregnation with HNO$_3$ solution, the number of acidic groups increased, resulting in an increase in the surface polar of PAC. The adsorption capacity of PAC-1 decreased because of the com-

petition of water molecules for polar organic compounds and partly destroy micropores. At the same time, the surface micropores PAC-2 increased after the subsequent sudden heat, and the new micropores and channels increased the adsorption capacity.

4 CONCLUSIONS

The results of Boehm titration and acid/base titrations for modified PACs show that when impregnated with HNO$_3$ solution, the surface acidity of PAC-1 and PAC-2 increased and the maximum amount of total surface acidic functional groups increased. The results of low-temperature nitrogen adsorption show that PAC-2 has the highest BET surface area of 660 m^2/g and a total pore volume of 0.577 cm^3/g, and surface areas and micropore volumes of PAC decreased after immersion but increased after the subsequent sudden heat. The decomposition of acid groups after the thermal method affects the surface structure of PAC and creates new micropores and channels for adsorption. The adsorption kinetic result is in agreement well with the pseudo second-order kinetic model and the best adsorption capacity of modified PAC was that of PAC-2, 30.67 mg/g COD$_{Cr}$, in salicylic acid methyl ester wastewater.

ACKNOWLEDGMENTS

This work was supported by the Science and Technology Development Program of Shandong province (2014GSF117025).

REFERENCES

Boehm H P, 1994.Some aspects of the surface chemistry of carbon blacks and other carbons, *Carbon.* 32: 759–769.

Liang X and Wang X, 2011.The technology of activated carbon modification and its application in wastewater treatment. *Technol. Water Treat.* 37: 1–6.

Mohammad S S, Wan M A, Wan D, 2011. Amirhossein H and Arash A N. Ammonia modification of activated carbon to enhance carbon dioxide adsorption: Effect of pre-oxidation. *Appl. Surf. Sci.* 257: 3936–3942.

Song J, Yue D and Nie Y, 2009.Adsorption properties of HNO$_3$-modified granular activated carbons on low molecular weight organic matter in leachate. *Environ. Chem.* 28: 788–792.

Sun Y, Yue Q, Mao Y, Gao B, Gao Y and Huang L, 2014. Enhanced adsorption of chromium onto activated carbon by microwave-assisted H$_3$PO4 mixed with Fe/Al/Mn activation *J. Hazard. Mater.* 265: 191–200.

Wu G, Sun X and Zhang Q, 2011. Review of surface oxidizing modification of activated carbon and influence on adsorption capacity. *J. Zhejiang A & F Univ.* 28: 955–961.

*Advances in Materials Science, Energy Technology
and Environmental Engineering – Patty & Zhou (Eds)*
© 2017 Taylor & Francis Group, London, ISBN 978-1-138-19668-1

Product remanufacturability evaluation based on the theory of LCC

Qinglan Han & Ben Hu
Business School of Central South University, Changsha, China

ABSTRACT: Remanufacturing completely retains the added value of waste products, which is important for the protection of the environment, recycling of resources, and development of recycling economy. In addition, it is especially significant to establish a scientific evaluate model to find whether the waste product is valuable to be remanufactured. Based on the theory of LCC, this paper establishes the product life cycle cost model to analyze the cost of original manufacturing and remanufacturing respectively. The environmental impact value is obtained using Eco-indicator 99 index by the LCA evaluation method. Then, integrate the data of LCC and LCA. Quantify the comprehensive evaluation index by the AHP method. Finally, the feasibility of this model is verified with the study of waste car engine, and decision support for remanufacturers is provided.

1 INTRODUCTION

With the rapid development of economy, the mechanical products increase quickly and scrapped cars amount to 16 million in 2020. Those scrapped cars will cause energy shortages and environmental pollution. The establishment of a scientific product remanufacturing assessment model to solve the problem of disposal of scrapped products becomes significantly obvious.

Lund R (1998) noted that remanufacturing retained the most valuable substances and summed up seven evaluation criteria of remanufacturing. Some assessed the energy consumption and emissions during the process of original manufacturing and remanufacturing by the LCA method, as in Kerr (2001), Vanessa (2004), DP (2007), and Liu (2014). Jorge (2010) quantified the environmental impact of a truck part during life cycle through comparing the value of EI99. Schau (2011) evaluated the LCC of three different materials of a generator remanufactured in different locations to obtain optimum scheme selection. Ferrer (1997) made a cost-benefit analysis and provided a basis for the remanufacturing company to make a decision. Some established an evaluation model for reconstructed electrical products with resource utilization, as in Mao G P (2009a and Mao G P (2009b). Some established a remanufacturing quantitative evaluation model of waste products, which considered several technical and economic indicators in Zhang Guo-qing (2005) and Zeng Shou-Jin (2012). Some established a manufacturing evaluation model using the fuzzy comprehensive evaluation method in Bras B (1996), Ghazalli (2011), and Hou Z Z (2009). Deng Qian-Wang (2014) quantified the green feature index of remanufactured products based on energy consumption, environmental emissions, and cost evaluation.

It can be observed that the study of recycled products' remanufacturing evaluation separated the environmental and economic factors. Foreign literature focused more on the environmental impact assessment, while domestic literature used the AHP method to combine the evaluation index, which is generally subjective. In this paper, the value of LCC and LCA attempts are integrated to evaluate the value of economy and ecology, and the evaluation model is established to assess waste products remanufacturability.

2 COMPREHENSIVE ASSESSMENT MODEL OF REMANUFACTURABILITY

This paper no longer considers the difference between the using and the final disposing stages when comparing the impact of original manufacturing and remanufacturing products from the perspective of product life cycle. The two main results of the cost and environmental impact with the application of the LCC and LCA methods are shown in Figure 1.

2.1 Life cycle cost model

According to the product life cycle cost theory, the total cost is the sum of multiple tasks costs. To complete each task, it needs to consume resources by Li Shi-Hui (2013).

2.1.1 Remanufactured products cost model
The total cost comprised: dismantling cost, cleaning cost, detecting cost, reprocessing cost, assembling

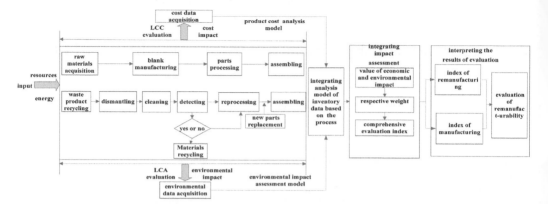

Figure 1. Comprehensive assessment of the remanufacturing process.

cost, and replacing cost. Renewable benefits of discarded parts have a deduction which cannot be remanufactured. The remanufactured product cost is calculated as follows:

$$C_{RM} = C_1 + C_2 + C_3 + C_4 + C_5 + C_6 + C_7 - C_8 \quad (1)$$

① Recycling cost C_1

$$C_1 = c_r \bullet w + w \bullet d \bullet c_d \quad (2)$$

c_r is the unit price of weight for the recycled product, w is the weight of the recovery product, d is the distance of transportation, and c_d is the unit price of transport.

② Dismantling cost C_2

$$C_2 = t_d \bullet s_{wd} + c_{eqd} \quad (3)$$

$$t_d = \sum_{k=1}^{m} N_{jk} \bullet t_{rk} + \sum_{i=1}^{n} t_{di} + t_a \quad (4)$$

$$t_a = k_a \bullet t_d \quad (5)$$

$$c_{eqd} = \sum_{i=1}^{n} c_{eqdi} \quad (6)$$

t_d is the total dismantling time, N_{jk} is the number of connected components that have a connection with k, t_{rk} is the time to remove the connecting piece of k, t_{di} is the separation time of the part i, n is the total number of dismantling parts, k_a is the proportion of basic time, s_{wd} is the wages of unit time, and c_{eqd} is the total cost of the tool loss, i.e. the sum of tool cost for each component.

③ Cleaning cost C_3

$$C_3 = c_m + t_c \bullet s_{wc} \quad (7)$$

c_m is the total cleaning media cost, t_c is the total cleaning time, and s_{wc} is the wages for unit cleaning time.

④ Detecting cost C_4

$$C_4 = c_{eqt} + t_{test} \bullet s_{wt} \quad (8)$$

c_{eqt} is the tool cost of detecting process, t_{test} is the total detection time, and s_{wt} is the wages for unit detecting time.

⑤ Reprocessing cost C_5

The decision variables are introduced between 0–1. $X = 1$ indicates that the parts can be reused, and $X = 0$ represents the parts for material regeneration (Xie Jia-Ping, 2003). In addition, the probability of the reuse of components can be expressed as $P_i(t) = e^{-\lambda it}$.

$$C_5 = \sum_{i=1}^{n} \sum_{k=1}^{l} X \bullet P_i(t) \bullet \left(t_k \bullet c_k + s_k \bullet t_k \bullet n_k \right) \quad (9)$$

$$c_k = I_k \bullet \tau_k \bullet h_k / H + p_k \bullet c_{kwh} + c_{aux,k} \quad (10)$$

t_k, c_k represent the time consumed and the cost of unit time in the processing operation k, respectively; s_k, t_k, n_k represent the wages of unit time, working hours, and the number of workers, respectively; I_k, τ_k, h_k represent the equipment investment cost, the annual depreciation rate, and working hours, respectively; H, p_k, c_{kwh} represent the equipment annual working hours, consumption, and the price of electricity, respectively; and $c_{aux,k}$ represent other energy costs.

⑥ Replacing cost C_6

$$C_6 = \sum_{j=1}^{Nn} q_{nj} \bullet p_{nj} \quad (11)$$

q_{nj}, p_{nj} represent the quantity and price of the new replacement parts j; and N_n represents the numbers required for new-type N.

⑦ Assembling cost C_7

$$C_7 = \sum_{l=1}^{k} \left(t_l \bullet c_l + s_l \bullet t_l \bullet n_l \right) \tag{12}$$

t_l, c_l represent time consumed and the assembly cost of process l; s_l, t_l, n_l represent the wages of unit time, working hours, and the number of workers for assembly process l.

⑧ Renewable benefits C_8

$$C_8 = R_{renew} - C_{renew} \tag{13}$$

$$R_{renew} = \sum_{i=1}^{n} \sum_{k=1}^{m} w_i \bullet \beta_{ik} \bullet \gamma_{ik} \bullet p_{ik}$$
$$\bullet \left[(1 - X_i) + (1 - P_i(t)) X_i \right] \tag{14}$$

$$C_{renew} = \sum_{p=1}^{k} \left(t_p \bullet c_p + s_p \bullet t_p \bullet n_p \right) \tag{15}$$

λ_i, t, w_i represent the degradation rate, the service time, and the weight of component i; β_{ik}, γ^k, p^{ik} indicate the proportion of renewable material, reproduction rates, and prices for the material k; t_p, c_p indicate time consumption and recovery cost of material recycling process p; s_p, t_p, n_p indicate the wages of unit time, working hours, and workers for process p.

2.1.2 *Manufactured products cost model*

Cost elements are input into the product manufacturing process, including materials, equipment, energy, and labor. Combining with the characteristics of each production stage, the building product manufacturing cost is shown as follows:

$$C_{OM} = C_m + C_w + C_a \tag{16}$$

$$C_m = \sum_{k=1}^{s} c_k \bullet q_k \bullet (1 + r_k) \tag{17}$$

$$C_{w/a} = \sum_{m=1}^{n} \left(t_m \bullet c_m + s_m \bullet t_m \bullet n_m \right) \tag{18}$$

C_m, C_w, C_a represent material costs, processing costs, and assembly costs; c^k, q^k, r^k represent the price, quantity, and scrap rate in the production process of the required material k; t_m, c_m represent the consumed time and the cost of process m; and s_m, t_m, n_m represent the wages of unit time, working hours, and the number of workers in the process m.

2.2 *Life cycle environmental impact assessment*

Environmental factors such as energy consumption must be considered in the product transferring process. The LCA method evaluates the environmental impact during the product life cycle by quantifying the product energy consumption

and waste emissions. The life cycle environmental impact will be classified into three categories as resource consumption, environmental damage, and damage to human health as in Goedkoop M (2000) and Liu Zheng (2013). Then, a single weighted value is obtained as EI99, which indicates the degree of environmental impact.

① Many kinds of environmental impact factors EIF_k are classified into 11 kinds of ecological problems IC_{ij} with the help of the characteristic factor. Its formula is given as follows:

$$IC_{ij} = \sum \alpha_k \bullet EIF_k \tag{19}$$

② Ecological problems are classified into three environmental impact categories as IC_i by dimensional normalization parameter. Therefore, the formula is given as follows:

$$IC_i = \sum \beta_i \bullet IC_{ij} \tag{20}$$

③ Obtain the ecological index EI by weighting the three categories, which can be based on the important degree among those.

$$EI = \sum w_i \bullet IC_i \tag{21}$$

2.3 *Comprehensive evaluation of remanufacturability*

The cost and environmental impact of different dimensions, EI, and LCC must be standardized. Then, the results of LCC and LCA were integrated. Finally, the comprehensive evaluation index was obtained, and the calculation is as follows:

$$LCC_k = \frac{\min[LCC_k]_1^N}{LCC_k} \tag{22}$$

$$EI_k = \frac{\min[EI_k]_1^N}{EI_k} \tag{23}$$

$$Score_k = \omega \bullet LCC_k + (1 - \omega) \bullet EI_k \tag{24}$$

where ω represents the economic weight, $(1-\omega)$ represents the environmental weight; LCC_k represents the life cycle cost standardization value, EI_k represents the ecological index standardization value; and if $Score_{rem} > Score_{om}$, it represents the waste products with good remanufacturability.

3 MODEL APPLICATION

This paper takes the common heavy truck engine WD615.87 as an example, with six cylinders and the total weight for 873.68 kg. The remanufacturability is evaluated by comparing the cost and environmental impact between the original manufacturing engine and the remanufacturing engine.

3.1 *Remanufacturing engine cost*

Detection is a necessary step before parts remanufacturing, and the waste engine cost will be divided into two parts:

Part one is the calculation of the cost from recycling to detecting stages.

① Recycling cost
It is mainly determined by the purchase price and transportation costs. Presently, the transaction price is between 2800 (RMB) and 3500 for unit ton[1]. This paper takes the middle price of 3150. Then, obtain the average recovery distance for each scrap engine about 200 km. Assume that the waste engine transported by trucks to the remanufacturing plant which weighs 10 tons and consumes fuel for diesel, then the fuel consumption amounts to 40 L per 100 km. The transportation cost is composed of the fixed cost and the variable cost. The former includes the driver's salary and the latter includes the fuel cost. The driver's average wage will be 200 per day and 5.5 for per liter diesel. Calculate the cost in the waste recovery stage of the engine as follows: The purchasing price is 2752.09, the transportation cost is 55.92 and the recycling cost is 2808.01.

Data and the remanufactured engine production time are shown as follows: dismantling time is 6.5h, cleaning time and detecting time is 4.5, processing time is 61.5, and assembly time is 2.5. The total time[2] is 79.5.

② Dismantling cost
Owing to the low level of remanufacturing automation, the process of dismantling, cleaning, and inspection is mainly done manually. Assume that unit labor cost of those procedures are equal ($s_{wd} = s_{wc} = s_{wt} = 0.55$ for per minute). The dismantling cost is calculated to be 214.5.

③ Cleaning cost
Assuming that the cost of cleaning media is not considered, the cleaning cost is 148.5.

④ Detecting cost
Assume that tool depreciation is not considered in the paper. Finally, we get the testing cost to be 148.5.

Another part is the calculation of the cost from reprocessing to completing the assembly stages. Collect and record the testing results of the engine

main parts, which have been serviced for 7 years before scrapped[3], as given in Table 1.

⑤ Reprocessing cost
About ten parts have to be remanufactured which need to be invested in machinery, energy, and labor elements when remanufacturing. According to the unit cost of machining and labor, the reprocessing cost is calculated to be 8175.3.

⑥ Replacing cost
Owing to severe wear and aging, four parts cannot be remanufactured. Thus, these parts need to be replaced. Calculate the replacement cost to be 825.8.

⑦ Renewable benefits
Costs of four parts for material recovery mainly considered the equipment cost and the labor cost, and the sales revenue is mainly decided by the amount of recycled materials and unit price. The renewable rates of cast iron and aluminum material are 85% and 90%, respectively. The work by Han Qing-Lan (2015) deals precisely with it. Estimate materials revenue, cost, and benefits separately as follows: 197.07, 33.21, and 163.86.

⑧ Assembling cost
All parts will be transported into the assembly line finally, using the formula (12) to calculate the assembly cost as 1750.6.

Then, according to formula (1), we get the remanufactured engine total cost as 13907.35.

3.2 *Original engine manufacturing cost*

Engine manufacturing cost includes the materials cost, machinery operating cost, and labor cost. Add up the weight available in Table 1 for main materials: cast iron, 511.47; alloy steel, 275.64; and aluminum alloy, 86.57 kg. The prices are 1.83, 9.06, and 22.78 respectively. In addition, the rate of waste material needs to be considered when calculating. Assume that the rate is 0.1 in the paper. Then, calculate the material cost.

In this case, the main engine components are manufactured by their own processing plants with purchasing materials. Collect the manufactured engine production time as follows: processing time is 108.55 and assembly time is 4.5. The total time is 113.05. Consider the main cost with the equipment cost and the labor cost in the manufacturing and assembling stage. Then, calculate those cost based on the cost of unit equipment operating time and worker time.

The material cost, processing cost, and assembly cost amount to the engine total cost, and the results are as follows: material cost is 5945.89, processing cost is 24166.12, and assembly cost is 2727.13. Total cost is 32839.14.

---

[1]Data from Chinese recovery trading network.
[2]Data from the publication of Remanufacturing and Recycling Economy.

[3]Data of the cast iron engine average age from the car market network.

Table 1. Test results of main engine parts.

Part name	Used directly	Need remanufacturing	Material recycling	Reprocessing time /min	Price of new/¥	Main material	Weight /kg
cylinder block	0*	1**	0	930		cast iron	298.52
crankshaft	0	1	0	870		alloy steel	199.02
link	0	1	0	405		alloy steel	29.85
flywheel	1	0	0			cast iron	39.8
cylinder head	0	1	0	480		cast iron	99.51
gear chamber	1	0	0			cast iron	29.85
air compressor	0	1	0	247		alloy steel	29.85
Intake pipe	1	0	0			aluminum	9.95
front exhaust manifold	1	0	0			cast iron	14.93
behind exhaust manifold	1	0	0			cast iron	14.93
turbocharger	0	1	0	255		aluminum	19.9
piston	0	0	1		598.5	aluminum	21.89
bolt	0	0	1		108.2	alloy steel	9.95
cylinder liner	0	0	1		86.4	cast iron	13.93
oil cooler cover	1	0	0			aluminum	4.98
spiracle	0	1	0	55		alloy steel	1.99
tappet	0	1	0	45		alloy steel	1.99
piston ring	0	0	1		32.7	alloy steel	1.00
injector	0	1	0	55		alloy steel	1.99
fuel pump	0	1	0	348		aluminum	29.85
	6	10	4	3690	825.8		873.68

** indicates "yes", while * indicates "no".

3.3 LCA inventory analysis and environmental impact assessment

First, using the EI99 evaluation system, environmental factors are classified into the following categories: fossil fuels, minerals, acidification, respirable inorganic, and climate change. Then, the five factors are classified as three kinds of environmental impact categories. Using the data of materials quantity in Table 1 as well as the environmental impact on production steel, iron, and aluminum in Liu Zhi-Chao (2013) and Chen Wei-Qing (2009), calculate the environmental impact of raw materials acquisition stage. Calculate the environmental impact of energy consumption and pollutant emissions in the process and transportation stage. Materials recycled can be as a deduction in the LCA inventory data. Ultimately, calculate the data of life-cycle environment impact between the original manufacturing and the remanufacturing processes as given in Table 2.

Characterize and dimensionally normalize the inventory data, respectively. Di Xiang-Hua (2005) determined the weight as 20%, 40%, and 40% of three environmental impact categories. Finally, calculate the value of EI99 as shown in Table 3.

Table 2. Inventory of the original manufacturing and the remanufacturing processes.

Category	Substance	OM	RM
Energy	electricity/MJ	8266.36	2786.44
	coal/kg	2703.74	707.71
	oil/kg	104.13	66.24
	natural gas/kg	24.81	7.17
Resources	iron ore/kg	2272.39	62.74
	aluminum ore/kg	406.93	57.29
Air emissions	CO_2/kg	4844.01	1250.33
	CO/kg	15.37	10.33
	NO_x/kg	11.83	4.72
	CH_4/kg	13.42	3.68
	SO_2/kg	14.44	11.43
	H_2S/kg	0.03	0.43
	HCL/kg	0.84	0.25

3.4 Evaluation of remanufacturability

Determine the economic weight as 60% using the AHP method. Standardize the results of LCC and EI and then calculate the evaluation index.

Consider the economy and ecology of waste engine remanufacturing from the perspective of product life cycle. The value is $Score_{RM} > Score_{OM}$, indicating that the waste engine has good

Table 3. Evaluation value of original and remanufactured engines.

	LCC	Standardized LCC	EI	Standardized EI	Score
OM	32839.14	0.42	315.53	0.39	0.41
RM	13907.35	1	121.82	1	1

remanufacturability, which is consistent with that the car is a more popular object for remanufacturing.

4 CONCLUSION

Based on the LCC theory, the product life cycle cost model is developed by analyzing the cost element inputs of each activity during the original manufacturing and remanufacturing process. The ecological impact is measured in the manufacturing process by applying the method of LCA and EI99 index. The original manufacturing and remanufacturing product life cycle cost is calculated, respectively, with the theory of LCC. Then, the indicator of LCC and EI99 are integrated to get the ultimate quantitative evaluation index. It can provide a decision for manufacturers about how to dispose waste products. Finally, the evaluation model is feasible through the application with waste engine. Compared with the original engine, the remanufactured engine reduces resources consumption and environmental emissions in the stages of obtaining materials and manufacturing, with great economic and environmental benefits.

ACKNOWLEDGMENTS

This work was financially supported by the National Natural Science Foundation (71172101). The project is Building and Application Knowledge based on the Theory of Life Cycle Cost.

REFERENCES

Bras B, Hammond R (1996). Towards Design for remanufacturing-metrics for assessing remanufacturability, Proceedings of the 1st international workshop on reuse, Eindhoven, The Netherlands.

Chen Wei-Qing, Wan Hong-Yan (2009). Life cycle assessment of aluminium and the environmental impacts of aluminium industry. L M. 5.3–10.

DP Adler, PA Ludewig (2007). Comparing Energy and Other Measures of Environmental Performance in the Original Manufacturing and Remanufacturing of Engine Components. Asme Intl Manuf Sci & Eng Conf. 851–860.

Deng Qian-Wang, Luo Jing-Zhi (2014).Green Evaluation of Retired Construction Machinery Products Remanufacturing. Chin Surf Eng. 3. 101–107.

Di Xiang-Hua (2005). Life cycle emission inventories for the fuels consumed by thermal power in China. Chin Environ Sci. 5. 632–635.

Ferrer Geraldo (1997). The Economics of Tire Remanufacturing Resour Conserv Recy. 19. 221–255.

Ghazalli Z, Murata A (2011). Development of an AHP-CBR evaluation system for remanufacturing: end-of-life selection strategy. Intl J Sustainable Eng. 4(01). 2–15.

Goedkoop M, Spriensma R (2000). The Eco-indicator'99: A damage oriented method for Life Cycle Impact Assessment. Zoetermeer.

Han Qing-Lan, Zhang Yang (2015). Materials Selection Evaluation Based on LCC Theory from Eco-design Perspective. Sci Tech Mgt Res. 7.180–184.

Hou Z Z (2009). The Evaluation of Remanufacturing Design Scheme about Electromechanical Product. Dvpt & Innovation Machr & Elec Prods. 2. 20–22.

Jorge Amaya, Peggy Zwolinski (2010). Environmental benefits of parts remanufacturing: the truck injector case. Hefei, ANHUI. Chin.

Kerr W, Ryan C (2001). Eco-efficiency gains from remanufacturing: A case study of photocopier remanufacturing at Fuji Xerox Australia. J Clean Prod. 9. 75–81.

Li Shi-Hui, Han Qing-Lan (2013). Construction of Knowledge-Database Based on Life Cycle Cost Management. A Res. 7. 35–42.

Liu Z, Li T (2014). Life Cycle Assessment-based Comparative Evaluation of Originally Manufactured and Remanufactured Diesel Engines J Ind Eco. 18. 567–576

Liu Zheng (2013). Research on Environmental Consciousness Design and Evaluation Method of Electromechanical Products. ZheJiang U.

Liu Zhi-Chao (2013). Life Cycle Assessment Methodology of Original Manufacturing and Remanufacturing of an Engine. Dalian U Tech.

Lund. R (1998). Remanufacturing: an American resource. Rst Inst Tech, NY, USA. Jun 16–17.

Mao G P (2009a). A remanufacturability assessing model for waste electromechanical product. Mod Manuf Eng. 6. 114–118.

Mao G P (2009b). A Study on the Environmental Influence Comparison between Engine Manufacturing and Remanufacturing. Auto Eng. 31. 565–568.

Schau EM, Traverso M (2011). Life Cycle Costing in Sustainability Assessment—A Case Study of Remanufactured Alternators. Sustainability. 3. 2268–2288.

Shao Xin-Yu, Deng Chao (2008). Integration and optimization of life cycle assessment and life cycle costing for product design. J Mech Eng. 9. 13–20.

Vanessa M Smith, G A K (2004). The Value of Remanufactured Engines: Life-Cycle Environmental and Economic Perspectives. J Ind Eco. 8. 193–221.

Xie Jia-Ping, Chen Rong-Qiu (2003). The Cost-Benefit Analysis Model of Recovery Processing in Reverse Logistics of Assembly Products. Chin Circ Econ. 17. 25–28.

Zeng Shou-Jin, Liu Zhi-Feng (2012). Green remanufacturing comprehensive assessment method and its application of electromechanical products based on Fuzzy AHP. Mod manuf Eng. 7.1–6.

Zhang Guo-qing, Jing Xue-dong (2005). The Assessment Method and Model of Remanufacturability. J Shanghaijiaotong U. 39. 1431–1436.

Advances in Materials Science, Energy Technology and Environmental Engineering – Patty & Zhou (Eds)
© 2017 Taylor & Francis Group, London, ISBN 978-1-138-19668-1

Structure-dependent mechanical property of carbon nanotube fibers

Jingdong Zhu

School of Material Science and Engineering, Tianjin University, Tianjin, P.R. China

ABSTRACT: The mechanical behavior of Carbon Nanotube (CNT) fibers depends on their fabrication process. In this study, CNT fibers were synthesized by chemical vapor deposition process, and the mechanical properties of the CNT fibers with different structures were measured. The results indicate that the arrangement of CNT bundles within a fiber determines the mechanical behavior.

1 INTRODUCTION

Carbon Nanotube (CNT) fibers have attracted much interest as novel materials for various applications. In particular, the mechanical property of CNT fibers is interesting (Lu, 2012). Currently, CNT fibers are synthesized by four methods: 1) by spinning from a CNT solution (Ericson, 2004), 2) by spinning from an aligned CNT array (Zhang, 2004), 3) by spinning from a CNT aerogel formed by Chemical Vapor Deposition (CVD) (Li, 2004; Koziol, 2007; Zhong, 2010), and 4) by twisting/rolling from a CNT film (Ma, 2009). The CNT fibers obtained by different fabrication processes show different mechanical behaviors. However, the reason for this is not known yet. In this study, a series of experiments were conducted to study the relationship between the structure and mechanical property of the CNT fibers obtained by CVD; the results indicate that the arrangement of CNT bundles within the fiber affects the mechanical behavior of the CNT fiber.

2 EXPERIMENTAL

CNT fibers were spun by CVD using acetone as the carbon source, containing ferrocene and thiophene. The reaction conditions were as follows: an injection rate of 8 mL/min, a reaction temperature of 1170°C, and a hydrogen flow of 1000sccm. The spinning velocity was 5 mL/min. The as-spun CNT fibers were pulled out from the reaction mixture and treated with acetone; a more detailed description of the procedure can be found elsewhere (Zhong, 2010). The obtained as-spun fibers have a hollow structure. Without acetone treatment, the CNT fibers were obtained as a solid. Twisted CNT fibers were obtained by twisting a 3-cm-long as-spun fiber using a homemade twisting machine with a controlled number of turns.

The cross-sectional and surface morphologies of the CNT fibers were characterized using a scanning electron microscope (JSM-6700F) and further characterized using a transmission electron microscope (Tecnai-G20 F20) and Raman spectrometer (Renishaw). The mechanical properties of the CNT fibers were evaluated using a tensile tester (XQ-1, Shanghai New Fiber Instruments, China, Shanghai). The extension rate was 3 mm/min, and the gauge length was 10 mm.

3 RESULT AND DISCUSSION

3.1 *Structures of CNT fibers*

The fibers were continuous and soft similar to cotton yarns. Three hierarchy structure s were observed in the CNT fibers. At the fundamental level, mainly double-wall CNTs were observed in the diameter range 5–10 nm; at the next level, the CNTs formed bundles of 10–100 nm in diameter. Then, the CNT bundles formed dog-bone cross-sections, thus producing stacks of parallel graphene sheets as shown in Figure 1a. At the microstructural level, the CNT bundles formed a continuous network orientated preferably along the fiber axis. However, three different CNT network structures were observed, namely, hollow structure, solid structure, and twisted structure, depending on the synthesis and treatment of the CNT fiber as explained below. The crystallization of the CNT fibers was characterized by Raman spectroscopy. An IG/ID value of 3.1 was obtained (Fig. 1b), indicating that the CNT fibers obtained in this study have less defect than those obtained by spinning from an aligned CNT array (Li, 2007).

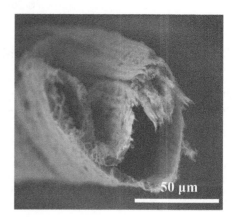

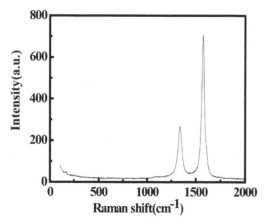

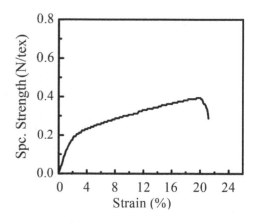

Figure 1. a) Transmission electron microscopy image of the cross-section of a CNT bundle. b) Raman spectrum of CNT fibers.

Figure 2. a) Cross-sectional image of a hollow CNT fiber, b) specific stress–strain curveof the hollow CNT fiber.

3.2 As-spun fiber

The as-spun CNT fibers have a hollow structure with a diameter in the range 50–100 μm. This can be attributed to the treatment with acetone; a similar phenomenon has been reported (Futaba, 2006). Sometimes, the CNT fiber walls collapse due to the flexibility of the fiber. The fiber walls comprise CNT bundles, in intimate contact with each other. A typical CNT fiber is shown in Figure 2a.

The mechanical property of the as-spun CNT fibers was measured. The CNT fiber strength was evaluated in terms of specific strength (GPa/(g/cm³), which was calculated by dividing the measured force in N with the linear density of the CNT fiber in tex (gram per kilometer fiber, g/km), because GPa/(g/cm³) is equivalent to (N/tex). Because the linear density of a fiber can be measured precisely, the specific strength thus obtained represents the actual strength of the CNT fibers as it does not include the fiber density in g/km, which is difficult to measure.

Figure 2b shows a "two-stage" curve: high-slope and low-slope stages, representing the elastic and plastic transformations of the CNT fiber, respectively. According to Li (2011), the elastic transformation originates from the rearrangement of the alignment with the axis of the CNT bundle within the fiber, whereas the plastic transformation originates from the sliding of CNT bundles against each other.

3.3 Solid-core fiber

The CNT fibers obtained without the treatment with acetone have a solid structure with a diameter in the range 200–250 μm. A typical CNT fiber is shown in Figure 3a. The CNT fiber is filled with loose CNT bundles, and the CNT bundles interconnect to form a loose web with a porosity of up to 98.1%, in contrast to the dense hollow structure of the as-spun CNT fibers.

The mechanical tensile property of the CNT fibers was measured as shown in Figure 3b. A simple relationship was established between the specific

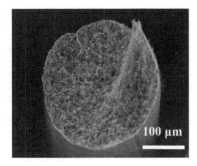

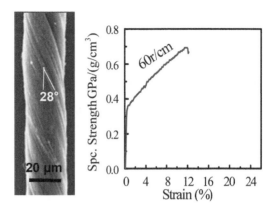

Figure 4.   a) SEM image of the surface of a twisted fiber, b) specific stress–strain curveof the twisted fiber.

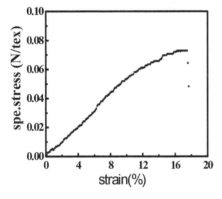

Figure 3.   a) Cross-sectionalimage of a solid CNT fiber, b) specific stress–strain curveof the solid CNT fiber.

stress and strain of the CNT fibers. The initial modulus of the CNT fibers was calculated to be 0.52 N/tex, much lower than that of the as-spun fiber (10.34 N/tex), indicating that a porous CNT structure resulted in a lower modulus compared to the dense CNT structure. What is the reason for this? According to Ma (2009), the modulus of a CNT web structure originates from the transformation of curved CNT bundles aligned with the fiber axis. This is probably because the structure with an enormous porosity has a smaller steric hindrance for the transformation of CNT bundles. Therefore, it is easier to change the structure of a porous CNT bundle than a dense CNT bundle. Furthermore, the typical porous CNT structure has a lower tensile strength (0.07 N/tex) than the dense CNT structure (0.39 N/tex), probably because the porous web structure causes a poor stress transfer than the dense CNT structure. Notably, the contact between the adjacent CNT bundles plays a key role in the stress transfer.

3.4   *Fiber with twisting treatment*

The as-spun CNT fibers with a hollow structure can be mechanically twisted to form twisted CNT fibers. The twisting was carried out by the uniform wisting of the as-spun CNT fibers, resembling the twisting of a paper roll. The twisted CNT fibers have a round cross-section with a diameter of ~30 μm. A typical twisted fiber is shown in Figure 4a with a twisted degree of 60 r/cm, and thus the corresponding twisted angle was 28°. The mechanical tensile property of the CNT fibers was measured, and the result is shown in Figure 4b.

Clearly, the curve exhibits a two-stage profile, similar to that of the as-spun CNT fibers. The plastic transformation caused by the sliding of the CNT bundles was also observed. However, the modulus of the twisted CNT fibers was 86.8 N/tex, much higher than that of the as-spun CNT fibers, indicating that the twisting significantly increased the modulus of CNT fibers. The twisting generated stress within the fiber, and the curved CNT bundles within the fiber were extended under the stress, thus reducing the extension space for the subsequent tensile measurement. Moreover, the specific stress (0.69 N/tex) of the twisted CNT fibers was higher than that of the as-spun CNT fibers (0.39 N/tex), probably because twisting increased the compactness of the CNT bundles within a fiber, thus increasing the stress transfer.

4   CONCLUSION

In this study, CNT fibers were synthesized by CVD, and the mechanical property was investigated. The results indicate that the CNT fibers with a porous structure have a lower specific stress and specific modulus than the CNT fibers with a dense structure, which originated from a large transformation of the CNT bundles within the CNT fibers and poor stress transfer, respectively. Moreover, the specific stress and specific modulus of CNT fibers could be increased by a twisting treatment because of the rearrangement of the CNT bundles within the fiber.

ACKNOWLEDGEMENT

Author is grateful to Prof. Ya-Li Li at Tianjin University for providing carbon nanotube fibers and also indebted to Liu Yakun for her support in experiments.

REFERENCES

Ericson LM, Fan H, & Peng HQ, et al. Macroscopic, neat, single-walled carbon nanotube fibers. Science 2004; 305(5689): 1447–1450.

Futaba DN, Hata K, & Yamada T, et al. Shape-engineerable and highly densely packed single-walled carbon nanotubes and their application as super-capacitor electrodes Nat. Mat. 2006; 5: 978–994.

Koziol K, Vilatela J, & Moisala A, et al. High-Performance carbon nanotube fiber. Science 2007; 318: 1892–1895.

Li QW, Li Y, Zhang XF, & Chikkannaavar SB et al. Structure-dependent electrical properties of carbon nanotube fibers. Adv. Mater. 2007; 19(20): 3358–3363.

Li Q, Kang YL, & QIU W, et al. Deformation mechanisms of carbon nanotube fibres under tensile loading by in situ Raman spectroscopy analysis. Nanotechnology, 2011; 22(22): 5704.

Li YL, Kinloch IA, & Windle AH, Direct spinning of carbon nanotube fibers from chemical vapor deposition synthesis. Science 2004; 304(5668): 276–278.

Lu WB, Zu M, & Byun JH, et al. State of the art of carbon nanotube fibers: opportunities and challenges. Adv. Mat. 2012; 24: 1805–1833.

Ma WJ, Liu LQ, & Yang R, et al. Monitoring a micromechanical process in macroscale carbon nanotube films and fibers. Adv. Mater. 2009; 21(5): 603–608.

Zhang M, Atkinson KR, & Baughman RH, Multifunctional carbon nanotube yarns by downsizing an ancient technology. Science 2004; 306(5700): 1358–1361.

Zhong XH, Li YL & Liu YK, et al. Continuous multilayered carbon nanotube yarns. Adv. Mater. 2010; 22(6): 692–696.

Advances in Materials Science, Energy Technology
and Environmental Engineering – Patty & Zhou (Eds)
© 2017 Taylor & Francis Group, London, ISBN 978-1-138-19668-1

In-situ DRIFTS study on photocatalytic reduction of CO_2 over Cu_2O-P25 composites

C.Y. Yan, W.T. Yi, J. Xiong & J. Ma
College of Chemistry, Chemical Engineering and Material Science, Zaozhuang University, Zaozhuang, China

ABSTRACT: Cu_2O-P25 (TiO_2) nano-heterostructures with different mass ratios were synthesized via a wet chemical precipitation and hydrothermal method, and were characterized by X-Ray Diffraction (XRD), Field-Emission Scanning Electron Microscopy (FESEM), UV-vis Diffuse Reflectance Spectra (DRS), Fourier-Transform Infrared Spectra (FTIR) and N_2 adsorption-desorption isotherms. In-situ Diffuse Reflectance Infrared Fourier Transform Spectroscopy (DRIFTS) study on absorption and photocatalytic reduction of CO_2 over Cu_2O-P25 was performed under UV-vis light. The weak signals of reaction intermediates such as formic acid, formaldehyde, and methoxy were found on the IR spectra, which may be due to the slow photocatalytic CO_2 reduction on the photocatalyst.

1 INTRODUCTION

The reduction of CO_2 for producing fuel like chemicals has attracted much attention under the current background of the depletion of fossil resources and the increase of CO_2 emissions. However, the activation of CO_2 is one of the biggest challenges in chemistry. In the long run, the photocatalytic reduction of CO_2 using solar energy, i.e. the artificial photosynthesis, is the most attractive route for converting CO_2 to fuels and chemicals.

Undoubtedly, TiO_2 is the traditional and most commonly used photocatalyst since 1972. However, a large band gap of TiO_2 (3.2 eV for anatase) restricts its use only to the narrow light-response range of ultraviolet. And the low quantum efficiency is another problem for TiO2, which is due to the recombination of electrons and holes in the photocatalytic process. In order to utilize the solar energy effectively, several approaches for TiO_2 modification have been proposed in the past decade, such as noble metal deposition, metal ions doping, semiconductor coupling, dye sensitization, etc. Among which, semiconductor coupling is an effective strategy to realize the efficient photocatalytic activity of TiO_2. In the heterostructures formed between two semiconductors, the photo-induced electrons and holes can be separated more easily and effectively due to the difference of energy levels of the two semiconductors, hence the promotion of the quantum efficiency of TiO_2. In recent years, the p-type Cu_2O with a band gap of 2.0 eV has been reported as a good candidate for visible light photocatalysis and exhibits huge potential for solar light utilization. Cu_2O-TiO_2 heterojunction

prepared as an efficient photocatalyst has attracted much attention (Cheng 2014, Han 2015). Cu_2O-TiO_2 heterostructures showed much higher degradation ability for organic waste water (Santamaria 2014, Tusui 2014), H_2 evolution in comparison with pure TiO_2 (Xi 2014, Li 2015). However, to the best of our knowledge, the preparation of Cu_2O/TiO_2 heterostructure by hydrothermal method has never been reported. In this paper, the hydroxylamine hydrochloride was used as the reductant to prepare Cu_2O/TiO_2 composites via a wet chemical precipitation and hydrothermal method. And photocatalytic reduction of CO_2 over Cu_2O-P25 was performed under UV-vis light.

2 EXPERIMENTAL

2.1 *Synthesis of the materials*

All reagents were of analytical grade and used as it without further purification. Cu_2O-TiO_2 composites were prepared via a wet chemical precipitation and hydrothermal method. In a typical procedure, $Cu(NO_3)_2 \cdot 2.5H_2O$ was firstly dissolved into 25mL distilled water. Then, a certain amount of P25 was dispersed in above solution under magnetic stirring for 30 min to obtain a uniform suspension. Subsequently, NaOH solution with two times the mole amount of $Cu(NO_3)_2 \cdot 2.5 H_2O$ was added into above suspension, and then the hydroxylamine hydrochloride was added as the reductant during strong magnetic stirring. The suspension was transformed into a 50 mL Teflon-lined autoclave after stirring for 2 h. The autoclave was sealed and maintained at 150 °C for 10h, and then cooled down to room temperature spontaneously. The pricipitate was collected and washed with distilled water 3 times by centrifugation. The solid composites were then dried under vacuum at 80°C. Samples were denoted as x-Cu_2O-P25, where x represents the mass ratio of Cu_2O (0%, 2% and 5%).

2.2 Sample characterization

The XRD patterns were recorded by a 6100 model X-ray diffractometer using Cu Kα radiation as the X-ray source (30 KV, 20 mA). The morphologies of the samples were characterized using a JSM-7800F scanning electron microscope. Nitrogen adsorption measurements were carried out at liquid nitrogen temperature with an ASAP 2020HD88 surface area and porosity analyzer. The diffused reflectance UV-visible spectra (DRS) of the samples were recorded by an UV-2401pc spectrometer with a diffuse reflectance accessory using $BaSO_4$ as the reference at room temperature. The FTIR spectra were recorded on a VERTEX 70 spectrometer.

2.3 In situ DRIFTS study

In situ Diffuse and Reflectance Infrared Fourier Transform Spectroscopy (DRIFTS) experiments were carried out using a VERTEX 70 spectrometer equipped with a liquid N_2 cooled MCT detector, and a three window DRIFTS cell. Two ZnSe windows allowed IR transmission, and a third (Quartz) window allowed the introduction of UV-vis light into the cell. CO_2 (99.99%) was used as received. Prior to illumination, the photocatalyst was fed into the cell, heated up to 120°C and maintained at this temperature for 30 min, followed by cooling down to 30°C in a dry He stream of 30 mL · min^{-1} in order to remove physic-sorbed water. Before recording a background spectrum of the catalyst, CO_2 (10 vol% in He, 20 mL · min^{-1}) was purged for 30 min. For experiments involving water vapor, CO_2 was bubbled through a saturator at room temperature, which added approximately 4 vol% water vapor to the CO_2 feed for saturation of the surface by exposure. A 100W Hg lamp (λ: 250~600 nm) was used as the light source. During illumination, reactants were held stationary in the cell at room temperature. In-situ IR signals were thus recorded in determined time intervals under UV/Vis light irradiation (100 Watt Hg lamp, λ: 250–600 nm), against a background of the pre-treated catalyst.

3 RESULTS AND DISCUSSION

Figure 1(a) shows the XRD patterns of Cu_2O-P25 composites with different mass ratios. The pure P25 and Cu_2O were also characterized for comparison.

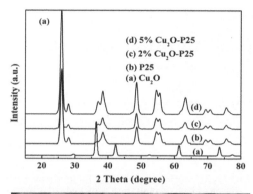

Figure 1.   (a) XRD patterns of P25, Cu_2O and Cu_2O-P25; (b) SEM image of 5% Cu_2O-P25.

In Figure 1(a), for Cu_2O, the diffraction peaks at 29.2°, 36.5°, 43.1°, 61.8°, 74.2°, 77.6° were assigned to (110), (111), (200), (220), (311), (222) planes of cubic Cu_2O. The P25 consists anatase and rutile phases of TiO_2, matching well with the specification of the product provided by the Degussa Company. For Cu_2O-P25 composites, the diffraction peak positions referred to the peaks of P25 except the peak at 36.5°, which is the characteristic peak of Cu_2O. The peak height at 36.5° increased gradually with the increase of mass ratio of Cu_2O. Moreover, the peak positions of TiO_2 move to the low angle slightly with the increase of Cu_2O content, indicating the change of the crystal lattice of P25.

The morphology of 5% Cu_2O-P25 measured by Scanning Electron Microscopy (SEM) is shown in Figure 1(b). The composites are composed of particles with sizes from few tens to several hundred nanometers. It is noticed that particles with uniform spheroidal shape are nearly homogeneously distributed. The sub-micro powder are partially covered with aggregated nano-particles with average agglomerate size about 70–100 nm.

The FTIR spectra of the samples are shown Figure 2(a). The composite samples show similar FTIR spectra with P25, indicating the similar

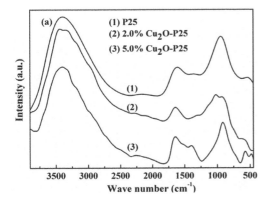

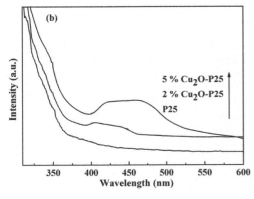

Figure 2. (a) FTIR spectra; (b) DRS spectra of Cu$_2$O-P25 samples.

The small twin-bands located in the range of 3550–3700 are the combined tones of CO$_2$ (g) and adsorbed CO$_2$, which may be rendered by the oxygen vacancy on the surface of the material. A strong adsorption band at 2400–2300 cm^{-1} represents the gas phase CO$_2$, which can be clearly reflected after adsorption of CO$_2$.

BET surface area and pore size distribution of 5% Cu$_2$O-P25 were measured by N$_2$ adsorption-desorption at liquid N$_2$ temperature. The isotherm of 5% Cu$_2$O-P25 was of typical type IV according to the IUPAC classification as shown in Figure 3(b), indicating the mesoporous structure of the composite. The BET area of 5% Cu$_2$O-P25 was measured to be 185.2 m^2/g, and the pore size distribution was in the range of 5–30 nm, with average pore size of 13 nm.

For photocatalytic reduction of CO$_2$, the background of the photocatalysts (P25 and 5% Cu$_2$O-P25) after adsorbing CO$_2$ and before illumination was collected. The time-profiled DRIFTS spectra during the illumination in 120min are shown in Figure 4. Clearly, we could not get new informa-

surface composition with P25. The broad bands around 3800–2500 cm^{-1} and 1650–1610 cm^{-1} correspond to the stretching vibrations of O-H bond and bending vibrations of strongly adsorbed water coordinated to Ti^{4+} on the surface of the samples respectively. The broad band at 950 cm^{-1} was assigned to the Ti-O-Ti bending mode. The main peaks at 650–550 cm^{-1} and 495–436 cm^{-1} were ascribed to absorption bands of Ti-O and O-Ti-O flexion vibration, respectively. The characteristic peak at 620 cm^{-1} corresponds to the Cu-O stretching vibration, which may be overlapped by the Ti-O absorption band.

Figure 2(b) shows the UV-vis diffuse reflectance spectra of P25, and Cu$_2$O-P25 composites. It is obvious that P25 just absorbs UV light with wavelength shorter than 400 nm. The Cu$_2$O-P25 composites show strong absorption both in the UV and visible light regions shorter than 500 nm, indicating the promoted light absorption ability of the composites.

FTIR spectra of 5% Cu$_2$O-P25 before and after thermal treatment/CO$_2$ adsorption against background spectra of KBr are shown in Figure 3(a).

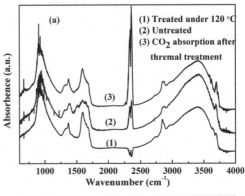

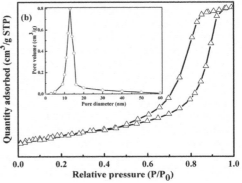

Figure 3. (a) FTIR spectra of 5% Cu$_2$O-P25 before and after thermal treatment/CO$_2$ adsorption; (b) Nitrogen adsorption–desorption isotherm and pore size distribution of 5% Cu$_2$O-P25.

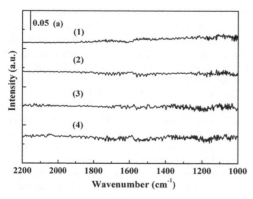

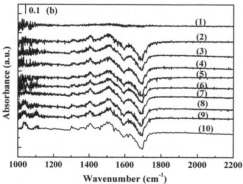

Figure 4. (a) Time-profiled DRIFTS spectra of P25 in 120 min of illumination, (1) 5 min, (2) 30 min, (3) 60 min, (4) 120 min; (b) Time-profiled DRIFT spectra of 5% Cu_2O-P25 in 120 min of illumination, (1) 20 s, (2) 5 min, (3) 15 min, (4) 30 min, (5) 45 min, (6) 60 min, (7) 75 min, (8) 90 min, (9) 105 min, (10) 120 min.

tion from the FTIR spectra for P25 in Figure 4(a), for the spectra nearly did not change at all during illumination, indicating that P25 did not have the activity to reduce CO_2 to fuel like chemicals.

However, for 5% Cu_2O-P25 composite in Figure 4(b), things changed and new peaks appeared during illumination, indicating the formation of new chemicals. The absorption band of bicarbonate (1675, 1417 cm^{-1}) and carbonate (1335 cm^{-1}) increased with increasing irradiation time. The band at 1620 cm^{-1} was assigned to the formic acid. The absorption band at 1505 cm^{-1} corresponds to the bending vibration $\delta(CH_2)$ of formaldehyde (H_2CO). Furthermore, weak absorption bands at 1106 and 1053 cm^{-1} are signed to formaldehyde and methoxy, respectively, indicating the formation of new chemicals by photocatalytic reduction of CO_2, which is consistent with the literature (Wu 2010). The formation of formaldehyde and formic acid may be due to the enhanced electron transfer rate by Cu_2O-P25 heterojunction, especially Cu^+ on the surface of the composites.

4 SUMMARY

Cu_2O-P25 nano-composites with different mass ratios were synthesized via a hydrothermal method. In-situ Diffuse Reflectance Infrared Fourier Transform Spectroscopy (DRIFTS) study on photocatalytic reduction of CO_2 over Cu_2O-P25 was performed under UV-vis light. The adsorbed CO_2 progressively converts to formic acid, formaldehyde, and methoxy on the surface of the composites. The formation of the fuel like chemicals may be due to the enhanced electron transfer rate by Cu_2O-P25 heterojunction, especially Cu^+ on the surface of the composites.

ACKNOWLEDGMENTS

This work was financially supported by Natural Science Foundation of Shandong Province (No. ZR2014BQ025), the Science and Technology Research Program of Shandong Province (No. 2012GSF11703), and Science and Technology Research Program of Zaozhuang City (No. 201435).

REFERENCES

Cheng W., T. Yu, & K. Chao (2014). Cu_2O-decorated mesoporous TiO_2 beads as a highly efficient photocatalyst for hydrogen production, *Chem. Cat. Chem.*, 6, 293–300.

Han T.H., D.M. Zhou, H.G. Wang & X.M. Zheng (2015). The study on preparation and photocatalytic activities of Cu_2O/TiO_2 nanoparticles. *J. Environ. Chem. Eng.*, 3(4), 2453–2462.

Li Y.P., B.W. Wang, S.H. Liu, X.F. Duan & Z.Y. Hu (2015). Synthesis and characterization of Cu_2O/TiO_2 photocatalysts for H_2 evolution from aqueous solution with different scavengers. *Appl. Surf. Sci.*, 324, 736–744.

Santamaria M., G. Conigliaro, F. Di Franco, & F. Di. Quarto (2014). Photoelectrochemical Evidence of Cu_2O/TiO_2 Nanotubes Hetero-Junctions formation and their Physicochemical Characterization. *Electrochim. Acta*, 144, 315–323.

Tsui L.K. & Zangari G. (2014). Modification of TiO_2 nanotubes by Cu_2O for photoelectrochemical, photocatalytic, and photovoltaic devices. *Electrochim. Acta*, 128, 341–348.

Wu J. & C. Huang (2010). In situ DRIFTS study of photocatalytic CO_2 reduction under UV irradiation. *Front. Chem. Eng. China* 4, 120–126.

Xi Z., C. Li, L. Zhang, M. Xing & J. Zhang (2014). Synergistic effect of Cu_2O/TiO_2 heterostructure nanoparticle and its high H_2 evolution activity. *Int. J. Hydrogen Energy*, 39, 6345–6353.

*Advances in Materials Science, Energy Technology
and Environmental Engineering – Patty & Zhou (Eds)*
© 2017 Taylor & Francis Group, London, ISBN 978-1-138-19668-1

Influence of TiO_2 on drag reduction and shear degradation resistance of polyolefin in diesel

Xiaodong Dai & Huanrong Liu
Shengli College, China University of Petroleum, Dongying, Shandong, China

Yanzong Zhang
Sichuan Agricultural University, Chengdu, Sichuan, China

Jingmiao Li
PetroChina Pipeline Company, Langfang, Hebei, China

ABSTRACT: Polyolefin as a drag-reducing agent with long molecular chains would lose most of the drag-reducing capacity caused by shear degradation. Nanocomposites combining many unique physical and chemical properties showed a possible way to develop a DRA with better performance. The potential of polyolefin/TiO_2 nanocomposite as a DRA were investigated in our research. Surface-modified TiO_2 with organic chains could improve the interaction with the polymer. Drag reduction and shear degradation resistance were analyzed using a diesel experimental apparatus. It was shown by drag reduction and degradation test that a nanocomposite DRA presented significant improvement for shear degradation resistance, and more than 40% DR capacity was left after shear degradation. In addition, SCA chains had influence on performance, and the longer chains had influence on TiO_2 particle surface, the better shear resistance, caused by the stronger interaction between nanoparticles and polymer chains. Positive synergistic effect reinforced the application property after combination of nano TiO_2 and polyolefin.

1 INTRODUCTION

Drag reduction in flow by a Drag-Reducing Agent (DRA) had generated great interest because of energy saving and flow increase in engineering practice. Oil production, transportation in pipeline, heating, and cooling system had widely used a DRA to reduce the pumping power and increase the piping capacity. Polymer, surfactants, fibers, and micro-bubbles could be used as DRA in different solvent system (Abubakar A, 2014). In this paper, polyolefin commonly applied in oil pipeline as a DRA was discussed. By decades of development, polyolefin with more than $6*10^6$/m molecular weight, narrow MW distribution, and parts of branch could be more suitable for better performance (Guo, 2013; Witold Brostow,2013). While the major problem that affects the effectiveness of DRA during turbulent flows was the mechanical degradation. It would occur under high shear force by the interactions of turbulence or passing through a centrifugal pump. The long polymer chains were broken out permanently, and lost most of the drag-reducing rate. Therefore, it should add fresh DRA station by station to maintain the drag reduction rate (Li, 2013; Witold

Brostow, 2008). Nanocomposites were a kind of composited material and first put forward in the 1980s. Moreover, polymer-based nanocomposites referred to the nanocomposite of polymer as the matrix, and the nanomaterial as the dispersed phase. Nanomaterials with large surface area presented a specific surface effect. Therefore, a strong interface effect could be easily formed between polymers and nanomaterials, and could give themselves outstanding mechanical performance, and thermal, electrical, and other properties (Xu, 2010; Tan, 2007). Because of the mechanical degradation of polyolefin, we had reported the effect of adding nanoparticle SiO_2 to improving the anti-shearing performance, and shown better shear resistance than polyolefin, in our former work(Dai, 2016).

In this work, we evaluated the potential of polyolefin-based nano TiO_2 composites with better shear resistance ability. By modifying the surface properties of TiO_2, it could reinforce the distribution of nanoparticle in polymer matrix. The influences of different coupling agents on surface modification and TiO_2 content were studied by testing diesel circuit. The drag reduction and shear resistance performance were also evaluated.

2 EXPERIMENTAL

2.1 *Preparation of nanocomposites*

First, anatase nano TiO_2 with a surface area of 70 m^2/g, average particle size of 30 nm (Fig. 1), was dried at 120°C for 24h. Then, nano TiO_2 was dispersed in ethanol with ultrasonic wave for 30mins. A certain amount of a silane-coupling agent (SCA) was dissolved in ethanol with intensive stirring, and, if required, deionized water was added for prehydrolysis. Silane solvent was slowly added into a TiO_2/ethanol beaker, keeping at 75–80°C for 5 h with reflux for sufficient modifying reaction. After that, the product was purified by centrifugation to remove the by-products and unreacted materials. Then, the power product was dried at vacuum for 10h, that was surface modification nano TiO_2. The SCA used in experiments were KH-550, KH-560, KH-570, and A-172.

Polyolefins and ethanol were under ultrasonic wave for 0.5 h. Then, the dried maleic anhydride (at 80°C for 10 h) was added into the solution, and then under ultrasonic wave for 10–30mins, and the grafted polyolefins with anhydride were obtained as the interfacial compatibilizer.

The above as-obtained compatibilizer was added to polyolefin slurry, which was then added to the modified nano TiO_2. The slurry was slowly stirred at 80°C for 2 h, and the composite DRA was obtained.

2.2 *Experiment loop test*

The experimental apparatus to evaluate the performance of DRA is shown in Figure 2. 0# diesel was used as a test fluid. Temperature was controlled by an air conditioner at 25°C. DRA was first

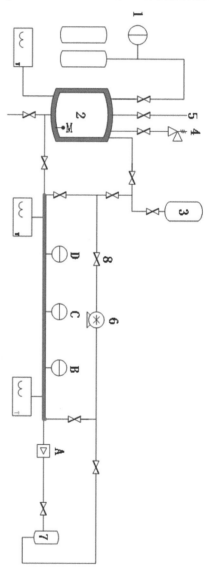

Figure 2. Experimental loop for the performance test of DRA. 1, N2; 2, pressure surge tank; 3, dilution tank; 4, safety relief valve; 5, air relief valve; 6, centrifuge pump; 7, reflux accumulator; 8, ball valve; A, flow sensor; B,C,D, pressure sensors.

Figure 1. Anatase nano TiO_2 particle used in the experiment.

dissolved in diesel at a concentration of 1/1000, and then added into the diesel tank(2). Condensed N_2 was used as power to maintain the flow pressure. Different kinds of DRA with different modified TiO_2 contents, and different TiO_2 contents were investigated in this study. The calculation procedure is as follows:

1. Measuring the flow rate before (Q_0) and after (Q_1) addition of DRA
2. Calculating the flow increase rate: $TI = (Q_1 - Q_0)/Q_0 \times 100\%$
3. Calculating the drag-reducing rate: $DR = [1 - 1/(1 + TI)^{1.79856}] \times 100\%$
4. Investigating the shearing degradation resistance performance: The DR results from the unsheared and sheared DRA, and unsheared and sheared composite DRA were plotted, and the shearing degradation resistance was evaluated according to the tendency and slope of the curve.

3 RESULTS AND DISCUSSION

3.1 *Influence of silane coupling agent*

Four kinds of SCA were used for surface modification of nano TiO_2. The flow increase rate and drag-reducing rate were investigated as shown in Figure 3.

It has been observed that synthesized the polymer nanocomposite presented slightly higher DR and flow increase effects than the commonly used polyolefin. Through the synthesis process, the modified TiO_2 was not dispersed in polyolefin slurry, but in its diesel solution. With the recombination reaction, polyolefin would dissolve well, and present slightly good performance. As different SCAs have different lengths of chains, it would influence the interaction with the polymer chain. Therefore, the longer the chain, the better the performance.

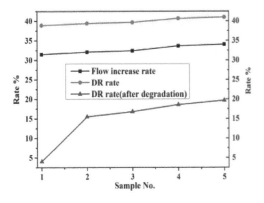

Figure 3. Influence of SCAs on DR performance.

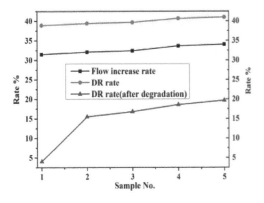

Figure 4. Combination between modified TiO_2 and polyolefin chains.

After shear degradation by pump, it was found that polyolefin left about 10% DR capacity. While, by combination of TiO_2 with polyolefin, it presented a greater increase in the shear resistance capacity.

The combination between modified TiO_2 and polymer chain is shown in Fig. 4. It has been observed that modified TiO_2 particles with organic chain on surface would interact with polyolefin chains and reinforce the rigidity of polymer chain. Therefore, under shear stress, more DRA could remain and perform the DR effect. Moreover, the longer the organic chain on the TiO_2 surface, the better the shear resistance. It was thought that longer chain functional groups had more interaction points with the polymer chain. Sample 5 has the best drag reduction, flow increase, and shear resistance performance. In the next section, SCA of KH-570 was selected to modify the TiO_2 surface property.

3.2 *Influence of nano TiO_2 content*

From the previous section, it has been concluded that TiO_2 with SCA modification could improve the shear resistance of DRA. The influence of TiO_2 content is discussed in this section. TiO_2 contents were 0.5%, 1%, 2%, and 4% (w). All samples undertook centrifugal pump mechanic shear for one time. It is shown in Fig. 5, that polyolefin rarely had shear resistance capacity. With the increase in TiO_2 content to 1%, the shear resistance of the polymer was enhanced by the combination effect. After that, with the increase in TiO_2 content from 2% to 4%, there was no significant influence on the shear resistance of the polymer. It could conclude that 1% TiO_2 modified by KH-570 processed the best effect.

All results indicated that DRA from nanoparticles mixed with polyolefin composites had a

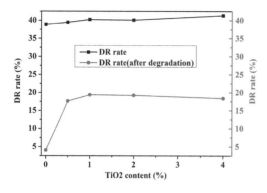

Figure 5. Influence of TiO_2 content on DR performance.

much better shear degradation resistance, due to the TiO_2-polymer matrix synergetic effect. Similarly, our former reports [8,9] about modified nano SiO_2 combined with polymer had demonstrated the potential of shear degradation resistance. Our studies demonstrated that nanoparticles with a modified surface combined with polyolefin matrix would be a promising way to improve the DR and shear resistance. In addition, it needs further research to strength nanoparticle dispersion and improve application property.

4 CONCLUSIONS

From the results, it can be observed that the combination of modified nano TiO_2 and polyolefin could not obviously enhance the drag-reducing effect and flow increase performance. Only slight improvement of application properties could attribute to efficient dissolution. While, for the nanocomposite properties, many researches had reported the synergetic mechanism. In our study, surface modified TiO_2 by SCA combined with polyolefin can significantly improve the shear degradation resistance. After one-time degradation by pump, more than 40% of DR capacity remained. Nano TiO_2 was hydrophilic, while polyolefin was hydrophobic. Hence, the anhydride-grafted polyolefin was selected as an interfacial compatibilizer, surface

modified TiO_2 combined with organic chains shows hydrophobic property, which improves its comparability with polyolefin, and thereby increasing the TiO_2-polyolefin interfacial binding strength. Meanwhile, the longer the organic chains on TiO_2 surface, the better the shear degradation resistance, because of more interaction between organic chains and polymer chains. Through the analysis of the TiO_2 content, the sample with 1% TiO_2 shows the best performance. Adding more TiO_2 could cause aggregation of particles, and reduce the interaction. Finally, based on the synergistic effect of TiO_2 and polymer matrix, nanocomposite DRA shows excellent shear degradation resistance performance.

REFERENCES

Abubakar A., Al-Wahaibi T., Al-Wahaibi Y., et al. 2014. Roles of drag reducing polymers in single—and multiphase flows. *Chemical engineering research and design, 92:2153–2181.*

Dai Xiaodong, Li Guoping, Guo Xu, et al. 2015. Potential of SiO_2 nanocomposite as DRA of oil. International Conference on Materials Engineering and Information Technology Applications. *925–928.*

Dai Xiaodong, Guo Xu, Li Guoping, et al. 2016. Preparation and performance test of nanocomposites as drag reducing agent for oil. *Oil & Gas storage and transportation, in press.*

Guo Xu, Dai Xiaodong, Song Linhua, et al. 2013. Analysis on polymeric nano-composites as drag reduction agent for oil. *Oil & Gas storage and transportation, 32(10):1037–1042.*

Li Guoping. 2006. Theory research and application technology of DRA. *Shandong, Shandong University.*

Tan Xiumin, 2007. Functionalization of polyolefin and nanocomposites. *Tianjin, Tianjin University.*

Witold Brostow, Hamide Ertepinar, R. P. Singh. 1990. Flow of dilute polymer solutions: Chain Conformations and degradation of drag reducers. *Macromolecules, 23:5109–5118.*

Witold Brostow, 2008. Drag reduction in flow: Review of applications, mechanism and prediction, *Journal of industrial and engineering chemistry, 14:409–416*

Xu Lixin. 2010. Design, Preparation and Application of polyolefins-modified nanoparticles. *Doctoral thesis of Zhejiang University.*

*Advances in Materials Science, Energy Technology
and Environmental Engineering – Patty & Zhou (Eds)
© 2017 Taylor & Francis Group, London, ISBN 978-1-138-19668-1*

Adsorption of aqueous heavy metal ions by a poly(*m*-phenylenediamine)/attapulgite composite

Y.B. Ding, J.J. Wang & Y.C. Xu

*Department of Polymer and Coating, College of Chemistry and Chemical Engineering,
Jiangxi Science and Technology Normal Universitys, Nanchang, P.R. China*

ABSTRACT: A novel poly(*m*-phenylenediamine)/attapulgite (P*m*PD/ATP) composite was synthesized and investigated for adsorption of three kinds of heavy metal ions (Cu(II), Cr(IV), and Pb(II)) from aqueous solutions. The effects of *m*PD/ATP mass ratio on yield and copper ion adsorption were investigated to evaluate the removal capacity of three kinds of heavy metal ions. Then, adsorption time and adsorption kinetics were studied. The results indicated that at least 96. 1% of heavy metal ions could be removed from the solution by P*m*PD/ATP composite when *m*PD/ATP mass ratio is 1:0.6. Kinetic study indicated that the three kinds of heavy metal ions adsorption by P*m*PD/ATP composite fitted a pseudo-second-order kinetic model, indicating that the adsorption process of three kinds of heavy metal ions was predominantly controlled by a chemical process.

1 INTRODUCTION

With the rapid development of modern industry and agriculture, water pollution caused by toxic heavy metal ions (such as Cu, Cr, Pb, Hg, etc.) has become a serious environmental problem. These toxic metal ions, even at low concentrations, have deteriorated water resources and easily accumulated in the human body throughout the food chain, causing a variety of diseases and disorders (Samiey, 2014). Therefore, the removal of heavy metal ions from wastewaters has been one of the urgent environmental issues. Different methods have been developed for the removal of toxic metal ions (Fu, 2011).

Adsorption is a well-known separation method and has been recognized as one of the efficient and economic methods for wastewater treatment. Recently, many polymer adsorbents have been used for the treatment of wastewater containing heavy metal ions due to its simplicity and convenience (Cui, 2012; Wang, 2016; Liao, 2014; Huang, 2014). Among these polymers, poly(*m*-phenylenediamine) (P*m*PD) is used as nitrogen atoms of P*m*PD have a high affinity for heavy metal ions and the cost is low (Huang, 2006). However, P*m*PD particles are easily aggregated, which results in relatively low surface area, therefore limiting its adsorption capacity and practical application.

Attapulgite (ATP, or palygorskite) is a crystalline hydrated magnesium silicate absorbent mineral with a fibrous morphology. It has a large specific surface area (Chen, 2007). To reduce costs and improve the comprehensive water-absorbing properties of the material, fabricating a composite that consists of polymer and attapulgite micropowder was considered a priority.

In this study, a poly(*m*-phenylenediamine)/attapulgite (P*m*PD/ATP) composite was synthesized and used to remove Cu(II), Cr(IV) and Pb(II) from aqueous solutions. A series of batch experiments were conducted to investigate the adsorption of Cu(II), Cr(IV), and Pb(II) with P*m*PD/ATP composite adsorbent. The pseudo-first order and the pseudo-second order kinetic models have been determined.

2 EXPERIMENTAL METHODS

2.1 Materials and reagents

Ammonium persulfate, *m*-phenylenediamine, copper nitrate, potassium dichromate, lead nitrate, bis(cyclohexanone)oxalyl dihydrazone, acetone, phenylanilineurea, disodium EDTA, absolute ethyl alcohol and ammonia water were purchased from Sinopharm Chemical Reagent Co., Ltd, concentrated H_2SO_4 was purchased from Xilong Chemical Co., Ltd, attapulgite was purchased from Changzhou new Mstar Technology Ltd, and all reagents and materials were used as received without further purification. All reagents used in this study were of analytical grade.

2.2 Preparation of PmPD/ATP composite

The PmPD/ATP composite was synthesized by oxidative polymerization. The procedure described below is a modification of the procedure described by Huang et al. (Huang, 2006). ATP (0.5585 g) and mPD (3.2442 g) were dispersed in 240 mL of H_2O, and then 60 mL of 0.5 mol/L $(NH_4)_2S_2O_8$ solution was slowly added to the mixture under vigorous stirring over a period of 6 h at 0°C. Finally, the resulting PmPD/ATP was separated and washed with ultrapure water and vacuum-dried at 323 K until reaching a constant weight. The resultant PmPD/ATP composite was powdered in a mortar and stored.

2.3 Adsorption experiments

Adsorption experiments were conducted in batch equilibrium mode. All experiments were conducted by mixing 20 mL of heavy metal ions solutions with 0.05 g of the PmPD/ATP composite in a 50 mL Erlenmeyer flask. The Erlenmeyer flask was then transferred to an incubator shaker and vibrated at 150 rpm for 1 h to ensure the equilibrium adsorption. For kinetic studies, 100 mg/L was chosen as the initial concentration of heavy metal ions solution.

After adsorption, the mixture was filtered with a syringe filter of 0.45 μm. Heavy metal ions concentrations in the supernatants were determined by a UV spectrometer (TU-1810PC, Beijing Purkinje General Instrument Co., Ltd, China). The adsorbed amounts were then calculated by the following equation:

$$A_e = (C_0 - C_e)/C_0 \qquad (1)$$

$$Q_e = (C_0 - C_e)V/M \qquad (2)$$

where A_e is the adsorptivity of heavy metal ions on PmPD/ATP composite, C_0 is the initial concentration of heavy metal ions (mg/L), and C_e is the equilibrium concentration of heavy metal ions in solution (mg/L).

Q_e is the adsorption capacity of heavy metal ions on PmPD/ATP composite (mg/g), m is the mass of adsorbent used (g), and V is the volume of heavy metal ions solution (L).

3 RESULTS AND DISCUSSION

3.1 Effect of mPD/ATP mass ratio on yield

Figure 1 gives the results of mPD/ATP mass ratio on yield. From figure 1, it can be observed that the yield is the lowest for mPD/ATP mass ratio which is 1/0, as mPD is not completely converted into

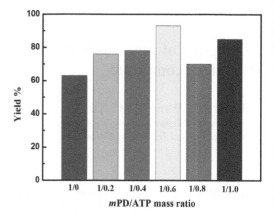

Figure 1. Effect of mPD/ATP mass ratio on yield.

PmPD (Huang, 2006). Yield increases as mPD/ATP mass ratio decreases from 1/0.2 to 1/0.6, as high surface area of ATP is favorable to improve the yield. However, when the mass ratio of mPD/ATP is 1/0.8 and 1/1.0, there is a slight decline on the yield, mainly because excessive ATP prohibits the polymerization reaction of mPD.

3.2 Effect of mPD/ATP mass ratio on copper ion adsorption

Figure 2 shows the results obtained from mPD/ATP mass ratio on copper ion adsorption. As can be seen from Figure 2, the adsorptivity is the lowest for ATP (81.99%), while the adsorptivity is almost the same for other PmPD/ATP composites (≥ 94.98%), and this is attributable to PmPD with many nitrogen atoms having a high affinity for copper ion (Huang, 2012).

3.3 Effect of adsorption time on heavy metal ion adsorption

For the yield and adsorptivity reasons, PmPD/ATP composite with mPD/ATP mass ratio of 1/0.6 has been applied to study the adsorptivity with time. As can be seen from Figure 3, the composite has a high adsorptivity and a high rate of adsorption for the three kinds of heavy metal ions (Cu(II), Cr(IV), and Pb(II)), which provided evidence for the adsorption mechanism of complexation between Cu(II), Cr(IV), and Pb(II) and functional groups (Cui, 2012). The adsorptivity of Cu(II), Cr(IV), and Pb(II) maintained at a constant level after 50 minutes contact time. The quick adsorption process is favorable for the application of PmPD/ATP in the removal of Cu(II), Cr(IV), and Pb(II) from large volume of solutions.

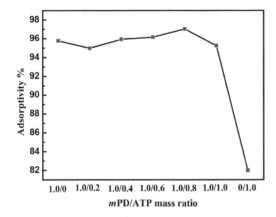

Figure 2. Effect of mPD/ATP mass ratio on copper ion adsorption.

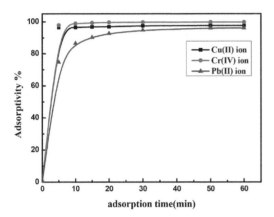

Figure 3. Effect of adsorption time on heavy metal ion adsorption.

3.4 Adsorption kinetic

In order to evaluate the controlling mechanism of adsorption processes, pseudo-first-order and pseu-do-second-order kinetic equations were used to test the experimental data. For pseudo-first-order kinetics, the adsorption process can be described by Largergren's rate equation (Largergren, 1898):

$$\ln(Q_e - Q_t) = \ln Q_e - k_1 t$$

The pseudo-second-order kinetics (Ho, 1999) based on adsorption capacity can be expressed as follows:

$$t / Q_t = 1 / k_2 Q_e^2 + t / Q$$

where Q_e (mg/g) is the adsorption amount, Q_t (mg/g) is the adsorption amount at time t, and k_1 is the equilibrium rate constant of the pseudo-first-

order adsorption (min^{-1}), which is determined from the slope of plot of $\ln(Q_e - Q_t)$ versus t (Fig. 4a). k_2 is the equilibrium rate constant of the pseudo-second order adsorption($g\ mg^{-1}\ min^{-1}$), which is similarly determined from the slope of plot of t / Q_t versus t (Fig. 4b). Simulated parameters based on pseudo-first-order and pseudo-second-order kinetic equations are given in Table 1.

As illustrated in Table 1, the correlation coefficient values of the pseudo-first-order model are low and the calculated equilibrium adsorption capacity does not agree with the experimental values, indicating that the pseudo-first-order kinetic model poorly fist the adsorption processes of PmPD/ATP composite for three kinds of heavy metal ions. However, for pseudo-second-order kinetic equation, the correlation coefficients are found to be higher than 0.99, and its calculated equilibrium adsorption capacities fit well with the experimental data. These suggest that the adsorption data are well represented by pseudo-second-order kinetics, which indicates that the adsorption process was controlled by chemical complexation (Cui, 2012).

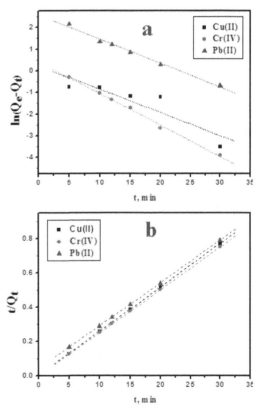

Figure 4. Kinetic models of three kinds of heavy metal ions adsorbed onto PmPD/ATP. (A) Pseudo-first-order model and (B) Pseudo-second-order model.

Table1. Fitting parameters of three kinds of heavy metal ions adsorbed onto PmPD/ATP using the pseudo-first-order and pseudo-second-order kinetic models.

Heavy metal ions	Q_e(exp)(mg/g)	First-order kinetics				Second-order kinetics			
		k_1 (1/min)	R_1^2		Q_e(calc) (mg/g)	k_2 (g/(mg min))	R_2^2		q_e(calc) (mg/g)
Cu(II)	39.04	0.10749	0.89822		1.27	0.2131	0.9999		39.09
Cr(IV)	39.89	0.1461	0.99816		1.51	0.2122	1.0000		40.03
Pb(II)	38.43	0.1107	0.99522		12.81	0.0151	0.9999		40.06

4 CONCLUSIONS

In this study, PmPD/ATP composite was prepared by in situ oxidative polymerization and three kinds of heavy metal ions(Cu(II), Cr(IV), and Pb(II)) adsorbed onto the composite adsorbent were investigated. The yield of PmPD/ATP composite is up to 93% and at least 96.1% of heavy metal ions could be removed from the solution by PmPD/ATP composites when the mPD/ATP mass ratio is 1:0.6. The kinetic study indicated that the three kinds of heavy metal ions adsorbed by PmPD/ATP composite fitted a pseudo-second-order kinetic model, indicating that the adsorption process of the three kinds of heavy metal ions was predominantly controlled by a chemical process. The present results imply that the PmPD/ATP composite is a suitable candidate for the adsorptive removal of three kinds of heavy metal ions in wastewater treatment.

REFERENCES

Chen, H. & Wang, A. (2007). Kinetic and isothermal studies of lead ion adsorption onto palygorskite clay. J Colloid Interf Sci. 307, 309–316.

Cui, H., Qian, Y., Li, Q., Zhang, Q. & Zhai, J. (2012). Adsorption of aqueous Hg (II) by a polyaniline/attapulgite composite. Chem. Eng.J. 211, 216–223.

Fu, F. & Wang, Q. (2011). Removal of heavy metal ions from wastewaters: a review. J Environ Manage. 92, 407–418.

Ho, Y.S. & McKay, G. (1999). Pseudo-second order model for sorption processes. Process Biochem. 34, 451–465.

Huang, M., Peng, Q. & Li, X. (2006). Rapid and effective adsorption of lead ions on fine poly (phenylenediamine) microparticles. Chem-Eur J. 12, 4341–4350.

Huang, M.R., Lu, H.J. & Li, X.G. (2012). Synthesis and strong heavy-metal ion sorption of copolymer microparticles from phenylenediamine and its sulfonate. J Mater Chem. 22, 17685–17699.

Huang, S., Min, C., Liao, Y., Du, P., Sun, H., Zhu, Y. & Ren, A. (2014). Intrinsically conducting polyaminoanthraquinone nanofibrils: interfacial synthesis, formation mechanism and lead adsorbents. RSC Adv. 4, 47657–47669.

Largergren, S. (1898). Zur theorie der sogenannten adsorption geloster stoffe. Kungliga Svenska Vetenskapsakademiens. Handlingar. 24, 1–39.

Liao, Y., Cai, S., Huang, S., Wang, X. & Faul, C.F.J. (2014). Macrocyclic Amine-Linked Oligocarbazole Hollow Microspheres: Facile Synthesis and Efficient Lead Sorbents. Macromol Rapid Comm. 35, 1833–1839.

Samiey, B., Cheng, C.H. & Wu, J. (2014). Organic-inorganic hybrid polymers as adsorbents for removal of heavy metal ions from solutions: A review. Materials. 7, 673–726.

Wang, X., Lv, P., Zou, H., Li, Y., Li, X. & Liao, Y. (2016). Synthesis of Poly(2-aminothiazole) for Selective Removal of Hg(II) in Aqueous Solutions. Ind Eng Chem Res. 55, 4911–4918.

*Advances in Materials Science, Energy Technology
and Environmental Engineering – Patty & Zhou (Eds)*
© 2017 Taylor & Francis Group, London, ISBN 978-1-138-19668-1

Effect of Mg content on the microstructure and properties of a Zn-Al-Mg alloy

Xiaodong Hao, Chongfeng Yue, Sheming Jiang, Qifu Zhang, Yanan Bai & Lingjun Li
China Iron and Steel Research Institute Group, Beijing, China

ABSTRACT: For the development of Zn-Al-Mg alloy coating, the microstructure and corrosion resistance of Zn-Al-Mg alloy with different contents of Mg (1~3 wt%) were researched. The corrosion resistance of experimental Zn-Al-Mg alloy was 2 to 3.3 times of the alloy with GI coating. The corrosion resistance and hardness of Zn-Al-Mg alloy both increased with the increase in the Mg content. The solidification structure of Zn-Al-Mg alloy was mainly composed of Zn-rich phase, Mg_2Zn_{11} phase, and Al-rich phase; part of the Zn-rich phase and Mg_2Zn_{11} phase form binary eutectic phase, and part of the Zn-rich phase, Mg_2Zn_{11} phase and Al-rich phase form ternary eutectic phase. With the increase in Mg content, the Mg_2Zn_{11} phase was significantly increased.

1 INTRODUCTION

Steel is the largest output and the most widely used metallic material in the world. The steel material has a variety of excellent performances such as good toughness, high strength, and cost effectiveness, but its biggest drawback is its susceptibility to corrosion in use. Hot-dip galvanizing is the most cost-effective way to protect steel, and 10% to 12% of the steel is protected through galvanizing in the world (Yan, 1996; Lu, 1997; Hideloshi Shindo, 1999; Nishimura K, 2003; Rincon O). In recent years, due to the shortage of zinc, Zn-Al-Mg alloy coating with excellent corrosion resistance attracted widespread attention. The corrosion resistance of the coating can be significantly improved with the addition of Mg (Dutta M, 2010; Liang, 1997; Notowidjojo, 1998). The microstructure and corrosion resistance of alloy coating are determined by the coating composition, the alloy coating and alloy ingot with same components having similar tissue compositions. The corrosion resistances of alloy coating and alloy ingot have consistency by influence of its composition. Zn-Al-Mg alloy coatings were studied preliminary by analyzing the organization and performance of Zn-Al-Mg alloy with different Mg contents in this paper (Liang, 1998; Eliot, 1993; Liu, 2005).

2 MATERIALS AND METHODS

The performance of Zn-Al-Mg alloy ingots with 1.5 wt% Al, 1~3 wt% Mg were compared with GI coating (Zn–0.2 wt% Al), ZAM coating (Zn-6 wt% Al-3 wt% Mg), and Super Dyma coating (Zn-11 wt% Al-3 wt% Mg) alloy ingots. Experimental alloy composition is shown in Table 1.

The ingots were cut into cylindrical pellets as Φ44 × 5 mm. The edges of NSS samples were sealed as shown in Figure 1.

The FEI Quanta 650 Scanning Electron Microscope (SEM) and Energy-Dispersive Spectroscopy (EDS) were used for alloy organization

Table 1. Composition of experimental Zn-Al-Mg alloy ingot (wt%).

Element	Zn0.2 Al	1.5 Al 1 Mg	1.5 Al 1.5 Mg	1.5 Al 2 Mg
Al	0.2	1.5	1.5	1.5
Mg	0	1	1.5	2
Zn	Bal.	Bal.	Bal.	Bal.

Element	1.5 Al 2.5 Mg	1.5 Al 3 Mg	ZAM	SD
Al	1.5	1.5	6	11
Mg	2.5	3	3	3
Zn	Bal.	Bal.	Bal.	Bal.

Figure 1. Zinc-Aluminum-Magnesium alloy samples.

observation and EDS analysis. Philips Analytical X'Pert Primped X-Ray Diffraction (XRD) was used for analyzing the phase structure of the alloy. FQY-025 salt spray test box was used for analyzing neutral salt spray test. HVS-1000Z using a micro Vickers hardness tester was used for metal hardness testing, and the loading force is 100 g and holding time is 15 s.

3 EXPERIMENTAL RESULTS AND ANALYSIS

3.1 *Effect of Mg on the corrosion resistance of Zn-Al-Mg alloy*

Different Mg contents of the zinc-aluminum-magnesium alloy samples were tested by the neutral salt spray test, and samples of alloy of GI, ZAM, and SD coating composition were used as references. The ingot surface morphology after different times of neutral salt spray test is given in Table 2. With the increase in neutral salt spray corrosion time, the surface white rust of ingot alloy with each component increased. After 1440 h (60 days) of the neutral salt spray test, the alloy ingot surface of GI coating composition had the most white rust.

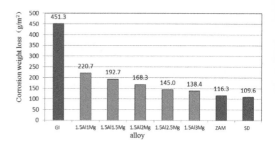

Figure 2. Mass loss after the neutral salt spray test (60 days) by corrosion of Zinc-Aluminum-Magnesium alloy with different Mg contents.

After 60 days of the neutral salt spray test, the corrosion products on the surface of the alloy ingots were removed, and the mass loss was calculated; the results are shown in Figure 2. Under the neutral salt spray test conditions, the zinc-aluminum-magnesium coating composition alloy having better corrosion resistance than the GI composition alloy; the corrosion resistance of experimental zinc-aluminum-magnesium coating composition alloy was 2.0 to 3.3 times of the GI coating composition alloy. The corrosion resistance of the alloy increased with the increase in Mg content, and ZAM and SD composition alloy has better corrosion resistance.

3.2 *Effect of Mg on the hardness of Zn-Al-Mg alloy*

The wear resistance and formability were affected by the hardness of zinc-aluminum-magnesium coating hardness. In order to study the influence of Mg content on the hardness of the zinc-aluminum-magnesium alloy, the hardness of various Mg content zinc-aluminum-magnesium alloy was tested. Vickers hardness test results are shown in Figure 3. The Al content is 1.5 wt% unchanged, as the Mg content is 1.5 wt%. Zinc-aluminum-magnesium vertical sectional phase diagrams are shown in Figure 4. The final phase of zinc-aluminum-magnesium alloy was constituted by Zn-rich phase, Al-rich phase, and Mg_2Zn_{11} phase. There is a eutectic point at 341.2 °C, and the eutectic reaction was reacted the alloy composition of Zn-1.5 wt% Al-0.6 wt% Mg.

$$L \leftrightharpoons (Zn) + (Al) + Mg2Zn11 \qquad (1)$$

3.3 *Solidification structure of Zn-Al-Mg alloy*

The SEM and EDS analysis of Zn-1.5 wt% Al-1.5 wt% Mg alloy's balance solidification structure are shown in Figure 5, and the XRD analysis of the alloy's balance solidification structure is shown

Table 2. The surface morphology of Zinc-Aluminum-Magnesium alloy after the neutral salt spray.

Component	0h	240h	720h	1440h
GI				
1.5Al1Mg				
1.5Al1.5Mg				
1.5Al2Mg				
1.5Al2.5Mg				
1.5Al3Mg				
ZAM				
SD				

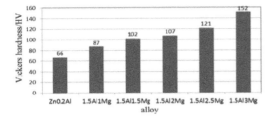

Figure 3. Comparison of the hardness of Zinc-Aluminum-Magnesium alloy with different Mg contents.

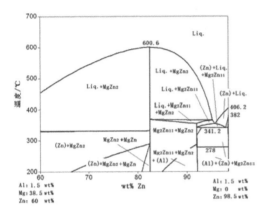

Figure 4. Zinc-Aluminum-Magnesium vertical sectional phase diagram.

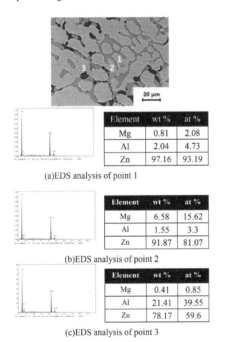

(a)EDS analysis of point 1

Element	wt %	at %
Mg	0.81	2.08
Al	2.04	4.73
Zn	97.16	93.19

(b)EDS analysis of point 2

Element	wt %	at %
Mg	6.58	15.62
Al	1.55	3.3
Zn	91.87	81.07

(c)EDS analysis of point 3

Element	wt %	at %
Mg	0.41	0.85
Al	21.41	39.55
Zn	78.17	59.6

Figure 5. SEM and EDS analysis of Zn-1.5 wt% Al-1.5 wt% Mg alloy.

in Figure 6. It can be observed from figure 5 that the white striped phase is a Zn-rich phase, the gray phase is Mg_2Zn_{11} phase, and the lack aggregation phase is Al-rich phase.

The zinc-aluminum-magnesium alloy microstructure with Mg contents of 1 wt%, 1.5 wt%, 2 wt%, 2.5 wt%, and 3 wt% were comparatively researched. The SEM morphology of zinc-aluminum-magnesium alloy with different Mg contents is shown in Figure 7. When the Mg content is 1 wt%, the the alloy is mainly composed of Zn-rich phase and a pale gray Mg_2Zn_{11} phase, and small amount of black massive Al-rich phase enriched

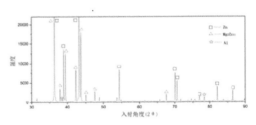

Figure 6. XRD analysis of Zn-1.5 wt% Al-1.5 wt% Mg alloy.

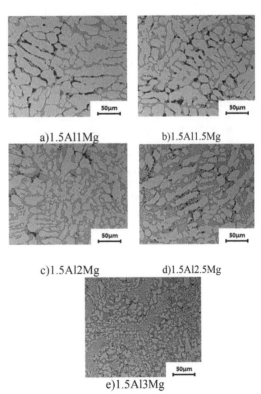

Figure 7. SEM micrographs of Zn-1.5 wt% Al-x wt% Mg alloy with different Mg contents.

407

in Mg_2Zn_{11} intermediate phase. When the Mg content increased to 1.5 wt%, the alloy were still consisting of Zn-rich phase, Mg_2Zn_{11} phase, and Al-rich phase, Zn-rich phase was significantly reduced, and Mg_2Zn_{11} phase ratio was increased. When the Mg content increased to 2 wt%, Zn-rich phase ratio was further reduced, Mg_2Zn_{11} phase ratio was further increased, and part of the discrete small block was distributed in Mg_2Zn_{11} phase. When the Mg content increased to 2.5 wt%, the Zn-rich phase was further reduced, and Mg_2Zn_{11} phase ratio was further increased. When the Mg content increases to 3 wt%, the Zn-rich phase particle size is significantly smaller, Zn-rich phase ratio decreases, part of Mg_2Zn_{11} appear with obvious dendrite morphology, forming $Zn\text{-}Mg_2Zn_{11}$ binary eutectic organizations, and Al-rich phase dispersed in the gap of the particles of Zn-rich phase and a Mg_2Zn_{11} phase.

4 CONCLUSIONS

1. Neutral salt spray test results show that, in the experimental composition range, the corrosion resistance of zinc-aluminum-magnesium alloy is 2 to 3.3 times of the GI coating composition alloy.
2. With the Al content of 1.5 wt% and Mg content within the range of 1~3 wt%, the corrosion resistance greatly affected by the Mg content, and the corrosion resistance of the alloy significantly increased with the increase in Mg content.
3. With the Al content of 1.5 wt% and Mg content within the range of 1~3 wt%, the hardness of zinc-aluminum-magnesium alloy has a great effect of Mg content, and significantly increased with the increase in Mg content.
4. The solidification structure of zinc-aluminum-magnesium alloy is mainly composed of Zn-rich phase, Mg_2Zn_{11} phase, and the Al-rich phase; part of Zn-rich phase and $Mg_2 Zn_{11}$

phase formed the binary eutectic phase, and part of Zn-rich phase, Mg_2Zn_{11} phase, and Al-rich phase formed the ternary eutectic phase.

REFERENCES

Dutta M., Halder A.K., Singh S.B, in: Morphology and properties of hot dip Zn-Mg and Zn-Al-Mg alloy coatings on steel sheet, Surface and Coatings Technology, 2010, 25(10): 2578–2584.

Eliot S.A, in: Method for metallographically revealing intermetallic formation at Galfan/steel interfaces, Material Charaterization, 1993, (30): 295–297.

Hideloshi Shindo, Kaznmi Nishimura, in: Developments and Properties of Zn-Mg Galvanizied Steel Sheet "DYMAZINC" Having Excellent Corrosion Resistance. Nippon steel technical report, 1999, (79): 63–67.

Liang, Chen H., Chang S.L, in: A thermodynamic description of the Al-Mg-Zn system, Metallurgical and materials Transactions, 1997, 28: 103–108.

Liang P., Tarfa T., Robinson J.A. et al, in: Experimental investigation and thermodynamic calculation of the Al-Mg-Zn system, Themochimica Acta, 1998, 314(1): 87–110.

Liu Yan, Xu Binshi, Zhu Zixin, et al, in: New pattern Zn-Al-Mg-RE coating technics for steel structure sustainable design, 2005, 12(10): 211–214.

Lu-zhu, in: Sustainable Development and corrosion Protection Technology, Crorosion & Prothection, 1997, 18(2): 51–54.

Nishimura K., Shindo H., Nomura H, et al, in: Corrosion resistance of hot dip Zn-Mg galvanized steel sheet, Tetsu to Hagane, 2003, 89(1): 174–179.

Notowidjojo, in: Possible souce of Dross Formation in Zinc-0.1% Nickel Galvanizing Process, Material Forum, 1989, (13): 73–75.

Rincon O., Rincon A., Sanchez M., et al, in: Evaluating Zn, Al and Al-Zn coatings on carbon steel in a special atmosphere, Construction and Building Materials, 23(3): 1465–1471.

Yan-hong, in: Hot-dip coating corrosion resistance and application. Chemical Corrosion and [1] YAN-hong, in: Hot-dip coating corrosion resistance and application. Chemical Corrosion and Protection, 1996, 16(1): 49–52.

*Advances in Materials Science, Energy Technology
and Environmental Engineering – Patty & Zhou (Eds)*
© 2017 Taylor & Francis Group, London, ISBN 978-1-138-19668-1

Beam manipulation by a slit film structure containing metal/dielectric composite materials

Kai Zhang & Zhijun Wang
Chinaunicom Institute, Beijing, China

ABSTRACT: A three-dimensional full-vector nonlinear Finite Difference Time Domain (FDTD) method is performed to investigate the beam deflection phenomenon of a three-slit film structure containing silver/silicon dioxide (Ag:SiO$_2$) composite materials. The simulation results show that the deflection angle can be turned easily by controlling the concentration of Ag particles, which were embedded in SiO$_2$ mix. The beam deflection phenomenon can be explained by the surface plasmons excitation and enhancement of the local field near the Ag particles and this can also be used in telecom operators.

1 INTRODUCTION

A three-dimensional full-vector nonlinear Finite Difference Time Domain (FDTD) method is performed to investigate the beam deflection phenomenon of a three-slit film structure containing silver/silicon dioxide (Ag:SiO$_2$) composite materials. The simulation results show that the deflection angle can be turned easily by controlling the concentration of Ag particles, which were embedded in SiO$_2$ mix. The beam deflection phenomenon can be explained by the surface plasmons excitation and enhancement of the local field near the Ag particles.

2 PRINCIPLE AND SIMULATION METHOD

When light passes through dielectric, it will be influenced by the refractive index of the dielectric. It has been proved that the propagation constant β grows steadily with the increasing dielectric constant ε_d when the slit width w is equal to the value in Shi (2005):

$$\tanh\left(\sqrt{\beta^2 - k_0^2 \varepsilon_d}\, w/2\right) = -\frac{\varepsilon_d \sqrt{\beta^2 - k_0^2 \varepsilon_m}}{\varepsilon_m \sqrt{\beta^2 - k_0^2 \varepsilon_d}} \quad (1)$$

where k_0 is the wave vector of light in free space, ε_m and ε_d are the relative dielectric constants for the metal and the materials in the slit, respectively, and w is the slit width. The imaginary part of β represents the decibel loss coefficient per unit length, which is usually ignorable for light propagation in short slit. We mainly focus on the Re(β/k0), representing the effective refractive index in the slit

and determining the phase retardation. From the relationship between Re(β/k0) and ε_d reported in Min (2007), we can conclude that Re(β/k0) will be increased as ε_d grows.

Obviously, the dispersion relation in equation (1) implies a potential way of phase modulation by varying the dielectric constant ε_d, which can be implemented with metal/dielectric composite materials. It is well known that dielectric constant ε_d in nonlinear media depends on the third-order nonlinear susceptibility:

$$\varepsilon_d = \varepsilon_l + \chi^3 |E|^2 \quad (2)$$

where ε_l is the linear dielectric constant and χ^3 is the third-order nonlinear susceptibility. When we use a laser with a certain wavelength and intensity, ε_d can be modulated by changing χ^3. Many articles have reported that the third nonlinear susceptibility value depends on the noble metal concentration of nonlinear composite materials (Liao, 1998; Zhou, 2003). The relationships between χ^3 and noble metal concentration in Au:SiO$_2$ composite films and in Ag:Bi$_2$O$_3$ composite films, respectively, are shown in Fig. 1 and Fig. 2. It can be explained by two effects: local field enhancement and Mie resonance of the particles (Liao, 1998). From the above, we can conclude that the third nonlinear susceptibility χ^3 can be enhanced by increasing the concentration of the noble metal. As the χ^3 increased, the Re(β/k0) which relates to the effective refractive index will be increased too. Therefore, the speed of light can be slowed down when it passes through metal/dielectric composite material. Hence, we can manipulate the phase retardation as well as output beam by varying the concentration of the noble metal.

A three-dimensional full-vector Nonlinear Finite Difference Time Domain (NFDTD) method has

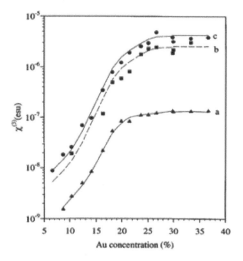

Figure 1. Concentration dependence of χ^3 in Au:SiO$_2$ composite films.

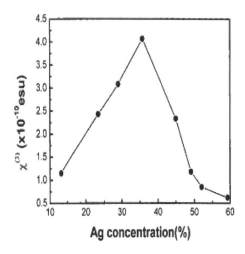

Ag concentration(%)

Figure 2. Concentration dependence of χ^3 in Ag:Bi$_2$O$_3$ composite films.

been used in our work (Taflove, 2000). The Drude dispersion model is used to simulate the metallic film. The Perfectly Matched Layer (PML) (Fujii, 2005) has been applied to boundaries of the simulated area.

3 SIMULATION RESULTS AND DISCUSSION

We investigate the deflection phenomenon of the beam by simulating a three-slit film structure, which is shown in Fig. 3. A 600-nm-thick silver film is the main board of the structure. Three slits with the same widths of 100 nm are chiseled in the film and the interspacing is 400 nm. Each slit is

filled with Ag:SiO$_2$ composite nonlinear materials with different concentrations of the Ag particle. For idealization, where the Ag particle we used is spherical and be arranged symmetrically. The Ag particles concentrations are 2.1%, 2.6%, and 3.1% in the slits from top to bottom, respectively. Then, the structure is illuminated by a TM polarized plane wave of 562 nm wavelength.

The results of FDTD simulations of electric-field intensity are shown in Fig. 4, Fig. 5 and Fig. 6. Fig. 4 is pure SiO$_2$ with no Ag nanoparticles as the dielectric. We can see that the output light

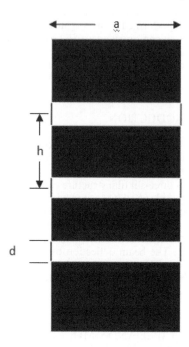

Figure 3. Schematic geometry of the three-slit film structure under study. The parameters are as follows: the thickness of the Ag film is 600nm, the distance between two silts, h, is 400nm, and each slit width, d, is 100 nm. A TM-polarized plane wave (562 nm wavelength) is incident from the left side of the slit array.

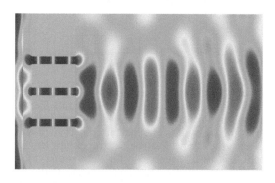

Figure 4. Structure with pure SiO$_2$ as the dielectric.

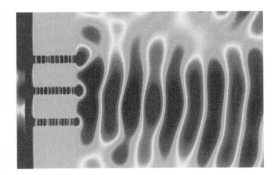

Figure 5. The Ag concentrations are 2.1%, 2.6%. and 3.1% from top to bottom.

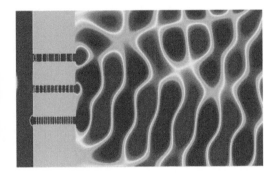

Figure 6. The Ag concentrations are 1.6%, 2.6%. and 4.0% from top to bottom.

is remained and deflection phenomenon does not occur. Fig. 5 is the structure with nanocomposites with different concentrations of Ag nanoparticles in each silt. We can find that the output light is deflected with a deflection angle about 10°. To further investigate the relationship between the deflection angles and the concentrations of Ag nanoparticles, we change the Ag concentration of each slit, make them 1.6%, 2.6%, and 4.0% from top to bottom. From Fig. 6, we can see that the deflection angle is enhanced to 30°.

The result can be explained by the principle of phase retardation. When we embed Ag particles into SiO_2, the effect of local field enhancement near Ag nanoparticles can increase the χ^3 of the composite materials which leads to a large effective refractive index. The more the Ag nanoparticles in SiO_2 the larger the effective refractive index in the slit. Therefore, light in the slit with fewer Ag nanoparticles will travel faster than the light in the slit with more Ag nanoparticles, which result in the deflection phenomenon. In Fig. 6, the light speed in the first slit is faster than that in Fig. 5, but the light in the third slit is slower than that in Fig. 6, that is why the deflection angle is bigger in Fig. 5 than in Fig. 6.

4 CONCLUSION

In this paper, embedding $Ag:SiO_2$ composite materials in metallic nano-slit array is proposed to be an active and new way to turn the angle of the output beam. The reason of beam deflection phenomenon through the nano-slit array is attributed to surface plasmons excitation and enhancement of local field which can lead to a large third nonlinear susceptibility χ^3. Through FDTD simulation, it has been clearly shown that the deflection angle can be controlled easily by embedding composite materials with different Ag concentrations into the slits. We believe that metallic nano-structures with metal/dielectric composite materials will lead to various new phenomena (such as four-wave mixing, bistability effect, and more actively controlled nano-optic devices) will be designed. This result can be used for improving the service efficiency of telecom operators.

REFERENCES

Battaglin G., Calvelli P., Cattaruzza E., Gonella F., Polloni R., Z-scan study on the nonlinear refractive index of copper nanocluster composite silica glass, Appl. Phys. Lett. 78(25), 2001, pp. 3953.

Fujii M., Koos C., Poulton C., Sakagami I., Leuthold J., Freude W., A simple and rigorous verification technique for nonlinear FDTD algorithms by optical parametric four-wave mixing, Microwave and Optical Technology Letters 48(1), 2005, pp. 88.

Huang W.Y., Qian W., El-Sayed M.A., Coherent Vibrational Oscillation in Gold Prismatic Monolayer Periodic Nanoparticle Arrays, Nano Lett. 4(9), 2004, pp. 1741.

Liao H.B., Xiao R.F., Wang H., Wong K.S., and Wong G.K.L., Large third-order optical nonlinearity in Au:TiO2 composite films measured on a femtosecond time scale, Appl. Phys. Lett. 72(15), 1998, pp. 1871.

Link S., Mohamed M.B., El-Sayed M.A., Simulation of the Optical Absorption Spectra of Gold Nanorods as a Function of Their Aspect Ratio and the Effect of the Medium Dielectric Constant, J. Phys. Chem. B 103(16), 1999, pp. 3073.

Min C., Wang P., Jiao X., Deng Y. Ming H., Beam manipulating by metallic nano-optic lens containing nonlinear media, Optics Express 15(15), 2007, pp. 5941.

Min C.,Wang X., Beam focusing by metallic nano-slit array containing nonlinear material, Appl. Phys. B 90(1), 2008, pp. 97.

Ning T, Chen C, Zhou Y, Lu H, Shen H, Zhang D, Wang P, Ming H, Yang G, Third-order optical nonlinearity of gold nanoparticle arrays embedded in a BaTiO3 matrix,Applied Optics 48(2), 2009, pp. 2.

Porto J.A.. Martin-Moreno J., Garcia-Vidal F.J., Optical bistability in subwavelength slit aperturescontaining nonlinear media, Phys. Rev. B 70(8), 2004, pp. 081402.

Prem Kiran P., Shivakiran-Bhaktha B.N., Narayana-Rao D., Nonlinear optical properties and surface-plasmon enhanced optical limiting in Ag–Cu nanoclusters co-doped in SiO2 Sol-Gel films, J.Appl. Phys. 96(11), 2004, pp. 6717.

Shi H., Wang C., Du C., Luo X., DongX., Gao H., Beam manipulatingby metallic nano-slits with variant widths, Optics Express 13(18), 2005, pp. 6815–6820.

Smolyaninov I., Quantum Fluctuations of the Refractive Index near the Interface Between a Metal and a Nonlinear Dielectric, Phys. Rev. Lett. 94(5), 2005, pp. 057403.

Taflove A., Hagness S., Computational Electrodynamics: The Finite-Difference Time-Domain Method Artech House, Boston 2000.

Wurtz G.A., Pollard R., and Zayats A.V., Optical Bistability in Nonlinear Surface-Plasmon Polaritonic Crystals, Phys. Rev. Lett., 97, 2006, pp. 057402.

Yang G., Zhou Y., Long H., Li Y., Yang Y., Optical nonlinearities in Ag/BaTiO3 multi-layer nanocomposite films, Thin Solid Films 515(20), 2007, pp. 7926.

Zhou P., You G.J., Li Y.G., Han T., Li J., Wang S.Y., Chen L.Y., Linear and ultrafast nonlinear optical response of Ag:Bi2O3 composite films, Appl. Phys. Lett. 83(19), 2003, pp. 3876.

Advances in Materials Science, Energy Technology
and Environmental Engineering – Patty & Zhou (Eds)
© 2017 Taylor & Francis Group, London, ISBN 978-1-138-19668-1

Discussion on the mineralogical characteristics and formation causes of Lizhu iron polymetallic ore deposits in Zhejiang Province

Delong Jia
Development and Research Center, China Geological Survey, Beijing, China

Yanfang Feng, Wanyi Zhang, Xiaofeng Yao, Yongsheng Li & Zhihui Zhang
Development and Research Center, China Geological Survey, Beijing, China

ABSTRACT: The predecessors have rarely made researches on the Lizhu iron polymetallic ore deposits in Zhejiang Province and the formation causes are controversial. Through studies on mineralogical characteristics of deposit, it has been considered in this paper that the Lizhu iron polymetallic ore deposit is typical skarn deposit generated by magmatic hydrothermal replacement.

1 INTRODUCTION

The geotectonic position of Lizhu iron polymetallic ore deposit is at northwest side of northeast section of Zhejiang Province of Qinzhou-Hangzhou metallogenic belt. The region is strong in structural fold with abnormal development of fracturation. During the Indo-China Period, and the Yanshanian Period, the acidic magma activities were frequent, which has caused the common stratum alteration in this region and, meanwhile, brought enrichment mineralization. The degree of scientific research on Lizhu iron ore is low. Fan Xiaoren et al. (2009) studied the metallogenic characteristics and ore-controlling factors of east ore belt of Lizhu Iron Deposit and mentioned in the research contents that the genetic type of ore deposit is skarn deposit related to magmatic intrusion. After research, Cao Shuying (1984, 1986) considered that the formation of this deposit experienced long-term and complicated process. The deposit was initially formed at the Sinian Xifengsi Period. With the formation of Dengying magnesium calcium carbonates, ferruginous deposition occurred and stripped iron-containing stratum with low taste formed. Until the Yanshanian period, strong regional fold and fracturation resulted in the intrusion of magma, and with the thermal force and hydrothermal process, heat recrystallization, hydrothermal metasomatism recrystallization and skarnization, sulfide mineralization and ferruginous leaching re-enrichment, and other series of superposition and reconstruction occurred in the original iron-containing stratum, which has formed deposition hydrothermal reconstruction magnet ore deposit with lean ore taking the dominant position. Through the studies on ore mineral characteristics of the deposit, this paper has discussed the genetic types of ore deposit.

2 CHARACTERISTICS OF THE ORE MINERAL

The metallic minerals are composed of magnetite, molybdenite, pyrite, and galena. There are a great variety of nonmetallic minerals in the mine lot, mainly including garnet, diopside, phlogopite, idocrase, chlorite, chondrodite, epidote, hornblende, and calcite, among which, garnet, diopside, and other skarn minerals have close relationship with the mineralization process in time, space, and genesis (as shown in Figure 1).

2.1 Metallic mineral

1. Magnetite
According to the structural characteristics of magnetite, the magnetite in this mine lot can be divided into early and late periods. Early period magnetite mainly presents black gray, subhedral-xenomorphic fine-grained texture, disseminated strip, or banded structure, and the grain size is generally between 0.01 and 0.2 mm, which is the main component of iron ore in this deposit. In this period, magnetite is mainly in mineral symbiosis with phlogopite and other degraded and altered minerals and the phenomenon of metasomatism skarn mineral can be observed. In addition, it can be observed that part magnetite is filled in grains of skarn mineral, for example, it can be seen that magnetite is filled in grains of diopside. In the late period, magnetite mainly presents xenomorphic, medium-fine-grained texture, vein, or nodular structure, and the

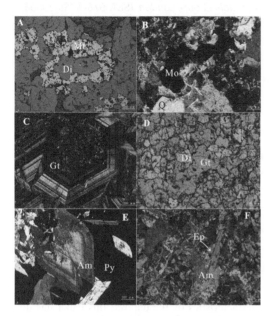

Figure 1. Mineral characteristics of Lizhu iron ore deposit. A, the magnetite (Mt) replaced diopside (Di); B, molybdenite (Mo) in the quartz (Q) vein; C, euhedral garnet shows clear oscillatory zoning (Gt); D, Garnet (Gt) and diopside (Di); E, euhedral amphibole (Am); F, amphibole (Am) and epidote (Ep).

phenomenon of vein magnetite crossing the surrounding rock and skarn can be observed.

The electronic probe result shows that the SiO_2 content in the magnetite of this mine lot is 0.59~2.46%, TiO_2 content is 0.01~0.03%, and Al_2O_3 content is 1.12~1.15%, TFeO content is 88.57~90.31%, MgO content is 0.08~0.19%, and CaO content is 0.27~0.37%, possessing the typical component characteristics of contact metasomatism-type magnetite mine.

2. Molybdenite

Molybdenite is less distributed in this mine lot, but mainly distributed in east and west regions. The molybdenite of molybdenum ore in east region is mainly in flaky form with a grain size of 0.01~0.07 mm, which is mainly distributed in quartz vein or skarn mineral in star, dot, and stringer shapes, in which, the molybdenite granularity in quartz vein is coarse. The molybdenite of molybdenum ore in west region is mainly in flaky form and clintheriform crystalline form, has idiomorphic-subhedral fine-grained texture, and the majority is disseminated structure with minority of micro vein structure. In addition, it can be observed that part molybdenite and nickel minerals present collection disseminated body.

3. Pyrite

Pyrite in this mine lot is well developed and is mainly pale yellow and in idiomorphic-subhedral medium coarse grained texture, which is widely distributed in ore body, skarn, and surrounding rocks. Pyrite is late-staged hydrothermal alteration mineral, which is mainly generated in vein or scattered shape and part pyrite has symbiotic relationship with degraded altered minerals. For example, medium and pyrite contains idiomorphic medium-coarse-grained hornblende.

4. Galena

Zinc blende in this mine lot is mainly in xenomorphic-subhedral-grained texture and the grain size is generally from 0.01 to 0.07 mm, mainly presenting in the disseminated form. Zinc blende in mine lot often manifests solid solution exsolution texture and the phenomenon of copper pyrite's distribution in zinc blende by presenting emulsion droplet can be observed.

2.2 *Nonmetallic minerals*

1. Garnet

According to the structural characteristics of garnet, it can be divided into early and late periods in this mine lot. Early garnet presents sandy beige, Kelly, and minority in puce. It is in subhedral-idiomorphic medium-coarse-grained texture and the grain size is generally from 0.02 to 2.5 mm in zonal structure development; it is tawny under plane-polarized light and positive relief under polar altitude and part abnormal interference color under cross-polarized light, which can reach I-grade gray. Garnet in this period is mainly in symbiosis with diopside, which is the main component mineral of skarn in this mine lot. In the late period, garnet is less distributed and mainly presents brown and Kelly, subhedral-xenomorphic fine-grained texture, presenting vein in the surrounding rock or skarn.

As shown in Table 1, electronic probe result shows that the endmember component of garnet of Lizhu iron ore is dominated by andradite endmember (64.53~91.60%, on average 77.56%) and second is grossularite endmember (2.34~32.82%, on average 19.46%), basically consistent with other skarn-type iron deposit garnet components (Einaudi et al., 1981; Einaudi and Burt, 1982; Meinert, 1989) in the world, which is typical andradite-grossularite series. Calculate the molecular formula of garnet by 12 oxygen atoms as standard and the garnet of Lizhu iron ore can be expressed as:

$$Ca_{(2.85-2.96)}Mg_{(0-0.02)}Mn_{(0.05-0.08)}Fe^{2+}_{(0-0.06)}$$
$$Fe^{3+}_{(1.28-1.98)}Al_{(0-0.7)}Ti_{(0-0.06)}Si_{(2.96-3.04)}O_{12}$$

2. Diopside

According to the structural characteristics of diopside, the diopside in this mine lot can be divided into early and late periods. Early diopside is the main component mineral in skarn, mainly presenting sage green and in short columnar crystalline

Table 1. Results of analysis of Lizhu iron mine garnet electron probe (wt%) and the end members.

No.	LZB-1	LZB-2	LZB-3	LZB-4
SiO_2	35.370	36.680	37.110	37.048
TiO_2	0.181	0.014	0.040	0.016
Al_2O_3	1.664	6.944	6.761	5.902
FeO	26.316	20.632	20.866	21.655
MnO	0.708	0.933	0.917	0.881
MgO	0.101	0.032	0.040	0.070
CaO	32.935	33.330	33.673	33.639
Si	2.9680	2.9787	2.9871	2.9944
Ti	0.0114	0.0009	0.0024	0.0010
Al	0.1646	0.6646	0.6414	0.5622
Cr	0.0000	0.0000	0.0000	0.0000
Fe^{3+}	1.8468	1.3490	1.3656	1.4409
Fe^{2+}	0.0000	0.0523	0.0391	0.0228
Mn	0.0503	0.0642	0.0625	0.0603
Mg	0.0126	0.0039	0.0048	0.0084
Ca	2.9612	2.9001	2.9041	2.9131
And	91.60	66.99	68.04	71.93
Pyr	0.420	0.130	0.160	0.280
Spe	1.660	2.120	2.080	2.010
Gro	6.310	29.02	28.42	25.02

No.	LZB-5	LZB-6	LZB-7	LZB-8
SiO_2	37.451	36.299	35.910	37.019
TiO_2	0.104	1.049	0.119	0.048
Al_2O_3	7.379	4.214	0.961	5.283
FeO	19.175	22.669	27.293	21.774
MnO	0.883	0.963	0.684	0.786
MgO	0.009	0.066	0.126	0.028
CaO	33.416	33.562	32.603	32.516
Si	3.0240	2.9644	3.0011	3.0408
Ti	0.0063	0.0645	0.0075	0.0030
Al	0.7022	0.4056	0.0947	0.5114
Cr	0.0000	0.0000	0.0000	0.0000
Fe^{3+}	1.2776	1.5483	1.8996	1.4594
Fe^{2+}	0.0173	0.0000	0.0079	0.0364
Mn	0.0604	0.0666	0.0484	0.0547
Mg	0.0011	0.0080	0.0157	0.0034
Ca	2.8910	2.9368	2.9194	2.8618
And	64.53	77.12	95.25	74.05
Pyr	0.04	0.27	0.52	0.12
Spe	2.03	2.21	1.62	1.85
Gro	32.82	20.40	2.34	22.75

No.	LZB-10	LZB-11	LZB-12	LZB-13
SiO_2	36.269	36.152	36.281	36.462
TiO_2	0.018	0.000	0.040	0.000
Al_2O_3	3.908	3.372	4.884	4.282
FeO	23.684	24.298	22.874	23.621
MnO	0.809	0.642	0.933	1.094
MgO	0.088	0.085	0.052	0.043
CaO	32.687	32.609	32.459	33.365
Si	3.0043	3.0087	2.9972	2.9811
Ti	0.0011	0.0000	0.0025	0.0000

(*Continued*)

Table 1. Results of analysis of Lizhu iron mine garnet electron probe (wt%) and the end members. (*Continued*).

No.	LZB-10	LZB-11	LZB-12	LZB-13
Al	0.3815	0.3307	0.4755	0.4126
Cr	0.0000	0.0000	0.0000	0.0000
Fe^{3+}	1.6149	1.6635	1.5247	1.6000
Fe^{2+}	0.0259	0.0277	0.0556	0.0151
Mn	0.0568	0.0453	0.0653	0.0758
Mg	0.0109	0.0106	0.0064	0.0052
Ca	2.9011	2.9078	2.8730	2.9228
And	80.89	83.41	76.23	79.50
Pyr	0.36	0.35	0.21	0.17
Spe	1.90	1.51	2.18	2.51
Gro	15.99	13.79	19.53	17.32

form, idiomorphic-subhedral medium-fine-grained texture and the grain size is generally between 0.01 and 0.5 mm; it is in weak pleochroism and positive relief under plane-polarized light; colorful blue, green, purple and, other bright interference colors can be seen under cross-polarized light, which is of typical pyroxene cleavage. In the late period, diopside presents idiomorphic coarse-grained texture, usually constituting stringer crossing ore with serpentine and carbonate minerals.

Electronic probe analysis result shows that: the variation range of endmember component of diopside of pyroxene is among 36.5~92.9%, on average 74.2%; the variation change in endmember component of hedenbergite is between 4.9 and 59.4%, on average 23.2%; the endmember component content of johannsenite is from 1.0 to 6.0%, on average 2.6%, showing that pyroxene in the mine lot is typical diopside-hedenbergite solid solution series (as shown in Table 2). The endmember component content of diopside is basically consistent with diopside content range of other domestic (Di50~90; Zhao Bin et al. 1987) and overseas (Di20~80; Einaudi et al., 1981; Meinert, 1989) skarn-type iron deposits. Calculate the molecular formula of diopside by six oxygen atoms and four positive ions as standard and the pyroxene of Lizhu iron ore can be expressed as:

$$Na_{(0-0.02)}Ca_{(0.55-1.09)}Mg_{(0.46-0.93)}Fe^{2+}_{(0.02-0.78)}Fe^{3+}_{(0-0.10)}Al_{(0.0-0.13)}Si_{(1.93-2.07)}O_6$$

3. Hornblende

Hornblende in this mine lot mainly presents black green and black color, has idiomorphic-subhedral medium-coarse-grained texture and part in pegmatitic texture and well-developed banded structure and part is in symbiosis with epidote.

The electronic probe analysis results show that the SiO_2 content in hornblende is 39.61~40.93%, TiO_2 content is 0.006~0.169%, Al_2O_3 content is

Table 2. Results of analysis of Lizhu iron mine diopside electron probe (wt%) and the endmembers.

No.	LZC-1	LZC-2	LZC-3	LZC-4
SiO_2	56.145	52.991	50.831	52.013
TiO_2	0	0.119	0.014	0.004
Al_2O_3	0.026	1.727	1.29	1.135
FeO	1.611	3.997	23.096	6.546
MnO	0.702	0.624	1.585	0.643
MgO	17.078	14.614	8.032	14.299
CaO	27.55	27.968	12.96	23.012
Na_2O	0.009	0.03	0.19	0.045
K_2O	0.022	0.022	0.181	0.017
Si	1.99004	1.92834	2.02271	1.97412
Al (iv)	0	0.07166	0	0.02588
Al (vi)	0	0.0024	0.0605	0.02489
Fe^{3+}	0.03061	0.09801	0	0.00735
Fe^{2+}	0.01703	0.02263	0.77666	0.20031
Mn	0.02108	0.01923	0.05342	0.02067
Mg	0.90239	0.79279	0.47647	0.80905
Ti	0	0.00326	0.00042	0.00011
Ca	1.04627	1.09047	0.55256	0.93581
Na	0.00062	0.00212	0.01466	0.00331
K	0.00099	0.00102	0.00919	0.00082
Di	92.92	85.00	36.47	77.99
Hd	4.91	12.94	59.44	20.02
Jo	2.17	2.06	4.09	1.99

No.	LZC-5	LZC-6	LZC-7
SiO_2	51.593	54.084	54.823
TiO_2	0.041	0.196	0.161
Al_2O_3	0.527	1.24	0.932
FeO	14.174	2.486	2.949
MnO	1.776	0.401	0.316
MgO	7.734	16.9	16.391
CaO	18.891	22.439	19.443
Na_2O	0	0.018	0.039
K_2O	0	0.099	0.007
Si	2.9680	2.9787	2.9871
Al (iv)	2.07329	1.99829	2.06128
Al (vi)	0.02496	0.05229	0.0413
Fe^{3+}	0	0	0
Fe^{2+}	0.48694	0.07735	0.09474
Mn	0.06045	0.01255	0.01006
Mg	0.46332	0.93086	0.91873
Ti	0.00124	0.00545	0.00455
Ca	0.81339	0.88832	0.78327
Na	0	0.00129	0.00284
K	0	0.00467	0.00034
Di	45.84	91.19	89.76
Hd	48.18	7.58	9.26
Jo	5.98	1.23	0.98

Table 3. Results of analysis of Lizhu iron mine hornblende electron probe (wt%) and the endmembers.

No.	LZD-1	LZD-2	LZD-3	LZD-4
SiO_2	40.937	39.61	40.924	40.806
TiO_2	0.006	0.069	0.064	0.169
Al_2O_3	14.851	14.267	11.455	13.349
FeO	11.319	17.143	21.654	14.679
MnO	0.726	0.772	0.954	0.709
MgO	13.188	10.038	11.356	7.512
CaO	12.653	12.965	12.884	12.587
Na_2O	2.299	2.118	0.922	2.117
K_2O	1.152	1.394	1.311	1.247
SiT^*	6.10	6.02	6.37	6.19
Al_T	1.90	1.98	1.63	1.81
Al_C	0.71	0.58	0.47	0.58
Fe^{3+}_C	0.00	0.00	0.15	0.00
Ti_C	0.00	0.01	0.01	0.02
Mg_C	2.93	2.27	1.74	2.57
Fe^{2+}_C	1.37	2.14	2.63	1.83
Mn_C	0.00	0.00	0.00	0.00
Fe^{2+}_B	0.04	0.04	0.03	0.03
Mn_B	0.09	0.10	0.13	0.09
Ca_B	1.86	1.86	1.84	1.88
Na_B	0.00	0.00	0.00	0.00
Ca_A	0.16	0.25	0.31	0.16
Na_A	0.66	0.62	0.28	0.62
K_A	0.22	0.27	0.26	0.24

ing to the classification of hornblende by Leake et al. (1997), the hornblende in this region can be divided into pargasite and ferropargasite.

4. Epidote

Epidote in this mine lot is grass-blade yellow in color, in long column or granular crystal form, and has idiomorphic-subhedral medium-coarse-grained texture; it has obvious pleochroism and positive relief under plane-polarized light; it has red, green, and other bright II-III interference colors under cross-polarized light. Epidote is the product of hydrothermal alteration and the phenomenon of symbiosis with hornblende mineral can be observed, which is usually distributed among early period skarn minerals presenting interstitial materials or cut cross surrounding rock presenting vein shape. It is also discovered in altered diorite-porphyrite in the mine lot.

3 CONCLUSIONS

1. Through the research on ore mineral characteristics, it has been considered that the Lizhu iron polymetallic ore deposit is typical skarn deposit generated by magmatic hydrothermal replacement.
2. The formation of deposit has experienced, – early skarn stage: in this stage, garnet, diopside, and

11.46~14.85%, TFeO content is 11.32~21.65%, MgO content is 7.51~13.19%, CaO content is 12.59~12.97%, and Na_2O content is 0.92~2.30%, as shown in Table 3, possessing the component characteristic of calciferous amphibole. Accord-

416

other silicate minerals are widely developed, but hardly there is generation of magnetite. α late skarn stage: it is the important mineralization stage, when magnetite, phlogopite, hornblende, epidote, chlorite, serpentine, and other degraded and altered minerals are mainly generated. β quartz-sulfide stage: in this stage, quartz, calcite, pyrite, chalcopyrite, galena, sphalerite, molybdenite, and other low-temperature hydrothermal solution minerals are mainly generated.

ACKNOWLEDGMENTS

I would like to express my heartfelt thanks to Luo Bangyue, Senior Engineer of Zhejiang Lizhu Iron Ore Group Co., Ltd., Li Runhao, Director of Zhejiang Geological Survey Bureau of Non-ferrous Metals as well as Doctor Yao Lei, Master He Pengfei, and Master Liu Peng of China University of Geosciences (Beijing) who have helped me a lot in finishing this paper.

REFERENCES

Cao Shuying. Applied Mathematical Statistics Distribution Characteristics Study on the Factors of StrataboundIron Ore of Zhejiang Lizhu. Journal of Chengdu College of Geology, 1986, 04: 31–35.

Cao Shuying. Geologic Feature and Factor Discussion of Zhejiang Lizhu Iron Deposit. Journal of Chengdu College of Geology, 1984, 02: 1–11.

Einaudi M.T. and Burt D.M. Introduction Terminology, Classification and Composition of Skarn Deposit. Econ. Geol, 1982, 77: 745–754.

Einaudi M.T. Meinert L.D. and New berrv R.J. Skarn Deposits. Econ. Geol, 1981, 75: 317–391.

Fan Xiaoren, Xia Guoqiang, Xie Jirong et al. Metal— logenic Characteristics and Ore-controlling Factors Analysis of East Ore of Lizhu Iron Ore. Mineral Resources and Geology, 2009, 23 (6): 538–541.

Meinert L.D. Gold Skarn Deposits Geology and Exploration Critera. In: Keays R, Ramsay R, Groves D, ed. The Geology of Gold Deposits. Monogr: Econ. Geol, 1989, 6: 537–552.

Leake B.E, Woolley A.R, Arps C.E.S, et al. Nomenclature of Amphibole: Report of the Subcommittee on Amphiboles of the International Mineralogical Association, Commission on New Minerals and Mineral Names. Am Mineral, 1997, 82: 1019–1037.

Zhao Bin, M.D. Barton. Component Characteristics of Garnet and Pyroxene in Contact Metasomatism Skarn Deposit and the Relationship with Mineralization. Journal of Mineral, 1987, 01: 1–8.

*Advances in Materials Science, Energy Technology
and Environmental Engineering – Patty & Zhou (Eds)*
© 2017 Taylor & Francis Group, London, ISBN 978-1-138-19668-1

Effects of biochar pyrolysis temperature on its characteristics and heavy metal adsorption

Lifang Zhao, Qingsong Liu, Shiyue Liu, Wencong Zhao, Haoming Li & Bing Han
China University of Geosciences, Beijing, China

ABSTRACT: Pyrolysis temperature is one of the most important factors that affect the properties of biochar. Considering the environment pollution caused by heavy metal, Cornstalk-based Biochars(CB) are produced at a temperature range of 250~450°C and their characteristics on the adsorption of Cr(VI) were studied. The results showed that the best corn stalk pyrolysis temperature was 250°C. The adsorption of Cr(VI) agreed better with Langmuir model. In terms of adsorption kinetic, adsorption of Cr(VI) followed pseudo second order kinetic model. The substantial acidic functional groups on CB surface played a crucial role on the adsorption of Cr(VI) . The characteristics of Cr(VI) adsorption was closely related to the pyrolysis temperature.

1 INTRODUCTION

Biochar is a pyrogenic black carbon that has attracted increased attention in both political and academic areas (Verheijen et al., 2009). A number of studies have suggested that application of biochar could effectively sequester carbon in soils and thus mitigate global warming, increased soil nutrients, water holding capacity and reduced emissions of other greenhouse gases from soils (Yao, 2009; Lehmann, 2008). Several research efforts have been made to evaluate biochar as an adsorbent in water treatment applications for removing various contaminations, including heavy metals, nutrients, and organic compounds (Beesley, 2010; Inyang, 2012; Yao, 2012).

Various types of biomass, such as trees, grasses (Mukherjee, 2011), industrial waste (Bernd, 2014), are used to produce biochars and majority of them are agricultural and forestry waste. The advantages of adsorption of heavy metals by biochar are low cost, high efficiency and cause no secondary pollution. Recently, biochar has attracted widespread attention due to its adsorption of heavy metals from solution.

The main purpose of this study was to investigate the adsorption characteristics of Cr(VI) ions from aqueous solutions using biochars produced from cornstalk at a temperature range of 250~450°C. Adsorption experiments were carried out to investigate the effects of solution pH, adsorbent concentration, contact time. To understand the adsorption process, the adsorption kinetics, isotherms and thermodynamics were further studied using data obtained from these experiments. The mechanisms responsible for the Cr(VI) removal by these four biochars were characterized using X-ray photoelectron spectroscopy, Fourier Transform Infrared Spectroscopy (FTIR), Boehm titration and Cation Exchange Capacity (CEC).

2 MATERIALS AND METHODS

2.1 Materials

Corn stalks were collected from a farm in Beijing, China. It was cleaned by distilled water and dried for 24 h with oven at 80°C. and then cut into pieces of approximately 2–3 cm in length. These pieces were smashed and sieved through 20-80 mesh sieve. The ground sample was tightly placed in a ceramic pot, covered with a fitting lid, and then pyrolyzed at 250°C, 350°C, 400°C and 450°C in a muffle furnace under oxygen-limited conditions for 2h. These four biochar samples were designated as CB250, CB350, CB400 and CB450.

2.2 Biochar characterization

Elemental analysis was determined using X-ray photoelectron spectroscopy. FTIR was used to identify functional groups on the biochar surfaces. To obtain the observable absorption spectra, the samples were prepared by mixing biochar with spectroscopic grade KBr and then compressing the mixture into pellets. FTIR spectra were recorded between 4,000 and 400 cm^{-1}. Boehm titration was used to identify functional groups' quantity. Sodium ion concentration was measured to determine the cation exchange capacity of biochars.

2.3 Batch adsorption experiments

For adsorption kinetics of heavy metals, 0.1 g of biochars was mixed with 40 mg · L^{-1} of K$_2$Cr$_2$O$_7$ solution at pH of 3. The suspensions were agitated on a shaker at 250 rpm and 25°C for certain sampling time(between 0~72 h) intervals, after which 5 mL of the solutions was intermittently sampled and filtered. Ion concentration determination of total Cr ion concentration was measured.

For isotherms experiment, 0.1 g of biochars was contacted with 10, 20, 40, 60, 80, 100 mg·L^{-1} of K$_2$Cr$_2$O$_7$ solution at pH of 3. The suspensions were agitated on a shaker at 250 rpm and 25°C for 72 h, after which 5 mL of the solutions was intermittently sampled and filtered. Ion concentration determination of total Cr ion concentration was measured.

In the experiment examining the pH edge, 0.1 g of biochar was mixed with 40 mg/L K$_2$Cr$_2$O$_7$ solution (100 mL) at final pH of 2–7. The suspensions were agitated on a shaker at 250 rpm and 25°C for 7 d, after which 5 mL of the solutions was intermittently sampled and filtered. Ion concentration determination of total Cr ion concentration was measured.

3 RESULTS AND DISCUSSION

3.1 Elemental analysis

Surface elemental composition of biochar was showed as Table 1. Pyrolysis temperature increased from 250°C to 450°C, the carbon content of CB biochar was gradually reduced, which decreased from 77.74% of CB250 to CB450 of 67.59%. With the pyrolysis temperature increasing, oxygen content was decreased first and then increased. This indicated that during pyrolysis oxygen was lost more than carbon. Nitrogen content was least of CB450, which demonstrated nitrogen was decomposed beyond 450°C. Phosphorus content of CB350 was the most. With the pyrolysis temperature increasing, Phosphorus content was increased first and then decreased. From the above results, elements was closely associated with the pyrolysis temperature.

Table 1. Surface elemental composition of CB biochar series, at/%.

Samples	C	N	P	O
CB250	77.74	1.20	0.27	19.61
CB350	75.71	1.10	0.39	19.55
CB400	71.11	1.73	0.36	20.74
CB450	67.59	0.32	0.29	23.07

3.2 FTIR

It can be seen in Figure 1 which was FTIR spectra of the biochars. Main absorption peaks of four biochars were very similar. Peak of 3400~3420 cm^{-1} was corresponded to the −OH stretching vibration. At the peaks of 2850~2847 cm^{-1} and 1700~1710 cm^{-1}, there were −CH and −COOH appeared respectively. With the pyrolysis temperature increasing, the peak of −COOH turned weak.

Strong adsorption peak in the range of 1580~1620 cm^{-1} was corresponded to the C = O vibration (Chen, 2013). At higher pyrolysis temperatures, new adsorption peaks such as −CH$_2$ appeared and had the trend to moving to lower waves, which indicated that the aromatic of biochars was gradually enhanced.

3.3 Boehm titration and CEC

Acidic functional group and cation exchange capacity of biochars were showed as Table 2. With the pyrolysis temperature increasing, Acidic functional group decreased from 3.17 mmol · g^{-1} of CB250 to 1.70 mmol · g^{-1} of CB450. Cation exchange capacity had the same trend, CB250 was the largest and CB450 was the least.

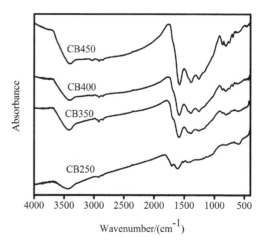

Figure 1. FTIR spectra of the biochars.

Table 2. Acidic functional group and cation exchange capacity of biochar.

Biochar	Acidic functional group (mmol · g^{-1})	Cation exchange capacity
CB250	3.17	130
CB350	2.58	97.5
CB400	1.95	64.9
CB450	1.70	57.1

3.4 Adsorption kinetics of Cr

Total Cr was rapidly adsorbed onto biochars (Figure 2). Nearly 80% of the Cr adsorption capacity was reached within 24 h, and then, Cr adsorption increased steadily to reach equilibrium in about 72 h except CB250. In order to investigate the mechanism of adsorption and potential rate-controlling steps, the pseudo-first-order and pseudo-second-order models have been employed to fit the experimental data, and the kinetic parameters for Cr adsorption are listed in Table 3.

The calculated q_e values from the pseudo-first-order linear plots did not agree well with the experimental ones, indicating that the adsorption of Cr onto biochar probably did not follow the pseudo-first-order kinetic models. Compared with r^2 and q_e values, kinetics data for the adsorption of Cr from aqueous solution were in good agreement with pseudo-second-order kinetic models.

3.5 Adsorption isotherms

The amount of total Cr adsorbed by biochar increased with increasing initial Cr concentrations (Figure 3). In the low concentration region (<20 mg/L), Adsorption amount of Cr

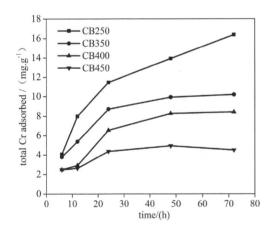

Figure 2. adsorption kinetics of Cr(VI).

Table 3. Kinetic parameters for Cr adsorption on biochar.

Adsorbent	Pseudo-first-order			Pseudo-second-order		
	q_e/mg/g	K_1	R^2	q_e/mg/g	K_a	R^2
CB250	16.16	0.05	0.98	21.31	0.002	0.98
CB350	10.33	0.07	0.98	12.26	0.008	0.98
CB400	9.04	0.05	0.96	11.97	0.003	0.91
CB450	4.74	0.09	0.88	5.16	0.028	0.97

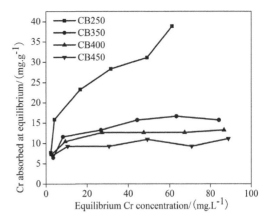

Figure 3. Adsorption isotherms of Cr(VI).

Table 4. Isotherms parameters for Cr adsorption on biochar.

Adsorbent	Freundlich			Langmuir		
	K_f	n	R^2	q_{max}	K_L	R^2
CB250	6.79	2.40	0.91	41.46	0.09	0.95
CB350	5.43	3.75	0.84	17.12	0.19	0.99
CB400	6.70	6.10	0.88	13.92	0.34	0.99
CB450	6.52	8.92	0.68	10.76	0.39	0.97

concentration increased rapidly. However, when the equilibrium concentration was more than 40 mg/L, Cr adsorption capacity did not nearly change. With the pyrolysis temperature increasing, Cr adsorption capacity was decreased. It had close trend with Acidic functional group and cation exchange capacity of biochar. Meanwhile, under acidic conditions, Cr^{6+} in the aqueous solution was reduced to Cr^{3+}, which was easily adsorbed by biochar.

Freundlich and Langmuir equations were fitted to the Cr adsorption data. The kinetic parameters for Cr adsorption are listed in Table 4. The calculated q_{max} and r^2 values from Langmuir equations were in good agreement with the experimental ones, indicating that the adsorption of Cr onto biochar followed the Langmuir models.

3.6 pH edge effect on Cr adsorption

pH is an important parameter affecting adsorption of Cr. Figure 4 demonstrates that the adsorption of Cr depends mainly on the pH of the aqueous solutions. The Cr adsorption decreased with increased solution pH. This is likely because the surface of the biochar bears more negative charge as the pH increases, causing a greater electrostatic repulsion

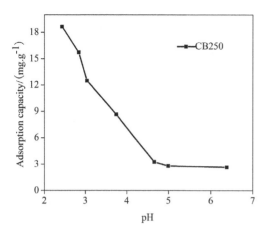

Figure 4. Effect of final pH on Cr(VI) adsorption.

toward the more negatively charged $Cr_2O_7^{2-}$. In addition, under strong acid conditions, Cr^{6+} is reduced to Cr^{3+}, the adsorption capacity increases, meanwhile, Cr^{3+} ion radius than Cr^{6+} ion', is more conducive to the occurrence of surface microporous adsorption (Bao, 2013).

4 CONCLUSIONS

This study investigated the ability of biochars produced from cornstalk-based biochar (CB) at a temperature range of 250~450°C to adsorb Cr(VI) from aqueous solution. The adsorption capacity and removal efficiency were controlled by solution pH, contact time, initial ion concentration. The adsorption kinetics and isotherm experiments showed that pseudo-second-order and Langmuir models can describe Cr(VI) adsorption on the biochars. XPS, FTIR Boehm titration and CEC indicated that the substantial acidic functional groups on CB surface played a crucial role on the adsorption of Cr(VI). The characteristics of biochar adsorption of Cr(VI) was closely related to the pyrolysis temperature.

ACKNOWLEDGMENTS

This research was financially supported by Hubei Hanjiang Watershed water security key technology research and demonstration (No. 2012ZX07205002-02-02)

REFERENCES

Beesley L, Moreno-Jimenez E, Gomez-Eyles J, 2010. Effects of biochar and greenwaste compost amendments on mobility bioavailability and toxicity of inorganic and organic contaminants in a multi-element polluted soil. Environmental Pollution. 158(6): 2282–2287.
Bernd W, Ernst A. Elmar S, et al. 2014. Phosphorus bioavailability of biochars produced by thermo-chemical conversion. J. Plant Nutr. Soil Sci. 84–90.
Guodan C, Qing H, Kai-song Z. Effect of temperature and duration of pyrolysis on properties of bio-dried sludge biochar. Chinese Journal of Environmental Engineering, 7(3): 1133–1138 (in Chinese).
Hanfeng B, Weiwei Y, Liqiu Z. 2013. efficiency and kinetics of heavy metals to remove the sludge based activated carbon. China Environmental Science. 33(1): 69-74.
Inyang M, Gao B, Yao Y, et al. 2012. Removal of heavy metals from aqueous solution by biochars derived from anaerobically digested biomass. Bioresource Technology. 50–56.
Lehmann J, Skjemstd J, Sohi S, et al. 2008. Australian climate-carbon cycle feedback reduced by soil black carbon. Nature Geoscience. 1(12): 832–835.
Mukherjee A, Zimmerman A, Harris W. 2011. Surface chemistry variations among a series of laboratory-produced biochars. Geoderma. 163(3-4): 247–255.
Verheijen F, Jeffery S, Bastos A.C, et al, 2009. Biochar application to soils–a critical scientific review of effects on soil properties processes and functions. EUR 24099 EN Office for the Official Publications of the European Communities, Luxemburg, 149.
Yao Y, Gao B, Chen H, et al. 2012. Adsorption of sulfamethoxazole on biochar and its impact on reclaimed water irrigation. Journal of Hazardous Materials. 209–210, 408–413.

*Advances in Materials Science, Energy Technology
and Environmental Engineering – Patty & Zhou (Eds)*
© 2017 Taylor & Francis Group, London, ISBN 978-1-138-19668-1

A hierarchical analysis of performances of powder extinguishing agents

Shuai Wang
Tianjin Fire Research Institute, Tianjin, China

ABSTRACT: Powder extinguishing agents are considered as one of the most widely applied fire extinguishing agents used currently all over the world. This paper introduces classical classifications of powder extinguishing agents and elaborates performance requirements of powder extinguishing agents based on the compilation of current national, regional, and international standards and most recent research achievements. The paper also generalized and recommends relevant test methods and pros and cons accordingly with a view to offer readers more valuable resources for future research and development.

1 INTRODUCTION

Powder extinguishing agent is regarded as one of the most classical and effective extinguishing agents in fire protection. This paper aims at elaborating different test methods currently available based on distinct chemical properties and performances with a view to help prove fire performance as a final collective purpose.

This paper begins with a brief introduction of powder extinguishing agent as a whole, and explain in detail test methods from different prospectives of chemical performances of powder extinguishing agents.

2 INTRODUCING A POWDER EXTINGUISHING AGENT

Powder extinguishing agent is a kind of fine powder which is dry and easy to flow and is used to extinguish fire. The powder extinguishing agent is based on inorganic ions, which have fire extinguishing performance and have additives with special physical properties of hydrophobic property, caking, and fluid modified property. Most classical powder extinguishing agent is applied in daily lives, which is classified as BC powder extinguishing agent and ABC powder extinguishing agent. The BC powder extinguishing agent is used to extinguish class B (liquid) and class C (gas) fire while ABC powder is to extinguish class B, class C, and class A (solid) fire. According to the circumstances in China specifically, the main base material of BC powder extinguishing agent is sodium bicarbonate, while monoammonium phosphate and/or ammonium sulfate are/is the main base material(s) of the ABC powder extinguishing agent.

3 MAIN CHEMICAL COMPONENTS

3.1 General components

Generally, powder extinguishing agents are compounded and combined with main fire extinguishing agents (base materials) and other additives with physical properties to improve the physical properties of powder extinguishing agents, i.e. dewatering, resistance to caking and lumping, and fluidity (Yang Liang, et al., 2015).

3.2 Base materials

Base materials play a decisive role in fire performances in powder extinguishing agents. There are some most common base materials currently used in powder agent, e.g. ammonium phosphate, sodium bicarbonate, potassium sulfate, potassium bicarbonate, potassium chloride, ammonium sulfate, sodium chloride, barium chloride, etc. Ammonium phosphate and ammonium sulfate have properties to extinguish Class A fire and could be applied as base materials of ABC powder, while other compounds are applied in BC powder. In accordance with recent research outcomes, some inorganic compounds (e.g. aluminum salt, copper salt, sodium salt, potassium salt, etc.) have high fire extinguishing performances but cannot come into actual application due to many realistic considerations of cost, toxicity, strong moisture absorption, etc. (ZHOU Wenying, et al., 2004).

3.3 Additives

Powder extinguishing agents are solid powder with small particle size. The particles and gas are combined and pressurized to spray towards flame in fire extinguishing process. In accordance with researches, the researcher finds out that long time storage, impact, and strong shock will influence the physical properties of powder, which will directly lead to a harmful impact on the fire performance of powder extinguishing agents. Additives are applied under such circumstances with a view to improve the fluidity of the powder, to reduce the impact of moisture to powder, and to develop the resistance to caking and lumping of powder agents.

4 PERFORMANCES AND TEST METHODS

4.1 *General*

Based on the research and analysis on the chemical and physical compounds and combination in powder extinguishing agents, relevant performances and test methods have developed and standardized to evaluate the general characteristics of powder agents.

4.2 *Main chemical components*

Main chemical components are referred to as the dominant effective fire extinguishing component. Currently, different standards have succinct requirements about the components of the main chemical components. In China's GB standard, the general requirement may be 75% or more of the total content. ISO regulates to have 90% or more of the total content (GB4066.2:2004).

About test methods, China standardized to use the combustion method to test components of BC powder, and regulates to use the precipitation method to test monoammonium phosphate (GB4066.1:2004). EU regulates to use the titration method. ISO working group of powder extinguishing agents preferred to use the titration method as an informative method to be added in ISO 7202, based on the consideration of international circumstances and various methods available in the world (ISO 7202:2012).

4.3 *Moisture absorbance*

Moisture absorbance refers to the absorbance of the sample powder extinguishing agent exposed in certain test temperature and comparative humid environment. Moisture absorbance is an important index to test anti-humidity of powder and also directly influence the other performances of powder extinguishing agents, such as resistance to caking and lumping and electrical insulation.

Ordinarily, there are two test methods, the constant temperature and humidity oven test and the saturated ammonium chloride method.

The test method of constant temperature and humidity oven test is to locate powder sample in a

Table 1. Test method of MAP in powder extinguishing agents (translated from China's national standard).

Reagent
Sodium molybdate, citric acid, nitric acid, distilled water, quinolin, aceton, nitric acid solution

Solution preparation
Solution A—Place 70 g sodium molybdate into 400 ml beaker and solute it with 100 ml distilled water;
Solution B—Place 60 g citric acid into 1000 ml beaker, solute it with 100 ml distilled water, and add 85 ml nitric acid;
Solution C—Add Solution A into Solution B and mix;
Solution D—Mix 35 ml nitric acid and 100 ml distilled water in 400 ml beaker, and add 5 ml quinoline;
Add **Solution D** into **Solution C** and mix them. Place the solution for one night, filter with the filter paper or cotton, and add 280 ml aceton into the filtrate, dilute it to 1000 ml with the distilled water, then store the solution in the brown volumetric flask in dark place.

Test method
a) Weigh 1 g precisely of test sample, accurate to 0.0002 into 100 ml beaker. Add 2 ml acetonum in the beaker and stir completely.
b) After volatilization of aceton, add a little (60–70)_ distilled water to beaker, solute and filter. Use 250 ml distilled water to wash the insolubles, and collect the filtrate and washing solution into 500 ml volumetric flask. Dilute it to 500 ml with the distilled water and shake it up. The 500 ml diluted solution is Sample Solution A.
c) Move 25 ml Sample Solution A to 400 ml beaker with the pipet, add 10 ml nitric acid solution, dilute it to 100 ml with the distilled water and heat it to nearly boiled. Add (40–45) ml quinoline Mo lemon ketone reagent, cover the watch glass, boil for 1 min on the closed electric furnace, and place the beaker at room temperature and let the liquid cool down gradually.
d) The crucible filter should be dried for 45 min at $(180 \pm 2)°C$. Filter the top clear liquid first and wash the sediment with 100 ml distilled water. Place the sediment and crucible filter together in dry oven for 45 min at $(180 \pm 2)°C$. Move the dried item to desiccator for 45 min. After cooled down, weigh the item.

Result
Calculate the ammonium dihydrogen phosphate content of the sample X1(%) from the equation (A.1):

$$X_1 = m_1 \times 1.0396 \times 100/m_0 \dots\dots\dots\dots\dots\dots\dots(A.1)$$

where:
m_0—mass of the sample, g;
m_1—mass of the quinoline phosphomolybdate sediment, g.

test oven with the environment of $(21 \pm 3)°C$ and 78% comparative humidity for 24 hours and calculate moisture absorbance accordingly. This method could test many samples at the same time, but will result in comparatively large errors due to different

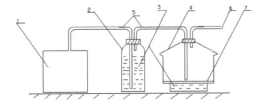

key

    1 air supply

    2 bottle

    3 saturated ammonium chloride solution;

    4 Φ250 mm humidistat;

    5 Φ6 mm glass tube;

    6 air outlet;

    7 pore plate

Figure 1.   Saturated ammonium chloride test facility.

locations with succinct humidity in the oven. The test method of saturated ammonium chloride is to put sample in the test oven of saturated ammonium for 24 hours with constantly injecting air with condition of 5 L/min and 78% comparatively humidity. Currently, standards from China, USA, and Canada are developed to use this method.

4.4   Resistance to caking and lumping

This performance is to test the cohesion ability of the sample after experience of moisture absorbance and dryness cycle. Based on the current ISO standard regulation, the procedure is to place an excess of powder in the Petri dish, place the Petri dish in the desiccators at $(20 \pm 5)°C$ for (21 ± 3) h, and then place them in the oven for (24 ± 1) h. Cover the Petri dish and allow it to cool at (20 ± 5) °C for (60 ± 10) min, and then remove the cover and overturn the Petri dish on to a clean sheet of paper. Let the powder slide into a sieve in a way that formed lumps are not crushed. Shake the sieve with revolving movements horizontally to separate the formed lumps from free flowing powder without crushing the lumps. Using a spatula to lift the lumps, drop them from a height of (200 ± 10) mm onto a clean sheet of paper, which is placed on a hard surface. Let the powder slide onto the sieve again and shake with revolving movements horizontally to separate the formed lumps from free flowing powder as described earlier.

4.5   Bulk density

Bulk density refers to the ratio of mass and volume of the sample powder when the powder is at natural loose status. This performance is an important requirement for filling powder in extinguishers. Generally, the bulk density value can be more than 0.85 g/cm³ for BC powder and more than 0.80 g/cm³ for ABC powder based on the current ISO standard of powder.

The test method is to place 100 g ± 0,1 g of the powder in a clean, dry, 250 ml stoppered glass measuring cylinder. Secure the stopper in the cylinder. Rotate the cylinder end over end for ten complete revolutions, at approximately one revolution every 2 s. immediately after the ten revolutions have been completed, set the cylinder upright on a level surface and allow the powder to settle for 180 s. Calculate the bulk density afterwards.

4.6   Water repellence

Water repellence is to test the repellence ability of powder extinguishing agent in contact with liquid water. This performance is to test the particle size of dewatered white carbon black and technique of silication in producing the powder agent. There are many methods currently available, and China has regulated to use the "dripping method". The procedure is to drip 3 drops water in the Petri dish filled with the smooth surfaced sample powder. Place the Petri dish in the oven under $(20 \pm 5)°C$ and 75% comparatively humidity for 1 hour. Slip off 3 drops of water afterwards and see if there are any differences on the contact surface between the powder and water drops.

4.7   Electrical insulation value

This performance is to evaluate breakdown voltage of powder in experiencing required number of cycles of drop test. Electrical insulation directly influences the safety in handling a fire extinguisher in extinguishing class E (electrical fire) fire.

A test cup can be used in the test. The procedure is as follows: fill the test cup with the extinguishing powder and compact it by dropping the cup 500 times at a frequency of 1 Hz through a height of 15 mm using an impact machine. The impact machine submits the cup to repetitive shock pulses by dropping it from the height given above onto a solid surface. The dropping operation shall be guided and have acceleration approaching free fall. The cup may, if desired, be clamped in a suitable protective casing during this procedure. Using the transformer, apply an electrical potential to the electrodes, increasing the potential at a uniform rate until breakdown occurs as indicated by a continuous discharge across the gap between the electrodes. Record the voltage as the dielectric breakdown strength.

4.8 Sieve analysis

Sieve analysis is considered as one of the most important test indexes in test performances of powder extinguishing agents. It influences the performance and fluidity of the powder and will especially affect fire performance. In that case, this performance is always of high attention in developing the relevant standard. One of the most classical methods is the sieve method. The sieve method is to use a sieve-shaking device. Accurately weigh approximately 20 g $\pm$ 0,02 g of the powder into the top sieve. Assemble on the shaking device and shake for 10 min. Weigh the quantity of powder retained on each sieve and report as the cumulative percentage of the original sample mass retained (Zhao Ting, 2002).

4.9 Fire test performance

Fire test is the ultimate purpose in testing any performances of powder extinguishing agents. In China's standard, the fire test is evaluated in two fire models, Class B and Class C fire and Class A fire. BC powder can test Class B and Class C fire while ABC powder can test two fire models accordingly. The method to test fire extinguishing performance is to use 3 kg of manual extinguisher to extinguish oil fire. It is considered to be qualified to test Class B and Class C, if oil fire could be extinguished completely. It is also required to use 3 kg manual extinguisher to extinguish wood crib fire. If the wood crib fire could be extinguished within 15 minutes, the agent is qualified to extinguish Class A fire.

5 CONCLUSION

There are also some other performances which have received great attention in the field of powder extinguishing agents. The performances have been regulated in different international, regional, and national standards in the world. However, the core attention of the research is to try to evaluate the chemical content of the powder extinguishing agents, and to test if the fire performance of powder will be changed and differed after long time of storage, treatment, and shipment (Shu, 1999).

In order to polish the deficiencies of ordinary powder extinguishing agents, more new products are available in the market. Superfine powder is one of the most promising powder agents, offering an effective pathway to fire performances and also calls for the need for future standard development of the subject products.

REFERENCES

International Organization of Standardization. ISO 7202: 2012 *Fire protection—Fire extinguishing media—Powder.*

Shu Zhongjun, Opportunity and Challenges of powder extinguishing agents. *Annual paper book of CFPA (1999).*

Standardization Administration of China. GB 4066.1–2004 *Fire extinguishing media—Part 1: BC Powder.*

Standardization Administration of China. GB 4066.2–2004 *Fire extinguishing media—Part 2: ABC Powder.*

Yang Liang, et al., *Fire extinguishing agents and standardization.* Scientific and technical documentation press, 2015.

Zhou Wenying, et al., Introduction of new type of powder extinguishing agent. *Fire technology and products information,* Chapter 5 (2004).

Zhao Ting, Sieve analysis of powder extinguishing agents. *Fire technology and products information,* Chapter 1 (2002).

Advances in Materials Science, Energy Technology
and Environmental Engineering – Patty & Zhou (Eds)
© *2017 Taylor & Francis Group, London, ISBN 978-1-138-19668-1*

The exploration and control methods of mine dust particles

Z.T. Xu, J.Y. Wang & M.X. Han
*Key Laboratory of Clean Combustion for Electricity Generation and Heat-supply Technology of Liaoning
Province, Shenyang Institute of Engineering, Shenyang, China*

ABSTRACT: To eliminate the harm of coal mine dust prevention and control work, there are solutions: one is having the technical equipment and the other is the management measure. Under the current situation of Chinese coal mine dust technology and equipment, which has lagged behind, it is very necessary to establish a set of strong dust prevention and integrated management system, which will create good working conditions for the safe production of coal mines.

1 INTRODUCTION

With the constant increase in exploitation scale and strength in mining, the extraction, excavation, transportation, and other productive progresses of coal mines have increased dramatically the dust amount. Especially, the respirable dust concentration shows a substantial upward trend. In addition, when it comes to the issues which related to coal mine dust, it is only a simple analysis of the results to analyze the causes of excessive dust. If the causes` analysis and control measurements of exceeding standard only aim at the key control points, then it lacks systematic and intensive instructors to further improve the level of dust hazard prevention to coal mining enterprises.

2 THE HARM OF DUST PARTICLE

Weather experts and medical experts believe that the fine particulate matter caused by the haze hazards an even greater risk to human health than storms. Particle size of 10 microns or more can be blocked out of the nose; the particle size between 2.5 to 10 micron particles can enter the respiratory tract, but some may be excreted through sputum or get blocked by nasal hairs inside the barrier. The harm to human health is relatively small. However, the particle size of 2.5 microns or less fine particulate matter, equivalent to the diameter of a human hair 1/10 size, cannot easily be blocked. Directly after being inhaled into the bronchial, pulmonary gas exchange gets interfered with and can cause asthma, bronchitis, and cardiovascular and other diseases. On average, a person inhales about 10,000 liters of air each day. The dust that is inhaled into the alveoli can be absorbed quickly without liver detoxification and can be directly distributed to the body through blood circulation.

Table 1. Average concentrations of heavy metals.

Element	Size ≤ 2.0 µm	Size > 2.0 µm
Ca	2.328	3.761
Mg	2.082	3.123
Al	1.870	4.363
Mn	0.038	0.057
Fe	1.824	3.388
Zn	0.296	0.127
Pb	0.280	0.115
V	0.004	0.002
Cu	0.021	0.018
Co	0.008	0.009

Second, it can damage the ability of hemoglobin to transport oxygen and cause blood loss. For example, it can aggravate respiratory diseases, and lead to congestive heart failure and coronary heart disease amongst others. Briefly, these particles can also enter the bloodstream through the bronchi and alveoli. The harmful gases and heavy metals dissolved in the blood cause greater damages to human health. The physical structure of a human body decides the non-ability to filter and block PM 2.5. However, the harm of PM2.5 to human health exposes its terror gradually with the advances in medical technology. The main harmful metal content of PM 2.5 can be seen in Table 1.

3 DUST PREVENTION AND CONTROL WORK, STILL WITH PROBLEMS

3.1 *Lack of awareness in dust hazards*

Some grass-roots local officials and business leaders understand the importance of working safely in these environments, but lack awareness in the potential harm of pneumoconiosis, and therefore pay inadequate attention. Meanwhile, with not enough knowledge of dust hazards, the underground line workers generally work without wearing dust masks. Adding to the risk, their dust-proof equipment is often idle. All these negative factors certainly will endanger the physical health of workers and shorten the life of precision instruments. What is more, they will even cause dust explosions.

3.2 Insufficient understanding of the importance of dust prevention and control work

Owing to the latent dust hazards, business leaders focus more on emergency gas disasters and the management of dominant disasters. Industry-related policies and regulations on dust control are not perfect, one reason being inadequate staff training. This factor enables dust prevention to remain in a frozen state.

3.3 Imperfection of system construction in enterprises

Many coal mining enterprises do not establish a sound and integrated management system to prevent dust, and some companies do not even have a special dust-proof structure. This causes a "headless state" or "long state" chaos, and aids difficulties and risk to human health when working with dust.

3.4 A serious sufficient input of coal mine dust work

Since the enterprises bias to the awareness of dust prevention, the investment of funding and human resources has seen a severe shortage. Investment of resources is insufficient to introduce advanced dust equipment, and the available dust equipment lacks specialist management and maintenance, meaning that the effectiveness of dust work is relatively poor.

4 FEASIBILITY OF INTRODUCING PM2.5 DUST CONTROL STANDARDS

4.1 Theoretical feasibility

The concept of PM2.5 in the atmosphere has been applied. The new standard was introduced in 2006 by the United States. It reduced the 24 h concentration limit of PM2.5 from 65 $\mu g/m^3$ to 35 $\mu g/m^3$ and improved the PM2.5 concentrations limit. In recent years, China has also introduced the concept of PM2.5 into the governance standards of

atmosphere. Compared with the atmosphere, there is also coal mine PM2.5, but slightly different in terms of concentration and composition. Thus, in theory, the introduction of coal mine dust control PM2.5 is feasible.

4.2 PM2.5 detection feasibility

In coal mines, the total coal dust and respirable dust concentration detection system is the main way to detect underground dust in any country on the planet. There is no professional coal mine testing equipment for PM2.5. However, in recent years, atmosphere PM2.5 dust detection systems have been developed rapidly. Systems such as the LD310 aerodynamic particle size spectrometer and the handheld 3016IAQ can achieve real-time monitoring to PM2.5 concentrations. Compared with the atmospheric environment, the underground PM2.5 testing equipment needs to meet some special requirements, such as being explosion-proof and intrinsically safe. Therefore, PM2.5 dust detection system can be introduced to the coal mine with modifications. In fact, there are articles to describe PM2.5 testing. Nie, Baisheng adopted a DRX laser dust concentration instrument to measure the distribution of fully mechanized mining around PM2.5 in a Shanxi coal mine with other people. The results show that: the PM2.5 concentration reaches a maximum 380.74 mg/m^3 at the point of 5-meter length in the leeward side of a shearer and takes 69.2% in total dust concentration. Thus, it can be achieved to have real-time monitoring of PM2.5 in the coal mine. In conclusion, it is feasible to introduce technical PM2.5 detection in coal mines.

5 RESEARCH DIRECTION

The components, and their existing form and pathogenesis of coal mine PM2.5 research needs to be further strengthened. Compared with atmospheric dust, the coal mine dust components are more complex and have more harmful ingredients. This complexity and diversity of harmful ingredients in underground PM2.5 will affect the incidents of pneumoconiosis, and become one of the factors to standardize PM2.5.

The distribution of the coal mine PM2.5 concentration range needs further study. Currently, only a few researchers focus on underground PM2.5. The distribution and the concentration change of underground PM2.5 are not clear. Therefore, it is not feasible to copy the PM2.5 limits in the atmosphere compared with the coal mine. The relevant research needs to be further strengthened in order to make the key points and difficulties of underground dust governance explicit and provide the basis for the future development of standardizing PM2.5.

The control technology and equipment for coal mine PM2.5 needs to be updated and improved. All along, the coal mine management has been focusing on particle size which is less than 5μm, and respirable. However, to the dust particle size less than 2.5μm, not enough attention has been paid. After the introduction of PM2.5 to the underground dust control, it is bound to bring higher requirements to the existing dust control technology and equipment. On the one hand, the existing technology and equipment needs to be updated and transformed. On the other hand, it needs to constantly introduce a new technology of dust proof and fall, such as magnetic dust removal, electro-coagulation dust removal, ultrasonic nebulization technique of dust removal, and other new technologies to enhance the removal of PM2.5.

6 INTEGRATED MANAGEMENT OF EXCESSIVE DUST

6.1 *Management of unsafe behaviors*

To eliminate the harm of coal mine dust prevention and control work, two solutions were presented: one is having the technical equipment and the other is the management measure. The management of unsafe behaviors has been systematically described in "Coal mine safety risk and pre-control management system standards". It contains the links of crew access management, the identification and conclusion of unsafe behaviors, the standard of post, the control and formulation of unsafe behaviors, the staff training and educating, the supervision of crew behaviors, and the establishment of employee files amongst others. It forms a PDCA closed-loop system of crew management.

6.2 *Organization and management system*

The technical specification of laws and regulations, which related to coal occupational disease prevention has explicitly stipulated that the coal mining enterprises should establish sound occupational hazard prevention and mechanism management. They should also equip a full-time manger to be responsible for the daily management of occupational hazard prevention control work. "The Mine workplace occupational hazard prevention and control (Trial)" clearly requires that coal mining enterprises should establish and improve occupational hazard prevention responsibility systems, occupational hazard prevention plans, implement programs, occupational hazards notification systems, occupational hazards prevention publicity

and education systems, occupational hazard protection facilities management system, and others. In total, 14 occupational hazard prevention and control systems need attention. The establishment and improvement of the system and mechanism are the effective guarantees to proper and well-organized dust prevention work.

7 CONCLUSIONS

PM2.5 is the major risk factor which causes pneumoconiosis. In China, each link of coal mine work can produce a large amount of PM2.5. Furthermore, the high concentration of free metal ions in dust increases greatly the danger of PM2.5. It affects seriously the health of mine workers. Additionally, due to the lack of restrictions which related to the management standard and no specific detectors, it is not sure if common measures of dust removal are effective to PM2.5 dust. Therefore, it is necessary to introduce PM2.5 standard to coal mine dust control. According to the requirements of 'rule' NO.73, with the construction of a perfect system, relying on scientific management, increasing investment, the development and introduction of advanced dustproof technology and equipment, and the improvement of technological content of the equipment, it will achieve the scientization of dust prevention management and automation of integrated dust proof, and therefore, be able to prevent the occurrence of accidents.

REFERENCES

Boffetta P, Jourenkova N, Gustavsson P. 1997. Cancer risk from occupational and environmental exposure to polycyclic aromatic hydrocarbons. *Cancer Causes & Control.* 8(3): 444–472.

Cai Qy, Mo Ch, Wu Qt. 2008. The Status of Soil Contamination by Semivolatile Organic Chemicals (SVOCs)in China: A Review. *Science of the Total Environment.* 389(2/3): 209–224.

Cao, Y.Z. 2012. Composition and content analysis of polycyclic aromatic hydrocarbons in surface soil in China. *Journal of Environmental Science.* 32(1): 197–203.

Hu, J. 2005. Study on polycyclic aromatic hydrocarbons (PAHs) in air water soil environment of Guiyang City. *Institute of geochemistry*, Chinese Academy of Sciences.

Li, J. 2005. Preliminary estimate of the economic loss of air pollution in Lanzhou City. *Gansu Science and Technology.* 21(9): 15–17.

Li, Y.S. 2014, Mining of coal mine dust control analysis. *Modern Science.* 05:107.

Li, Q. 2013. Using laser radar and satellite remote sensing to obtain the concentration distribution of suspended particulate matter in urban land. Journal of Peking University. *Natural Science Edition.* 49(4): 673–682.

Liu, Y.J. 2010. Characteristics of atmospheric particulate matter PM2.5 and black carbon concentration in spring of 2009 in Beijing City. *Powder technology in China*. 16(1): 18–22.

Pu, W.W. 2011. Influence of meteorological factors on PM2.5 pollution in late summer and early autumn in Beijing area. *Journal of Applied Meteorology*. 22(6): 716–723.

Wang, C. 2015. Characteristics and sources of polycyclic aromatic hydrocarbons in airborne particulate matter in Beijing, Tianjin and Hebei. *China Environmental Science*. 35(1): 1–6.

Yan, X.Y. 2007. Distribution characteristics of airborne particulate matter and heavy metal pollution in Shenyang City. *Science of environmental protection*. 33(3): 20–22.

Zhao, Q. 2009. Regional pollution characteristics of atmospheric particulate matter in Beijing and its surrounding areas. *Environmental science*. 30(7): 1873–1880.

Zhou, J.2013. The significance and approach of PM2.5 pollution control in urban atmosphere. *Environmental Research and Monitoring in Gansu*. 16(1): 29–31.

*Advances in Materials Science, Energy Technology
and Environmental Engineering – Patty & Zhou (Eds)*
© 2017 Taylor & Francis Group, London, ISBN 978-1-138-19668-1

Comprehensive treatment of wastewater derived from enzymatic preparation of L-methionine

Y. Mei, C. Hu & J. Zhang
School of Chemical and Environmental Engineering, Wuhan Polytechnic University, Wuhan, Hubei, China

ABSTRACT: High-concentration organic wastewater is derived from the production process of L-methionine by enzyme catalysis. It not only pollutes the environment, but also wastes the resource if discharged directly. In this study, the pH value of wastewater was adjusted to 3.0, and 46 g substances were recovered from 1 L wastewater, including methionine and N-acetyl methionine. Subsequently, the mixed wastewater was treated by A/O biological process, and the water quality of the effluent wastewater matched the second-degree requirements of the overall National Sewage Discharge Standard, when the retention time of anoxic and aerobic tanks and nitrification liquid reflux ration were 4 h, 8 h, and 200%, respectively.

1 INTRODUCTION

Methionine is an important amino acid, which is mainly produced in a large scale by oxidizing propylene to acrolein. The product is chiral DL-methionine and often used as an additive in feed industry. L-Methionine produced through enzymatic racemization is widely used in the pharmaceutical industry, which is an important raw material for drugs and precursor substances. However, this method often brings large volume of amino acid wastewater with high concentrations of COD_{Cr}, BOD_5 and nitrogen. If the wastewater is discharged directly, it not only wastes resources, but also causes eutrophication in the receiving water, harms to agriculture, fishery, tourism, and other areas, as well as pose a great threat to the safety of drinking water and food. Therefore, much focus has been given to the treatment of amino acid wastewater comprehensively.

Liang & Wei (1994) used chemical oxidation—biochemistry—flocculation to deal with the sewage coming from the preparation of L-methionine by the acrolein Hein chemical method. Xie & Huang (2001) proposed the recycling utilization and disposal of amino acid wastewater. Chen (2010) studied the pretreatment– hydrolysis—CASS—biological aerated filter system, which treats the amino acid wastewater from Yichang pharmaceutical factory. After being treated, the water quality of the effluent could comply with the national requirements of Fermentative Pharmaceutical Industrial Wastewater Discharge Standard (GB 21903–2008) (2008). Han et al (2014) treated the sewage by the new process of pretreatment—IC—

A/O—advanced catalytic oxidation and the water quality of effluent reached the first-degree requirements of GB 8978–1996 (1996). The purpose of this study is to solve the wastewater problem in the process of producing L-methionine by enzymatic preparation, which is able to recycle many useful substances and match the standards of wastewater discharge.

2 MATERIALS AND METHODS

2.1 Materials

Wastewater was collected from Hubei L-methionine manufacturer. Reagents were purchased from Sinopharm Chemical Reagent Co., Ltd (Shanghai, China), and were of analytical grade.

2.2 Methods

2.2.1 Determination of the COD_{Cr} of wastewater
Using the rapid digestion spectrophotometry method (HJ/T 399–2007) (2007), chemical oxygen demand of the water was determined.

2.2.2 Determination of the BOD_5 of wastewater
Determination of the five-day biochemical oxygen demand referred to the water quality standard (HJ 505–2009) (2009).

2.2.3 Determination of the concentration of total nitrogen in wastewater
Using the alkaline potassium persulfate digestion ultraviolet spectrophotometry method

(GB 11894 – 89) (1989), the total nitrogen content of the water was determined.

2.2.4 Determination of the concentration of ammonia nitrogen in wastewater

Using the Nessler's reagent spectrophotometry method (HJ 535–2009) (2009), ammonia nitrogen content of the water was determined.

2.2.5 Recycling amino acid and its derivatives

1 L of high-concentration methionine wastewater was taken and the pH was adjusted to acidity by adding hydrochloric acid. After being static for 12 hours, the wastewater was filtrated, the precipitate was dried at 80°C–85°C, and the filtrate was collected to be processed. Then, the dried precipitate was extracted with 1.5 L 95% ethanol, and separated by suction filtration to get the precipitate and the filtrate. The filtrate was distilled under reduced pressure to get the powdery solid, which was extracted with anhydrous ethanol. The insoluble was named solid A. The substances extracted with anhydrous ethanol were named solid B after being distilled under reduced pressure. The solids A and B were analyzed by infrared. The filtrate was distilled under reduced pressure on a rotary evaporator RE-52 A (Shanghai YaRong Biochemical Instrument Factory, Shanghai, China). The technological process is listed in Fig. 1.

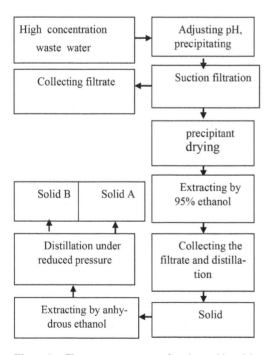

Figure 1. The recovery process of amino acid and its derivative.

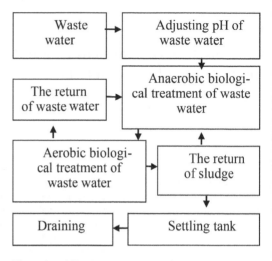

Figure 2. A/O treatment process of wastewater.

2.2.6 Infrared analysis of recycling products

Samples were prepared according to references [9]. Infrared spectroscopy Model FTIR-650 (Tianjin GangDong Sci. & Tech. Development Co,. Ltd) was used for this analysis.

2.2.7 Treatment of wastewater with A/O process

Wastewater (Materials and Methods 2.2.5) was treated with A/O listed in Fig. 2. The seeding sludge was collected from the settling tank of Wuhan Shahu Sewage Treatment Plant, and then the sludge was cultured and domesticated 30 days. During this period, the temperature was maintained at 21°C-26 °C. The sludge return ratio was 50% and the DO of the aerobic tank was 3–4 mg/L.

3 RESULTS AND ANALYSIS

3.1 Adjusting the pH value of wastewater and recovering amino acid and its derivatives

1 L of high-concentration wastewater was taken and the pH was adjusted to 3.0 using hydrochloric acid, and then, 185 g of the precipitate was collected. To further investigate the composition of precipitate and recover the useful substance, the precipitate was extracted with 95% ethanol, because the methionine and acetyl methionine were freely soluble in 95% ethanol. The insoluble substance was 139 g in weight and the filtrate weighed 46 g after being distilled and dried (at 80°C–85°C). It was found that 46 g of the solid should be the mixture of methionine and N-acetyl methionine through analysis of the preparation process of L-methionine. According to the physical characteristics of methionine and N-acetyl methionine, the mixture

was re-extracted with anhydrous ethanol, and the extract was obtained by the extraction of insoluble matter and anhydrous ethanol. After being filtrated, the insoluble matter was dried and weighed; it was 4.8 g, named as solid A. Then, the filtrate was distilled and dried, named B, it was 41.2 g.

3.2 Infrared analysis of solids A and B

Two kinds of solid samples were prepared for infrared analysis. The results of infrared spectra are shown in Figs 3 and 4. From the IR spectra peak of Fig. 3 and combined with organic functional groups standard

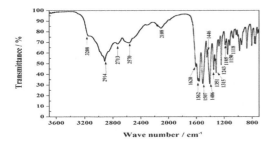

Figure 3.   IR spectra of solid A.

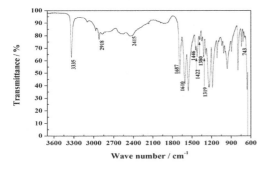

Figure 4.   IR spectra of solid B.

infrared spectrum, the following conclusions were drawn: 2914 cm^{-1} was the stretching vibration peak of -CH and -CH$_2$; 1446 cm^{-1} and 1150 cm^{-1} were the bending vibration peak of -CH2; 1351 cm^{-1} was the bending vibration peak of -CH; 1118 cm^{-1} and 1150 cm^{-1} were the stretching vibration peak of C-O; 3208 cm^{-1} was the stretching vibration peak of O-H, 1446 cm^{-1} was the bending vibration peak of O-H in carboxylic acid; 1590–1600 cm^{-1} and 1409 cm^{-1} were stretching vibration peak of C = O; 1185 cm^{-1} was the stretching vibration peak of O-C = O; 1580–1650 cm^{-1} was bending vibration of -NH$_2$; 1243 cm^{-1} was the stretching vibration peak of C-N; 1507 cm^{-1}, 1562 cm^{-1}, and 2108 cm^{-1} were the bending vibration peak for N-H; and 2570 cm^{-1} was the vibration peak for -S-H. According to the peak that represented the functional groups, the solid A was methionine.

From the IR spectra peak of Fig. 4 and combined with organic functional groups standard infrared spectrum, the following conclusions were drawn: 3335 cm^{-1} was the stretching vibration peak of -OH and N-H; 2451 cm^{-1} was the vibration peak for—S-H; for carboxyl functional groups, -C-O, -C = O and -OH separately appeared on 1289 cm^{-1}, 1687 cm^{-1} and 3335 cm^{-1}; 1687 cm^{-1} was the stretching vibration peak of C = O; 1610 cm^{-1} and 743 cm^{-1} was the bending vibration peak of -N-H; 2918 cm^{-1} was the stretching vibration peak of -CH$_3$; 1380 cm^{-1} and 1446 cm^{-1} were the bending vibration peaks of -CH$_3$; 1422 cm^{-1} and 1319 cm^{-1} were the bending vibration peaks of -CH$_2$; and 1319 cm^{-1} and 1380 cm^{-1} were the bending vibration peaks of -CH. According to the peak that represented the functional groups, the solid B was N-acetyl methionine.

3.3 Treatment of wastewater with A/O process

After recovery of useful substances and desalination from the high-concentration amino acid wastewater, the filtrate was mixed with other wastewater and the terminal COD$_{Cr}$ ranged from

Table 1.   Results of the treatment of amino acid mixed wastewater.

Influent quality (mg/L)			Treatment time (h)		Reflux ratio	Effluent quality (mg/L)				
COD	BOD	TN	Anoxic	Aerobic		pH	COD	BOD	TN	NH$_3$-N
1260	869.4	75.8	3	6	100%	7.1	261	83	20	7
1260	869.4	75.8	3	6	200%	6.8	247	70	16	6
1260	869.4	75.8	4	8	100%	7.0	102	45	3	1
1260	869.4	75.8	4	8	200%	7.1	83	27	3	1
1980	1375	132	3	6	100%	7.2	307	158	60	19
1980	1375	132	3	6	200%	7.1	285	136	45	18
1980	1375	132	4	8	100%	7.1	210	102	30	11
1980	1375	132	4	8	200%	7.3	181	76	21	11

1260 to 1980 mg/L, BOD_5 from 869 to 1375 mg/L, and Total Nitrogen (TN) from 76 to 132 mg/L. The mixed wastewater had good biodegradability performance ($BOD_5/COD_{Cr} > 0.3$, $BOD_5/TN > 3.5$). According to the characteristics of wastewater, A/O biological treatment process was designed (Fig. 2) and the results are shown in Table 1. When the water ingressed with a low concentration of COD_{Cr} (1260 mg/L), HRT in anoxic and aerobic tanks were 4 h and 8 h, respectively; the reflux ratio of nitrification liquid was 200%; the COD_{Cr}, BOD_5, TN, and NH_3-N of the quality of effluent were 83 mg/L, 27 mg/L, 3 mg/L, and 1 mg/L, respectively, when with high concentration of COD_{Cr} (1980 mg/L), the COD_{Cr}, BOD_5, TN, and NH_3-N were 181 mg/L, 76 mg/L, 21 mg/L, and 11 mg/L, respectively. According to the results above, it has been demonstrated that the drainage after being treated was able to achieve the secondary discharge standard of Integrated Wastewater Discharge Standard issued by the Ministry of Environmental Protection (GB 8978–1996) (1996).

4 CONCLUSIONS

Although progresses have been made in treatment of amino acids wastewater, due to that the characteristics of wastewater was different, subsequently, the treatment of wastewater was different. This study designed the process to recover useful substances from high-concentration amino acid wastewater derived from the enzymatic preparation of L-methionine, and treat the filtrate via A/O process. The results showed that 46 g of useful substance was recovered from 1 L of high-concentration wastewater. Using the A/O biological treatment process, we could remove most of COD and nitrogen from wastewater, and the quality of effluent was able to achieve the secondary discharge standard of Integrated Wastewater Discharge Standard issued by the Ministry of Environmental Protection (GB 8978–1996) (1996).

REFERENCES

Chen, L. 2010. Treatment processes of amino acids wastewater. *Wuhan: Wuhan University of Science and Technology* 15–20.

Han, Z.F., Wang, M., Li, H.S., Mai, W.L. & Du, J.X. 2014. Treatment of amino acids wastewater with IC-A/O-advanced catalytic oxidation process pretreatment. *Technology of Water Treatment* 40(11): 125–128.

Liang, Y.Q. & Wei, Q.W. 1994. Treatment of waste water derived from preparing of methionine. *Environmental Protection of Chemical Industry* 14(2): 66–71.

Lu, Y.Q. & Deng, Z.H. 1989. *Practical infrared spectrum analysis.* Beijing: Publishing House of Electronics Industry.

Ministry of Environmental Protection of the People's Republic of China. 2007. HJ/T 399–2007, water quality-Determination of the chemical oxygen demand-Fast digestion-Spectrophotometric method. Beijing: Standards Press of China.

Ministry of Environmental Protection of the People's Republic of China. 2009. HJ 505–2009, water quality-Determination of biochemical oxygen demand after 5 days (BOD5) for dilution and seeding method. Beijing: Standards Press of China.

Ministry of Environmental Protection of the People's Republic of China. 2009. HJ535–2009, water quality-Determination of ammonia nitrogen-Nessler's reagent spectrophotometer. Beijing: Standards Press of China.

Ministry of Environmental Protection of the People's Republic of China. 1996. GB 8978–1996, integrated wastewater discharge standard. Beijing: Standards Press of China.

Ministry of Environmental Protection of the People's Republic of China. 1989. GB 11894–89, water quality determination of total nitrogen-alkaline potassium persulfate digestion UV spectrophotometric method. Beijing: Standards Press of China.

Xie, S.X. & Huang, G.H. 2001. Recovery and treatment of amino acid wastewater. *Industrial Water Treatment* 21(2): 38–39.

Advances in Materials Science, Energy Technology and Environmental Engineering – Patty & Zhou (Eds)
© 2017 Taylor & Francis Group, London, ISBN 978-1-138-19668-1

Synthesis of an ESA/AMPS copolymer and evaluation of its scale inhibition performance

Zhenfa Liu, Yuelong Xu, Lihui Zhang, Haihua Li & Yuhua Gao

Institute of Energy Resources, Hebei Academy of Science, Shijiazhuang, China
Hebei Engineering Research Center for Water Saving in Industry, Shijiazhuang, Hebei, China

ABSTRACT: A polyepoxysuccinic acid copolymer (ESA/AMPS) was synthesized from 2-Acrylamido-2-Methylpropane Sulfonic acid (AMPS) with Epoxysuccinic Acid (ESA) prepared from Maleic Anhydride (MA). The ESA/AMPS copolymer was characterized by FT-IR, NMR, and elemental analyzer. The performances of scale inhibition against $CaCO_3$ were studied. The results showed that when the concentration of ESA/AMPS copolymer was 28 mg/L, the scale inhibition rate against $CaCO_3$ was 74.9%. The influence of ESA/AMPS copolymer on the morphology and crystal phase of $CaCO_3$ scales was studied by SEM and XRD. The results showed that ESA/AMPS copolymer distorted calcium carbonate crystallite more remarkably, and made the crystal phase of $CaCO_3$ transferred from calcite to aragonite, and made the calcium carbonate crystallite much finer and more dispersive.

1 INTRODUCTION

With the problem of scale in the industrial water equipment appearing, the research and application of scale inhibitor are more and more concerned by researchers (Zhang, 2013; Roweton,1997; Liu, 2011; Abdel-Gaber, 2011). Polyepoxysuccinic Acid (PESA) is a kind of biodegradable green water treatment agent with non-nitrogenous and non-phosphorus features, which has become a new focus in the field of water treatment because of its higher efficiency of scale inhibition, dispersion, and biodegradability (Suharso, 2011; Liang, 2008; Sun, 2009; Chaussemier, 2015; Bu, 2015). Consequently, the modification of PESA was a desirable method in order to improve the scale inhibition performances of PESA. Meanwhile, it was of great academic value and practical significance to study the mechanism of scale inhibition of the modified sulfonic acid groups.

In this paper, a polyepoxysuccinic acid copolymer (ESA/AMPS) was synthesized and the performances of scale inhibition against $CaCO_3$ were studied. The scale inhibitory mechanism against $CaCO_3$ was also studied by by SEM and XRD.

2 EXPERIMENTAL PROCEDURE

2.1 Materials

The chemical reagents used are analytical-grade maleic anhydride, sodium hydroxide, hydrogen peroxide (30%), sodium tungstate dehydrate, sodium molybdate, sodium orthovanadate, ammonium ferrous sulfate, calcium chloride, potassium dihydrogen phosphate, ammonium iron (II) sulfate, sodium bicarbonate, and ethanol. 2-Acrylamido-2-Methylpropane Sulfonic acid (AMPS) is of industrial grade, and Polyepoxysuccinic Acid (PESA) was synthesized by ourselves.

2.2 Synthesis and characterization of ESA/AMPS copolymer

For this experiment, 24.50 g of maleic anhydride and 40 mL deionized water were added to four 250 mL flasks, and the mixture was stirred until completely dissolved. Then, 50 mL of NaOH solution (10 mol/L) was slowly dropped to the solution, after finishing the mixed solution was heated to 55°C. The catalyst weighing 0.6 g was added to the mixed solution, and 10 ml H_2O_2 (30%) was uniformly added to the solution in 30 min. Then, the solution was incubated for 0.5 h at 65°C to complete the epoxidation to obtain the ESA. A certain amount of AMPS was added to the solution, ammonium sulfate solution was intermittently added to the mixed solution, and after that keeping the temperature some time for the copolymerization, a brown yellow viscous liquid was obtained. Then, absolute alcohol were mixed and kept still. A kind of yellow viscous material was precipitated out from the solution. The material was filtrated and dried at 65°C, and a brown yellow solid was obtained, which was the ESA/AMPS copolymer. Then, the ESA/AMPS copolymer was characterized by IR and NMR spectroscopy. The relevant synthesis reactions are expressed in scheme 1.

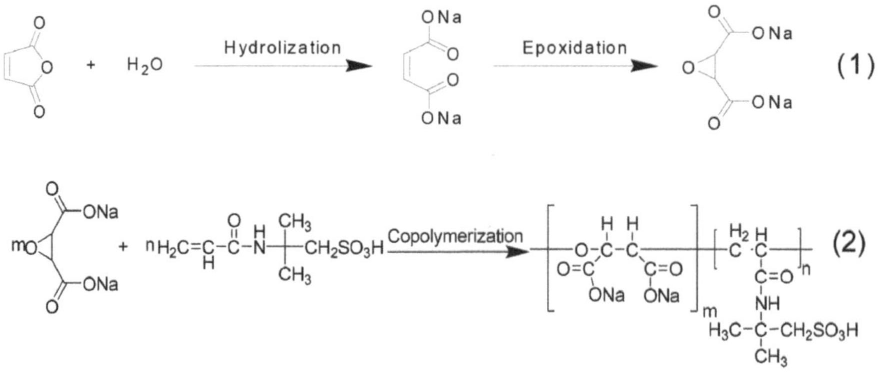

Scheme 1 Synthesis route of the ESA/AMPS copolymer.

2.3 Test of scale inhibition performance of ESA/AMPS copolymer against CaCO₃

The main steps were as follows: the water used in the experiment containing 600 mg/L Ca^{2+} and 1200 mg/L HCO_3^- calculating in terms of $CaCO_3$ was prepared from $CaCl_2$ solution and $NaHCO_3$ solution and the pH of water was adjusted to ≈ 9 by adding borax buffer. Following this, 500 ml of water containing a certain amount of ESA/AMPS, was kept in 80 °C thermostatic water bath for 10 h. After cooled to the room temperature, some supernatant liquor was taken and the concentration of Ca^{2+} in the water was measured by the EDTA titration method. Meanwhile, the blank without ESA/AMPS was conducted, and the scale inhibition efficiency ($X(Ca_{2+})$) was calculated using formula (1), as follows:

$$X(Ca_{2+}) = \frac{C_1 - C_0}{C_2 - C_0} \times 100\% \qquad (1)$$

where C_0 is the concentration of Ca^{2+} in the solution without ESA/AMPS copolymer after heating, mg/L; C_1 is the concentration of Ca^{2+} in the solution with ESA/AMPS copolymer after heating, mg/L; and C_2 is the concentration of Ca^{2+} in the solution without ESA/AMPS copolymer and before heating, mg/L.

3 RESULTS AND DISCUSSION

3.1 FT-IR analysis of ESA/AMPS copolymer

The FT-IR spectra of the ESA/AMPS copolymer are shown in Fig. 1a. From the spectrum, the peak at 3434.73 cm^{-1} is assigned to N-H from-C(=O)-NH-; the peaks at 1570.69 cm^{-1} and 1402.98 cm^{-1} are assigned to C = O from-COO-;

the peak at 1310.65 cm^{-1} is assigned to N-H from-C(= O)-NH-; the peak at 1214.57 cm^{-1} is assigned to C-N from-C(=O)-NH-; 1122.45 cm^{-1} is the characteristic absorption peak of the—COC-; the peak at 1047.67 cm^{-1} is attributed to-SO_3-; and the peak at 624.34 cm^{-1} is attributed to S-C. Above all, the peak at 1675 cm^{-1} -1645 cmcm^{-1} disappears, which are the characteristic peak of carbon-carbon double bond. It indicates that the copolymerization of monomers is completed. Hence, we concluded that the synthesized product was predominantly ESA/AMPS copolymer.

3.2 NMR spectrum of ESA/AMPS copolymer

Fig. 1b was solid nuclear magnetic resonance ^{13}C spectra of the ESA/AMPS copolymer. From the spectrum, the NMR signals, 180–170 ppm, were assigned to one acylamino carbon and two carboxyl carbon atoms, respectively; 77 ppm NMR signal was assigned to ether carbon; 74 ppm NMR signal was assigned to tertiary carbon that was connected with acylamino; 25 ppm NMR signal was assigned to the carbon that was connected with the sulfo group; and 17 ppm NMR signal was assigned to methyl carbon. This indicated that there were carboxyl group, amide group, ether group, and sulfo group in the copolymer, and therefore, it can be concluded that the synthesized product was ESA/AMPS copolymer. Moreover, there were not the signals of C = C between 105 and 140 ppm, that indicates that the copolymerization of monomers AMPS was completed and the result was consistent with that FT-IR analysis.

3.3 Scale inhibition performance of ESA/AMPS copolymer and PESA against CaCO₃

Fig. 2 shows the scale inhibition performance of ESA/AMPS copolymer and PESA against $CaCO_3$.

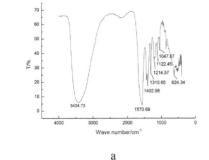

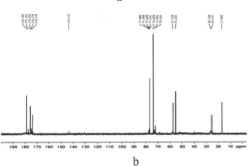

a

b

Figure 1. FT-IR and ^{13}C spectra of the ESA/AMPS copolymer.

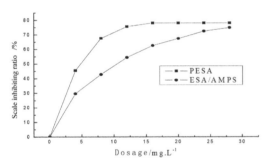

Figure 2. Scale inhibition performance of ESA/AMPS and PESA against CaCO$_3$.

When the dosage was 16 mg/L, the scale inhibition efficiency was 78.1% and kept stable. The values slowly increased with that of ESA/AMPS copolymer, when the dosage was 28 mg/L, and the scale inhibition efficiency was 74.9%. Therefore, when the dosage was at low concentration, the scale inhibition efficiency of ESA/AMPS copolymer against CaCO$_3$ scale was poorer than that of the PESA, but in the high-concentration range, the scale inhibition efficiency of ESA/AMPS copolymer against CaCO$_3$ scale was close to that of PESA.

4 CHARACTERIZATION OF CACO$_3$ SCALE

4.1 SEM micrographs of CaCO$_3$ scale

Fig. 3 is the SEM micrographs of CaCO$_3$ scales from the scale inhibition experiment at different dosages of ESA/AMPS copolymer. As shown in Fig. 5-a, CaCO$_3$ scales are composed of dense rhombohedron crystals that have a regular shape featuring well-defined cubes and glossy surface. It can be observed that when the dosages of the copolymer were 10 mg/L, 20 mg/L, and 30 mg/L added to the water samples, CaCO$_3$ scales changed obviously, edges and corners disappeared from Fig. 5-b, 5-c, and 5-d, and this kind of structure was loose. With the increase in the dosage, the deformation of the particles was more serious, and the size of the particles decreases obviously. It suggests that the copolymer molecular chains absorb on the activity growth-sites of the particles which result in the serious distortion of the grains and the sizes smaller and less than 4 μm; this kind of morphology has no a obvious regular shape featuring and show a good dispersion state. Consequently, in water used in the industry, these scales were easy to wash away with the water rather than attached to the pipe surface.

4.2 XRD characterization of CaCO$_3$ scale

Fig. 4 is the XRD images of CaCO$_3$ scales formed in the water at different dosages of ESA/AMPS copolymer. As shown in Fig. 8-a, the diffraction peaks at 22.98° (102), 29.32° (104), 35.90° (110), 39.34° (113),

a-Without copolymer; b-Addition of 10 mg/L copolymer;

c- Addition of 20 mg/L copolymer; d- Addition of 30 mg/L copolymer

Figure 3. SEM micrographs of CaCO$_3$ scale.

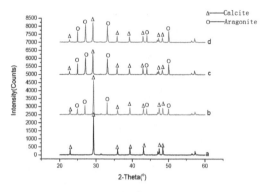

a - no copolymer;
b - Addition of 10 mg/L copolymer;
c - Addition of 20 mg/L copolymer;
d - Addition of 30 mg/L copolymer

Figure 4. XRD spectra of $CaCO_3$ scale.

43.10° (202), 47.42° (018), and 48.44° (116) are the characteristic peaks of calcite crystal in the absence of copolymer, which indicate that $CaCO_3$ deposits are mainly calcite. As shown in Fig. 8-c, 8-b and 8-d, the morphology of $CaCO_3$ deposits shows mainly calcite crystal at the different dosages of addition of the copolymer, but the characteristic peaks of aragonite crystal appear in the XRD images, and with the increase in the copolymer dosages, the diffraction peak strengths of the calcite crystal greatly reduce; meanwhile, the diffraction peak strength of the aragonite crystal is stronger. This suggested that the copolymer resulted in the crystal distortion and the unstable aragonite crystal stable existence. Meanwhile, the solubility and dispersion of $CaCO_3$ in water are increased, which was consistent with the results of SEM analysis.

5 CONCLUSION

1. ESA/AMPS copolymer scale inhibitor was synthesized with maleic anhydride and AMPS, FT-IR analysis showed that the molecular chain of the copolymer contained some effective scale inhibition and dispersion groups, such as carboxylic group, amide group, and sulfo group.
2. ESA/AMPS copolymer had good scale inhibition and dispersion performance for $CaCO_3$, and when the concentration of ESA/AMPS copolymer was 28 mg/L, the scale inhibition rate against $CaCO_3$ was 74.9%.

ACKNOWLEDGMENTS

This work was financially supported by the National Natural Science Foundation of China (21376062).

REFERENCES

Abdel-Gaber, A.M., B.A. Abd-El-Nabey and E. Khamis: Desalination Vol. 278 (2011), p. 337.
Bu, Y.Y., Y.M. Zhou and Q.Z. Yao: J. Appl. Polym. Sci. 2015, DOI: 10.1002/APP.41546.
Chaussemier, M., E. Pourmohtasham and D. Gelus: Desalination Vol. 356 (2015), p. 47.
Liang B.F. and Y. Yang: Petrochemical Technology & Application. Vol. 26 (2008), p. 39.
Liu, Z.Y., Y.H. Sun and X.H. Zhou: J. Environ. Sci. Vol. 23 (2011), p. S153.
Roweton, S., S.J. Huang and G. Swift: J. Environ. polymer degradation Vol. 5 (1997), p. 175.
Suharso, Buhani and B. Syaiful: Desalination Vol. 265 (2011), p. 102.
Sun, Y.H., W.H Xiang and Y. Wang: J. Environ. Sci. 21 (2009), p. S73.
Zhang, B., D.P. Zhou and X.G. Lv: Desalination Vol. 327 (2013), p. 32.

*Advances in Materials Science, Energy Technology
and Environmental Engineering – Patty & Zhou (Eds)*
© 2017 Taylor & Francis Group, London, ISBN 978-1-138-19668-1

Research on a new polyepoxysuccinic acid copolymer

Yuelong Xu, Zhenfa Liu, Lihui Zhang, Haihua Li & Yuhua Gao
Hebei Academy of Science, Institute of Energy Resources, Shijiazhuang, China
Hebei Engineering Research Center for Water Saving in Industry, Shijiazhuang, Hebei, China

ABSTRACT: A new polyepoxysuccinic acid copolymer (ESA/AMPS) was prepared from 2-Acrylamido-2-Methylpropane Sulfonic acid (AMPS) with Epoxysuccinic Acid (ESA). The performances of scale inhibition against $Ca_3(PO_4)_2$, dispersion of Fe_2O_3, and biodegradability of the ESA/AMPS copolymer were studied. The results indicated that when the concentration of ESA/AMPS copolymer was 35 mg/L, the scale inhibition rate against $Ca_3(PO_4)_2$ was 100%, and 61% of the copolymer could be biodegraded within 28 days.

1   INTRODUCTION

The problem of scale in the industrial water equipment attracts more and more attention in the water treatment field (Liu, 2015; Demadisa, 2015; Gao, 2015; Chaussemier, 2015; Bu, 2015; Feng, 2006). Polyepoxysuccinic Acid (PESA) as a kind of biodegradable green water treatment agent with non-nitrogenous and non-phosphorus features has become a new focus in the field of water treatment because of its higher efficiency of scale inhibition, dispersion, and biodegradability. Although PESA has a good performance of scale inhibition against $CaCO_3$, there were almost no performances of scale inhibition against calcium phosphate, calcium sulfate, and calcium silicate, and dispersion of ferric oxide because of its structure mainly containing a carboxylic group. When it is used in high-rigidity water, its performance of scale inhibition is worse than the commercially available scale inhibitors that contain organic phosphate compound (Zhang, 2010; Zhang, 2008; Chen, 2014; Zhang, 2015). However, those water treatment agents containing phosphorus are not environmentally friendly and difficult to biodegrade. Therefore, it is very important to develop biodegradable environmentally friendly water treatment agents. In order to improve the properties of PESA, the researchers have made more studies about the modification of PESA, and achieved some progress.

Although some progresses have been presently made on the modification of PESA, the scale inhibition performance of the modified product is not obvious and we should use larger dosage in order to get the good performances of scale inhibition against $Ca_3(PO_4)_2$. In this paper, the performances of ESA/AMPS copolymer against $Ca_3(PO_4)_2$, dispersion of Fe_2O_3, and biodegradability of the ESA/AMPS copolymer were studied.

2   EXPERIMENTAL PROCEDURE

2.1   *Materials*

The chemical reagents used are analytical-grade maleic anhydride, sodium hydroxide, hydrogen peroxide (30%), sodium tungstate dehydrate, sodium molybdate, sodium orthovanadate, ammonium ferrous sulfate, calcium chloride, potassium dihydrogen phosphate, ammonium iron (II) sulfate, sodium bicarbonate, and ethanol. 2-Acrylamido-2-Methylpropane Sulfonic acid (AMPS) is of industrial grade, and polyepoxysuccinic acid (PESA) was synthesized by ourselves.

2.2   *ESA/AMPS copolymer*

2.2.1   *Synthesis of ESA/AMPS copolymer*

For this experiment, 24.50 g of maleic anhydride and 40 ml deionized water were added to four 250 ml flasks, and then was magnetic stirred until completely dissolved. Then, 50 ml of NaOH solution (10 mol/L) was slowly dropped into the solution; after dripping is finished, the mixed solution was heated to 55°C. Following this, 0.6 g of 1:1 ratio composite catalyst was added to the mixed solution, and 10 ml H_2O_2 (30%) was uniformly added to the solution for 30 min. Then, the solution was incubated at 65°C for 0.5 h to complete the epoxidation to get the ESA. A certain amount of AMPS was added to the solution, ammonium sulfate solution was intermittently added to the mixed solution, and after that keeping the temperature some time for the copolymerization, a brown yellow viscous liquid was obtained. The relevant synthetic routes are expressed in scheme 1.

2.2.2 Purification of ESA/AMPS copolymer

The ESA/AMPS copolymer solution and anhydrous alcohol were mixed and kept still. A kind of yellow viscous material precipitated out from the solution. The material was filtrated and dried at 65°C, and a brown yellow solid was obtained, which was the ESA/AMPS copolymer.

2.3 Test of ESA/AMPS copolymer scale inhibition performance

2.3.1 Test of scale inhibition performance against $Ca_3(PO_4)_2$

The test of scale inhibition performance about (against) $Ca_3(PO_4)_2$ was conducted according to the Chinese National Standard method (GB/T 22626-2008), Specific steps were as follows: the water used in the experiment containing 250 mg/L Ca^{2+} in terms of $CaCO_3$ and 5 mg/L PO_4^{3-} was prepared from $CaCl_2$ solution and KH_2PO_4 solution and the pH of water was adjusted to ≈ 9 by adding borax buffer. Then, 500 ml with a certain amount of ESA/AMPS was kept in 80°C thermostatic water bath standing for 10 h. After cooled to the room temperature, some supernatant liquor taken and the concentration of PO_4^{3-} in the water was measured by UV spectrophotometry. Meanwhile, the blank without ESA/AMPS was conducted, and the scale inhibition efficiency (X_{PO4}^{3-}) was calculated using formula (1), as follows:

$$X(PO_4^{3-}) = \frac{C_1 - C_0}{C_2 - C_0} \times 100\% \qquad (1)$$

where C_0 is the concentration of PO_4^{3-} in the solution without ESA/AMPS copolymer after heating in to-be-tested solution (control test), mg/L; C_1 is the concentration of PO_4^{3-} in the solution with ESA/AMPS copolymer after heating in to-be-tested solution, mg/L; and C_2 is the concentration of PO_4^{3-} in the solution without ESA/AMPS copolymer and is not heated in to-be-tested solution, mg/L.

2.3.2 Test of the dispersion performance of ferric oxide

Some $CaCl_2$ solution, $(NH_4)Fe(SO_4)_2$ solution, borax buffer (pH ≈ 9) and a certain amount of ESA/AMPS were mixed to blend the water used in this experi-

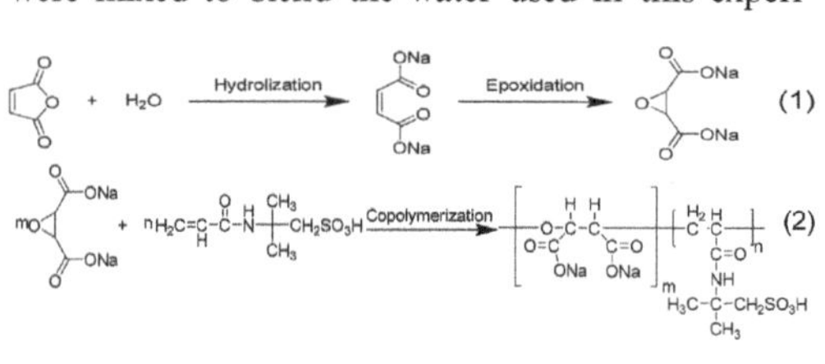

Scheme 1. Synthesis route of the ESA/AMPS copolymer.

ment. It was found that 500 mL of solution contained 150 mg/L Ca^{2+} in terms of $CaCO_3$ and 10 mg/L Fe^{3+}. After stirring for 15 min, the water was kept in 50°C thermostatic water bath standing for 5 h. After being cooled to room temperature, the transmittance at 450 nm of the supernatant liquor was measured using a 724-spectrophotometer and the distilled water was used as the reference (T = 100%). The lower the transmittance was, the higher the efficiency of the copolymer on the dispersion of Fe_2O_3 was.

2.3.3 Test of the biodegradability of ESA/AMPS copolymer

The biodegradability of ESA/AMPS copolymer was studied by the shaking-bottle incubating test. Specific steps were as follows: 500 mL solution with a certain amount of copolymer and 1 mL inoculum was prepared in a conical flask, and sealed by a cotton plug. Then, this flask and a flask with the blank water were placed on a shaking-bottle incubating device at 20°C at the same time, the Chemical Oxygen Demands (COD) was detected at 1, 7, 14, 21, and 28 days according to the Chinese National Stand Method of GB/T 15456-2008, respectively. The degradation rate of ESA/AMPS was calculated using formula (2):

$$\text{Degradation Rate} = (1 - \frac{C_t - C_{bt}}{C_o - C_{b0}}) \times 100\% \qquad (2)$$

where C_0, C_t, C_{b0}, and C_{bt} represent the initial values of COD in the inoculated solution with the copolymer, the COD value at time t in the inoculated solution with the copolymer, the initial value of COD in the blank solution, and the COD value at time t in the blank solution, respectively.

3 RESULTS AND DISCUSSION

3.1 Scale inhibition performance of ESA/AMPS copolymer and PESA against $Ca_3(PO_4)_2$

Fig. 1 shows the results of the scale inhibition experiments of ESA/AMPS copolymer and PESA against $Ca_3(PO_4)_2$. From Fig. 1, we know that when the dosage was less than 10 mg/L. The scale inhibition performances of ESA/AMPS copolymer and PESA against $Ca_3(PO_4)_2$ were not noticeable. While the dosage was more than 20 mg/L, the scale inhibition performances of ESA/AMPS copolymer against $Ca_3(PO_4)_2$ were dramatically reached, and the maximum inhibition efficiency was 100% at the concentration of 35 mg/L. For PESA, the scale inhibition performance has no obvious change throughout the measured concentration range. It can be known that the scale inhibition performance of ESA/AMPS copolymer against $Ca_3(PO_4)_2$ was better than that of PESA.

3.2 Dispersion performance of ESA/AMPS and PESA on Fe₂O₃

The results of Fe_2O_3 experiments of ESA/AMPS copolymer and PESA are shown in Fig. 2. From Fig. 2, we know that when the dosage was less than 20 mg/L, the dispersion performances of ESA/AMPS and PESA on Fe_2O_3 were not noticeable; when the dosage was more than 25 mg/L, the dispersion performance of ESA/AMPS noticeably increased. The maximum dispersion efficiency, 57%, was reached at the copolymer concentration of 35 mg/L. However, the dispersion of PESA is always poor in the tested concentration range. It can be known that when the dosage achieved a certain amount, the dispersion performance of ESA/AMPS was better than that of PESA.

3.3 Biodegradation performance of ESA/AMPS and PESA

The biodegradability of ESA/AMPS copolymer and PESA were studied by shaking-bottle incubating test, and the results are shown in Fig. 3.

It can be observed that the degradation rates of PESA and ESA/AMPS copolymer were increased with the extension of time from Fig. 3. At the seventh day after the start of the experiment, the biodegradation rates of ESA/AMPS and PESA can reach up to 51% and 52%, respectively; and the

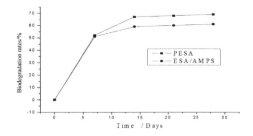

Figure 3. Biodegradation of ESA/AMPS copolymer and PESA.

biodegradation rates can also reach up to 61% and 69% in 28 d, respectively. The results show that the ESA/AMPS copolymer and PESA are both easily biodegradable and environmentally friendly.

4 CONCLUSION

1. The concentration of ESA/AMPS copolymer was 35 mg/L, and the scale inhibition rate against $Ca_3(PO_4)_2$ was 100%.
2. The ESA/AMPS copolymer is environmentally friendly and 61% of the copolymer can be biodegraded within 28 days.

ACKNOWLEDGMENTS

This work was financially supported by the National Natural Science Foundation of China (21376062) and the Natural Science Foundation of Hebei Province (E2013302037) and (B2015302008).

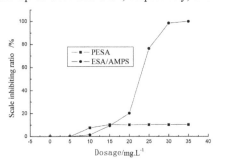

Figure 1. Scale inhibition performance of ESA/AMPS and PESA against $Ca_3(PO_4)_2$.

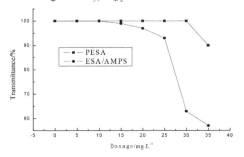

Figure 2. Dispersion performance of ESA/AMPS and PESA on Fe_2O_3.

REFERENCES

Bu, Y.Y., Y.M. Zhou and Q.Z. Yao, et al.: J. Appl. Polym. Sci. 2015, DOI: 10.1002/APP.41546.

Chaussemier, M., E. Pourmohtasham and D. Gelus, et al.: Desalination Vol. 356 (2015), p. 47.

Chen, Y.Y., Y.M. Zhou and Q.Z Yao, et al: Desalination and Water Treatment Vol. 41, (2014), p 88.

Demadisa K.D. and M. Prearia: Desalination and Water Treatment Vol.55 (2015), p. 749.

Feng, H.X., T. Zhang and Y. Wang, et al.: Modern Chemical Industry Vol. 26 (2006), p. 331.

Gao, Y.H., L.H. Fan and L. Ward, et al: Desalination Vol. 365 (2015) 220–226.

Liu, Z.F., S.S. Wang and L.H. Zhang, et al: Desalination Vol. 362 (2015), p. 26.

Zhang, J.M., D. Jin: Industrial Water Treatment Vol. 30 (2010), p 18.

Zhang, S.F., Y.J. Shu and S.M. Tang. China water & Wastewater Vol. 24, (2008), p 42.

Zhang, Y.Q., Y.M. Tang and J.Q. Xu, et al.: Journal of Solid State Chemistry, Vol. 231, (2015), p 7.

*Advances in Materials Science, Energy Technology
and Environmental Engineering – Patty & Zhou (Eds)
© 2017 Taylor & Francis Group, London, ISBN 978-1-138-19668-1*

High-performance and low-cost SERS substrate on a flexible plastic sheet

Q. Zhang, L. Liu & H.X. Ge

Nanjing University, Nanjing, Jiangsu, China

ABSTRACT: We demonstrated a simple and low-cost way to fabricate flexible Surface-Enhanced Raman Scattering (SERS) substrates over large area. A method based on combination of nanoimprint lithography and E-beam evaporation of gold was used to fabricate gold/PC nanopillar arrays. The gap between the gold-coated nanopillars as well as between the gold caps on the tops and the gold film at the bottom could be controlled to sub-10 nm. The effect on SERS enhancement was investigated by three-dimensional numerical simulations and experimental SERS measurements. The structure possessed a high SERS enhancement with an average enhancement factor of 4.4×10^8, a large-area uniformity with a relative standard deviation of 15%, and an excellent reproducibility with a relative standard deviation of 18%.

1 INTRODUCTION

Plasmonic nanostructure arrays of noble metals have recently attracted considerable interest in the fields of photonic (Luther et al. 2011), photovoltaics (FerryMunday & Atwater 2010), photocatalysis (LinicChristopher & Ingram 2011), and biological and chemical sensors (LeeJeon & Kim 2015) due to their extraordinary characteristics such as light scattering and absorption, and giant enhancement of the local electric fields (Fang & Zhu 2013). Of particular interest is the combination of plasmonic nanostructures and Raman spectroscopy that provides a powerful analytical technique, namely, Surface-Enhanced Raman Scattering (SERS) for the ultra-sensitive detection even to a single molecular level (Liu et al. 2011, Li et al. 2010). The capacity of SERS is generally attributed to two well-known mechanisms, the electromagnetic mechanism and chemical mechanism. The electromagnetic mechanism is induced by the excitation of the Localized Surface Plasmon Resonance (LSPR) of the noble metal nanostructures as well as their interactive coupling. The chemical mechanism arises from the charge transfer interaction between adsorbed molecules and noble metal surface. It has been understood that the electromagnetic effects contribute mainly to the Enhancement Factor (EF) of SERS (Pitarke et al. 2007).

The first plasmonic nanostructures used for SERS were random structures of noble metals such as colloidal aggregates of Ag or Au nanoparticles and roughened metal surfaces. It has been demonstrated that the hot spots in SERS (regions of the highest local electric field) are associated with the sharp edges (Lu et al. 2009) or the small gaps (Jiwei et al. 2013, Liu et al. 2011) of either roughened surfaces or aggregates of nanoparticles. They can provide a Raman EF of up to 10^{11} (Nie 1997), sufficient for single-molecule detection. However, the sparse hot spots and lack of the uniformity and reproducibility in EF for these random nanostructures limit their practical application (Zhou et al. 2014). Alternatively, ordered nanostructure arrays have been suggested to improve the reliability and reproducibility of SERS measurement. Various types of nanostructures have been designed and fabricated as SERS substrates, such as nanopillars, nanorods, nanotips, nanowires, nanocones, nanorings, nanobowls, and nanocrescents, to name a few (Gong et al. 2012). As nanofabrication techniques have been rapidly developing, a variety of methods have been used to fabricate metallic nanostructures. E-Beam Lithography (EBL) and Focus Ion Beam (FIB) offer high fidelity and high controllability in terms of design and prediction (Altissimo 2010, Yu and Golden 2007), but they suffer from low throughput and high cost. On the other hand, self-assembly approaches, such as nanosphere lithography (Willets & Van Duyne 2007) and block copolymer lithography (Wang et al. 2009) can produce high-resolution nanostructures over a large area with a low cost, but it is difficult to obtain highly uniform and controllable structures of arbitrary symmetries across the whole area.

The above issues of conventional fabrication techniques can be solved by Nanoimprint Lithography (NIL), which is a high-throughput fabrication technique capable of replicating nanopatterns from a master mold over a large area with nanometer precision at low cost (Chou et al. 1997). Various nanostructures of noble metals for SERS have been successfully fabricated by NIL. For example, a molecular trap structure of flexible polymer nanofingers is formed by UV-curing NIL. The gold-coated flexible nanofingers can be driven together to trap molecules by capillary force and at the same time create reliable hot spots at the fingertips for detecting the trapped molecule based on SERS (Hu et al. 2010). A novel plasmonic architecture, termed a 'disk-coupled dots-on-pillar antenna array' (D2PA), is fabricated by NIL in combination with guided self-assembly and self-alignment. This architecture couples a dense Three-Dimensional (3D) cavity nanoantenna array, through nano-gaps, with dense plasmonic nanodots (Li et al. 2011). These SERS substrates show significant improvements in both enhancement and large-area uniformity over previous works. Nevertheless, it is still a challenging task to achieve large-area nanoimprint molds with dense and high-resolution nanostructures, which usually require slow and expensive top-down lithographic methods (e.g., EBL or FIB).

Anodic Aluminum Oxide (AAO) templates with hexagonal array of nanopores have two significant advantages, namely (1) their large working areas (> 100 cm^2) and (2) narrowly distributed pore size, which can be varied from 5 to 500 nm by changing the oxidation condition. Owing to these merits, AAO templates have been applied as nanoimprint mold to produce nanodot arrays in various fields, such as data storage (Chen et al. 2013), organic photovoltaics (Kim et al. 2010) and plasmonic devices (Liu et al. 2014). The major challenge faced by the AAO template-based imprint method is that the AAO templates are not flat, and it causes large area defects in conventional thermal imprint because of the nonconformal contact between template and flat sample substrate.

To date, a practical SERS substrate with high EF, uniformity and reproducibility, large area, and low cost is still strongly desired. In this paper, we present a simple and low-cost method to fabricate gold-coated nanopillar arrays on flexible plastic sheets. In this approach, nanopillar arrays are formed on polycarbonate sheets by thermal nanoimprint technique with AAO template. A thin gold layer is deposited on the nanopillar arrays. An average EF of trans-1,2-Bis (4-Pyridyl) Ethylene (BPE), 4.4×10^8, with an excellent uniformity, variation of 15%, is achieved on a plastic nanopillar array with a diameter of 90 nm, pillar spacing of 20 nm, and deposited gold thickness of 90 nm. The high SERS activity is attributed to the dense nanogaps smaller than 10 nm that are created not only between the gold-coated nanopillars but also between the gold caps on the tops of the nanopillars and the gold film at the bottom of the nanopillar arrays.

2 METHODS

2.1 Materials

AAO templates with a nanopore diameter (d) of 90 nm and pore spacing (s) of 20 nm were purchased from Shanghai Shangmu Technology Co. Ltd. Polycarbonate (PC) sheets with 0.3 mm thickness were purchased from DuPont. Trichloro (1H,1H,2H,2H-perfluorooctyl) silane was purchased from Gelest, Inc.

2.2 Fabrication of gold-coated nanopillars

The surface of the AAO template was first cleaned by O$_2$ plasma, and then coated with a self-assembled monolayer of trichloro (1H,1H,2H,2H-perfluorooctyl) silane by vapor-phase deposition. Figure 1 shows the schematic of the process of preparing the gold-coated nanopillar arrays. The PC sheet was thermally imprinted by the AAO template ($d = 90$ nm, $s = 20$ nm) at 165 °C under a pressure of 0.6 MPa for 5 min using an air-pressure imprinter (ImprintNano, China). After removing the AAO template, a gold layer with 90 nm thickness was deposited on the imprinted PC sheet by e-beam evaporation. The resulting nanopillar arrays were ready to be used for SERS measurement.

2.3 Characterization

All the Scanning Electron Microscopic (SEM) images were detected using a field-emission scanning electron microscope (ZEISS ULTRA-55). Contact angle measurements were performed

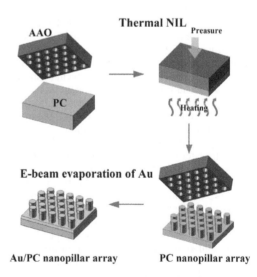

Figure 1. Schematic process of the preparation steps of gold/PC nanostructures.

using a contact angle meter (JC2000 CS). To illustrate the performance of our prepared SERS-active substrate, the Raman spectra of the samples were acquired using a Renishaw inVia Raman microscope system with the laser excitation at a wavelength of $\lambda = 633$ nm with 0.6 mW laser output power during just one-second collection time, and the Raman signals were collected via a 50× magnification objective. The laser spot size was approximately 2 µm². Before Raman measurements, all the samples were soaked in the prepared BPE solutions with a known concentration for 1 h, and subsequently dried in air as well.

3 RESULTS AND DISCUSSION

Before thermal imprinting, it is crucial to modify the AAO template surface with an anti-stick agent trichloro (1H,1H,2H,2H-perfluorooctyl)silane to reduce the surface energy for mold separation. The O_2 plasma was used to clean the AAO surface and expose the hydroxyl groups on it. Trichloro (1H,1H,2H,2H-perfluorooctyl) silane was used because it can covalently bond to the AAO template via the reaction of a chlorosilane group with the hydroxyl group on the surface of aluminum oxide. Figure 2a and b shows the water contact angle measurement of the AAO template before cleaning and after surface modification, respectively. The contact angle increased from 79.0° to 135.3°, which indicated that the AAO template was much more hydrophobic after modification.

The simple fabrication process of gold-coated PC nanopillar arrays only involved thermal-imprinting the PC sheet with AAO templates, followed by depositing gold on it via e-beam evaporation. PC widely used in engineering is a readily available, strong, tough, and optically transparent material. It has a glass transition temperature of about 145 °C, and is easily thermoformed and molded. A major application of polycarbonate is the production of optical discs by injection molding. In our experiment, PC was thermally imprinted at the temperature of 165°C.

Although the nonflatness of the AAO templates may cause large area detects in thermal imprint due to the nonconformal contact between the template and conventional hard substrates such as silicon wafer, the flexible PC sheet allows a conformal contact with the nonflat AAO template under a uniform air pressure. Figure 2c and d show the photographs of an AAO template (20×20 mm²) with a nanopore diameter (d) of 90 nm and pore spacing (s) of 20 nm, and its corresponding imprinted PC sheet after deposition of 90 nm gold layer, respectively. The imprinted area on PC

Figure 2. Water contact angle of the AAO template (a) before cleaning, and (b) after surface modification. Photographs of (c) an AAO template and (d) corresponding imprinted PC sheet after deposition of 90 nm gold layer.

exhibited a uniform dark color without obvious defects. An AAO template with pore diameter $d = 90$ and spacing $s = 20$ nm were employed as imprint mold. Figure 3a shows the top view of SEM images of the template. Figure 3b and c are the SEM image of the top view and cross-section of its imprinted PC replica, respectively. As shown in Figure 3, the diameter of nanopillar (d) and the average distance between the neighboring nanopillar (i) were identical with the size of the nanopores and distance between the nanopores on the master AAO templates. It was difficult to break the PC sheet cooled in liquid nitrogen for taking SEM image of the cross-section due to its flexibility and toughness. Instead, the cross-section of the PC nanopillar array were obtained from the 180 nm thick PC films, which were spin-coated on silicon wafers by a PC/chlorobenzene solution and imprinted by the same AAO template.

3.1 SEM characterization of the structures

Followed by the physical deposition process as shown in Figure 1, a gold/PC nanopillar array could be expected. A gold layer with 90 nm thickness was deposited on the above-mentioned PC nanopillar array by e-beam evaporation. Figure 4 shows the SEM image of the gold/PC nanopillar array with 90 nm gold thicknesses (H) from well-defined PC template. Gold layers were deposited both on the tops of the nanopillars and at the bottom of the nanopillar array.

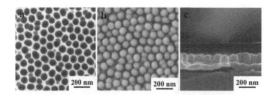

Figure 3. (a) Top-view SEM image of an AAO template with d = 90 nm, s = 20 nm. (b) Top-view SEM image of the imprinted PC replica (d = 90 nm, i = 20 nm). (c) Corresponding cross-section SEM image of the imprinted PC replicas.

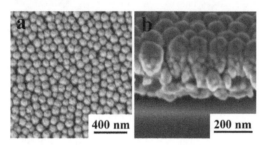

Figure 4. (a) Top-view SEM image of the gold/PC nanopillar array with d = 90 nm, i = 20 nm, H = 90 nm. (b) Cross-sectional SEM image of the gold/PC nanopillar array with d = 90 nm, i = 20 nm, H = 90 nm.

Two types of metallic nanogaps were formed between the neighboring nanopillars and between the metal-cap on the top of nanopillar and the gold film at the bottom. The size of both nanogaps got smaller because of the increase in the thickness of the metal at the bottom and expanding of the nanopillar due to this physical deposition. Dense sub-10 nm nanogaps could be achieved.

3.2 SERS measurement

The nanogaps between the gold-coated nanopillars as well as between the gold caps on the tops of the nanopillars and the gold film at the bottom of the nanopillar arrays can be considered as the hotspots for enhancing Raman intensity. As we know, the SERS EF could be expressed approximately as EF $\propto | E / E_0 |^4$, where $|E|$ is the amplitude of the enhanced local electric field and $|E_0|$ is the amplitude of incoming electromagnetic wave. To well evaluate the SERS enhancement effect of the gold/PC nanopillar array, three-dimensional numerical simulations were performed using the commercial software package (COMSOL Multiphysics) based on the finite-element method. The calculation domain comprised one complete and four-quarter gold/PC nanopillars in the x-y plane defined as Figure 5. The periodic boundary conditions were applied to the four sides of the rectangular simulation domain

to fit the periodicity of the whole structure. In the simulation, a plane wave with linear polarization was illuminated normally on the modeled structure. The refractive index of the PC was set at 1.59 and the permittivity of gold was described as a dispersive medium with the complex dielectric parameters taken from experimental data by Johnson and Christy (Johnson and Christy, 1972). Figure 5a shows the schematic of the gold/PC nanopillar array used for simulations. The nominal thickness of gold, which actually could be described as the thickness of gold on the PC bottom was 90 nm. The absorptance (A) spectra, which were defined as A = 1 – Transmittance (T) – Reflectance (R) were plotted in Figure 5b. An absorption peak is obvious and the absorption peak could be tuned to 633 nm which is the typical exciting laser wavelength used in our experiment. Figure 5c and d shows the electric field distributions along the x-y plane and x-z plane under the excitation of 633 nm for the sample.

It can be seen that dramatically field enhancement in the gaps between the neighboring metallic nanopillars as well as between the gold caps on the tops of the nanopillars and the gold film at the bottom of the nanopillar array are formed, which stems from the near-field coupling of the Localized Surface Plasmon (LSP) supported by the adjacent individual gold nanoparticles.

Using trans-1,2-Bis (4-Pyridyl) Ethylene (BPE) as the probe molecular, the SERS effect enhanced by the gold/PC nanopillar array was characterized. In our experiment, 18.2 g BPE was dissolved in 100 mL ethanol to form 10^{-1} M solution followed by stepwise diluting to lower concentrations (10^{-2} M-10^{-10} M).

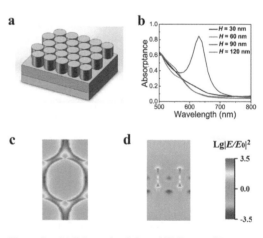

Figure 5. (a) Schematic of the gold/PC nanopillar array used for simulations. (b) Absorptance (A) spectra of the gold/PC nanopillar array. Corresponding maps of the electric field distributions along (c) the x-y plane, and (d) the x-z plane.

To demonstrate the SERS performance of our proposed metallic microstructure, we first soaked the substrate in ethanol solution of trans-1,2-Bis (4-Pyridyl) Ethylene (BPE) with concentrations from 10^{-6} M to 10^{-10} M for 1 h, and subsequently dried them in air. For comparison, a similar treatment to a flat PC sheet was also performed using an ethanol solution of BPE (10^{-1} M) for 1 h. All of the Raman spectra were measured under the same conditions. A helium-neon (He-Ne) laser was used as the excitation source with the wavelength of 633 nm. The acquisition time was 1 s and the laser power was 0.6 mW, as shown in Figure 6.

It is worthwhile to note that the Raman signal can still be clear when the concentration is 10^{-10} M. To quantify the SERS activity of the structure, we calculated Enhancement Factor (EF) by the formula: $EF = (I_{SERS}/N_{SERS})/(I_R/N_R)$, where I_{SERS} and I_R are the measured Raman intensity on the SERS-active metallic microstructure and the reference substrate (the flat PC sheet), respectively, and N_{SERS} and N_R are the number of probe molecules contributing to the SERS and the bulk Raman signal. In our work, the equation can be written as $EF = (I_{SERS}/I_R)/(C_R/C_{SERS})$, where C_R and C_{SERS} are the concentrations of the BPE solutions on the SERS substrate and the reference substrate. Taking the 1200 cm^{-1} peak into account, the EF is calculated to be about 4.4×10^8. It indicates that the nanostructures we obtained exhibit strong SERS EF, showing a fairly high SERS activity. Discrepancy between the measured and calculated SERS EFs can also be attributed to some other reasons, such as the additional contribution of chemical effects not considered in the simulation. The simulated structures are simplified as shown in Figure 5, and the actually structures are much more complex, for example, the pillars are not so regular

and the gold surface is not as smooth as that of the simulated structures.

For the application of an SERS-active substrate, the uniformity and reproducibility of the Raman signal over a very large area are of great importance, which are associated with the uniformity of the SERS hot-spots. Six positions were selected randomly on the sample and experimental SERS spectra of 10^{-7} M BPE were collected corresponding to these points, as shown in Figure 7a. Then, we randomly selected five different samples and randomly selected one position on each sample. Figure 7b shows SERS spectra of 10^{-7} M BPE collected from the different samples. We used the Relative Standard Deviation (RSD) of SERS peak around 1200 cm^{-1} to estimate the uniformity and reproducibility of the SERS signals. The RSD value is calculated to be about 15% and 18%, respectively. The intensity of Raman signals from different positions of

a

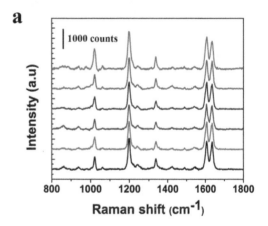

b

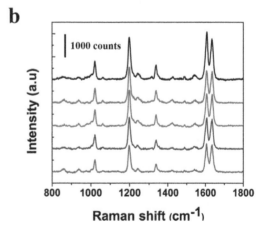

Figure 7. (a) SERS spectra of 10^{-7} M BPE from six randomly selected positions of one sample. (b) SERS spectra of 10^{-7} M BPE from five randomly selected samples.

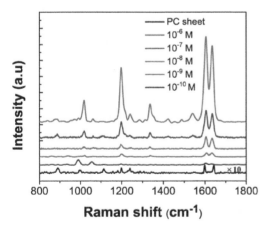

Figure 6. Experimental SERS spectra of BPE with different concentrations.

447

one sample and from different samples were almost same. Thus, the gold/PC nanopillar array we fabricated demonstrates excellent SERS signal uniformity and reproducibility.

4 CONCLUSION

In conclusion, we proposed a fairly simple, fast, and low-cost way to fabricated SERS substrates, which is based on the combination of AAO templating and nanoimprint lithography, followed by a physical deposition process to fabricate gold/PC nanopillar arrays. The gap between the gold-coated nanopillars as well as between the gold caps on the tops of the nanopillars and the gold film at the bottom of the nanopillar array was smaller than 10 nm. The average SERS EF could reach 4.4×10^8. In addition, the gold/PC nanopillar array demonstrated excellent SERS signal uniformity and reproducibility. The high SERS performance is contributed by the sub-10 nm nanogaps, bringing plenty of hotspots. In addition, since the gold/PC nanopillar array was supported on flexible plastic sheets, physical forces such as bending and twisting of the SERS substrate can provide anther degree of freedom to tune the SERS properties, and that may bring a wider use of the SERS substrates.

REFERENCES

Altissimo, M. 2010. E-beam lithography for micro-nanofabrication. *Biomicrofluidics*, 4.

Chen, X., Q. Li, X. Chen, X. Guo, H. Ge, Y. Liu & Q. Shen 2013. Nano-Imprinted Ferroelectric Polymer Nanodot Arrays for High Density Data Storage. *Advanced Functional Materials*, 23, 3124–3129.

Chou, S.Y., P.R. Krauss, W. Zhang, L.J. Guo & L. Zhuang 1997. Sub-10 nm imprint lithography and applications. *JOURNAL OF VACUUM SCIENCE & TECHNOLOGY B*, 15, 2897–2904.

Fang, Z. & X. Zhu 2013. Plasmonics in Nanostructures. *Advanced Materials*, 25, 3840–3856.

Ferry, V.E., J.N. Munday & H.A. Atwater 2010. Design Considerations for Plasmonic Photovoltaics. *Advanced Materials*, 22, 4794–4808.

Gong, X., Y. Bao, C. Qiu & C. Jiang 2012. Individual nanostructured materials: fabrication and surface-enhanced Raman scattering. *Chemical Communications*, 48, 7003.

Hu, M., F.S. Ou, W. Wu, I. Naumov, X. Li, A.M. Bratkovsky, R.S. Williams & Z. Li 2010. Gold Nanofingers for Molecule Trapping and Detection. *Journal of the American Chemical Society*, 132, 12820–12822.

Jiwei, Q., L. Yudong, Y. Ming, W. Qiang, C. Zongqiang, W. Wudeng, L. Wenqiang, Y. Xuanyi, X. Jingjun & S. Qian 2013. Large-area high-performance SERS substrates with deep controllable sub-10-nm gap structure fabricated by depositing Au film on the cicada wing. *Nanoscale Res Lett*, 8, 437.

Johnson, P.B. & R.W. Christy 1972. Optical Constants of the Noble Metals. *Phys. Rev. B*, 6, 4370–4379.

Kim, J.S., Y. Park, D.Y. Lee, J.H. Lee, J.H. Park, J.K. Kim & K. Cho 2010. Poly (3-hexylthiophene) Nanorods with Aligned Chain Orientation for Organic Photovoltaics. *Advanced Functional Materials*, 20, 540–545.

Liu, H., Zhang, L., Lang, X., Yamaguchi, Y., Iwasaki, H., Inouye, Y., Xue, Q. & Chen, M. 2011. Single molecule detection from a large-scale SERS-active Au79Ag21 substrate. *Scientific Reports*, 1, 1–5.

Lee, M., H. Jeon & S. Kim 2015. A Highly Tunable and Fully Biocompatible Silk Nanoplasmonic Optical Sensor. *Nano Letters*, 15, 3358–3363.

Li, J.F., Y.F. Huang, Y. Ding, Z.L. Yang, S.B. Li, X.S. Zhou, F.R. Fan, W. Zhang, Z.Y. Zhou, D.Y. Wu, B. Ren, Z.L. Wang & Z.Q. Tian 2010. Shell-isolated nanoparticle-enhanced Raman spectroscopy. *Nature*, 464, 392–395.

Li, W.D., F. Ding, J. Hu & S.Y. Chou 2011. Three-dimensional cavity nanoantenna coupled plasmonic nanodots for ultrahigh and uniform surface-enhanced Raman scattering over large area. *Opt Express*, 19, 3925–36.

Linic, S., P. Christopher & D.B. Ingram 2011. Plasmonic-metal nanostructures for efficient conversion of solar to chemical energy. *Nature Materials*, 10, 911–921.

Liu, X., Y. Shao, Y. Tang & K. Yao 2014. Highly Uniform and Reproducible Surface Enhanced Raman Scattering on Air-stable Metallic Glassy Nanowire Array. *Scientific Reports*, 4.

Lu, X., M. Rycenga, S.E. Skrabalak, B. Wiley & Y. Xia. 2009. Chemical Synthesis of Novel Plasmonic Nanoparticles. In *Annual Review of Physical Chemistry*, 167–192..

Luther, J.M., P.K. Jain, T. Ewers & A.P. Alivisatos 2011. Localized surface plasmon resonances arising from free carriers in doped quantum dots. *Nature Materials*, 10, 361–366.

Nie, S. 1997. Probing Single Molecules and Single Nanoparticles by Surface-Enhanced Raman Scattering. *Science*, 275, 1102–1106.

Pitarke, J.M., V.M. Silkin, E.V. Chulkov & P.M. Echenique 2007. Theory of surface plasmons and surface-plasmon polaritons. *Reports On Progress in Physics*, 70, 1–87.

Wang, Y., M. Becker, L. Wang, J. Liu, R. Scholz, J. Peng, U. Go sele, S. Christiansen, D.H. Kim & M. Steinhart 2009. Nanostructured Gold Films for SERS by Block Copolymer-Templated Galvanic Displacement Reactions. *Nano Letters*, 9, 2384–2389.

Willets, K.A. & R.P. Van Duyne. 2007. Localized surface plasmon resonance spectroscopy and sensing. In *Annual Review of Physical Chemistry*, 267–297..

Yu, Q. & G. Golden 2007. Probing the Protein Orientation on Charged Self-Assembled Monolayers on Gold Nanohole Arrays by SERS. *Langmuir*, 23, 8659–8662.

Zhou, Y., X. Zhou, D.J. Park, K. Torabi, K.A. Brown, M.R. Jones, C. Zhang, G.C. Schatz & C.A. Mirkin 2014. Shape-Selective Deposition and Assembly of Anisotropic Nanoparticles. *Nano Letters*, 14, 2157–2161.

Advances in Materials Science, Energy Technology and Environmental Engineering – Patty & Zhou (Eds)
© 2017 Taylor & Francis Group, London, ISBN 978-1-138-19668-1

Studies on novel multiferroic materials

Daoyang Wei
Department of physics, Tamkang University, Taiwan

ABSTRACT: Multiferroic refers to a kind of material that is simultaneously ferroelectric and ferromagnetic (or antiferromagnetic). This material has two or more kinds of arrangement at the same time, such as spin and polarization double arrangement structure. Owing to the coexistence of electric and magnetic fields, this kind of material is also composite. This functional material has shown great application prospect in the field of new magnetoelectric devices, spin electronic devices, high-performance information storage, and processing. It will push the development of several categories in condensed matter physics, charge, spin, orbit, lattice, etc., as soon as it is realized. Herein, we review some methods to synthesize double perovskite ion oxide $YBaNiFeO_5$ and put forward some feasible experimental operation applications.

1 INTRODUCTION

Multiferroic materials with the coexistence of at least two ferroic order. [ferroelectric, (anti-)ferromagnetic, and ferroelectric] have recently drawn ever-increasing interest due to their potential for applications as multifunctional devices. Multiferroic materials have dual functions of electricity and magnetism because of the magnetoelectric coupling phenomenon. It can cause the change in electric susceptibility by adding magnetic field or change the magnetic susceptibility by electric field in turn, as shown in Figure 1 (Shuai, 2010; Kundys, 2009; Lai, 2015).

Among them, the coexistence of ferroelectricity and ferromagnetism is highly desired. Only their

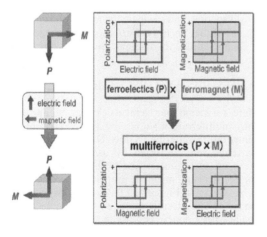

Figure 1. Multiferroic materials and their electromagnetic control.

coexistence is not enough; of most importance is to require a strong coupling interaction between two ferroic orders. In multiferroic materials, the coupling interaction between the parameters of different orders can produce additional functionalities, such as a Magneto Electric (ME) effect, an effect discovered more than a century ago.

From the view of material constituents, multiferroic ME materials can be divided into two types: single-phase and composite. The ME effect has been observed as an intrinsic effect in some natural material systems, which have been under intensive study recently, motivated by potential applications in information storage, spintronics and multiple-state memories. Thus far, over ten different compound families have been widely investigated as multiferroic ME materials such as well-known $BiFeO_3$ (BFO) and rare earth manganates. Although the intrinsic ME effect exists in the single-phase compounds, most multiferroic compounds exhibit low Curie temperatures (below room temperature), and a high inherent ME coupling (especially above room temperature) has not yet been found in the single-phase compounds. Among them, $BiFeO_3$ is unique with high Curie and Neel temperatures far above room temperature, and thus most widely investigated in recent few years.

In 2000, Hill published a paper entitled "why magnetic ferroelectrics so rare", describing the natural mutual exclusion of ferromagnetism and ferroelectricity which should be understood the formation of completion between the covalent bond and the coulomb repulsion. Covalent bonds leads to ion deviating from the center, thus exhibiting ferroelectricity. When the coulomb repulsion is stronger, the ions remain in the center, demonstrating ferromagnetism (Kundys, 2009).

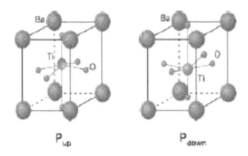

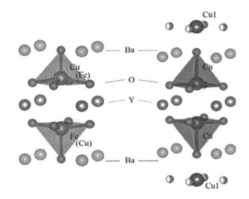

Figure 2. Space atomic distribution map of TiBaO3.

Figure 3. Space atomic distribution map (Left: YBa-CuFeO5, Right: YBa2Cu3O7).

Herein, as shown in Figure 2, we take $TiBaO_3$ for an example to explain the principle of multiferroic of perovskite structure. Ti:[Ar]$4s^2 3d^2$, Titanium ion provides ferromagnetic property. Ba: [Xe]$6s^2$, a lone pair, contributes ferroelectric property.

As it is all known, $YBaCuFeO_5$ has weak ferromagnetic properties at low temperature for successfully synthesized $YBaCuFeO_5$ at present. Here were placed the Cu by the same transition metal Ni with the expectation of obtaining a kind of multiferroic material at room temperature (Lai, 2015).

2 EXPERIMENTAL PROCEDURE

The method to $YBaNiFeO_{5+\delta}$ are divided into two categories: physical and chemical methods. Physical method is based on the method of $YBaCuFeO_5$ to study; chemical method is to burn $YBaFe_2O_5$ and doped Ni elements as the research method, and then get the $YBaNiFeO_{5+\delta}$.

2.1 The physical synthetic method

Step 1: Chose four compounds that contain Y, Ba, Ni, and Fe, respectively. Calculate the weight of each compound according to the mole ratio (1:1:1:1).

Step 2: Fully mix all the chemical compounds above in a mortar and crush.

Step 3: Calcine the powder grounded in a tubular furnace with the temperature of 900°C for the first firing.

Step 4: Characterize the sample with X-ray diffraction when the temperature drops down to room temperature.

Step 5: Press the powder into bulks for the second calcination.

Step 6: Characterize the bulk sample with X-ray diffraction, and compare the map with the standard one, and, if they are not consistent, change the second temperature until obtaining the right sample.

2.2 The chemical synthetic method

Step 1: Mix Y_2O_3 and citric acid monohydrate in a dry environment, add a little water to assist the dissolution at the same time.

Step 2: Add Fe and Ni powder into HNO_3 solution, pour the solution into the solution above step by step. Heat until the nitrogen gases stop producing.

Step 3: When the temperature drop to room temperature, add $BaCO_3$ and mix fully with a grinding machine, successively heat at 400°C for the first calcinations, and 800°C for the second calcinations.

Step 4: After two times calcinations, characterized with X-ray diffraction, when it is cooled down to room temperature.

Step 5: Press the powder into bulks with tablet compressing machine for the third calcination at 1020°C, and then feed oxygen to oxidize when the temperature drops to 600°C to make up the lack of oxygen.

Step 6: Characterize the bulk sample with X-ray diffraction, and compare the map with standard one, if they are not consistent, change the second temperature until obtaining the right sample.

3 RESULTS AND DISCUSSION

Here, we merge four elements (Y, Ba, Ni, and Fe) into multiferroic materials by the way of high-temperature sintering. X-Ray Diffraction (XRD) patterns were measured by using a Bruker-D8 Advance spectrometer, and we have also detected its electrical and magnetic properties.

As shown in Figure 4, $YBaCuFeO_5$ displays the phase of multiferroic material, belonging to ion oxide with a double perovskite structure, and has complex charge and electron spin structure.

As is reported, the basic principle of multiferroic $YBaCuFeO_5$ lies in: Fe: [Ar] 3d6 $4s^2$, Fe^{3+}

contributes d track with ferromagnetic properties and Cu:[Ar]3d^{10}4s^1,Cu^{2+} contributes a lone pair rack with ferroelectric property (Kundys, 2009).

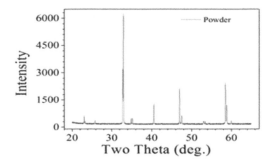

Figure 4.   XRD pattern of YBaCuFeO$_5$.

YBaCuFeO$_5$ is confirmed multiferroic with a low magnetoelectric effect at low temperature. However, most multiferroic materials are troubled with the weakness of low electrode strength, week magnetic field, and low operating temperature. Improving the shortcomings is the development trend of multiferroic materials. What I am doing is synthesizing other multiferroic materials, and here I take YBaNiFeO$_5$ as an example and I will try other elements such as YBaMnFeO$_5$, YFeO$_3$, and so on.

YBaNiFeO$_5$ is sintered at 900°C for 24 hours the first time, then 1100°C for 48 hours without the existence of oxygen gas. The raw chemical compounds are composed of Y$_2$O$_3$, NiO, Fe$_2$O$_3$, and BaCO$_3$.

While YBaNi$_2$O$_5$ is sintered at 900°C for 24 hours for the first time, and then at 1100°C for 24 hours without the adding of oxygen gas. The raw chemical compounds consist of Y$_2$O$_3$, Ni powder and BaCO$_3$.

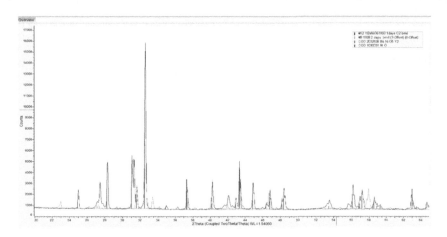

Figure 5.   XRD patterns of YBaNiFeO$_5$ and YBaNi2O$_5$.

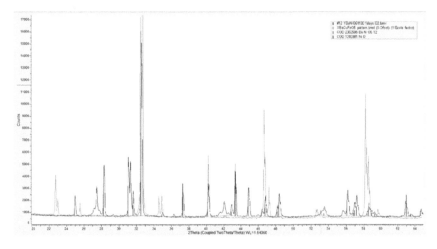

Figure 6.   XRD patterns of YBaCuFeO$_5$ and YBaNi$_2$O$_5$.

The pattern of YBaNiFeO$_5$ in Figure 5 is the best one after many trials. However, the pattern of YBaNi2O$_5$ is closer to the standard multiferroic materials. Therefore, it can be figured out that it is very difficult to burn the pure YBaNiFeO$_5$ compounds, and I decide to prepare the pure YBaNi$_2$O$_5$ compounds and figure out the reaction condition of pure YBaNiFeO$_5$ compounds according to the best reaction condition of YBaNi$_2$O$_5$ compounds (Er-Rakho, 1988; Karen, 2004).

As shown in Figure 6, there are two peaks, 33° and 40°, respectively, which are consistent with the standard one, meaning formation of the structure. Also, there are two weak peaks at around 47° and 59°, and it may not be due to the proper condition.

I will try higher temperature or prolong the sintering time with the addition of oxygen so as to increase the degree of crystallization.

4 CONCLUSION

Overall, this special material has a great potential of development in the area of information storage, microwave field, as well as electronic devices (sensors, converter, etc.). Meanwhile, multiferroic materials have the rich connotation of physical due to the spin-magnetic, electric-lattice coupling, attracting great attention of many researchers and have become a new hot spot. The breakthrough in basic and applied aspects will cause important significance.

REFERENCES

Er-Rakho L, Michel C, Lacorre P, et al. YBaCuFeO$_5$+ δ: A novel oxygen-deficient perovskite with a layer structure [J]. Journal of Solid State Chemistry, 1988, 73(2): 531–535.

Karen P, Suard E, Fauth F, et al. YBaMnCoO$_5$: Neither valence mixed nor charge ordered[J]. Solid state sciences, 2004, 6(11): 1195–1204.

Kundys B A. Maignan ChSimon. Multiferroicity with high-TC in ceramics of the YBaCuFeO$_5$ ordered perovskite [J]. arXiv preprint arXiv, 2009, 0903. 1790.

Lai, Yen Chung. Self-adjusted flux for the traveling solvent floating zone growth of YBaCuFeO$_5$ crystal[J]. Journal of Crystal Growth 413 2015: 100–104.

Shuai D, Liu J M. Multiferroic materials: past, present and future [J]. Physic, 2010, 39.10.

Advances in Materials Science, Energy Technology and Environmental Engineering – Patty & Zhou (Eds)
© *2017 Taylor & Francis Group, London, ISBN 978-1-138-19668-1*

Theoretical analysis of the separation of water load magnetic particles in a high-gradient magnetic field

Zuoming Zhu, Jinxing Cheng, Youpeng Wu, Chunan Wang, Fengtao Zhao & Weiwei Wen
Kadinuo Science and Technology (Beijing) Co. Ltd., Beijing, China

Jinxing Cheng
Institute of Nuclear and New Energy Technology, Tsinghua University, Beijing, China

ABSTRACT: Separating radioactive wastewater by using a high-gradient magnetic separator is one of the new technologies. In this paper, theoretical analysis of separation of water load magnetic particles in a high-gradient magnetic field was performed, and the mathematical physical model was developed and verified. The theoretical analysis results are concordant with the FLUENT simulation.

1 INTRODUCTION

With the progress of science and technology, technologies for radioactive wastewater treatment are becoming more and more advanced, and separating radionuclide in liquid by using a high-gradient magnetic separator is one of those advanced technologies. High-gradient magnetic separation is performed by adding a small amount of coagulant, magnetic particles, etc. to the wastewater, which flocculate and integrate with the pollutants. They remove the pollutants in the water by efficient sedimentation and magnetic filtering, and the magnetic particles will be recycled and reused by a magnetic drum separator, the basic working principle of which is production of high-gradient magnetic fields on the surface of magnetic media under an external magnetic field for capturing the magnetic particles nearby (Dong et al. 2009, Zheng et al. 2015, Zou 2013).

The purpose of this paper is to perform theoretical analysis of separation of water load magnetic particles in a high-gradient magnetic field, and establish a reasonable mathematical physical model and verify it, ultimately providing support for engineering applications of this technology.

2 STRESS ANALYSIS OF MAGNETIC PARTICLES IN THE WATER

The separation interval of a high-gradient magnetic separator is filled with numerous ferromagnetic assembled magnetic media—steel wool. In order to understand the process of capturing the magnetic particles in the separating interval, it is necessary to simplify complex issues, which means to observe the process that a single piece of ferromagnetic steel wool captures magnetic particles. First, we establish a simple model to describe capturing features of a single piece of ferromagnetic steel wool to particles, based on which the capturing process occurred in the whole separation space that should be considered. With reference to the relevant literature, the research model (Geoffrey et al. 2004) is shown in Figure 1.

Particles are affected by magnetic force, gravity, buoyancy and drag force of water flow, particle inertia force, the Brown force, diffusion force, etc. (Wang et al. 2010), in the system. Particles involved in this paper are of submicron sizes (0.1μm-1μm). Extraction of magnitude particles of this magnitude is dependent on the magnetic force attracting magnetic particles on the steel wool and the drag force of water flow to particles. Gravity, buoyancy, and other forces are ignored herein as they are relatively small.

2.1 Fluid drag force

Fluid drag force F_D affecting particles can be presented as (Paul 2011):

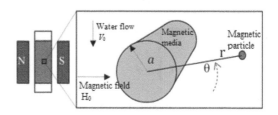

Figure 1. Model of a single piece of magnetic medium to capture magnetic particles.

$$F_D = 6\pi u V R_{core} \tag{1}$$

where u is the kinematic viscosity of the fluid media; R_{core} is the radius of the magnetic particle; and V is the movement speed of the particle relative to the fluid, $V = V_g - V_f$.

We assume that the steel wool is a standard cylinder. Although there is no analysis method for the low Reynolds number flow through the cylinder on the surface area, the speed of water around the outer layer of the steel wool can be approximately calculated as (Paul, 2011):

$$V_r = -V_0 G^- \sin\theta \tag{2}$$

$$V_\theta = -V_0 G^+ \cos\theta \tag{3}$$

where $G^\pm$ is the geometric factor.

The axial and angular flow resistances can be obtained by the following formula (Paul, 2011):

$$F_{vr} = -6\pi u R_{core} \left(\frac{dr}{dt} + V_0 G^- \sin\theta \right) \tag{4}$$

$$F_{v\theta} = -6\pi u R_{core} \left(r\frac{d\theta}{dt} + V_0 G^+ \cos\theta \right) \tag{5}$$

2.2 Gradient magnetic field force

The magnetic field gradient refers to the increasing or decreasing degree of an inhomogeneous magnetic field in a certain direction. A high-gradient magnetic field area will occur on the surface of filling magnetic media between the two poles of the uniform magnetic field. Forces affecting particles in the magnetic field near the magnetic media (Geoffrey et al. 2004) can be calculated as:

$$F_m = \mu_0 V_p M_{core} \Delta H \tag{6}$$

where μ_0 is the space permeability; V_p is the particle volume; M_{core} is the particle magnetization; and H is the magnetic field intensity of the space where the particles are.

The magnetic force on particles is determined by the magnetic field intensity first, $H(r, \theta)$.

$$H_r = \left(\frac{M_{wire}a^2}{2r^2} + H_0 \right) \cos\theta \tag{7}$$

$$H_\theta = \left(\frac{M_{wire}a^2}{2r^2} - H_0 \right) \sin\theta \tag{8}$$

where H_0 is the external magnetic field intensity, and M_{wire} is the magnetic rate, i.e. the magnetization of steel wool. The magnetic force affecting particles is calculated by formula (6).

In addition, the magnetization and magnetic field intensity of paramagnetic material particles have a linear relationship: $M_{core} = \chi H$. For paramagnetic material particles of submicron sizes, the magnetization can be calculated by the following formula (Takayasu et al. 1983):

$$M_{core} = M_{sp} \left\{ \coth\left(\frac{VM_{sp}H_0}{KT} \right) - \frac{KT}{VM_{sp}H_0} \right\} \tag{9}$$

For particles of submicron sizes and with a saturation magnetization of 80 emu/g, $VM_{sp} \gg kT$, M_{core} can be presented as (Takayasu et al. 1983):

$$M_{core} = \begin{cases} M_{sp} & (H_0 \geq M_{sp}/3) \\ 3H_0 & (H_0 \leq M_{sp}/3) \end{cases} \tag{10}$$

The axial and angular magnetic field forces are (Geoffrey et al. 2004):

$$F_{mr} = -\frac{4\pi\mu_0 M_{wire} M_{core} a^2 R_{core}^3}{3r^3} \left(\frac{M_{wire}a^2}{2H_0 r^2} + \cos 2\theta \right) \tag{11}$$

$$F_{m\theta} = -\frac{4\pi\mu_0 M_{wire} M_{core} a^2 R_{core}^3}{3r^3} \sin 2\theta \tag{12}$$

3 PARTICLE ADSORPTION MODEL

The purpose of this paper is to study and establish an adsorption model to describe the particles when the magnetic field direction, steel wool, the flow field move vertically, the radius and angle for a single magnetic medium to adsorb magnetic particles, and use the FLUENT software for verification.

The following conditions must be met for a magnetic particle to be adsorbed to the steel wool of the magnetic medium (Xiang et al. 1997): ① the resultant force of radial forces under the polar coordinate must point to the magnetic medium and ② the resultant force of tangential forces is large enough to put particles to a position under larger radial force.

Assume that magnetic particles have been static on the surface of steel wool. Let $F_m + F_v = 0$, $r_a = r/a$, and we can get:

$$\frac{dr_a}{dt} = -\tau_v G^- \sin\theta - \tau_m \left(\frac{M_{wire}}{2H_0 r_a^5} + \frac{\cos 2\theta}{r_a^3} \right) \tag{13}$$

$$r_a \frac{d\theta}{dt} = -\tau_v G^+ \cos\theta - \tau_m \frac{\sin 2\theta}{r_a^3} \tag{14}$$

wherein, $\tau_v = \frac{v_0}{a}$; $\tau_m = \frac{2\mu_0 M_{wire} M_{core} R_{core}^2}{9ua^2}$.

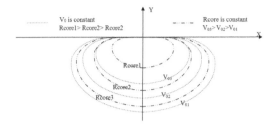

Figure 2. Schematic diagram of adsorption radius r_a.

As $R_{core} \ll a$, the geometric factor $G^{\pm}$ can be detached as:

$$G^- = 0, \quad G^+ = \frac{2R_{core}/a}{2 - \ln Re}$$

where Re is the turbulent Reynolds number.

If particles keep static after being adsorbed, then $dr_a/dt = 0, d\theta/dt = 0, r_a > 1$. By substituting this into formulas (13) and (14), we can get:

$$r_a = \sqrt{-\frac{\tau_m(2 - \ln Re)a\sin\theta}{\tau_v R_{core}}} \tag{15}$$

We can get from Formula (15) and Figure 2 that as the particle size decreases and the fluid velocity increases, the maximum adsorption radius of steel wool decreases. With the increasing magnetic field intensity, the magnetization of the particle M_{core} increases gradually to saturation, the magnetic force increases gradually to a fixed value, and the adsorption radius of steel wool also increases to saturation. Moreover, the adsorption angle $\theta < 0$. These conclusions will be verified in the FLUENT simulation.

4 ESTABLISHMENT OF MATHEMATICAL PHYSICS MODELS

Using the FLUENT simulation calculation for liquid-solid two-phase fluid with a volume fraction less than 12%, the liquid phase is set as continuous media, the solid phase is set as discrete media, and the liquid phase adopts the standard k-ξ turbulence model. In the FLUENT simulation calculation, for the movement of a single particle in the discrete-phase model, the trajectory of discrete-phase particles is solved by the differential equation of working forces on particles under the integral Laplace coordinate system. Under the Cartesian coordinate system, it can be expressed as Yan (1989):

$$\frac{du_p}{dt} = F_D(u - u_p) + \frac{g_x(\rho_p - \rho)}{\rho_p} + F_x$$

where $F_D(u - u_p)$ is the fluid drag item and F_D is the resistance coefficient. For simulation calculation of particles with sizes smaller than micron magnitude, Stokes' resistance coefficient can be used. Here, $g_x(\rho_p - \rho)/\rho_p$ is the gravity and buoyancy item. F_x refers to other forces, such as the Brown force, diffusion force, and the Saffman lift force. The magnetic field force is also involved here, which is embodied in the form of UDF program.

Magnetic particles are driven into the high-gradient magnetic field by the liquid-phase fluid. Being affected by the combination of flow field and magnetic field in the container, some particles are captured by the magnetic media. Owing to the complicated movement of particles in the container, in order to simplify the operation, the following assumptions are made for modeling:

1. Water flow belongs to incompressible fluid, and hence, controls the exit flow rate to achieve a steady flow of the fluid.
2. To reduce the calculation amount, select a limited area near the steel wool in a large container as the area for calculation. Study the adsorption properties only, and select only one piece of steel wool, whose diameter is far smaller than the space between steel wools. Considering the changing nature of the magnetic field, the mutual influence between steel wools can be ignored at the moment.
3. The assembled magnetic medium steel wool is a cylinder, with a diameter of 300 μm and a length much longer than the calculated area, and therefore, the model is built as a 2D model, and the steel wool is in the middle of the calculated area.

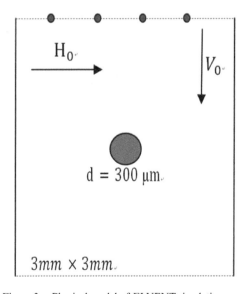

Figure 3. Physical model of FLUENT simulation.

455

4. When using the FLUENT software to calculate the discrete-phase particle itself, the particle size and volume should be considered. However, after being displayed via images and adsorbed to the steel wool, the discrete-phase particle is calculated regardless of the particle volume and changes in the steel wool magnetic field before and after the adsorption of magnetic particle.

The physical model is shown in Figure 3.

5 FLUENT NUMERICAL SIMULATION AND ANALYSIS

Set parameters as follows: set the diameter of steel wool as 300 μm, the saturation magnetization of steel wool as 871 mT, the saturation magnetization of iron oxide particles as 80 emu/g, the particle diameter as 1 μm, the magnetic field intensity as 1T, and the water flow rate as 1 cm/s. The particle density determined by particle measurement is 5 g/cm³. The adsorption effect is shown in Figure 4.

From Figure 4, we can get that the particles are adsorbed and captured by the steel wool when flowing through the steel wool in the water flow, and the closer the distance is, the more easily the particle is adsorbed. Particles being far away from the steel wool will also approach it when passing by, but due to the water flow and other forces, the particles leave the magnetic area, and flow to the downstream direction. Moreover, the adsorption angle is less than 0 and particles are mostly concentrated at the lateral side of steel wool, which is concordant

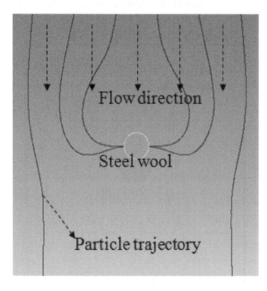

Figure 4. Schematic diagram of adsorption effect by FLUENT simulation.

with the result of the analytical method mentioned, and the experiment results of Paul C. Owings. In order to figure out effects of factors on the process of high-gradient magnetic field capturing magnetic particles, the single factor analysis method is adopted to analyze the adsorption effect of magnetic media and the adsorption radius, and the adsorption radius r_a is obtained by adding particles to the steel wool in the model with a distance of r.

6 CONCLUSIONS

In this paper, theoretical analysis of separation of water load magnetic particles in a high-gradient magnetic field was carried out, and the mathematical physical model was established and verified. The theoretical analysis results match well with the FLUENT simulation. The results of this study can provide support for the engineering application of this technology.

REFERENCES

Dong Huifang, Li Dagong, Jiang Huili & Liu Yanzhu (2009). Application and research of magnetic separation technology in water treatment. *Journal of Hebei Institute of Architecture and Civil Engineering.* 27(3), 46–50.

Geoffrey D. Moeser, Kaitlin A. Roach, William H. Green, & T. Alan Hatton (2004). High-gradient magnetic separation of coated magnetic nanoparticles. *AIChE Journal.* 50(11), 2835–2848.

Paul C. Owings (2011). *High gradient magnetic separation of nanoscale magnetite.* Kansas State University.

Takayasu, M. Gerber, & R. Friedlaender (1983). Magnetic separation of submicron particles. *IEEE Transactions on Magnetics.* 19(5).

Wang Fahui & Tie Zhanxu (2010). Numerical simulation for high gradient magnetic field located single magnetic medium in entrapping magnetic particles. *Coal Preparation Technology.* 2, 20–23.

Wang Fahui, Liu Xiufang & Zhang Dan (2010). Numerical simulation for multiple magnetic medium in entrapping magnetic particles in high gradient magnetic field. *Metal Mine.* 2, 103–106.

Xiang Fazhu, He Pingbo & Chen Jin (1997). Mathematical model for high gradient magnetic separation: its general review and development tendency. *Mining And Metallurgical Engineering.* 17(1), 42–46.

Yan Jici (1989). *Electromagnetism.* Higher Education Publishing Press.

Zheng Libing, Tong Juan, Wei Yuansong & Wang Jun (2015). State of arts of magnetic separation technology in water treatment. *Acta Scientiae Circumstantiae.* DOI: 10.13671/j.hjkxxb. 2015. 0775.

Zou Zhaolong (2013). Field application and exploration of magnetic separation technology in wastewater treatment technology. *Private Science and Technology.* 12, 69.

Advances in Materials Science, Energy Technology
and Environmental Engineering – Patty & Zhou (Eds)
© 2017 Taylor & Francis Group, London, ISBN 978-1-138-19668-1

Author index